Springer Series on
SIGNALS AND COMMUNICATION TECHNOLOGY

Grid Enabled Remote Instrumentation
F. Davoli, N. Meyer, R. Pugliese, S. Zappatore
ISBN 978-0-387-09662-9

Usability of Speech Dialog Systems
Th. Hempel
ISBN 978-3-540-78342-8

Handover in DVB-H
X. Yang
ISBN 978-3-540-78629-0

Multimodal User Interfaces
D. Tzovaras (Ed.)
ISBN 978-3-540-78344-2

Wireless Sensor Networks and Applications
Y. Li, M.T. Thai, W. Wu (Eds.)
ISBN 978-0-387-49591-0

Passive Eye Monitoring
R.I. Hammoud (Ed.)
ISBN 978-3-540-75411-4

Digital Signal Processing
S. Engelberg
ISBN 978-1-84800-118-3

Digital Video and Audio Broadcasting Technology
W. Fischer
ISBN 978-3-540-76357-4

Satellite Communications and Navigation Systems
E. Del Re, M. Ruggieri (Eds.)
ISBN 978-0-387-47522-6

Three-Dimensional Television
H.M. Ozaktas, L. Onural (Eds.)
ISBN 978-3-540-72531-2

Foundations and Applications of Sensor Management
A.O. Hero III, D. Castañón, D. Cochran, and K. Kastella (Eds.)
ISBN 978-0-387-27892-6

Human Factors and Voice Interactive Systems, Second Edition
D. Gardner-Bonneau and H. Blanchard
ISBN 978-0-387-25482-1

Wireless Communications: 2007 CNIT Thyrrenian Symposium
S. Pupolin
ISBN 978-0-387-73824-6

Adaptive Nonlinear System Identification: The Volterra and Wiener Model Approaches
T. Ogunfunmi
ISBN 978-0-387-26328-1

Wireless Network Security
Y. Xiao, X. Shen, and D.Z. Du (Eds.)
ISBN 978-0-387-28040-0

Satellite Communications and Navigation Systems
E. Del Re and M. Ruggieri
ISBN 0-387-47522-2

Wireless Ad Hoc and Sensor Networks
A Cross-Layer Design Perspective
R. Jurdak
ISBN 0-387-39022-7

Cryptographic Algorithms on Reconfigurable Hardware
F. Rodriguez-Henriquez, N.A. Saqib, A. Díaz Pérez, and C.K. Koc
ISBN 0-387-33956-6

Multimedia Database Retrieval
A Human-Centered Approach
P. Muneesawang and L. Guan
ISBN 0-387-25627-X

Broadband Fixed Wireless Access
A System Perspective
M. Engels and F. Petre
ISBN 0-387-33956-6

Distributed Cooperative Laboratories
Networking, Instrumentation, and Measurements
F. Davoli, S. Palazzo and S. Zappatore (Eds.)
ISBN 0-387-29811-8

The Variational Bayes Method in Signal Processing
V. Šmídl and A. Quinn
ISBN 3-540-28819-8

Topics in Acoustic Echo and Noise Control
Selected Methods for the Cancellation of Acoustical Echoes, the Reduction of Background Noise, and Speech Processing
E. Hänsler and G. Schmidt (Eds.)
ISBN 3-540-33212-x

(continued after index)

Franco Davoli • Norbert Meyer •
Roberto Pugliese • Sandro Zappatore
Editors

Grid Enabled Remote Instrumentation

Springer

Editors

Franco Davoli
DIST-University of Genoa
CNIT, University of Genoa
Research Unit
Via Opera Pia 13
16145 Genova, Italy

Norbert Meyer
Poznań Supercomputing and
Networking Center (PSNC)
Institute of Bioorganic Chemistry
Polish Academy of Sciences
ul. Noskowskiego 12/14
61-704 Poznań, Poland

Roberto Pugliese
Information Technology Department
Sincrotrone Trieste S.C.p.A.
S.S. 14, km 163.5
Area Science Park
34012 Basovizza (Trieste), Italy

Sandro Zappatore
DIST-University of Genoa
CNIT-University of Genoa
Research Unit
Via Opera Pia 13
16145 Genova, Italy

ISSN: 1860-4862
ISBN: 978-1-4419-3506-9 e-ISBN: 978-0-387-09663-6

Softcover reprint of the hardcover 1st edition 2009

Printed on acid-free paper

springer.com

TABLE OF CONTENTS

Part V. Applications

PREFACE

Performing measurement tasks and experiments on complex tele-laboratory structures is gaining momentum as a relevant aspect in a number of areas, from scientific and technological research, to environmental monitoring, to distance learning and training. The diffusion of e-Science requires real-time interaction of researchers' teams worldwide; this, in turn, relies on a number of functionalities, which include, among others, the widespread use of collaborative tools and the possibility to access and control remote instrumentation.

The Grid architecture (or, more generally, the service-oriented architecture, SOA), viewed as a tool for the integration of distributed resources, plays a significant role, not only as manager of computational resources, but increasingly as aggregator of measurement instrumentation and pervasive large-scale data acquisition platforms (e.g., sensor networks). In this context, the functionality of a service-oriented architecture allows managing, maintaining, and exploiting heterogeneous and geographically sparse instrumentation and acquisition devices in a unified way, by providing standardized interfaces and common working environments to their users. This is achieved through the properties of isolation from the physical network and from the peculiarities of the instrumentation, granted by standard middleware, together with secure and flexible mechanisms to seek, access, and aggregate distributed resources.

Though the notion of cooperative sharing of laboratory instrumentation, data and computing power in integrated virtual structures (collaboratories) dates back to 1989, successive advances in Grid middleware have mainly emphasized the aspect of distributed computation, with respect to the access to and control of real laboratory instrumentation and measurement devices. More recently, a growing interest has developed toward such infrastructures. However, making real instruments become full members of the Grid requires a number of advances in research and development of additional functionalities, to be integrated in existing middleware architectures. Noticeable efforts in this direction have been, among others, the CIMA (Common Instrument Middleware Architecture) project, funded by the National Science Foundation (NSF) in the United States, and the GRIDCC (Grid Enabled Remote Instrumentation with Distributed Control and Computation) and RINGrid (Remote Instrumentation in Next-generation Grids) projects, funded by the European Commission (EC) in Europe.

It is worth noting in this context that the issues that are relevant to the above-mentioned full integration of remote instrumentation services cover a wide range of inter-disciplinary research areas. Besides middleware functionalities and architectural aspects, we can mention the following: (i) communications and networking, in particular as regards quality of service (QoS) provisioning, cross-layer interaction with the service-oriented layers and wireless sensor networks (WSNs); (ii) instrumentation and measurement, including the study of the metrological problems connected with distributed measurement systems; (iii) collaborative tools and advanced data visualization techniques.

This book stems from the 2nd Workshop on Distributed Cooperative Laboratories – Instrumenting the Grid (INGRID 2007), held in Santa Margherita Ligure, Italy, from April 16 to 18, 2007. All chapters are based on presentations held at the workshop, either as keynote addresses, invited, or regular contributions. The first workshop in this series was held in Sorrento, Italy, in July 2005, within the framework of the International Tyrrhenian Workshop on Digital Communications (TIWDC), and originated a previous publication in edited book form by Springer on this streamline.

The INGRID Workshop and this book would not have been possible without the existence and activity of a number of European and National projects that participated in the organization. In the first category, the already-mentioned GRIDCC (http://www. gridcc.org) and RINGrid projects (http://www.ringrid.eu) have played a major role. As regards national projects, a Polish and an Italian effort have been involved, namely, VLAB (Virtual LABoratory; http://vlab. psnc.pl) and CRIMSON (Cooperative Remote Interconnected Measurement Systems Over Networks; http://www.prin-crimson.org). In addition, a follow-up effort was the creation of the OGF (Open Grid Forum) research group called RISGE (Remote Instrumentation Services in Grid Environment).

The book is structured into five parts, which contain each a number of chapters on the theme of the part. Part I, Sensor Networks as Data Acquisition Devices, covers aspects regarding the integration of WSNs into the Grid, data-centric storage, decentralized detection, multi-hop WSNs, and consensus protocols.

Part II, Networking and the Grid, touches some points of interaction between the service layers and networking, the impact of biologically inspired approaches on computer networking and on nanoscale biological networks, network monitoring, and the optimization of specific networking software, also in relation to Grid applications.

Part III is titled Instrumentation and Measurement and is dedicated to tele-measurement platforms, instruments' abstractions, metrological aspects, and collaborative tools for e-Science, with the use of remote instrumentation.

Part IV, Grid Resource Allocation, QoS, and Security, deals with brokering techniques, QoS-aware workflows and QoS guarantees, knowledge spaces, approaches to Grid security, support of interactive applications; it also provides a thorough overview and future directions of the CIMA middleware architecture and ends with a glimpse on Web 2.0 and a comprehensive review of several research activities and applications of this topic.

Part V is devoted to Applications and reports activities in a large number of different fields: support of interactivity and its use in plasma physics visualization; network performance monitoring in support of Grid applications; Grid support for network emulation; remote operation of food engineering equipment, automated irrigation systems, telecommunication devices, and particle accelerators; 3D data analysis; parallel image processing; electronic very large baseline interferometry (*e*-VLBI); digital libraries; workflow management in biomedicine.

The book complements the previous volume in this series,[1] by emphasizing the role of Grid middleware and services in the domain of remote instrumentation. In this respect, it is a first contribution to the discussion that initiated in the framework of the OGF, with the already-mentioned recent creation of the RISGE Research Group (http://forge.ogf.org/sf/projects/risge-rg). Its aim is to set the basis for further development and, at the same time, to offer an up-to-date and comprehensive overview of this field.

Franco Davoli
Norbert Meyer
Roberto Pugliese
Sandro Zappatore

[1] F. Davoli, S. Palazzo, S. Zappatore, Eds., *Distributed Cooperative Laboratories: Networking, Instrumentation, and Measurements*, Springer, New York, NY, 2006; ISBN 0-387-29811-8.

ACKNOWLEDGMENTS

We gratefully acknowledge the support of the many persons and institutions that contributed to the success of the 2nd Workshop on Distributed Cooperative Laboratories (INGRID 2007) and to this book. The European Commission, through the Directorate General for Information Society and Media (DG-INFSO), and in particular the Research Infrastructures Unit, has funded the GRIDCC and RINGrid projects, has deeply involved in the organization of the workshop, and has provided the presence of Dr. Maria Ramalho-Natario as one of the keynote speakers. Our thanks go to her and to the other keynote speakers, Prof. Tatsuya Suda, from the University of California at Irvine, and Prof. Geoffrey Fox, from Indiana University, for their excellent presentations, to the Program Committee members, to the participants from the GRIDCC, RINGrid, VLAB and CRIMSON projects, to the many people involved in the organization of the demonstrations, and to the staff of the CNIT (Italian National Consortium for Telecommunications) Laboratory of Savona, Italy, and of the CNIT National Laboratory for Multimedia Communications, Naples, Italy. We also acknowledge the following companies and institutions, which have given their sponsorship to the organization of the workshop: Elettronica Industriale Telecommunication Technology – Gruppo Mediaset, Italy; Eutelsat, France; MBI – Mediterranean Broadband Infrastructure, Italy; Regione Liguria, Italy; Selex Communications, Italy; SIR – Soluzioni In Rete, Italy. A special thanks goes to the International Institute of Communications (IIC), Genova, Italy, and in particular to Clotilde Fertini-Canepa and Kaji Foscoli, for the much-appreciated logistic organization, as well as to Stefano Vignola and Carlo Caligaris for the repeated formatting of the manuscript, and to Caitlin Womersley from Springer for her constant assistance. Finally, we appreciate the participation and interest in our workshop of all presenters and authors of the chapters. The success of the event would have not been possible without them and the high-quality contributions we received.

Part I

Sensor Networks as Data Acquisition Devices

DATA-CENTRIC STORAGE IN NON-UNIFORM SENSOR NETWORKS

M. Albano[1], S. Chessa[1,2], F. Nidito[1], and S. Pelagatti[1]

[1] *Computer Science Department, University of Pisa, Largo Pontecorvo 3, 56127 Pisa, Italy,*
[2] *Istituto di Scienza e Tecnologie dell'Informazione, Area della Ricerca, CNR di Pisa, S.Cataldo, 56100 Pisa, Italy*

Abstract Dependable data storage in wireless sensor networks is becoming increasingly important, due to the lack of reliability of the individual sensors. Recently, data-centric storage (DCS) has been proposed to manage network-sensed data. DCS reconsiders ideas and techniques successfully proposed in peer-to-peer systems within the framework of wireless sensor networks. In particular it assumes that data are uniquely named and data storage and retrieval is achieved using names instead of sensor nodes addresses. In this chapter, we discuss the limitations of previous approaches, and in particular of geographic hash tables (GHT), and introduce DELiGHT, a protocol that provides fine QoS control by the user and ensures even data distribution, also in non-uniform sensor networks. The merits of DELiGHT have been evaluated through simulation in uniform and Gaussian-distributed sensor networks. The simulation results show that the protocol provides a better load balancing than the previous proposals and that the QoS is ensured without appreciable overhead.

1. INTRODUCTION

A sensor is a micro-system, which comprises a processor, one or more sensing units (*transducers*), a radio transceiver and an embedded battery. A wireless sensor network (*WSN*) is a network formed by a large number of tiny and inexpensive sensors that cooperate to perform measurement tasks [4]. Sensors are spread in an environment (*sensor field*) without any predetermined infrastructure and cooperate to execute common monitoring tasks, which usually consist in sensing environmental data. The sensed data are collected by an external sink node, when it is connected to the network. The sink node, which could be either static or mobile, is in turn accessed by the external operators to retrieve the information gathered by the network. In a WSN, data can be accessed according to different paradigms, which define sensor-to-sensor and sensor-to-sink communications.

Recent paradigms [7,9,14] see the WSN as a distributed database. The WSN is then "programmed" by the sink by sending queries to the sensors. Sensors in turn reduce and filter the sensed data locally before sending an answer.

In this scenario, dependable in-network data storage is becoming increasingly important, due to the lack of reliability of sensor nodes [3, 5, 11, 12].

In WSNs the content of data is generally more important than the address of the sensor that has gathered the data. This naturally leads to data-centric storage (DCS) [12], in which data are univocally named and the node in which data are stored is determined by the name associated with them.

GHT [12] implements data-centric storage using geographic hash tables. Each datum has a unique *meta-datum* (or *name*), which is hashed uniformly as a coordinate in the sensing area, represented as a two-dimensional plane. GHT implements two operations: `put`, which stores data, and `get`, which retrieves them. In the `put` operation, the name of data to be stored is first hashed into a location (x,y) in the sensing field. Then, GHT selects the closest sensor to (x,y), which becomes the *home node* for that data. The home node is selected using the GFG routing protocol [6]. GFG uses two operation modes: greedy and perimeter. Each packet starts in the *greedy mode*, in which it is routed progressively closer to its destination at each hop. When a packet reaches a node s_i whose neighbors are all farther than s_i to the destination, GFG switches to the *perimeter mode* and the packet is forwarded using the *right-hand rule* (the packet is forwarded on the *next* edge clockwise from the edge from which the packet has been received). As soon as the packet reaches a node closer to destination than the previous ones, it returns to the greedy mode. If the destination (x,y) does not correspond to any sensor, GHT uses the perimeter mode of GFG to locate all the sensors surrounding (x,y) (called the *perimeter* of (x,y)). The closest sensor in the perimeter becomes the *home node* for (x,y). GHT stores a copy of the data in the home node as well as in all the sensors belonging to the perimeter. Storing on all the perimeter is essential to guarantee data persistence also in presence of node faults. Data retrieval uses a `get` operation. The name is first hashed into the destination (x,y), then GFG is used to route the request to (x,y). When the request reaches a node in the perimeter of (x,y), the data are returned back to the sender. Replicating all data on the perimeter of (x,y) is a simple choice, which allows to use GFG with almost no changes and which can work quite well on very large sensing fields with uniformly distributed sensors. However, this choice has two important drawbacks. The first one is that in non-uniformly distributed sensor networks the number of data stored on each sensor can be very different. This leads to load unbalance (a few nodes need to answer many `gets`) and to rapid consumption of sensor batteries. The second problem is related to dependability and QoS. In GHT, the user has no means to ask for better dependability for a really important datum. This is because the number of replicas of each datum depends only on the GFG perimeter around the home node of its meta-datum.

Other data-centric storage for sensor networks [3,5,11] differ from GHT in terms of algorithms used to compute the set of sensors that should store a given data, but like GHT they fail to address QoS in data dependability and they do not address issues related to sensors distribution.

Our work originates from GHT. In some previous work, we proposed Q-NiGHT [1,2], in which we implemented QoS using geographic hash functions. In Q-NiGHT, a user can specify an extra parameter for put operations which tell the system the level of QoS (i.e., the number of replicas) required for a given datum. Then, the protocol takes care of spreading exactly this number of replicas in the surroundings of the home node. Results show that Q-NiGHT was actually able to provide QoS with small extra overhead with respect to original GHT puts and gets. However, to reach these results, Q-NiGHT needed to know exact sensor distribution function in advance, showing poor results and load unbalance when the actual sensor distribution was different.

In this chapter we propose DELiGHT (Density Estimation for Load balancing in Geographic Hash Tables), a novel DCS protocol, which moves from GHT and Q-NiGHT in order to provide QoS control and good load balance among sensors in WSNs of any (possibly changing) distribution. DELiGHT uses a strategy similar to the *rejection method* [10] to build a hash function biased with sensor distribution. This spreads data more evenly among nodes. In addition, DELiGHT can provide QoS with different redundancy techniques. We detail the protocol using pure replication, allowing the user to choose the number of replicas required for a given datum. We conduct detailed simulations of DELiGHT , Q-NiGHT and GHT and compare the results obtained with respect to the load of each sensor (i.e., the number of data stored in each node) and the number of messages needed for data storage and retrieval. Results show the good performance of DELiGHT on different sensors distributions in terms of both protocol costs and load balance.

The chapter is organized as follows. Section 2 discusses the problems of load unbalance in GHT in more detail, presenting the results of simulations carried on with uniform and Gaussian distributions. Section 3 presents the density estimation technique used in the protocol. Then, Section 4 details DELiGHT and discusses how different redundancy techniques can be incorporated in the protocol. Finally, Section 5 reports on the simulation of our protocol and compares its performance with plain GHT. Conclusions are presented in Section 6.

2. LOAD UNBALANCE IN GHT

GHT [12] implements data-centric storage using geographic hash tables and the GFG protocol. We have already discussed how the protocol works in Section 1. Here, we discuss the behavior of GHT, w.r.t. load balancing. In particular, we describe a simulation experiment which gives better insights into the protocol behavior and settle some motivations for our proposal.

Our experiment is organized as follows. In order to measure the degree of unbalance of GHT on realistic scenarios, we simulated a flat square sensing field, with a 400 m side. Each node has circular transmission range with 10 m radius. In this area, we simulated several WSNs ranging from 3600 to 20,000 sensors, which correspond to a mean network density ranging in $(8, 40)$. For each density, we randomly generated 100 networks with uniform distribution. For each network, we compute the mean and the variance of the number of nodes found in a GFG perimeter as follows. For each sensor network the simulator randomly selects 1000 points and, for each point, it computes the number of nodes in the perimeter surrounding the point. GFG needs to work with a planar graph in order for the perimeter mode to behave correctly. We assume to planarize graphs using the GG (*Gabriel Graph*) [8] algorithm. Figure 1a shows that, as the network density increases, the average number of nodes in a perimeter decreases. However, the actual number of nodes remains highly variable. This variance is partly due to the behavior of nodes in the outer part of the sensing area, since in that area the probability of having very long perimeters (i.e., following the whole border) is high. With low densities, the probability that a random point belongs to the exterior of the network (and thus it is associated to the external perimeter) is not negligible.

Perimeters on the border. In order to understand this *border effect*, we performed another set of simulations in which the sensing networks are generated

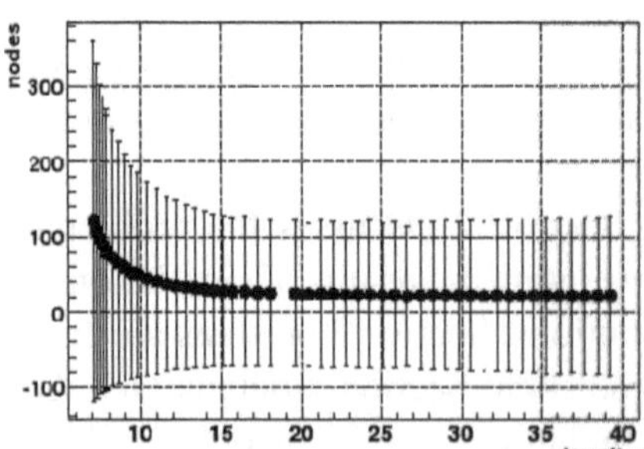

(a) GG with border

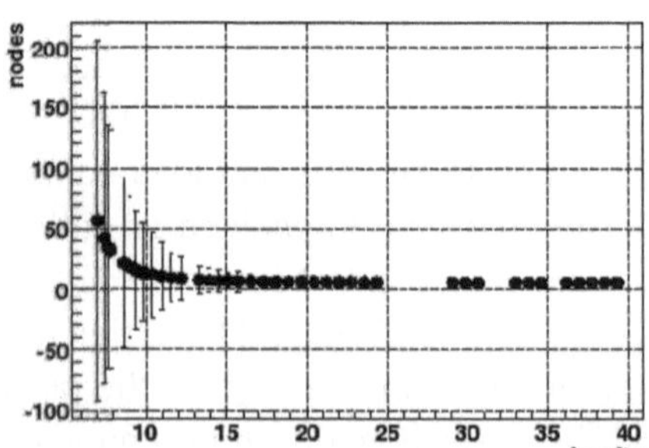

(b) GG without border

Figure 1. Mean and variance of perimeters (number of nodes) measured for different densities with GG planarization

in the same way as above but the external area is not used to store data. We "cut away" 5% of the area from each border, for a total of 19% of the total area. We randomly generate 1000 points in the white area and again measure the length of each perimeter and compute the mean and the variance. Figure 1b shows the results we computed. The mean and the variance improve if the border nodes are left out, but the standard deviation remains high, leading to a high load unbalance in the nodes.

Non-uniform sensor distribution. In order to understand the behavior of GHT with non-uniform sensor distribution, we repeated our experiments using a Gaussian function ($\sigma = 1$ with maximum on the center of the area) for distributing sensors. The function is structured to have the 99 percentile matching the area. The results are shown in Figure 2. The behavior is much worse than with uniform distribution because GHT uses a uniform hash function independently of the real distribution of the sensors. This brings to a pathological state of *load unbalance* that is due to the different quantity of data that must be managed by an equal number of sensors: A sensor on the border of the deployment area must manage a quantity of data that is larger than the quantity managed by a sensor in the center of the network.

Load unbalance and QoS. Another issue with GHT is that there is no way to control the QoS provided for each datum. Since the point (x,y) is obtained computing a hash function h on its associated meta-data M, the selection of the sensors candidate for storage is in practice independent of the importance of the datum. In principle this ensures the same treatment for each stored datum. However, if the meta-datum M is particularly popular and many sensors generate data described by M, the sensors located in the perimeter around $(x,y) = h(M)$ would be burdened with a high storage and communication load. For this reason the authors of GHT [12] introduce the technique of *structured*

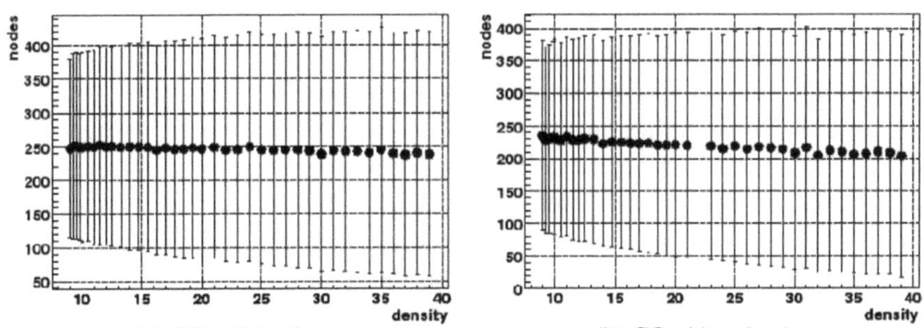

(a) GG with border (b) GG without border

Figure 2. Mean and variance of GHT perimeters for different WSNs densities, Gaussian sensor distribution

replication that replicates the same datum in different sub-areas. However, neither GHT nor the structured replication ensures that the level of redundancy associated to a data is related to the importance of the data itself: GHT assures only the same average treatment of each stored data. Another aspect is that, although the average level of redundancy of the meta-datum is constant, in practice it can vary significantly (due to the fact that each geographic point is surrounded by a different perimeter), even in case of uniform distribution of the sensors.

3. DISTRIBUTION ESTIMATION

GHT is designed to work on uniform networks and the hash function is a plain uniform hash function. When the network is randomly distributed the sensors can be distributed in a non-uniform way. This may bring to a pathological state of load unbalancing. To ensure data load-balancing DELiGHT must be able to spread data following a function fitting the distribution of the sensors in the environment.

In DELiGHT, this problem is addressed using a generic hash function (detailed in Section 4.2). This function takes two input parameters: The first one is the meta-datum to hash and the second one is the probability distribution function with which we want to spread the data in the network.

A solution that uses the actual distribution of the sensors is impracticable because this distribution can be unknown or it may vary with time, and measuring the number of neighbors of each sensor and broadcasting the information is too expensive in terms of energy consumption. Thus, to use the non-uniform hash function technique, DELiGHT must be able to approximate the actual distribution of the nodes during the lifetime of the network.

The DELiGHT architecture exploits the measurement of a limited number of sensors (*probes*) using an *estimation function*, in order to approximate the actual distribution function of the nodes.

At network startup, the sink chooses a number of geographical points in the network and sends them an initialization message. The message is routed by GFG and the sensors that are the closest to each of the selected geographical points detect their number of neighbors. Then they store this information in the network using the `put` operation of DELiGHT under the assumption of uniform nodes' distribution, using `Probed Distribution` (PD) as key.

Every node that needs to estimate the distribution of the sensors does a `get` operation to retrieve the measurements. Then, the sensor uses the estimation function to calculate the distribution. In front of frequent topology modifications, it is possible for the *probes* to detect periodically their number of neighbors to update the estimation function. The tradeoff between the communication cost of the distribution of *probe*'s measurements and the fidelity of the estimation is evaluated in Section 5.

Probes' selection algorithm chooses the geographical points in a regular pattern, as shown in Figure 3. The number of *probes* is k^2, and the Cartesian axis of the network area is divided in to $k+1$ segments. All the segments on the x-axis have the same length, except the first and the last, which have half length. The y-axis is divided in the same way. The geographical points are on the convergence of the delimiters of the segment. We set $k^2 = 25$, which proved a good tradeoff as shown in Section 5. Thus, the first and last segments of the x-axis are 10% of the x range, and each other segment is 20% of the range.

The estimation function is the weighted mean of the density measured by the three nearest *probes*, with the weight equal to 1 divided by the squared distance from the *probes*, or, in formulae, the density estimation for the geographical point (x, y) is

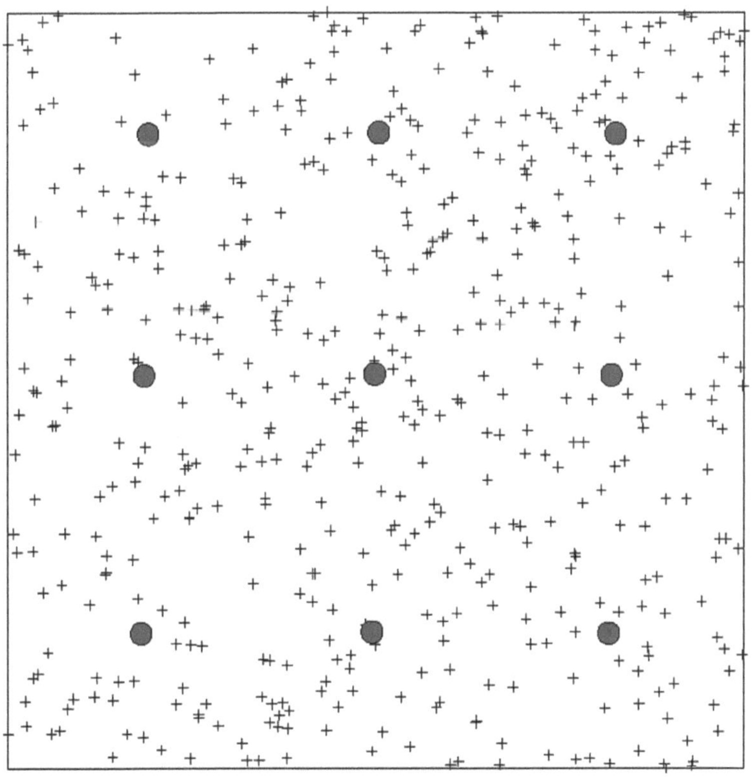

Figure 3. Geographical points chosen for nine *probes*

$$\rho(x,y) = \frac{\sum_{i=1}^{3} \frac{d_i}{(x-x_i)^2+(y-y_i)^2}}{\sum_{i=1}^{3} \frac{1}{(x-x_i)^2+(y-y_i)^2}} \tag{1}$$

with (x_i, y_i, d_i) probes, respectively, the coordinates and the number of neighbors of the three nearest *probes*.

4. DELIGHT: EXTENDING GEOGRAPHIC HASH TABLES WITH DENSITY ESTIMATION

In the initialization phase of DELiGHT the sink node selects the set of geographical points to be used to probe the network density and sends the initialization message to these points using GFG [6] as discussed in Section 3. Once the network density had been probed and stored in the network using the `put` operation of DELiGHT under the assumption of uniform distribution of the nodes, DELiGHT is full functional and can be used to store data by any node in the network taking into account the nodes' distribution.

To this purpose DELiGHT offers a modified version of the `put` and `get` operations. The first time a node performs either a `put` or a `get`, it preliminarily acquires PD (the `Probed Distribution`). The `put` and `get` operations are detailed next.

4.1 The `put` and `get` Operations

During the initialization, PROBED, a state variable which corresponds to meta-datum PD, is set to NULL. The `get` operation is used to retrieve a datum described by a meta-data (which could also be PD itself). It takes as parameter a key (the meta-data) and exploits the information about network distribution stored in PROBED, if available. If PROBED is NULL, then the `get` first acquires PROBED. This is achieved since the position of the nodes that stored PROBED can be computed using the function `RejectionHash` over metadata PD and assuming uniform distribution of the nodes (Figure 4).

Otherwise, if PROBED is available, it directly computes the location of the desired data using the function `RejectionHash` over the input meta-data and using PROBED as distribution of the nodes.

In DELiGHT the interface of the `put` includes, along with the meta-data K and the data D, a parameter Q expressing the desired QoS. `put` uses PROBED, a state variable, to estimate the distribution of the sensors. If PROBED is not available, it is first retrieved using a `get` operation with meta-datum PD. Then the algorithm calculates the geographical coordinates (x, y) and it routes the request to the home node for (x, y). In turn the home node will apply the dispersal protocol, which is presented in Section 4.3. The dispersal protocol stores the datum D, using Q as a parameter of the dependability required for the data, in terms of number of replicas required for the data (Figure 5).

```
get(K) {
    if (PROBED == NULL OR K == PD) {
        (x,y) ← RejectionHash(PD, UNIFORM)
        Route the request towards (x,y) using GFG
        PROBED is set to the received PD
    }
    if( K != PD ) {
        // now PD is available
        (x,y) ← RejectionHash(K, PROBED)
        Route the request towards (x,y) using GFG
        Return the received data associated to key K
    }
}
```

Figure 4. `get(K)` algorithm

```
put(K,D,Q) {
    if (PROBED == NULL )
        PROBED ← get(PD)
    // now PD is available
    (x,y) ← RejectionHash(K, PROBED)
    Route the request to the home node for (x,y) using GFG
    Store datum D using the dispersal protocol
}
```

Figure 5. `put(k,D,Q)` algorithm

4.2 Non-uniform Hashing

A hash function $h(k)$ is a kind of pseudo-random number generator: Starting from a seed (in our case the key k) it produces an output (in our case a value in R^2) such that for *near* values of the key, the hashed values must be *distant*. With this consideration in mind, we define a new hash function based on the *rejection method* [10], but with some differences. Rejection method is a technique used in random number generation to produce random numbers following any probability distribution, with limited dominion. The basic idea is the following: The probability function is boxed and we generate uniform random values in the box. If the value generated is below the distribution function, the value is accepted and returned. Otherwise we randomly generate new points in the box until a value is below the function. Notice that in principle there is a non-null

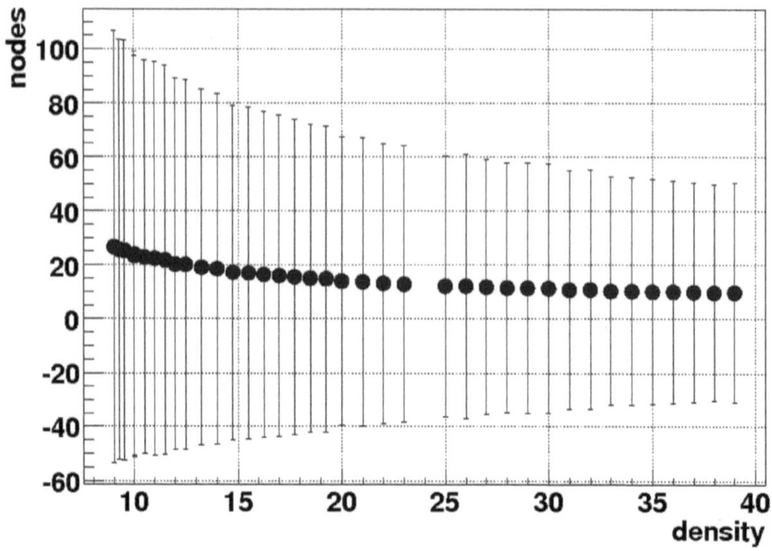

Figure 6. Mean and variance of GHT perimeters for different WSNs densities, Gaussian sensor distribution and non uniform hashing

probability of non-termination because any number of subsequent values can be generated all above the function. In practice, a good uniform hash grants to generate values in the box. The function `RejectionHash` returns a pair (x, y) of coordinates where to place data from its key `k`, belonging to the estimated PD. Instead of using uniform randomization, it uses random hashing on the key. At each iteration, if necessary, it changes lightly the key in a deterministic way. Figure 6 shows a good behavior of the non-uniform hash function. `RejectionHash` fits well the sensor distribution in the data dissemination strategy with a good global load balancing. These results are better than the results provided with uniform distribution and uniform hashing (Figure 1a and b). This is due to the Gaussian distribution of the nodes that does not have a border effect as evident as in the uniform distribution.

4.3 The Dispersal Protocol

As in GHT we call *home node* the sensor s_d (of coordinates (x', y')) geographically nearest to the destination coordinates. The home node naturally receives the packet as a consequence of applying GFG. Upon the reception of packet P_p, sensor s_d begins a *dispersal protocol*, which selects Q sensors to store a copy of `<M, D>`. The dispersal protocol is iterative and uses the concept of *ball*. Given a sensor s_d of coordinates (x, y), we denote with $B_{(x,y)}(\bar{r})$ the *ball* centered in (x, y) of radius $\bar{r}$, that is the set of sensors that are within a

Euclidean distance $\overline{r}$ from (x,y). n the first iteration s_{d} broadcasts a replica of D to all the sensors included in the ball $B_{(x',y')}(\overline{r})$. $\overline{r}$ is chosen in order to reach the Q sensors nearest to (x,y) with high probability. In simulations, to compute $\overline{r}$, we use a simple strategy based on the density estimation calculated as in Section 3. Using this heuristic, we have experimented that the method converges in no more than two iterations. Each sensor receiving a replica responds with an acknowledgment to s_{d}. Sensor s_{d} confirms the $Q-1$ acknowledgments received from the sensors geographically nearest to (x,y) and disregards the others. The confirmation requires an extra packet sent by s_{d}. Sensors which receive the confirmation keep the data while the other sensors will disregard the data after a timeout. If s_{d} receives $Q' < Q$ acknowledgments, then it executes another iteration of the dispersal protocol with $\overline{r} = 2\overline{r}$ in which it considers only the sensors in $B_{(x',y')}(2\overline{r}) - B_{(x',y')}(\overline{r})$. The dispersal protocol stops as soon as Q sensors have been hired or the outermost perimeter has been reached. When a node s_{g} of coordinates (r,z) executes `get(M)` it first computes $(x,y) = h(M)$ and sends a query packet $\mathtt{P}_{\mathrm{g}}$=<(x,y),<(r,z),M>> using the GFG protocol. In turn, packet $\mathtt{P}_{\mathrm{g}}$ will reach the perimeter surrounding (x,y) and it will start turning around the perimeter. Eventually, the packet will reach either the home node or another node containing a replica of the data D associated to M. This node will stop packet $\mathtt{P}_{\mathrm{g}}$ and will send the required data back to s_{g}. The complexity of the `put` protocol clearly depends upon the choice of $\overline{r}$ as this determines the number of iterations made to successfully place the Q replicas. However, if we know the distribution of sensors f, for any given (x',y'), coordinates of s_{d}, and Q, it is possible to fix $\overline{r}$ in such a way that, with high probability, at least Q sensors belong to the ball $B_{(x',y')}(\overline{r})$.

4.4 Behavior in Case of Faults

If some of the nodes holding the replicas of <M, D> fail, our protocol ensures graceful degradation of data availability. Due to GFG protocol, any `get` with key M is routed to the node geographically nearest to $(x,y) = h(M)$ (home node of M). If the faulty node is not the home node, the protocol implicitly discards it. If this is not the case, the second nearest node in the perimeter is always included in the ball built by our protocol.

5. SIMULATIONS AND RESULTS

In this section, we discuss the results of our simulation. We simulated a square with a 400 m side, with sensor transmission range of a perfect 10 m radius circle. We assumed a density of 14 and performed 2000 `put` operations with randomly generated meta-data using both GHT and DELiGHT. In these trials, DELiGHT uses a pure replication QoS with 15 replicas for each datum.

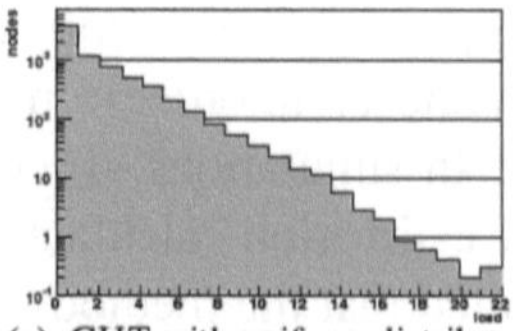

(a) GHT with uniform distribution

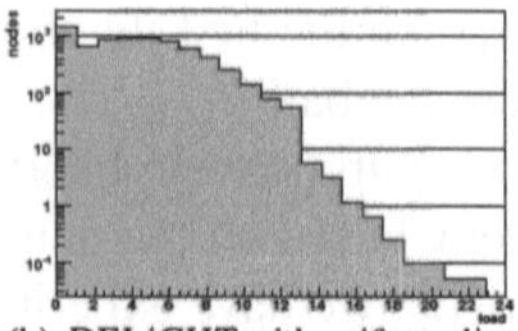

(b) DELiGHT with uniform distribution

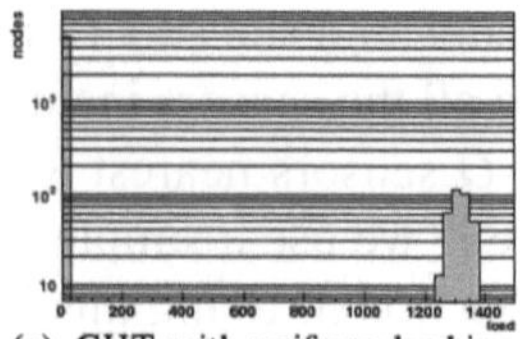

(c) GHT with uniform hashing and Gaussian distribution

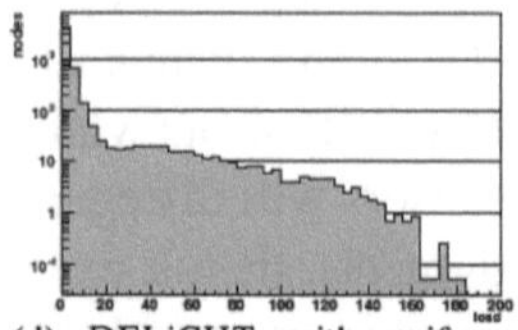

(d) DELiGHT with uniform hashing and Gaussian distribution

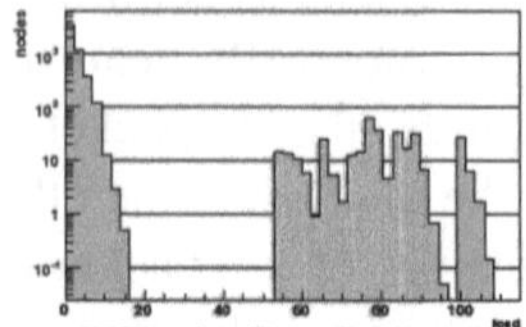

(e) GHT with Gaussian hashing and Gaussian distribution

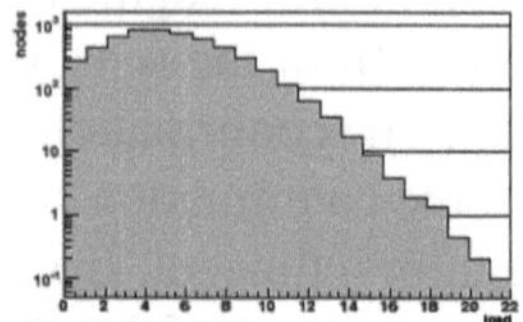

(f) DELiGHT with Gaussian hashing and Gaussian distribution

Figure 7. Average load of sensors

Figure 7 compares the behavior of GHT (graphics a,c,e) and DELiGHT (graphics b,d,f). Charts (a,b) consider uniform sensor distribution and uniform hashing, (c,d) Gaussian sensor distribution and uniform hashing and finally (e,f) Gaussian sensor distribution and Gaussian hashing. In all graphs, the x-axis shows the different load (e.g., number of data) on a node and the y-axis shows the number of nodes storing exactly this number of data. Values on the y-axis follow a logarithmic scale for better comprehension. We can notice that DELiGHT reaches better load balance even in the uniform case (graphs a,b), while from graphs (c,d) it is evident that DELiGHT reaches better load balance even in the uniform case. Figure 7c and d shows the average load of sensors in case of Gaussian sensor distribution and uniform hashing. GHT shows its usual unbalance problems, while DELiGHT is able to balance the load (despite uniform hashing) because it keeps replica distribution localized and avoids replication on long perimeters (which happens with GHT in low-density areas). This behavior is even more evident in Figure 7e and f in which we compare the load of GHT and DELiGHT in case of Gaussian distribution of sensors and Gaussian hashing.

Evaluating `puts` and `gets`. Figure 8 shows the mean and standard deviation of the cost of the basic `put` and `get` operations (number of hops needed to store a data). We performed 2000 `puts` and 2000 `gets` with randomly chosen meta-data. The QoS for DELiGHT is again pure replication with 15 replicas for each datum. In all graphs, the x-axis shows the sensor density in the network and the y-axis the operation cost measured. Graphs in the first row compare the cost of a `put` operation in GHT (a) and DELiGHT (b). The `put` is much more efficient in DELiGHT as it keeps the replicas localized in a ball without

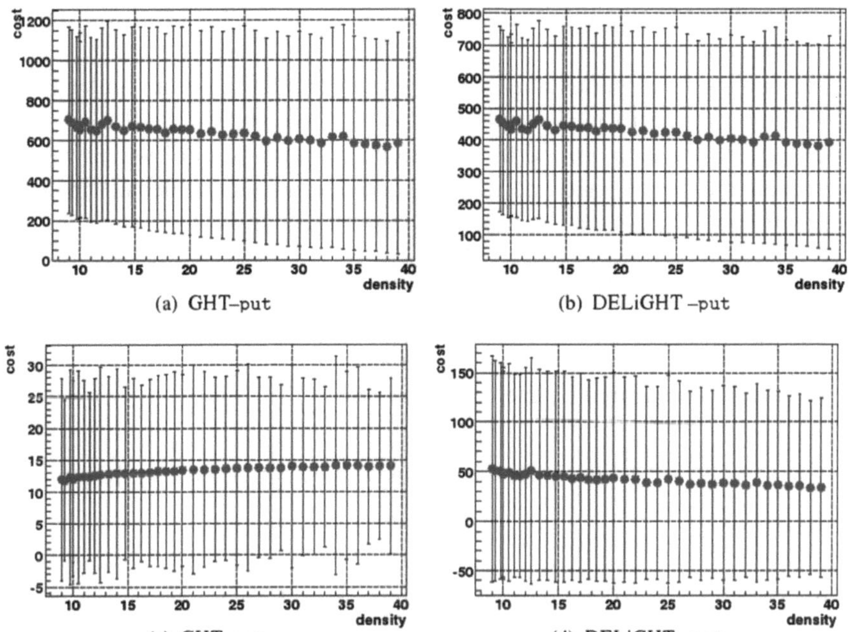

(a) GHT–put (b) DELiGHT –put

(c) GHT–get (d) DELiGHT –get

Figure 8. Mean and standard deviation of the costs (number of hops) of `put` and `get` with Gaussian distribution

following long perimeters across the network. On the other hand, Figure 8c and d shows the mean and standard deviation of the cost of a `get` operation with GHT (c) and DELiGHT (d). The cost of DELiGHT is greater than GHT. This is due to the fact that in GHT, as soon as a `get` request hits a node in the perimeter it immediately finds the data; on the other hand, using DELiGHT that request must travel until it reaches the replication ball, which may need more hops. In fact, if meta-data is hashed outside the external perimeter, a `get` operation could mean traversing the entire network before hitting the ball where the data was stored.

Note, however, that the higher costs incurred in the `get` of DELiGHT is largely counterbalanced by the much larger cost of the `put` of GHT, and that this additional cost also provides load balance and QoS in the network. Note also that the `put` operations are expected to be much more common than `gets` during the network lifetime.

Density estimation evaluation. In this section we show the results of the simulations that we performed to select an estimation function to approximate the sensor distribution.

Section 3 details the motivation to have an heuristic that estimates the density of the sensors.

The geographical points to place the *probes* are chosen as a regular pattern of k^2 points. We performed simulations with $k^2 = 9, 16, 25, 36, 49$.

Let $A_{x,y}$ be the area around the geographical point (x, y) of which we want to estimate the density, and (x_i, y_i, d_i) the coordinates and the density measured by the *probes*. The estimation functions that we test are

1. The mathematical mean of the densities measured by the *probes*
2. The theoretical density of the network
3. The density measured by the nearest *probe*
4. $\rho(x,y) = \frac{\sum_i \frac{d_i}{(x-x_i)^2+(y-y_i)^2}}{\sum_i \frac{1}{(x-x_i)^2+(y-y_i)^2}}$ with the summation performed on all the *probes* of the network
5. $\rho(x,y) = \frac{\sum_i^3 \frac{d_i}{(x-x_i)^2+(y-y_i)^2}}{\sum_i^3 \frac{1}{(x-x_i)^2+(y-y_i)^2}}$ with the summation performed on the three *probes* that are closer to (x, y).

The difference between heuristics 4 and 5 is that the latter performs the weighted mean only on the three *probes* of the PD (`Probed Density`) that are closer to (x, y) while the former uses the entire PD.

The other simulation parameters are

- Network area is a square with a 400 m side.
- Sensor transmission range is 10 m.
- The number of sensors is between 3600 and 20,000.
- Sensor distribution can be uniform or Gaussian.

The performance of the estimation is measured using the χ^2 of the estimation with a number of degrees of freedom equal to the number of nodes minus the number of *probes*. The χ^2 is divided by the number of degrees of freedom. As shown in [13], the χ^2 function should approach 1.

Note that the Gaussian distribution is approximated using polynomial functions. The motivation is that we look for a general solution to the estimation problem. Thus, we do not exploit the knowledge of the real distribution function.

The results of the simulations show that a uniform distribution of sensors is estimated equally well by every estimation function we proposed. More interesting are the results on Gaussian-distributed WSNs.

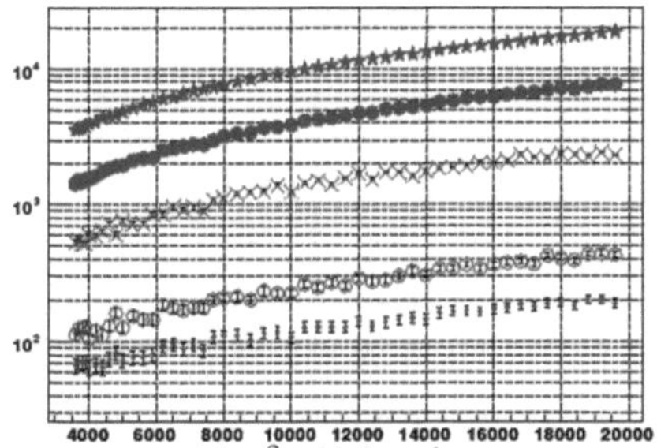

(a) χ^2 with 9 *probes*

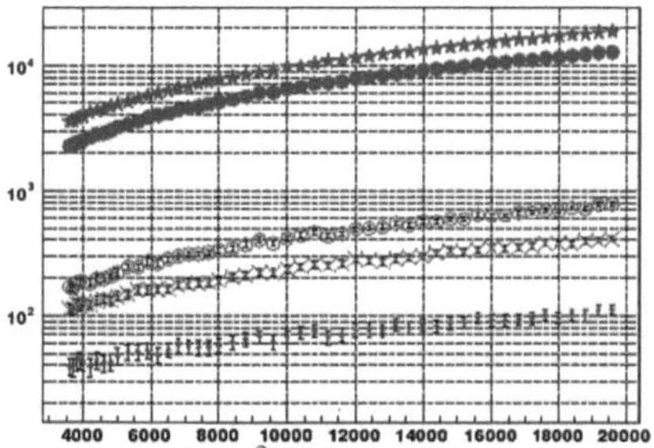

(b) χ^2 with 49 *probes*

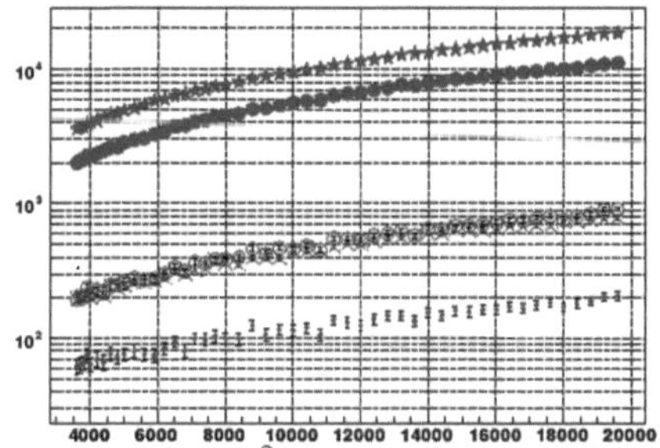

(c) χ^2 with 25 *probes*

Figure 9. χ^2 for different heuristics. From *top* to *bottom*, Heuristics 1, 2, 3, 4 and 5

The first two heuristics have a lousy performance. In fact, they calculate a global estimation, independent of the geographical point the density is estimated on.

Heuristic 3 performs slightly better, but it does not implement a real interpolation of the densities measured by the *probes*.

Heuristic 4 does a poor job, caused by the non-locality of the data that are used in the estimation.

Heuristic 5 significantly outperforms the other heuristics. The χ^2 cannot approach 1, because the estimation function we used is not a Gaussian.

Figure 9a shows a comparison of the different heuristics with 25 *probes*. Figure 9b shows the same comparison with 49 *probes*. The behaviors of the estimation functions are clearly visible, and it is seen that the quality of the approximation with more than 25 *probes* does not provide significant improvements, while with less than 25 *probes* the quality is poor.

6. CONCLUSIONS

In this chapter, we discussed the limitations of the GHT in practical WSNs and we have proposed a new protocol (DELiGHT) which overcomes these limitations and allows a fine QoS control by the user. DELiGHT corrects the ill-behavior of GHT in three ways: (1) It uses non-uniform hash tables, which allow a much more balanced distribution of data across the WSN when sensors are distributed in a non-uniform way. (2) The *dispersal protocol* used in DELiGHT allows the user to control the number of replicas of a given datum (using the quality of service parameters in the `put`). (3) Data replicas are placed in sensors which are "as close as possible" to the home node, which results in much more balanced load on all the WSNs. The merits of DELiGHT have been evaluated through simulation in uniform and Gaussian-distributed WSNs. The results show that the protocol performs a better load balancing with respect to GHT and has a smaller cost for `put` operations.

ACKNOWLEDGMENTS

This work was supported in part by the 6th EU framework programme projects SMEPP (contract number 033563), PERSONA (contract number 045459), and INTERMEDIA (contract number 38419).

REFERENCES

[1] Albano, M., Chessa, S., Nidito, F., Pelagatti, S.: "Geographic hash tables with QoS in non uniform sensor networks", *ACM Mobihoc 2006 (Poster Session)*, Firenze, Italy, 22–25 May 2006, p. 3.

[2] Albano, M., Chessa, S., Nidito, F., Pelagatti, S.: "Q-NiGHT: adding QoS to data centric storage in non-uniform sensor networks", *Technical report*, Dipartimento di Informatica, Università di Pisa, 2006.

[3] Araujo, F., Rodrigues, L., Kaiser, J., Liu, C., Mitidieri, C.: "CHR: a distributed hash table for wireless ad hoc networks", In: *Proc. of the 25th IEEE ICDCSW'05*, 2005.

[4] Baronti, P., Pillai, P., Chook, V., Chessa, S., Gotta, A., Hu, Y.F.: "Wireless sensor networks: a survey on the state of the art and the 802.15.4 and ZigBee standards", *Computer Communications*, 2007, doi: 10.1016/j.comcom.2006.12.020.

[5] Bian, F., Govindan, R., Schenker, S., Li, X.: "Using hierarchical location names for scalable routing and rendezvous in wireless sensor networks", In: *SenSys '04*, New York, 2004, pp. 305–306.

[6] Bose, P., Morin, P., Stoimenovič, I., Urrutia, J.: "Routing with guaranteed delivery in ad hoc wireless networks", *Wireless Netw., 7*(6), 2001, 609–616. Also in *DialM'99*, Seattle, August 1999, pp. 48–55.

[7] Intanagonwiwat, C., Govindan, R., Estrin, D.: "Directed diffusion: a scalable and robust communication paradigm for sensor networks", In: *Proc. of MobiCom 2000*, Boston, 2000, pp. 56–67.

[8] Jaromczyk, J., Toussaint, G.: "Relative neighborhood graphs and their relatives", *Proc. IEEE, 80*(9), 1992, 1502–1517.

[9] Maddenand, S., Franklin, M.J., Hellerstein, J.M., Hong, W.: "The design of an acquisitional query processor for sensor networks", In: *Proc. of the 2003 SIGMOD Conference*, San Diego, 2003, pp. 491–502.

[10] Neumann, J.V.: "Various techniques used in connection with random digits", In Taub, A.H., ed., *John von Neumann, Collected Works*, Volume 5, Pergamon Press, Oxford, 1951, 768–770.

[11] Newsome, J., Song, D.: "GEM: Graph EMbedding for routing and data-centric storage in sensor networks without geographic information", In: *Proc. of the First International Conference on Embedded Networked Sensor Systems*, Los Angeles, 2003, pp. 76–88.

[12] Ratnasamy, S., Karp, B., Shenker, S., Estrin, D., Govindan, R., Yin, L., Yu, F.: "Data-centric storage in sensornets with GHT, a geographic hash table", *Mob. Netw. Appl. (MONET), 8*(4), 2003, 427–442.

[13] Wilson, E.B., Hilferty, M.M.: "The distribution of chi-square", *Proc. Natl. Acad. Sci., 17*, 1931, 684–688.

[14] Yao, Y., Gehrke, J.: "The Cougar approach to in-network query processing in sensor networks", *SIGMOD Rec., 31*(3), 2002, 9–18.

EXTENDING INSTRUMENTATION GRIDS TO WIRELESS SENSOR NETWORKS

A. Lenis[1], M. Grammatikou[1], V. Maglaris[1], and S. Papavassiliou[1]

[1]*Network Management & Optimal Design Laboratory (NETMODE), School of Electrical & Computer Engineering, National Technical University of Athens (NTUA), Zografou 15773, Athens, Greece*

Abstract Grids of computing and storage resources have been extended with a new type of resources, the instruments, in what is referred to as instrumentation grids. Sensor grids can be viewed as an instance of instrumentation grids, whereby wireless sensor networks are integrated within a distributed computing/storage resource-sharing environment. They enable wireless sensor networks with computing power, high-capacity storage and standardized interaction with users and other sensor networks. In this chapter we propose a sensor grid based on the architecture and services of the instrumentation grid as developed within the GRIDCC project. We also present a sensor grid testbed and application that we deployed to study and improve our architecture.

1. INTRODUCTION

In recent years, grid computing has evolved as a platform for the coordinated sharing of distributed and heterogeneous resources over the network amongst dynamic virtual organizations. Today's existing implementations have focused on computational and data grids. They address the requirements for high computational power and data storage for applications with increasing demands in fields like high-energy physics, meteorology and biomedical computations.

Recently, along with computing elements and storage elements new types of resources – instruments – are joining grid infrastructures.

Instruments share common needs like the possibility to be accessed and monitored remotely and introduce requirements for real-time or highly interactive operation with computing grid resources. Instrumentation grids focus on the creation of a coherent collection of services that can allow the remote configuration of partitions and controls of a physical instrument and a better integration with the computational grid. Instrumentation grids introduce a new type of resource, the instrument element (IE) [7] that is equivalent to the computing elements (CE) and storage element (SE) of computational grids.

Inside the IE, a set of Web Service interfaces called VIGS (virtual instrument grid service) enables users to remotely access physical instruments. In addition the IE provides facilities for interactive co-operation between computing grid resources and applications that have real-time requirements or need fast interaction with CEs and SEs.

Wireless sensor networks are increasingly being deployed in many important applications that require interaction between computer systems and the physical world. They allow high-resolution (dense) observation of the physical environment and they increase the quantity and quality of real-world data for applications. Examples of sensor network applications include weather monitoring, healthcare monitoring of patients, security surveillance and smart home management [4,6].

Sensor grids extend the grid paradigm to add sensors as new types of resources in the network. A sensor grid is the result of the integration of multiple and isolated sensor networks with each other, in order to form a large overlay network, and with the existing grid resources and services.

Sensor networks greatly enhance their functionality by being organized according to the grid architecture and interworking with computational storage grids. First of all data from a larger number of sensors can be collected to increase the resolution of the observation space and to enable new distributed algorithms to be applied on disperse sensors data. Second, data can be redirected to storage and computing elements, complementing the sensor networks with computational power and storage capacity.

Sensor grid is a relatively new area of research. Design and implementation of such a system needs to address issues like interconnection of dissimilar devices, routing, scalability, availability, quality of service, security and scheduling.

We propose a sensor grid architecture based on the instrumentation grid platform to address these issues. The rationale behind this approach is that a sensor network, being a set of spatially distributed autonomous devices, has similar requirements for remote access, management and interaction with computing grid, as other instruments do.

The rest of the chapter is organized as follows. In Section 2, we present the instrumentation grid infrastructure as it is implemented in the GRIDCC project. In Section 3 we discuss how sensor networks can benefit from integration with the instrumentation grid in more detail. In Section 4 we describe the setup of our prototype sensor grid testbed, and finally Section 5 concludes the chapter.

2. THE GRIDCC PROJECT

"Grid enabled Remote Instrumentation with Distributed Control and Computation" (GRIDCC) [7] is a project funded by the European Community, aimed at providing access to and control of distributed complex instrumentation.

The GRIDCC system allows managing, maintaining and exploiting heterogeneous instrumentation and acquisition devices by providing the users with standardized interfaces and common working environments. This is achieved through the properties of isolation from the physical network and from the peculiarities of the instrumentation, granted by standard middleware, together with secure and flexible mechanisms to seek, access and aggregate distributed resources.

The core services of the system are shown in Figure 1 and can be summarized as follows:

1. Virtual control room (VCR): Provides a common set of collaboration tools and allows users to build complex workflows, which are then submitted to the execution services, and to directly monitor and control remote instruments in real time.

2. Compute elements (CE) and storage elements (SE): Provide access to computing and storage resources, which are common place in grid infrastructures.

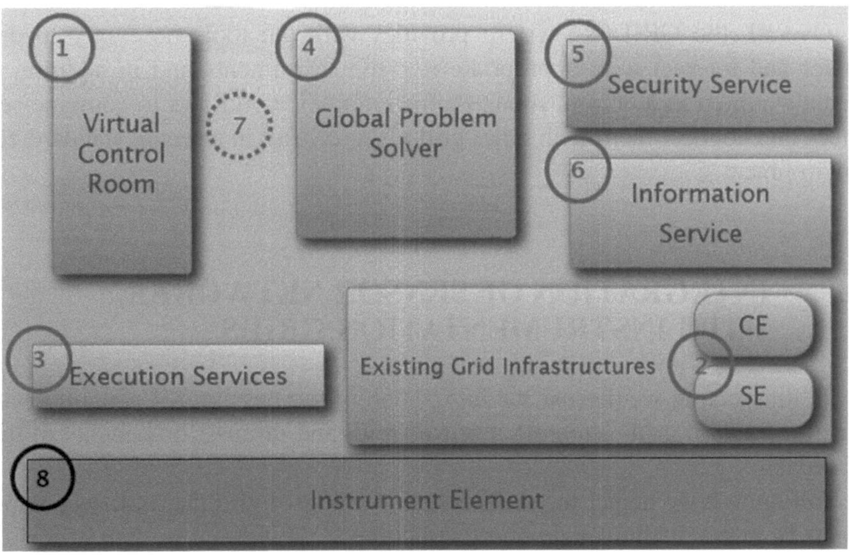

Figure 1. The GRIDCC services

3. Execution services (ES): Control the execution of the workflows defined by the user in the VCR, maintaining the status of the tasks that make up the workflow. Also, they support the advance reservation of resources. This information is passed to the user in the VCR upon request.

4. Problem solver (PS): Offers automated problem solving in a grid environment at two levels. The first, local to a given instrument element, allows solving problems related to functionalities of a given instrument. The second, that is global, allows solving system-wide problems.

5. Security services (SS): GRIDCC uses a split security system. When interacting with components of other grids the GSI security will be used and the users will be identified by their X.509 proxy certificate. When interacting with the IE the user will be identified by a Kerberos ticket to decrease the overhead of complex public key cryptography.

6. Information service (IS): Stores information about resources and provides such information to the ES and VCR to discover available and appropriate resources.

7. Monitoring service (MS): This service uses a publish/subscribe system to disseminate monitoring data.

8. Instrument element (IE): This is a unique concept to GRIDCC. It consists of a coherent collection of services which provide all the functionality to configure partition and control the physical instrument.

Overall, the GRIDCC system provides with the tools necessary to discover and interact with appropriate real-life instruments and in addition to visualize them in a control room, define complex workflows to combine services and connect them with other grid resources, all in a secure standardized environment.

3. INTEGRATION OF SENSOR NETWORKS WITH INSTRUMENTATION GRIDS

In this section we discuss the integration of wireless sensor networks with instrumentation grids, creating a sensor grid, and discuss the benefits of this approach.

In Figure 2 we depict the architecture of a sensor grid that utilizes components from the instrumentation grid structure.

The most important building block is the integration of the sensor network within the instrument element. The instrument element's internal structure

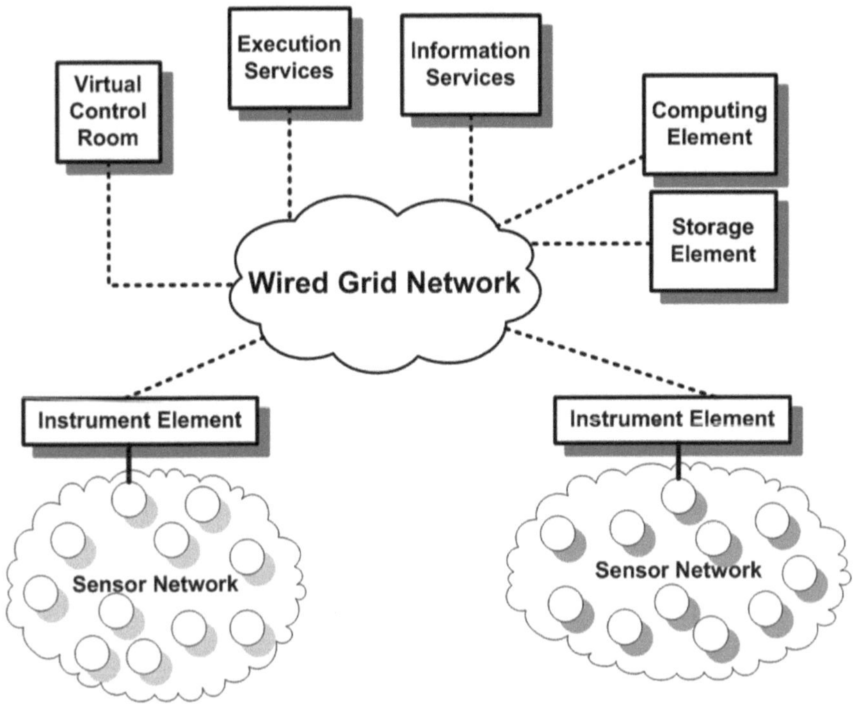

Figure 2. Sensor grid architecture

allows this kind of interaction by the implementation of a specific module, the instrument manager (IM). IMs are the internal modules of IEs that act as the mediators between the actual devices and the IE. In a wireless sensor network (WSN) the IM does not need to connect with all sensors. A single sensor, referred to as the *sink sensor*, can act as the cluster head, responsible for collecting information from the network and reporting it to higher layers for further processing.

The IM is composed of sub-modules responsible for keeping the state of the sensors in a state machine, publishing monitoring data and collecting and storing observation data. Especially for data movement, the IE can send all data directly to SEs with the data mover module. The IE can also detect problems or errors in the sensor network with the help of the local problem solver module

Once the IE connects to the WSN, it enables users of the wired grid network to access and exploit them as they would do with all other shared resources. Communications between the components on the wired part of the sensor grid are based on standard grid protocols (Web Services) to ensure interoperability and reusability with heterogeneous client applications.

3.1 Interconnection and Networking

Connectivity of wireless sensors is of an ad hoc nature (no pre-determined network hierarchy), exhibiting low bandwidth, high latency and susceptible to faults due to noise and signal degradation from environmental factors. On the other hand, grid connections are fast and reliable, based on high-quality Internet infrastructures. They use web standards to hide the low-level internal details of the resources to make applications technology agnostic.

Sensor grids require that sensor networks interconnect with the computing grid, so reliable networking and standardization protocols need to be introduced in the sensors. Since a sensor is by definition a lightweight power-limited device, it cannot have multiple interfaces to implement the complex functionality of reliable networking protocols.

The IE acts as a proxy that bridges the wireless sensor network with the wired grid network. Control messages, monitoring data and measurements are transformed from the binary protocols of the wireless devices to XML messages and vice versa. This approach exhibits several benefits. First, the extra load of communication with demanding protocols is handled outside the sensor network, thus preserving the battery power of the sensors for its main tasks. In addition, the interface of the IE, the virtual instrument grid service (VIGS), is based on standard Web Services specifications and is already compatible for deployment in a grid environment.

3.2 Scalability

A highly desirable feature in sensor applications is the ability to add more sensors to a single network or more sensor networks to a sensor grid without the need to update the software architecture. An application should be able to receive data from heterogeneous sensor networks, from different domains, and combine their abilities to create a large-scale sensor infrastructure in a dynamic nature.

Adding a sensor to a standalone network is simple. The sensor network detects the new node; its data are passed to the cluster head sensor that is connected to an IM, and from there to the IE and the grid network.

Adding a new sensor network is also a relatively straightforward procedure. First it is required that a new IE is instantiated with the appropriate IM that will interact with the new sensor network. The next step is to update the *information service* (a GRIDCC service) with the existence and capabilities of the new resource. After the resource and services catalogue is updated, client applications can discover and use this new resource.

With the help of the GRIDCC *execution service* users can create complex workflows to collect data from a large number of sensor networks dynamically and redirect them to storage and computing elements for further analysis. This process is not hard-coded and can be updated to include resources as they come and go in the grid network without the need to update software modules. From a *virtual control room* administrators can monitor the state and health of all the sensor networks and manage them accordingly.

3.3 Availability

Sensor nodes rely on limited battery power. Since the collected data can be critical, algorithms to increase the availability of the nodes are crucial.

The IE gathers monitoring data and, with the help of the local *problem solver* (*PS*), can react to changing conditions in the standalone network. If the IE detects a drop in the available power of the sensors it can reconfigure sensors to increase their lifespan (e.g., tune their power or switch them off). The PS supports rule-based algorithms and neural network, already developed within GRIDCC, but its modular architecture allows other algorithms for detection and reaction to be implemented and deployed to the system as plugins.

The *monitoring service* (*MS*) constantly updates with its publish/subscribe mechanism all interested parties, whether they are users at the *virtual control room*, the global *problem solver* or a custom application. The MS provides information on all sensor networks that participate in the grid, so decisions on increasing availability can be based on a global understanding of the state of all the resources and not on an isolated observation of a single resource.

3.4 Scheduling and Partitioning

Sensor grids allow multiple users and applications to share the same resources. To ensure that the sensors availability is not depleted by an overload of requests and that users do not interfere with each other, the platform has to provide scheduling and partitioning mechanisms. This applies to both the local scale, where a single resource can be partitioned to smaller parts, and the global scale where multiple resources from different domains can be combined.

A module of the IE, the *resource service*, provides users with the ability to partition a locally controlled network. A user can lock a subset of a sensor network for an application.

On a global scale, the *execution service* (*ES*) provides with the ability to lock a set of resources from more than one IE. The ES accepts reservation requests from users and aims at satisfying them in a single transaction, taking into account all reservations scheduled by other users.

3.5 Quality of Service

Quality of service (QoS) on a sensor grid determines whether sensor resources are provided efficiently, according to user requirements, using appropriate discovery and scheduling services.

The specification of the user's requirements should be described at a higher level, avoiding low-level technical details. A mechanism that maps these requirements to specific low-level configurations on the sensor networks is required. A user request will typically address parameters like amount of sensors, memory, bandwidth and latency. The service that interfaces with resources has to be able to export information on these parameters to external users and grid applications in general.

Once the appropriate resources are identified, a user must be able to reserve them in advance. This is tightly coupled to the scheduling and partitioning mechanisms. The QoS-enabled reservation service must utilize the existing mechanisms for scheduling and partitioning of resources, but at the same time needs to be adaptive in nature and be able to consider changes on the states of the resources. For example it must take into account changes in resource availability (power status), network topology, bandwidth and latency in order to select the resources that meet the requirements not only on the present time but also in the future with a high probability of success.

The IE has an embedded service that measures QoS parameters (e.g., latency) and provides this information to the other components of the grid network. The component that is more concerned with this information is the *execution service* (*ES*).

The ES receives as input a workflow accompanied with a desired level of quality of services participating in the workflow. This is based on standard XML formats like BPEL [1] and describes the QoS demands in a high level. The *virtual control room* has a module that helps users to write the workflow document, as BPEL is very complex in nature to write manually. The ES analyses the workflow and, using information on the state and abilities of the sensor resources, it identifies a set of resources that meet the desired requirements. The ES subsequently proceeds with the actual reservation using the scheduling mechanism of the IEs.

3.6 Security

Security is of utmost importance in any grid system. Organizations will not share their resources unless there is a guarantee that they will be protected in terms of authentication and authorization. An authentication and authorization infrastructure (AAI) enables a distributed community to form virtual organizations [3].

Wireless sensor networks are prone to security problems such as the compromising and tampering of sensor nodes, eavesdropping of sensor data and communication and denial of service attacks. Techniques to address these problems include sensor node authentication, efficient encryption, secure

MAC, etc. Some of these techniques are essential for the sensor network to filter the dataflow from and to authenticated-only nodes and the IE.

In sensor grids we also need to ensure that once a sensor network joins the grid fabric it is protected from malicious users. The key issue here is to allow access of an authorized and authenticated user to a sensor through a single controlled and secure channel. In our sensor grid architecture the instrument element is the only entry point for managing the sensor network resources. As this element is protected by well-established grid-secure mechanisms, it secures all instruments related to it.

4. TESTBED SETUP AND IMPLEMENTATION

Our wireless sensor testbed network at NTUA (NETMODE Laboratory) consists of 15 *Tmote Sky motes* [5]. These motes feature a 250 kbps 2.4 GHz IEEE 802.15.4 *Chipcon radio*, an integrated onboard antenna with 125 m range, integrated humidity and temperature sensors, ultra low current consumption, fast wakeup from sleep ($< 6\,\mu s$), hardware link – layer encryption and authentication and programming – data collection via USB.

At present we have deployed a sensor grid as shown in Figure 3 where sensors in different computer rooms are monitoring the room temperature, and

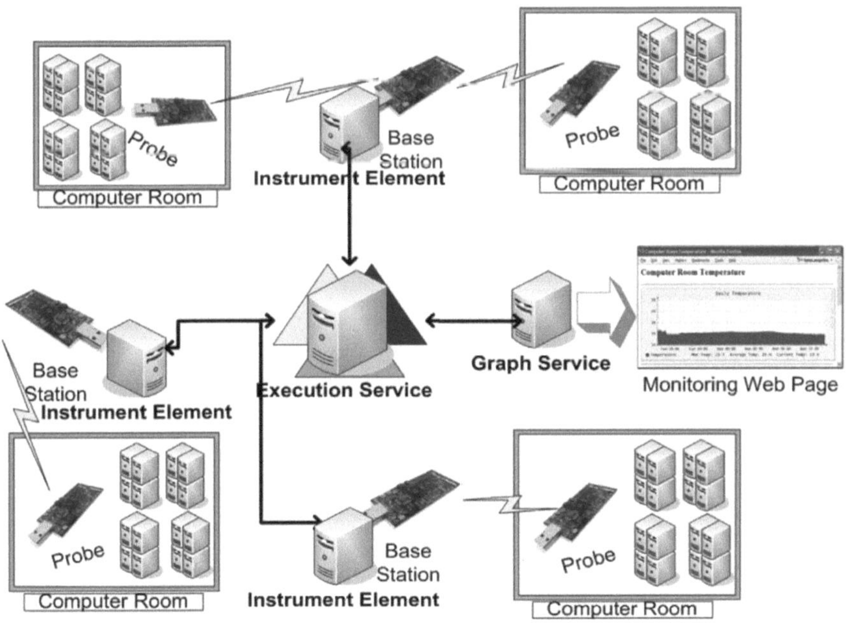

Figure 3. Testbed/application architecture

GRIDCC's IE and other services are collecting them and creating graphs for the users to view.

In more detail, we have installed the following components:

1. In each room we have deployed two sensors. One is acting as the monitoring probe and is placed inside the computer room. The other one is acting as the base station and collects the data from the probe. The location of the base station can be anywhere, limited to the probes transmission range. The base station is also connected to a PC via USB.

2. For each base station we have installed an instrument element that collects the data and makes them available to clients of the wired network.

3. One Web Service, the Graph Service, that takes as input monitoring data and creates MRTG graphs as output.

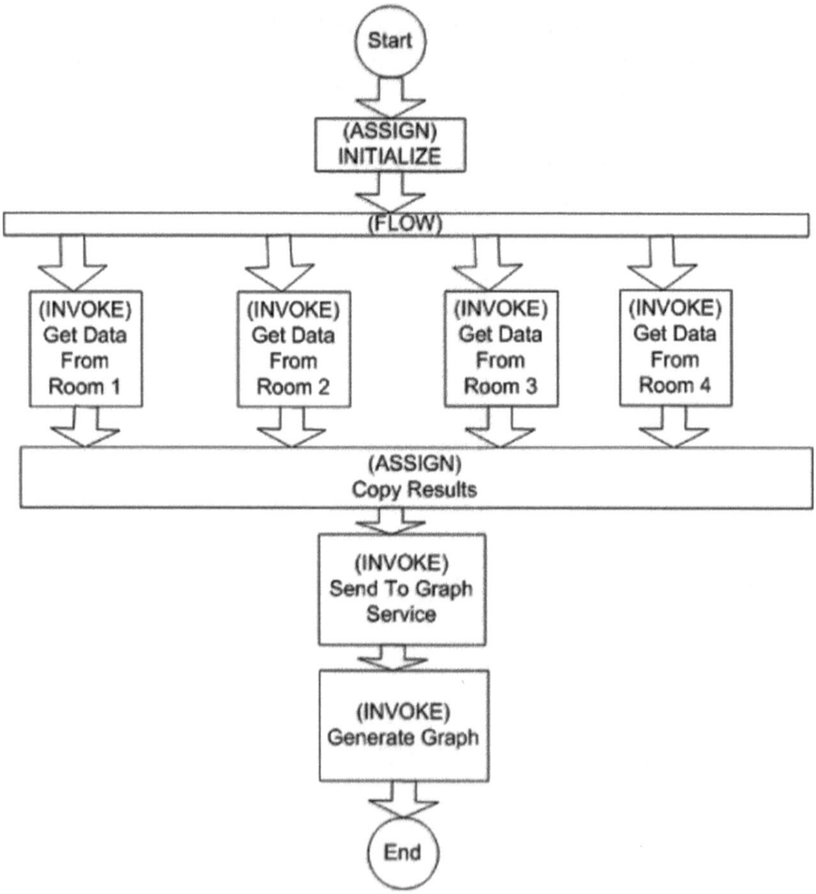

Figure 4. Application workflow

4. One execution service that orchestrates the flow of information in the wired network.

The application lifecycle is implemented as the workflow shown in Figure 4. First the execution service initializes some parameters like endpoints and local variables. It then invokes in parallel the instrument elements function that returns the values of the monitoring data. Next the values are copied to a variable that is the input for the Graph Service which is in turn invoked. Finally we run the recalculate graphs function of the service.

We have implemented the application workflow in BPEL and we are running it in circles of 5 min. The result of the workflow is the web page shown in Figure 5.

In another deployed workflow, the data from the sensors are collected from the corresponding IEs and are then stored in a storage element. Grid jobs are

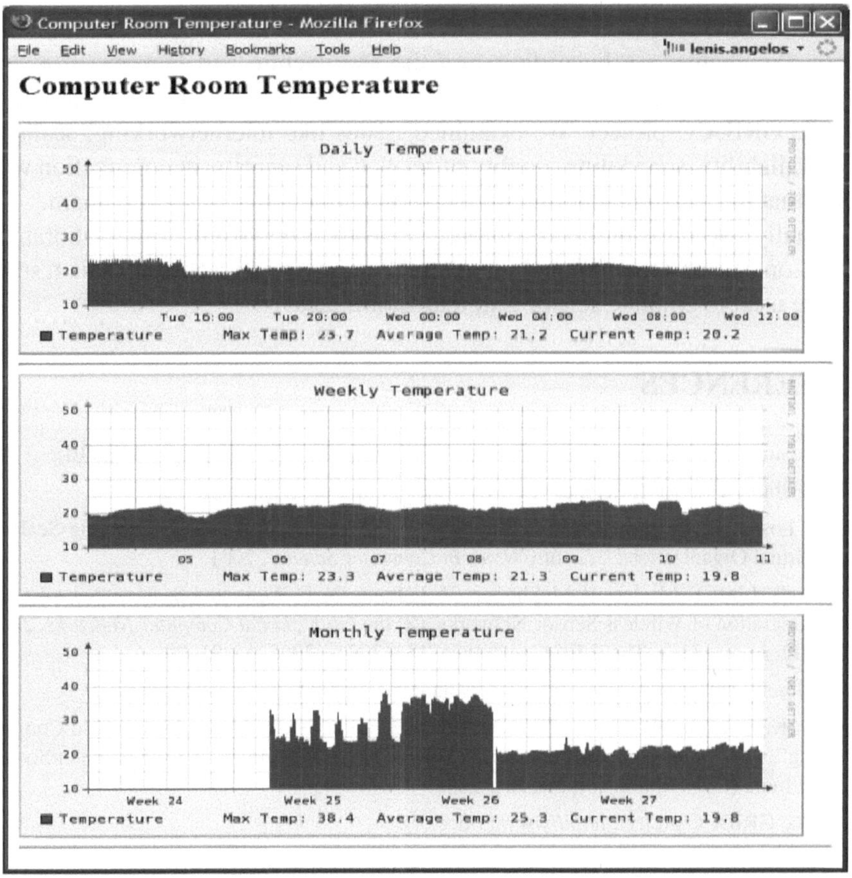

Figure 5. Web page with monitoring graphs

then submitted to a CE to analyze them as required. For this example we have used IASA's node, which is part of the EGEE grid [2], and consists of 44 HP DL140 servers, 4 HP DL360 servers, an HP MSA 1500 storage server and Cisco 3750 switches.

It is part of our current and future research plans to experiment more with the workflow and QoS mechanism of the GRIDCC infrastructure. In addition we will add more features on instrument element concerning power management, availability and error detection.

5. CONCLUSIONS

Wireless sensor networks and grid computing are technologies that can benefit greatly by converging with each other. This opens the path to an exciting new world of applications, as sensors become a key part of ambient, ubiquitous computing and the need for adding intelligence to sensor data is of paramount importance.

In this chapter we have discussed the architecture and features of a sensor grid implementation based on the instrumentation grid platform provided by the GRIDCC project. We examined issues like internetworking, scalability, availability, scheduling, quality of service and security, in conjunction with tools that the instrumentation grid platform should offer to address them.

Finally, we have outlined a prototype testbed deployed in our laboratory, based on sensor networks and computing/storage grid resources. This testbed can be used to emulate sensor grid realization ideas.

REFERENCES

[1] Business Process Execution Language, BPEL, http://en.wikipedia.org/wiki/BPEL.

[2] Enabling Grid for E-Science (EGEE) Project, http://public.eu-egee.org.

[3] I. Foster, C. Kesselman and S. Tuecke, "The Anatomy of the Grid: Enabling Scalable Virtual Organizations", *Lecture Notes in Computer Science*, 2001.

[4] H. B. Lim, Y. M. Teo, P. Mukherjee, V. T. Lam; W. F. Wong and S. See, "Sensor Grid: Integration of Wireless Sensor Networks and the Grid", *Local Computer Networks*, 2005, 30th Anniversary, The IEEE Conference, 15–17 Nov. 2005, pp. 91–99.

[5] Moteiv, Tmote Sky motes, http://www.moteiv.com/products/tmotesky.php.

[6] C.-K. Tham and R. Buyya, "SensorGrid: Integrating Sensor Networks and Grid Computing", *CSI Communications*, pp. 24–29, Vol. 29, No. 1, ISSN 0970-647X, Computer Society of India (CSI), Mumbai, India, July 2005.

[7] The GRIDCC project, http://www.gridcc.org.

REDUCED-COMPLEXITY DECENTRALIZED DETECTION OF SPATIALLY NON-CONSTANT PHENOMENA

G. Ferrari[1], M. Martalò [1], and M. Sarti [2]

[1] *Wireless Ad-hoc and Sensor Networks (WASN) Laboratory, Department of Information Engineering, University of Parma, Parma, Italy,*

[2] *Elettric 80 S.p.A., Viano, Reggio Emilia, Italy*

gianluigi.ferrari@unipr.it, martalo@tlc.unipr.it, sarti.m@elettric80.it

Abstract In this chapter, we study sensor networks with decentralized detection of a *spatially non-constant* phenomenon, whose status *might change* independently from sensor to sensor. In particular, we consider *binary* phenomena characterized by a *fixed* number of status changes (from state "0" to state "1") across the sensors. This is realistic for sensor networking scenarios where abrupt spatial variations of the phenomenon under observation need to be estimated, e.g., an abrupt temperature increase, as could be the case in the presence of a fire in a specific zone of the monitored surface. In such scenarios, we derive the minimum mean square error (MMSE) fusion algorithm at the access point (AP). The improvement brought by the use of quantization at the sensors is investigated. Finally, we derive simplified (sub-optimum) fusion algorithms at the AP, with a computational complexity lower than that of schemes with MMSE fusion at the AP.

1. INTRODUCTION

Sensor networks have been an active research field in the last years [1]. In particular, a lot of civilian applications have been developed on the basis of this technology, especially for environmental monitoring [14]. Several frameworks have been proposed for the analysis of sensor networks with a common binary phenomenon under observation [5, 15]. In [4], noisy communication links are modeled as binary symmetric channels (BSCs) and a few techniques, such as the use of multiple observations or selective decentralized detection, are proposed in order to make the system more robust against the noise.

While in [4] the focus is on decentralized detection of a binary phenomenon *common* for all sensors, it is of interest to extend this approach to a scenario where the status of the phenomenon may vary from sensor to sensor. In [10], the authors consider a scenario with a single phenomenon status change (denoted, in the following, as *boundary*) and propose a framework, based on a minimum mean square error (MMSE) detection strategy, to determine the position of the boundary. In [11], under the assumption of proper regularity of the observed boundary, a reduced-complexity MMSE decoder is proposed. In [2], the authors show that an MMSE decoder is unfeasible for large-scale sensor networks, due to its computational complexity, and propose a distributed detection strategy based on factor graphs and the sum–product algorithm. Finally, MMSE-based distributed detection schemes have also been investigated in scenarios with a common binary phenomenon under observation and bandwidth constraints [9].

In this chapter, we propose an analytical approach to the design of decentralized detection schemes in scenarios with *spatially non-constant* binary phenomena, i.e., phenomena with status (either "0" or "1") which may vary from sensor to sensor. First, we focus on phenomena with a *single* boundary, i.e., scenarios where there is a single (spatial) position in correspondence to which the phenomenon status changes. Then, we analyze scenarios with a *generic* number of boundaries, i.e., more realistic scenarios where the number of phenomenon status changes may be larger than one. For both these scenarios, we derive the MMSE fusion algorithms[1] at the AP, considering various quantization strategies at the sensors [8]. In order to make our approach practical, we derive simplified fusion algorithms with a computational complexity much lower than that of the MMSE fusion rules. Although heuristic, the simplified fusion algorithms guarantee a *limited* performance loss for sufficiently high values of the sensor SNR.

This chapter is structured as follows. In Section 2, we derive MMSE and simplified fusion rules in a scenario where the observed phenomenon is characterized by a single boundary. In Section 3, we extend our framework to the case with a spatially non-constant phenomenon with a generic number of boundaries. In Section 4, numerical results associated with the proposed fusion rules are presented. In Section 5, a simple computational complexity analysis, based on the number of operations required by the derived algorithms, is proposed. Finally, in Section 6 concluding remarks are given.

[1]Note that the proposed MMSE distributed detection schemes are *optimal* in the processing performed at the AP. However, the use of quantization at the sensors makes the overall schemes sub-optimal. Overall optimality holds only for a scenario with no quantization at the sensors.

2. PHENOMENA WITH A SINGLE BOUNDARY

2.1 MMSE Fusion Rule

Consider a network scenario where N sensors observe a (spatially) non-constant binary phenomenon characterized by a single status change across the sensors. The phenomenon status can be expressed as $\boldsymbol{H} = [H_1, H_2, \ldots, H_N]$, with

$$H_i \triangleq \begin{cases} 0 & \text{if } i < \alpha \\ 1 & \text{if } i \geq \alpha \end{cases} \qquad i = 1, \ldots, N$$

where the index α is the position of the *boundary* in correspondence to which the phenomenon status changes (from "0" to "1"). The position of the boundary is modeled as uniformly distributed across the sensors, i.e., $P(\alpha = \ell) = 1/N$, $\ell = 1, \ldots, N$.

The signal observed at the ith sensor can be expressed as

$$r_i = c_{\mathrm{E},i} + n_i \quad i = 1, \ldots, N$$

where

$$c_{\mathrm{E},i} \triangleq \begin{cases} 0 & \text{if } H_i = 0 \\ s & \text{if } H_i = 1 \end{cases}$$

and $\{n_i\}$ are additive observation noise samples. Assuming that the noise samples $\{n_i\}$ are independent with the same Gaussian distribution $\mathcal{N}(0, \sigma^2)$, the common signal-to-noise ratio (SNR) at the sensors can be defined as follows:

$$\mathrm{SNR}_{\mathrm{sensor}} = \frac{[\mathbb{E}\{c_{\mathrm{E},i} | H_i = 1\} - \mathbb{E}\{c_{\mathrm{E},i} | H_i = 0\}]^2}{\sigma^2} = \frac{s^2}{\sigma^2}$$

Each sensor quantizes the observed signal and the value sent by the ith sensor is denoted as $d_i \triangleq f_{\mathrm{quant}}(r_i)$, where the function $f_{\mathrm{quant}}(\cdot)$ depends on the specific quantization strategy. In the following, we consider (i) binary quantization, (ii) multi-level quantization, and (iii) absence of quantization (i.e., the observations $\{r_i\}$ are sent to the AP). Based on the messages sent by the sensors, the goal of the AP is to reconstruct, through an MMSE fusion strategy, the status of the distributed binary phenomenon $\boldsymbol{H}$. More precisely, in the considered setting the AP needs to estimate correctly the position of the boundary.

2.1.1 Binary Quantization

In this scenario, the ith sensor makes a decision comparing its observation r_i with a threshold value τ_i and computes a local decision $d_i \in \{0, 1\}$, i.e., $f_{\mathrm{quant}}(r_i) = U(r_i - \tau_i)$, where $U(\cdot)$ is the unit step function. In order to optimize the system performance, the thresholds $\{\tau_i\}$ need to be properly selected.

In this chapter, a common value τ at all sensors is considered. This choice is intuitively motivated by the fact that the sensor SNR is constant across the sensors. While in a scenario with a common binary phenomenon the relation between τ and s is well known [13], in the presence of a non-constant phenomenon the threshold τ needs to be optimized in order to minimize the probability of decision error at the AP. This optimization will be carried out for all the considered scenarios (see the beginning of Section 4 for more details).

The estimated boundary position, denoted as $\widehat{\alpha}$, can be reasonably chosen as the minimizing value for the mean square error (MSE), i.e.,

$$\widehat{\alpha} \triangleq \operatorname*{argmin}_{\tilde{\alpha}=1,\ldots,N} \mathbb{E}\left[|\alpha - \tilde{\alpha}|^2 |\boldsymbol{d}\right] \tag{1}$$

where $\boldsymbol{d} \triangleq [d_1, d_2, \ldots, d_N]$ is the vector of sensors' decisions. The solution of the MMSE problem (1) is well known [6, Ch.10]:

$$\widehat{\alpha} = \mathbb{E}[\alpha|\boldsymbol{d}] = \sum_{m=1}^{N} m P(\alpha = m|\boldsymbol{d}) \tag{2}$$

where each conditional probability $P(\alpha = m|\boldsymbol{d})$ can be expressed, using the Bayes formula and the total probability theorem [12], as

$$P(\alpha = m|\boldsymbol{d}) = \frac{P(\boldsymbol{d}|\alpha = m)P(\alpha = m)}{\sum_{\ell=1}^{N} P(\boldsymbol{d}|\alpha = \ell)P(\alpha = \ell)} \tag{3}$$

At this point, one has to calculate the probabilities $\{P(\boldsymbol{d}|\alpha = m)\}_m$ in (3). Since the noise samples are independent, conditionally on the value of the boundary position the sensors' decisions are also independent. Therefore, one obtains

$$P(\boldsymbol{d}|\alpha = m) = \prod_{k=1}^{N} P(d_k|\alpha = m) \tag{4}$$

where

$$P(d_k|\alpha = m) = \begin{cases} P\left(n_k \underset{d_k=1}{\overset{d_k=0}{\gtrless}} \tau\right) & k < m \\ P\left(n_k \underset{d_k=1}{\overset{d_k=0}{\gtrless}} \tau - s\right) & k \geq m \end{cases} \tag{5}$$

One should note that expression (5)—and, consequently, the estimated boundary position $\widehat{\alpha}$ in (2)—depends on the particular sequence $\boldsymbol{d}$ of sensors' decisions.

The computational complexity required for the evaluation of (2) is very high. For this reason, in the following, MMSE-based detection schemes will be applied only in scenarios with a (relatively) small number of sensors. In Section 2.2, we will derive a simplified fusion rule, in order to analyze scenarios with larger numbers of sensors.

2.1.2 Multi-level Quantization

Considering n_b quantization bits, we set the $2^{n_b}-1$ quantization thresholds as follows:

$$\frac{s}{2}, \frac{s}{2} \pm \Delta, \frac{s}{2} \pm 2\Delta, \ldots, \frac{s}{2} \pm \left(2^{n_b-1}-1\right)\Delta$$

where Δ is a system parameter which needs to be properly optimized—more details on its optimization will be provided in Section 4.1. The central threshold corresponds to $\tau = s/2$, so that if $n_b = 1$, one obtains the scenario with binary quantization [4]. In this case, the sensors' quantized decisions are taken according to the following rule:

$$d_i = \begin{cases} 0 & \text{if } r_i < s/2 - \left(2^{n_b-1}-1\right)\Delta \\ 1 & \text{if } s/2 - \left(2^{n_b-1}-1\right)\Delta \le r_i < s/2 - \left(2^{n_b-1}-2\right)\Delta \\ \cdots & \cdots \\ 2^{n_b}-2 & \text{if } s/2 + \left(2^{n_b-1}-3\right)\Delta \le r_i < s/2 + \left(2^{n_b-2}-1\right)\Delta \\ 2^{n_b}-1 & \text{if } r_i > s/2 + \left(2^{n_b-1}-1\right)\Delta \end{cases} \tag{6}$$

The MMSE fusion algorithm at the AP is similar to that in Section 2.1.1. However, one needs to properly compute the probabilities $\{P(\boldsymbol{d}|\alpha = m)\}_m$ in (4). According to (6), expression (5) can be generalized as follows:

$$P(d_k = j|\alpha = m) = \begin{cases} \Phi\left(\tau - \left(2^{n_b-1}-1\right)\Delta - s\cdot U(m-k+1)\right) & \text{if } j = 0 \\ \Phi\left(\tau - \left(\left(2^{n_b-1}-1\right)-j\right)\Delta - s\cdot U(m-k+1)\right) & \\ -\Phi\left(\tau - \left(\left(2^{n_b-1}-1\right)-j+1\right)\Delta - s\cdot U(m-k+1)\right) & \\ & \text{if } j = 1,\ldots,2^{n_b}-2 \\ 1-\Phi\left(\tau + \left(2^{n_b-1}-1\right)\Delta - s\cdot U(m-k+1)\right) & \\ & \text{if } j = 2^{n_b}-1 \end{cases} \tag{7}$$

where $\Phi(x) \triangleq \int_{-\infty}^{x} \frac{1}{\sqrt{2\pi}} \exp(-y^2/2)\mathrm{d}y$. Obviously, the computational complexity of the fusion algorithm at the AP increases with the use of multi-level quantization. In fact, from (7) one can conclude that 2^{n_b} probability values have to be evaluated—they reduce to $2^1 = 2$ values in the presence of binary quantization.

2.1.3 Absence of Quantization

In this case, the observations at the sensors are not quantized and a local likelihood value, such as the conditional probability density function (PDF) of the observable, is transmitted from each sensor to the AP. Obviously, this is not a realistic scenario, since an infinite bandwidth would be required to transmit a PDF value. However, investigating this case allows to derive useful information about the limiting performance of the considered decentralized detection schemes, since transmission of the PDF of the observables does not entail any information loss at the sensors.

The MMSE estimate of the boundary position can now be written as

$$\widehat{\alpha} = \underset{\tilde{\alpha}=1,\ldots,N}{\operatorname{argmin}} \mathbb{E}\left[|\alpha - \tilde{\alpha}|^2|\boldsymbol{r}\right] = \mathbb{E}\left[\alpha|\boldsymbol{r}\right] = \sum_{m=1}^{N} mP(\alpha = m|\boldsymbol{r}) \tag{8}$$

where $\boldsymbol{r} = [r_1, r_2, \ldots, r_N]$ is the vector of the observed signals (rather than the vector of decisions $\boldsymbol{d}$ as in (1)). The a posteriori probabilities in (8) can be expressed similarly to (3), i.e.,[2]

$$P(\alpha = m|\boldsymbol{r}) = \frac{p(\boldsymbol{r}|\alpha = m)P(\alpha = m)}{\sum_{\ell=1}^{N} p(\boldsymbol{r}|\alpha = \ell)P(\alpha = \ell)}$$

where, owing to the independence of the observations, $p(\boldsymbol{r}|\alpha = m) = \prod_{i=1}^{N} p(r_i|\alpha = m)$ and

$$p(r_i|\alpha = m) = \begin{cases} p_{\mathcal{N}}(r_i) & i < m \\ p_{\mathcal{N}}(r_i - s) & i \geq m \end{cases}$$

with $p_{\mathcal{N}}(n) \triangleq \frac{1}{\sqrt{2\pi\sigma^2}} \exp\left(-\frac{n^2}{2\sigma^2}\right)$.

2.2 Simplified Fusion Rule

Since the computational complexity of the MMSE fusion strategy rapidly increases with the number of sensors [10], in this section we derive a simplified low-complexity fusion algorithm. The key idea of this simplified algorithm consists in approximating the MMSE boundary estimate $\widehat{\alpha}$ in (2) (and, similarly, in (8)), which involves a *statistical* average, with a simpler *deterministic* expression. As in Section 2.1, various quantization levels at the sensors are considered. Note that the proposed approach relies on the fact that our goal is to estimate a *single* boundary. However, extensions of this approach to a scenario with multiple boundaries will be considered in Section 3.

[2]Note that the uppercase P is used to denote the probability of an event, whereas the lowercase p is used to denote the PDF of a random variable.

2.2.1 Binary Quantization

In this case, the boundary position can be estimated as follows:

$$\widehat{\alpha} = \underset{1\leq j\leq N}{\operatorname{argmin}} \left\{ \sum_{i=1}^{j-1} |d_i|^2 + \sum_{i=j}^{N} |d_i - 1|^2 \right\} \tag{9}$$

The intuition behind (9) is based on the fact that there is a single boundary: the initial sensors' decisions (from the first to the $(j-1)$th) are compared with "0," whereas the others (from the jth to the Nth) are compared with "1." The estimated boundary minimizes the simplified cost function $|\boldsymbol{d} - \boldsymbol{d}_j|^2$, where $\boldsymbol{d}_j \triangleq [0,\ldots,0,\underbrace{1}_{j\text{th position}},\ldots,1]$, over all possible values of j.

2.2.2 Multi-level Quantization

The approach proposed here can be extended to scenarios with multi-level quantization at the sensors. In particular, we propose the following decision rule at the AP:

$$\widehat{\alpha} = \underset{1\leq j\leq N}{\operatorname{argmin}} \left\{ \sum_{i=1}^{j-1} f_1(d_i) + \sum_{i=j}^{N} f_2(d_i) \right\} \tag{10}$$

where

$$\begin{aligned} f_1(d_i) &\triangleq \min\{|d_i|^2, |d_i - 1|^2, \ldots, |d_i - (2^{n_b-1} - 1)|^2\} \\ f_2(d_i) &\triangleq \min\{|d_i - 2^{n_b-1}|^2, |d_i - (2^{n_b-1} + 1)|^2, \ldots, |d_i - (2^{n_b} - 1)|^2\} \end{aligned}$$

The rationale behind (10) is the following. Assuming that the boundary is in the jth position, we compare the first $j-1$ decisions with the "low" half of the quantization levels (i.e., from 0 to $2^{n_b-1} - 1$), whereas the remaining $N - j + 1$ decisions are compared with the "high" half of the quantization levels (i.e., from 2^{n_b-1} to $2^{n_b} - 1$). In both groups, the quantization levels closest to the corresponding decisions are selected (through the functions f_1 and f_2).

As an aside, we remark that for sufficiently large sensor SNR the estimation strategy in (10) (and, as a special case, in (9)) leads to the same performance obtained with the MMSE fusion strategy.

2.2.3 Absence of Quantization

In this scenario, one can use the a posteriori probabilities of the two hypotheses at each sensor, conditionally on the observables, to derive the proper objective function. In this case, one can write[3]

$$\widehat{\alpha} = \operatorname*{argmax}_{1\leq j\leq N} \left\{ \sum_{i=1}^{j-1} P(H_i = 0|r_i) + \sum_{i=j}^{N} P(H_i = 1|r_i) \right\} \tag{11}$$

where, using Bayes formula and assuming $P(H_i = 0) = P(H_i = 1) = 1/2z, \forall i$, one has

$$P(H_i = \ell|r_i) = \frac{p(r_i|H_i = \ell)}{p(r_i|H_i = 0) + p(r_i|H_i = 1)} = \frac{p(n_i - \ell \cdot s)}{p(n_i) + p(n_i - s)} \qquad \ell = 0, 1$$

3. PHENOMENA WITH GENERIC NUMBERS OF BOUNDARIES

In this section, we focus on a network scenario where the status of the phenomenon under observation is characterized by a *generic* number, denoted as as N_{bs}, of boundaries.[4] This scenario is more realistic than that considered in Section 2, since a generic phenomenon (e.g., the humidity level) could change its status from "0" (e.g., low humidity) to "1" (e.g., high humidity), or vice versa, in correspondence to more than one sensor. Note that the case with $N_{\text{bs}} = 1$ corresponds to the previously investigated scenario with a single boundary. The following assumptions are expedient to simplify the derivation of the MMSE detection strategy:

- changes of the phenomenon status are not admitted in correspondence to the first and last sensors, i.e., $1 \leq N_{\text{bs}} \leq N - 2$;
- the phenomenon status is perfectly known at the first sensor ($H_1 = 0$) and there is no change at the last sensor (i.e., $H_N = H_{N-1}$).

According to the considered assumptions, the N_{bs} boundaries $\{\alpha_1, \ldots, \alpha_{N_{\text{bs}}}\}$ have to satisfy the following relation:

$$2 \leq \alpha_1 < \alpha_2 < \ldots < \alpha_{k-1} < \alpha_k < \ldots < \alpha_{N_{\text{bs}}} \leq N - 1.$$

[3]Note that in (11) the "argmax" function is used, instead of the "argmin" function used in (9) and (10), since the objective function needs to be maximized.

[4]We remark that throughout this chapter the status of the phenomenon will be supposed independent from sensor to sensor. The existence of *correlations* between sensors would require an extension of the derived algorithms. However, this extension goes beyond the scope of this chapter.

Therefore, between 1 and $\alpha_1 - 1$ the phenomenon status is "0," between α_1 and $\alpha_2 - 1$ the phenomenon status is "1," and so on. In order for the boundary distribution to be realistic, the following condition must necessarily hold:

$$\alpha_{k-1} < \alpha_k \leq (N-1) - (N_{\rm bs} - k) = N - N_{\rm bs} + k - 1 \qquad k = 2, \ldots, N_{\rm bs} \quad (12)$$

Condition (12) formalizes the intuitive idea that the kth boundary cannot fall beyond the $(N - 1 - N_{\rm bs} + k)$th position, in order for the successive (remaining) $N_{\rm bs} - k$ boundaries to have admissible positions.

3.1 MMSE Fusion Rule

3.1.1 Binary Quantization

Denoting as $\boldsymbol{\alpha}$ the sequence of boundaries $(\alpha_1, \ldots, \alpha_{N_{\rm bs}})$, the MMSE fusion strategy can be determined by directly extending the derivation in Section 2.1.1, obtaining $\widehat{\boldsymbol{\alpha}} = \mathbb{E}[\boldsymbol{\alpha}|\boldsymbol{d}]$. On the basis of the assumptions introduced at the beginning of this section, the generic term of the vector $\widehat{\boldsymbol{\alpha}}$ can be written as[5]

$$\widehat{\alpha}_k = \mathbb{E}[\alpha_k|\boldsymbol{d}] = \sum_{\alpha_k=1}^{N} P(\alpha_k|\boldsymbol{d}) = \sum_{\alpha_k=k+1}^{N-N_{\rm bs}+k-1} \alpha_k P(\alpha_k|\boldsymbol{d}) \qquad k = 1, \ldots, N_{\rm bs} \quad (13)$$

where the upper and lower bounds of the sum in the last term are properly modified in order to take into account the constraint (12). The computation of (13) can be carried out by extending (in a multi-dimensional sense) the approach in Section 2.1.1. The probability $P(\alpha_k|\boldsymbol{d})$ $(k = 1, \ldots, N_{\rm bs})$ can be obtained by marginalizing the joint probabilities of proper boundaries' sequences. By applying the Bayes formula and the total probability theorem [12], after a few manipulations one obtains

$$P(\boldsymbol{\alpha}|\boldsymbol{d}) = P(\boldsymbol{d}|\boldsymbol{\alpha})P(\boldsymbol{\alpha}) \left[\sum_{\alpha_1=2}^{N-N_{\rm bs}} \cdots \sum_{\alpha_k=k+1}^{N-N_{\rm bs}+k-1} \cdots \sum_{\alpha_{N_{\rm bs}}=N_{\rm bs}+1}^{N-1} P(\boldsymbol{d}|\boldsymbol{\alpha})P(\boldsymbol{\alpha}) \right]^{-1} \quad (14)$$

The first multiplicative term at the right-hand side of (14) can be written as

$$P(\boldsymbol{d}|\boldsymbol{\alpha}) = \prod_{i=1}^{N} P(d_i|\boldsymbol{\alpha}) = \prod_{i=1}^{\alpha_1-1} \underbrace{P(d_i|\boldsymbol{\alpha})}_{H_i=0} \prod_{j=\alpha_1}^{\alpha_2-1} \underbrace{P(d_j|\boldsymbol{\alpha})}_{H_j=1} \cdots \prod_{q=\alpha_{N_{\rm bs}}}^{N} \underbrace{P(d_q|\boldsymbol{\alpha})}_{H_q=0 \text{ or } 1} \quad (15)$$

[5]Note that for ease of notational simplicity, in (13) we use the same notation α_k to denote both the random variable (in the second term) and its realization (in the third and fourth terms). The same simplified notational approach will be considered in the remainder of Section 3.1. The context should eliminate any ambiguity.

where we have used the fact that the sensors' decisions are independent. Note that, in the last $N - \alpha_{N_{\rm bs}} + 1$ terms, $H_i = 0$ if $N_{\rm bs}$ is *even*, whereas $H_i = 1$ if $N_{\rm bs}$ is *odd*. As in Section 2.1.1, the component conditional probabilities at the right-hand side of (15) can be written as

$$P(d_i|\boldsymbol{\alpha}) = \begin{cases} P\left(n_i \underset{d_i=1}{\overset{d_i=0}{\lesseqgtr}} \tau\right) & \text{if } i \in \mathscr{I}_0(\boldsymbol{\alpha}) \\ P\left(n_i \underset{d_i=1}{\overset{d_i=0}{\lesseqgtr}} \tau - s\right) & \text{if } i \in \mathscr{I}_1(\boldsymbol{\alpha}) \end{cases} \tag{16}$$

where

$$\mathscr{I}_\ell(\boldsymbol{\alpha}) \triangleq \{\text{indexes } i \text{ such that } H_i = \ell|\boldsymbol{\alpha}\} \qquad \ell = 0,1 \tag{17}$$

The second multiplicative term at the right-hand side of (14), instead, can be written, using the chain rule [12], as

$$P(\boldsymbol{\alpha}) = \prod_{i=1}^{N_{\rm bs}} P(\alpha_i|\alpha_{i-1},\ldots,\alpha_1) = P(\alpha_1)\prod_{i=2}^{N_{\rm bs}} P(\alpha_i|\alpha_{i-1}) \tag{18}$$

where we have used the fact that the position of the ith boundary depends only on the position of the previous $(i-1)$th boundary. Each multiplicative term at the right-hand side of (18) can be written by observing that each boundary is uniformly distributed among the sensors according to the constraints introduced in (13). In particular, by using combinatorics, one obtains

$$\begin{aligned} P(\alpha_1) &= \frac{1}{N - N_{\rm bs} + 1} \\ P(\alpha_k|\alpha_{k-1}) &= \frac{1}{N - N_{\rm bs} + k - \alpha_{k-1}} \qquad k = 2,\ldots,N_{\rm bs} \end{aligned}$$

The last term at the right-hand side of (14) (i.e., the denominator) can be easily obtained by observing that it is composed by terms similar to those evaluated in (15) and (18).

Finally, the a posteriori probabilities of the boundaries' positions $\{P(\alpha_k|\boldsymbol{d})\}$ $(k = 1,\ldots,N_{\rm bs})$ in (13) can be obtained by proper marginalization of (14):

$$P(\alpha_k|\boldsymbol{d}) = \sum_{\sim\{\alpha_k\}} P(\alpha_1,\ldots,\alpha_{N_{\rm bs}}|\boldsymbol{d}) \qquad k = 1,\ldots,N_{\rm bs}$$

where the notation $f(y_i) = \sum_{\sim\{y_i\}} f(y_1,y_2,\ldots,y_n)$ $(i = 1,\ldots,n)$ means that the marginal function $f(y_i)$ is obtained from the joint function $f(y_1,y_2,\ldots,y_n)$ by summing over all variables $\{y_j\}$, with $j \neq i$ [7].

3.1.2 Multi-level Quantization

The derivation of the MMSE fusion algorithm for a scenario with multi-level quantization (with n_b quantization bits) at the sensors is the same as that provided in Section 3.1.1 for the case with binary quantization. However, as in Section 2.1.2, 2^{n_b} possible values for the decisions at the sensors are admissible (see (6)) and 2^{n_b} probabilities have to be computed (see (7)).

3.1.3 Absence of Quantization

Let us finally consider the scenario with no quantization at the sensors, i.e., with the sensors transmitting the PDFs of their observables. As in Section 2.1.3, the estimated boundaries can be written, according to the assumptions outlined at the beginning of Section 3, as

$$\widehat{\alpha}_k = \mathbb{E}\left[\alpha_k|\boldsymbol{r}\right] = \sum_{\alpha_k=k+1}^{N-N_{\text{bs}}+k-1} \alpha_k P\left(\alpha_k|\boldsymbol{r}\right) \qquad k = 1,\ldots,N_{\text{bs}} \tag{19}$$

The probabilities in (19) can be obtained, as in Section 3.1.1, through proper marginalization of joint conditional probabilities of the following type:

$$P(\boldsymbol{\alpha}|\boldsymbol{r}) = p(\boldsymbol{r}|\boldsymbol{\alpha})P(\boldsymbol{\alpha}) \cdot \left[\sum_{\alpha_1=2}^{N-N_{\text{bs}}} \cdots \sum_{\alpha_i+1}^{N-N_{\text{bs}}+i-1} \cdots \sum_{\alpha_{N_{\text{bs}}}=\alpha_{N_{\text{bs}}-1}+1}^{N-1} p(\boldsymbol{r}|\boldsymbol{\alpha})P(\boldsymbol{\alpha})\right]^{-1}$$

Since sensors' observations are independent, it holds that

$$p(\boldsymbol{r}|\boldsymbol{\alpha}) = \prod_{i=1}^{N} p(r_i|\boldsymbol{\alpha})$$

where, similar to Section 2.1.3,

$$p(r_i|\boldsymbol{\alpha}) = \begin{cases} p_{\mathscr{N}}(r_i) & \text{if } i \in \mathscr{I}_0(\boldsymbol{\alpha}) \\ p_{\mathscr{N}}(r_i - s) & \text{if } i \in \mathscr{I}_1(\boldsymbol{\alpha}) \end{cases}$$

where $\mathscr{I}_0(\boldsymbol{\alpha})$ and $\mathscr{I}_1(\boldsymbol{\alpha})$ are defined as in (17).

3.2 Simplified Fusion Rule

Obviously, the computational complexity of the MMSE distributed detection strategy in scenarios with an arbitrary number of phenomenon boundaries increases more rapidly than in scenarios with a single phenomenon boundary (for more details see Section 5). Therefore, the derivation of simplified fusion algorithms with low complexity (but limited performance loss) is crucial.

A first possible choice is a direct extension of the sub-optimal approach, presented in Section 2.2, for scenarios with a single phenomenon status change. However, this class of simplified fusion algorithms is not efficient in a scenario with multiple boundaries, since the number of comparisons with all possible sequences of boundaries rapidly increases with the number of sensors. Therefore, in the following we introduce another class of reduced-complexity fusion algorithms, which do not make use of these comparisons. As before, depending on the quantization strategy at the sensors, we distinguish three possible scenarios.

3.2.1 Binary Quantization

Define the following function:

$$f_{\mathrm{bq}}(k,\boldsymbol{d}_1^k) \triangleq \sum_{i=1}^{k} [P(H_i = 0|d_i) - P(H_i = 1|d_i)] \qquad k = 1,\ldots,N \qquad (20)$$

where $\boldsymbol{d}_1^k = (d_1,\ldots,d_k)$. The key idea of our approach is the following. The function $f_{\mathrm{bq}}(k,\boldsymbol{d}_1^k)$ is monotonically increasing (or decreasing), with respect to k, while the phenomenon does not change its status. In correspondence to each change of the phenomenon status, the function $f_{\mathrm{bq}}(k,\boldsymbol{d}_1^k)$ changes its monotonic behavior. More precisely, a phenomenon variation from "0" to "1" corresponds to a change, trend-wise, from increasing to decreasing; a phenomenon variation from "1" to "0" corresponds to a change, trend-wise, from decreasing to increasing. Therefore, by detecting the changes of the monotonic behavior of f_{bq} one can estimate the positions of the boundaries.

The probability $P(H_i = \ell|d_i)$ ($\ell = 0,1;\ i = 1,\ldots,N$) in (20) can be written, by applying the Bayes formula and following an approach similar to that in Section 2.2.3, as

$$P(H_i = \ell|d_i) = \frac{P(d_i|H_i = \ell)}{P(d_i|H_i = 0) + P(d_i|H_i = 1)}$$

where we have used the fact that $P(H_i = 0) = P(H_i = 1)$ and

$$P(d_i|H_i = \ell) = \begin{cases} P(s\cdot\ell + n_i < \tau) = P(n_i < \tau - s\cdot\ell) & \text{if } d_i = 0 \\ P(s\cdot\ell + n_i > \tau) = P(n_i > \tau - s\cdot\ell) & \text{if } d_i = 1 \end{cases}$$

3.2.2 Multi-level Quantization

A simplified fusion algorithm for a scenario with multi-level quantization at the sensors can be directly obtained from the one just introduced for the case with binary quantization, with the difference that the probability $P(d_i|H_i = \ell)$ ($\ell = 0,1; i = 1,\ldots,N$) can assume $2^{n_b} \geq 2$ values:

$$P(d_i|H_i=\ell) = \begin{cases} P(s\cdot\ell+n_i < \tau-(2^{n_b-1}-1)\Delta) & \text{if } d_i = 0 \\ P(\tau-(2^{n_b-1}-1) \leq s\cdot\ell+n_i < \tau-(2^{n_b-1}-2)\Delta) & \text{if } d_i = 1 \\ \vdots & \\ P(\tau+(2^{n_b-1}-2)\Delta \leq s\cdot\ell+n_i < \tau+(2^{n_b-1}-1)\Delta) & \\ & \text{if } d_i = 2^{n_b}-2 \\ P(s\cdot\ell+n_i \geq \tau+(2^{n_b-1}-1)\Delta) & \text{if } d_i = 2^{n_b}-1 \end{cases}$$

3.2.3 Absence of Quantization

In the absence of quantization at the sensors, one can use the probability $P(H_i = \ell|r_i)$ ($\ell = 0,1; i = 1,\ldots,N$) and introduce the following function:

$$f_{\mathrm{nq}}(k,\boldsymbol{r}_1^k) \triangleq \sum_{i=1}^{k}[P(H_i=0|r_i)-P(H_i=1|r_i)] \qquad k=1,\ldots,N$$

where $\boldsymbol{r}_1^k = (r_1,\ldots,r_k)$. The fusion algorithm at the AP is then the same as that presented in the case with binary quantization, but for the use of f_{nq} in place of f_{bq}.

4. NUMERICAL RESULTS

We now analyze, through Monte Carlo simulations, the performance of sensor networks using the decentralized detection algorithms previously described. We preliminary denote as $D(\boldsymbol{H},\widehat{\boldsymbol{H}})$ the quadratic distance between the observed phenomenon $\boldsymbol{H}$ and its estimate $\widehat{\boldsymbol{H}}$:

$$D(\boldsymbol{H},\widehat{\boldsymbol{H}}) \triangleq \left| < (\boldsymbol{H}\oplus\widehat{\boldsymbol{H}});(\boldsymbol{H}\oplus\widehat{\boldsymbol{H}}) > \right|^2 \tag{21}$$

where the notation $\oplus$ stands for bit-by-bit ex-or. Note that $\widehat{\boldsymbol{H}}$ is the estimated phenomenon, directly derived from the estimated boundaries' positions $\widehat{\boldsymbol{\alpha}}$. We will simply refer to D as "distance." We remark that (21) reduces to $D(\boldsymbol{H},\widehat{\boldsymbol{H}}) = |\alpha-\widehat{\alpha}|^2$ in the case of single-boundary phenomena.

The Monte Carlo simulation results are obtained through the following steps:

1. the number of boundaries and their positions are randomly generated (in the case of a single boundary, only its position is randomly generated);

2. the sensors' decisions (or the PDFs of the observables, according to the chosen quantization strategy at the sensors) are taken and transmitted to the AP;

3. the AP estimates the boundaries' positions through either MMSE or simplified fusion algorithms;

4. the distance D (between the true phenomenon and its estimate) is evaluated on the basis of the estimated sequence of boundaries;

5. steps 1–4 are repeated a sufficiently large number of times;

6. the average distance $\overline{D}$ is finally computed.

4.1 Single-Boundary Phenomena

In Figure 1, the distance $\overline{D}$ is shown, as a function of the decision threshold τ at the sensors, for three different values of the number N of sensors: (i) 8, (ii) 16, and (iii) 32. The sensor SNR is set to 0 dB. As expected, the optimum value of τ, which will be selected, corresponds to $s/2$ ($s = 1$ in the considered simulations). When the number of sensors is small, e.g., $N = 8$, the results in Figure 1 show that $\overline{D}$ depends on τ in a limited way, and there is not a well-pronounced minimum. The minimum (in correspondence to $\tau = 0.5$) becomes more pronounced when N increases—obviously, the larger is the number of sensors, the larger is the distance $\overline{D}$, since α can assume a wider set of values.

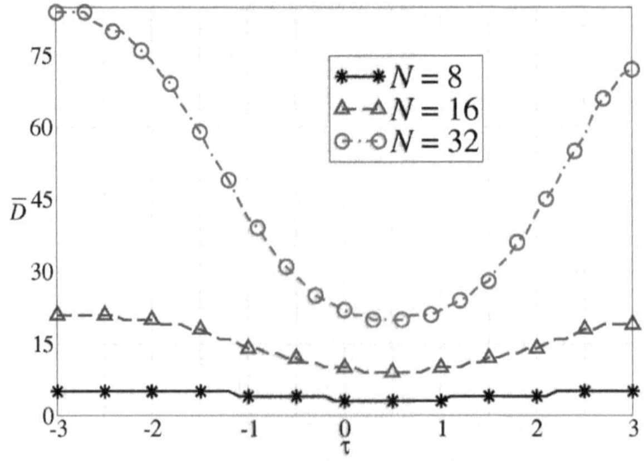

Figure 1. Distance, as a function of the decision threshold τ at the sensors, in a scenario with a *single boundary* phenomenon, binary quantization, MMSE fusion rule at the AP, and $SNR_{sensor} = 0$ dB. Three possible values of the number N of sensors are considered: (i) 8, (ii) 16, and (iii) 32

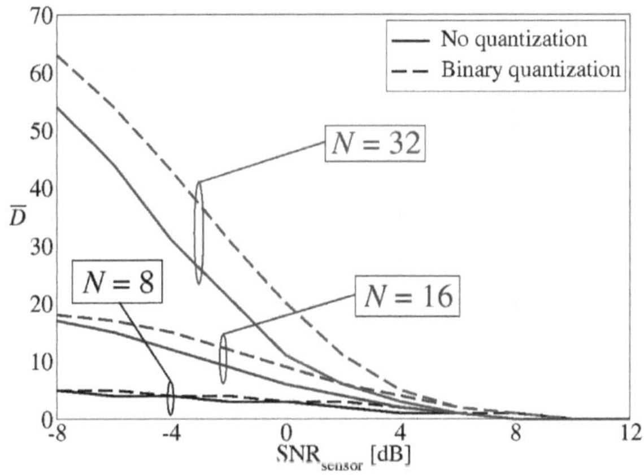

Figure 2. Distance, as a function of the sensor SNR, in a scenario with a *single boundary* phenomenon and MMSE fusion rule at the AP. Three possible values for the number N of sensors are considered: (i) 8, (ii) 16, and (iii) 32. *Solid lines* correspond to no quantization at the sensors, whereas *dashed lines* are associated with binary quantization

In all numerical results presented in the following, the threshold τ is optimized in order to minimize the probability of decision error.

In Figure 2, the distance $\overline{D}$ is shown, as a function of the sensor SNR, in scenarios with binary quantization at the sensors (dashed lines) and, for comparison, with no quantization (solid lines). Three possible values for the number N of sensors are considered: (i) 8, (ii) 16, and (iii) 32. As expected, the distance is a decreasing function of the sensor SNR. In fact, when the accuracy of the sensors' observations increases, the decisions sent by the sensors to the AP are more reliable and, consequently, the estimated phenomenon at the AP is closer and closer to the true phenomenon. Note also that the performance degradation incurred by the use of quantization, with respect to the unquantized case, increases for increasing number of sensors.

Let us now turn our attention to a scenario with multi-level quantization at the sensors. First, one needs to optimize the value of Δ in (6) for the multi-level quantization scheme. As for the decision threshold τ, the optimization is carried out by minimizing the distance $\overline{D}$. In Figure 3, the distance $\overline{D}$ is shown, as a function of Δ, in a scenario with MMSE fusion strategy at the AP and multi-level quantization at the sensors. The sensor SNR is set to 0 dB. As in Figures 1 and 2, three possible values for the number N of sensors are considered: (i) 8, (ii) 16, and (iii) 32. In each case, the performance with 2-bit quantization (solid lines) is compared with that with 3-bit quantization (dashed lines). One can observe that in the case with $N = 8$ sensors, the performance remains the same regardless of the value of Δ. As will be shown later, this is due to the fact that more than one quantization bit at the sensors do not lead

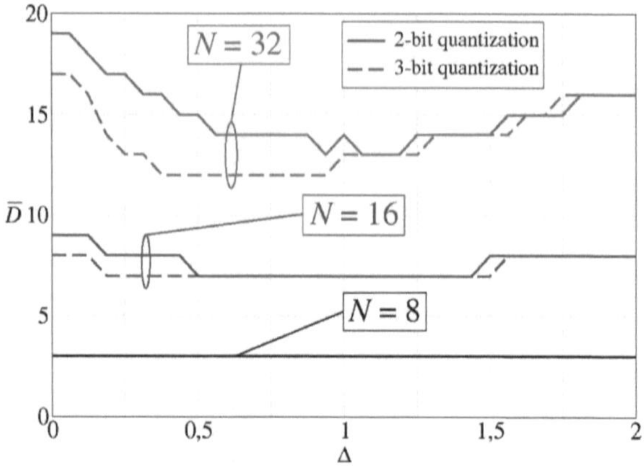

Figure 3. Distance, as a function of the quantization parameter Δ, in a scenario with a *single boundary* phenomenon, MMSE fusion rule at the AP, multi-level quantization at the sensors, and $\text{SNR}_{\text{sensor}} = 0$ dB. Three possible values for the number N of sensors are considered: (i) 8, (ii) 16, and (iii) 32. *Solid lines* correspond to 2-bit quantization at the sensors, whereas *dashed lines* are associated with 3-bit quantization

to any performance improvement when the number of sensors is too small. On the other hand, the minimum exists in the cases with $N = 16$ and $N = 32$. In the following, for any value of N the corresponding optimized value of Δ will be used. Note that a value of Δ slightly larger than 0.5 is practically the best for all considered values of N.

In Figure 4, the distance is shown, as a function of the sensor SNR, in a scenario with $N = 32$ sensors, MMSE fusion algorithm at the AP, and various quantization levels at the sensors. As expected, the larger is the number of quantization bits, the better is the performance. In fact, a larger amount of information about the observed phenomenon is collected at the sensors and, consequently, the reconstruction of the phenomenon at the AP is more reliable—obviously, the performance in a scenario with no quantization at the sensors represents a lower bound for the distance $\overline{D}$. Obviously, there is a price to pay in order to improve the performance through multi-level quantization. In fact, transmission of a larger number of bits leads to higher energy and/or bandwidth consumption.

We now focus on a decentralized detection scheme where the simplified fusion rule derived in Section 2.2 is applied to estimate the boundary position. In Figure 5a, the distance between the true boundary position and its estimate is shown, as a function of the sensor SNR, in scenarios with no quantization (solid lines) and binary quantization (dashed lines) at the sensors, respectively. Three different values for the number N of sensors are considered: (i) 8, (ii) 16, and (iii) 32. From the results in Figure 5a, one can observe that the simplified fusion rule leads to a performance loss with respect to the case with MMSE fusion

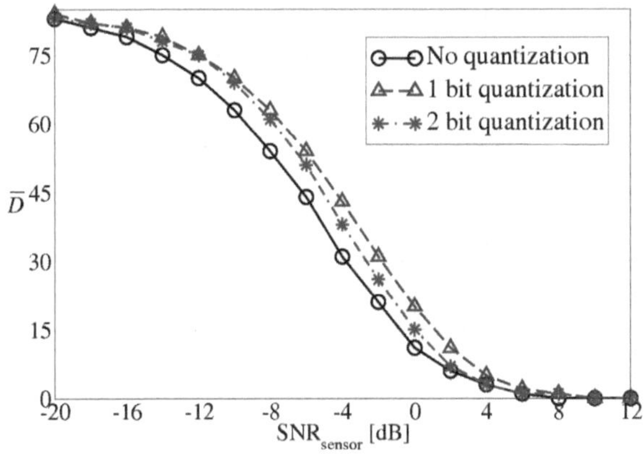

Figure 4. Distance, as a function of the sensor SNR, in a scenario with $N = 32$ sensors, *single boundary* phenomenon, and MMSE fusion rule at the AP. Two possible values for the number of quantization bits at the sensors are considered: (i) 1 and (ii) 2. For comparison, the performance in a scenario with no quantization at the sensors is also shown

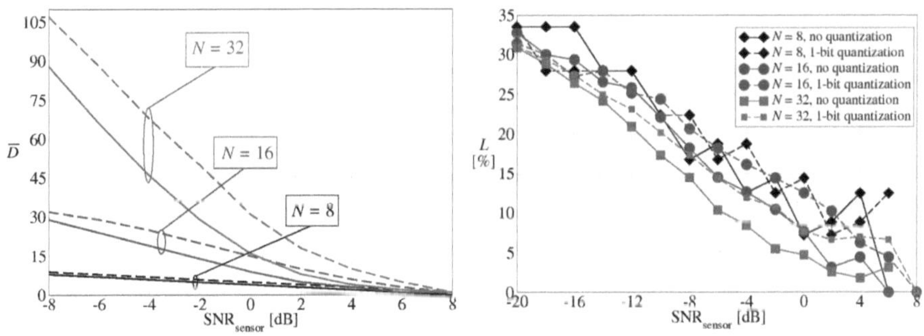

Figure 5. Performance, as a function of the sensor SNR, in a scenario with a *single boundary* phenomenon and simplified fusion algorithm at the AP: (**a**) distance and (**b**) percentage loss with respect to the MMSE fusion algorithm. Three different values for the number N of sensors are considered: (i) 8, (ii) 16, and (iii) 32. The performance in the presence of no quantization (*solid lines*) is compared with that using binary quantization at the sensors (*dashed lines*)

rule (compare the results in Figure 5a, for instance, with those in Figure 2). However, in the region of interest (SNR$_{\text{sensor}} > 0$ dB) the performance with the simplified fusion algorithm is close to that of the MMSE scheme. Moreover, the distance goes to zero with the same trend observed in Figure 2 for the MMSE fusion rule.

In order to evaluate the loss incurred by the use of the simplified fusion algorithm, define the following percentage loss:

$$L \triangleq \sqrt{\underbrace{\frac{D^{\text{simp}} - D^{\text{MMSE}}}{D^{\text{MMSE}}}}_{\text{Term}_1} \cdot \underbrace{\frac{D^{\text{simp}} - D^{\text{MMSE}}}{N^2}}_{\text{Term}_2}} \quad (22)$$

The intuition behind the definition of (22), given by the geometric average of Term_1 and Term_2, is the following. Term_1 represents the relative loss of the simplified fusion rule with respect to the MMSE fusion rule. However, using only this term could be misleading. In fact, for high sensor SNR, the terms D^{simp} and D^{MMSE} are much lower than N^2 (the maximum possible quadratic distance). Therefore, even if $D^{\text{simp}} > D^{\text{MMSE}}$ (for example, $D^{\text{simp}} = 4$ and $D^{\text{MMSE}} = 1$ with $N = 32$), both algorithms might perform very well. The introduction of Term_2 eliminates this ambiguity, since it represents the relative loss (between MMSE and simplified fusion algorithms) with respect to the maximum (quadratic) distance, i.e., N^2. In Figure 5b, the behavior of L is shown as a function of the sensor SNR. In the region of interest ($\text{SNR}_{\text{sensor}} \geq 0$ dB), one can observe that L is lower than 15%, i.e., the proposed simplified fusion algorithm is effective.

In Figure 6, we compare directly the performance of MMSE (solid lines) and simplified (dashed lines) fusion algorithms in a scenario with $N = 16$ sensors and multi-level quantization. Two possible quantization levels at the sensors are considered: (i) 2 bits (curves with circles) and (ii) 3 bits (curves with triangles). One can observe that, as in Figure 4, the performance improves

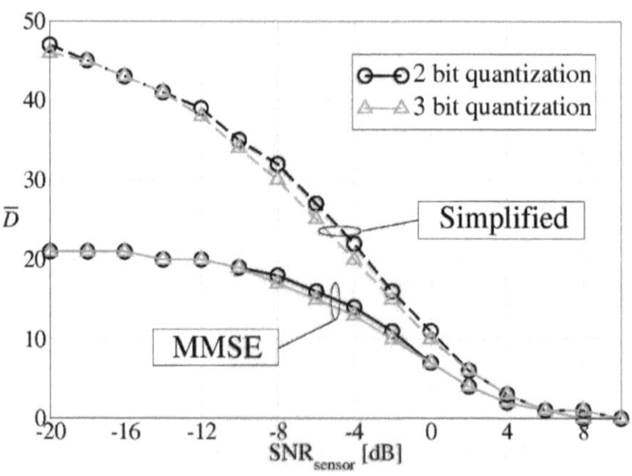

Figure 6. Distance, as a function of the sensor SNR, in a scenario with *single boundary* phenomenon and $N = 16$ sensors. Two possible values of quantization bits at the sensors are considered: (i) 2 bits (*circles*) and (ii) 3 bits (*triangles*). The performance with the MMSE fusion rule (*solid lines*) is compared to that with the simplified fusion rule (*dashed lines*)

for increasing number of quantization bits. Using a 3-bit quantization level is sufficient to achieve the performance limit corresponding to the absence of quantization—the improvement, with respect to a scenario with 2-bit quantization, is minor.

4.2 Phenomena with Multiple Boundaries

In Figure 7, the distance is shown, as a function of the sensor SNR, in a scenario with a multi-boundary phenomenon, for (a) $N = 8$ sensors and (b) $N = 32$ sensors. No quantization is considered at the sensors and the performance with the simplified fusion algorithm at the AP is compared directly with that obtained using the MMSE fusion rule. As expected, the distance $\overline{D}$ reduces to zero for increasing values of the sensor SNR and the performance with the MMSE fusion algorithm is better than that with the simplified fusion algorithm. We recall that the performance with the MMSE fusion rule is evaluated only with $N = 8$, since the computational complexity becomes unbearable for values of N larger than 8 (the simulations are too lengthy).

In order to investigate scenarios with larger numbers of sensors, the use of the reduced-complexity simplified fusion algorithms derived in Section 3.2 is mandatory. In Figure 7b, the distance is shown, as a function of the sensor SNR, in a scenario with $N = 32$ sensors and using the simplified fusion algorithm at the AP. Four different quantization scenarios at the sensors are considered: (i) no quantization, (ii) 1-bit quantization, (iii) 2-bit quantization, and (iii) 3-bit quantization. All curves overlap, i.e., the performance does not improve by using more than one quantization bit at the sensors. It remains to be investigated what is the relative loss of the simplified fu-

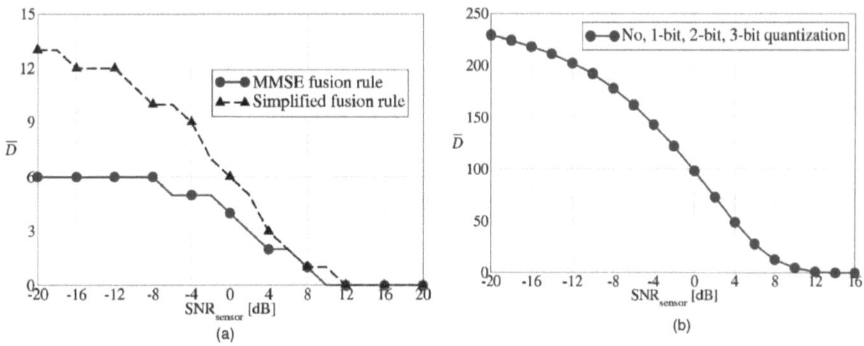

Figure 7. Distance, as a function of the sensor SNR, in a scenario with a *multi-boundary* phenomenon, considering (**a**) $N = 8$ sensors and absence of quantization (MMSE and simplified fusion algorithms at the AP are considered) and (**b**) $N = 32$ and multi-level quantization (only the sub-optimum fusion algorithm at the AP is considered)

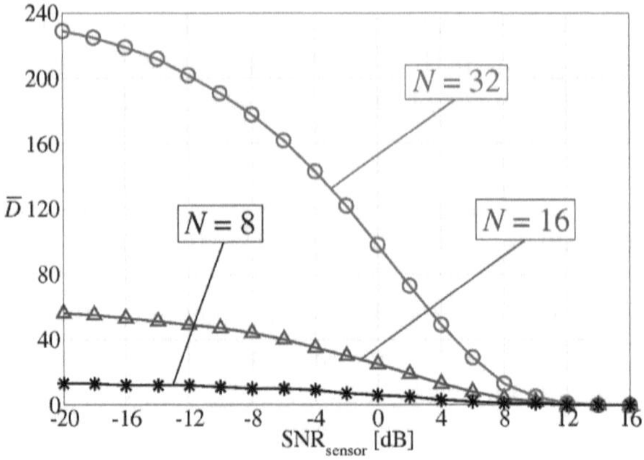

Figure 8. Distance, as a function of the sensor SNR, in a scenario with a *multi-boundary* phenomenon, simplified fusion algorithm at the AP, and 3-bit quantization at the sensors. Three different values for the number of sensors N are considered: (i) 8, (ii) 16, and (iii) 32

sion algorithm, with respect to the MMSE fusion algorithm, in scenarios with multi-boundary phenomena. The fact that the quantization strategy at the sensors has little impact suggests that this relative loss might not be negligible.

Finally, we investigate the impact of the number of sensors N on the system performance. In Figure 8, the distance is shown, as a function of the sensor SNR, considering three different values for N: (i) 8, (ii) 16, and (iii) 32. The simplified fusion algorithm at the AP and 3-bit quantization at the sensors are considered—we remark that similar results hold also for other quantization scenarios. As expected, the smaller is the number of sensors, the smaller is the distance between the observed phenomenon and its estimate. In fact, when the number of sensors increases, the number of phenomenon boundaries increases as well and, consequently, the distance can assume larger values. However, as expected, the distance reduces to zero for sufficiently high values of the sensor SNR.

5. COMPUTATIONAL COMPLEXITY

It is now of interest to evaluate the improvement, in terms of computational complexity with respect to the MMSE fusion rule, brought by the use of the simplified fusion algorithms introduced in Sections 2.2 and 3.2. As complexity indicators, we choose the numbers of additions and multiplications (referred to as N_s and N_m, respectively) required by the considered fusion algorithms.

In a scenario with *single-boundary* phenomena and MMSE fusion rule at the AP, by analyzing (2), (3), and (4), it is possible to show that $N_s^{opt} = \Theta(N^2)$ and

$N_m^{opt} = \Theta(N^3)$, where the notation $f(n) = \Theta(g(n))$ means that there exists an n_0 such that for $n > n_0$, $\exists c_1 \in (0,1)$, $c_2 > 1$ such that $c_1 g(n) \leq f(n) \leq c_2 g(n)$ [3]. In a scenario with the simplified fusion algorithm introduced in Section 2.2 (for single-boundary phenomena), instead, by analyzing (9), (10), and (11), it is possible to show that $N_s^{sub\text{-}opt} = \Theta(N^2)$ and $N_m^{sub\text{-}opt} = \Theta(N^2)$. Therefore, one can conclude that the complexity of this simplified fusion algorithm is lower than that of the MMSE fusion algorithm only in terms of the number of multiplications.

Let us now turn our attention to a scenario characterized by phenomena with *multiple boundaries*. By reasoning as in the cases with single-boundary phenomena, the number of operations (in terms of additions and multiplications) required by the MMSE fusion algorithm is $N_s^{opt} = \Theta(N^{2N_{bs}})$ and $N_m^{opt} = \Theta(N^{2N_{bs}+1})$, respectively—recall that N_{bs} is the number of boundaries. As described at the beginning of Section 4, in the considered simulation set-up the number N_{bs} of boundaries is chosen randomly between 1 and $N-2$. Therefore, one can assume that the phenomenon is characterized, on average, by $\frac{N-2}{2} = N/2 - 1$ boundaries. Under this assumption, the computational complexity of the MMSE fusion algorithm would be $N_s^{opt} = \Theta(N^{N-2})$ and $N_m^{opt} = \Theta(N^{N-1})$. On the other hand, the reduced-complexity fusion algorithm requires only N additions, since no multiplication has to be performed. Therefore, the complexity of the proposed simplified fusion algorithm is $N_m^{sub\text{-}opt} = 0$ and $N_s^{sub\text{-}opt} = N$, showing a significant complexity reduction with respect to the MMSE fusion algorithm—this also justifies the non-negligible performance loss for small values of the sensor SNR.

6. CONCLUSIONS

In this chapter, we have analyzed the problem of decentralized detection of spatially non-constant binary phenomena, i.e., phenomena with statuses characterized by single or multiple boundaries. An analytical framework has been developed to attack this problem, distinguishing between: (i) binary quantization at the sensors, (ii) multi-level quantization, and (iii) no quantization. In all cases, the MMSE fusion algorithm at the AP has been derived and the impacts of relevant network parameters (e.g., the decision threshold at the sensors, the interval of quantization, the sensor SNR, and the number of sensors) have been investigated. Then, we have turned our attention to low-complexity fusion rules. In particular, we have proposed a suboptimal fusion rule, based on a deterministic approximation of the MMSE strategy, in scenarios where the phenomenon has a single boundary. We have further simplified this approach in scenarios with multi-boundary phenomena. Our results show that the performance penalty introduced by the simplified fusion algorithms is asymptotically (for high sensor SNR) negligible. Finally, we have compared the computational complexities of MMSE and simplified fusion algorithms, in terms of required numbers of additions and multiplications. Our results underline that the simplified fusion algorithms allow to reduce the numbers of operations, especially in scenarios with multi-boundary phenomena.

REFERENCES

[1] I. Akyildiz, W. Su, Y. Sankarasubramaniam, and E. Cayirci, "A survey on sensor networks," *IEEE Commun. Mag.*, vol. 40, no. 8, pp. 102–114, August 2002.

[2] J. Barros and M. Tüchler, "Scalable decoding on factor trees: A practical solution for wireless sensor networks," *IEEE Trans. Commun.*, vol. 54, no. 2, pp. 284–294, February 2006.

[3] T. H. Cormen, C. E. Leiserson, R. L. Rivest, and C. Stein, *Introduction to Algorithms,* 2nd Edition, MIT Press, Cambridge, MA, USA, 2002.

[4] G. Ferrari and R. Pagliari, "Decentralized binary detection with noisy communication links," *IEEE Trans. Aerosp. Electron. Syst.*, vol. 42, no. 4, pp. 1554–1563, October 2006.

[5] I. Y. Hoballah and P. K. Varshney, "An information theoretic approach to the distributed detection problem," *IEEE Trans. Inform. Theory*, vol. 35, no. 5, pp. 988–994, September 1989.

[6] S. M. Kay, *Fundamentals of Statistical Signal Processing, Volume I: Estimation Theory*, Prentice-Hall, Upper Saddle River, NJ, USA, 1993.

[7] F. R. Kschischang, B. J. Frey, and H. A. Loeliger, "Factor graphs and the sum-product algorithm," *IEEE Trans. Inform. Theory*, vol. 47, no. 2, pp. 498–519, February 2001.

[8] W. M. Lam and A. R. Reibman, "Design of quantizers for decentralized estimation systems," *IEEE Trans. Commun.*, vol. 41, no. 11, pp. 1602–1605, November 1993.

[9] Z.-Q. Luo, "An isotropic universal decentralized estimation scheme for a bandwidth constrained ad hoc sensor network," *IEEE J. Select. Areas Commun.*, vol. 23, no. 4, pp. 735–744, April 2005.

[10] R. Nowak and U. Mitra, "Boundary estimation in sensor networks: Theory and methods," in *Proc. 2nd Int. Work. on Inform. Processing in Sensor Networks*, Palo Alto, CA, USA, April 2003.

[11] R. Nowak, U. Mitra, and R. Willett, "Estimating inhomogeneous fields using wireless sensor networks," *IEEE J. Select. Areas Commun.*, vol. 22, no. 6, pp. 999–1006, August 2004.

[12] A. Papoulis, *Probability, Random Variables and Stochastic Processes*, McGraw-Hill, New York, NY, USA, 1991.

[13] W. Shi, T. W. Sun, and R. D. Wesel, "Quasi-convexity and optimal binary fusion for distributed detection with identical sensors in generalized gaussian noise," *IEEE Trans. Inform. Theory*, vol. 47, no. 1, pp. 446–450, January 2001.

[14] S. N. Simic and S. Sastry, "Distributed environmental monitorning using random sensor networks," in *Proc. 2nd Int. Work. on Inform. Processing in Sensor Networks*, Palo Alto, CA, USA, April 2003, pp. 582–592.

[15] R. Viswanathan and P. K. Varshney, "Distributed detection with multiple sensors–part I: Fundamentals," *Proc. IEEE*, vol. 85, no. 1, pp. 54–63, January 1997.

PERFORMANCE ANALYSIS OF ZIGBEE WIRELESS SENSOR NETWORKS WITH RELAYING

G. Ferrari[1], P. Medagliani[1], and M. Martalò[1]

[1] *Wireless Ad-hoc and Sensor Networks (WASN) Laboratory, Department of Information Engineering, University of Parma, Parma, Italy*

gianluigi.ferrari@unipr.it, paolo.medagliani@unipr.it, martalo@tlc.unipr.it

Abstract In wireless sensor networks, transmitting information from sensors to a far access point (AP), out of direct transmission range, might make the use of *relaying* crucial. The goal of this chapter is to study the impact of relaying in Zigbee wireless sensor networks. In particular, we focus on Zigbee wireless sensor networks and analyze scenarios where the sensors transmit to an AP (or *coordinator*) (i) directly, (ii) through a relay node (or *router*), or (iii) through two relays. We study how the network performance (in terms of delay, transmission rate, and throughput) is influenced by the number of sensors, the traffic load, and the use of ACKnowledgment (ACK) messages. This performance analysis is carried out with simulations, analytical considerations, and experimental measurements. Our results show that the use of one or two relays, combined with the use of ACK messages (not efficiently managed by the upper layers of the network protocol stack), may cause a significant performance degradation. On the opposite, if ACK messages are not used, then the performance improves significantly. In addition, we also consider the impact of the network lifetime on the network transmission rate.

1. INTRODUCTION

Wireless sensor networks are an interesting research topic, both in military [1–3] and civilian scenarios [5, 17]. In particular, remote/environmental monitoring, surveillance of reserved areas, etc., are important fields of application of wireless sensor networking techniques. Typically, very low power consumption and low-cost hardware are required [13].

One of the latest standards for wireless sensor networking with low transmission rate and high energy efficiency has been proposed by the Zigbee Allianc [8, 21]. An experimental analysis of Zigbee wireless sensor networks, taking into account the impact of the most important system parameters (e.g., the received signal strength indication (RSSI), throughput, network transmission rate, and delay), is presented in [7, 8, 12].

In this chapter, we evaluate the impact of the presence of one or two *relay nodes* (also referred to as *routers*) on the network performance. Both experimental and simulation results, obtained using a wireless sensor network constituted by PICDEM Z nodes (produced by Microchip) [14] and the Opnet simulator [16], respectively, are presented. As performance indicators, we use *network transmission rate*, *delay*, and *throughput*. We evaluate the impact of the number of nodes and the packet length on the system performance. In addition, we investigate the impact of the variation of the ACKnowledgment (ACK) window duration (i.e., the time interval during which a node waits for an ACK message after transmitting its data). A simple-minded analytical framework is also proposed to validate the simulation results. Finally, we characterize the behavior of the network transmission rate as a function of the network lifetime given by the percentage of sensors' deaths which make the network dead (i.e., the maximum tolerable network death level), in the presence of *clustering*. A small number of required sensors' deaths (i.e., a short network lifetime) can be integrated as a stringent quality of service (QoS) condition.

2. ZIGBEE STANDARD OVERVIEW

The Zigbee standard is suited for the family of low-rate wireless personal area networks (LR-WPANs), allowing network creation, management, and data transmission over a wireless channel with the highest possible energy savings. Three different types of nodes are foreseen by the Zigbee standard: (i) *coordinator*, (ii) *router*, and (iii) *end device*. In the absence of a direct communication link from an end device to the coordinator, the router is employed to relay the packets toward the correct destination. The coordinator, in addition to being able to relay the packets itself, can also create the network, exchange the parameters used by the other nodes to communicate (e.g., a network IDentifier (ID), a synchronization frame), and send network management commands. The router and coordinator are referred to as *full function devices* (FFDs), i.e., they can implement all the functions required by the Zigbee standard in order to set up and maintain communications. The end devices, which are also referred to as *reduced function devices* (RFDs), can only collect data from sensors, insert these values into proper packets, and send them to destination nodes.

The Zigbee standard is based, at the first two layers of the ISO/OSI stack, on the IEEE 802.15.4 standard [10], which employs a *non-persistent* carrier sense multiple access with collision avoidance (CSMA/CA) medium access control (MAC) protocol and operates in the 2.4 GHz band (similarly to the IEEE 802.11 standard [9]). In addition, the IEEE 802.15.4 standard provides an optional ACK message to confirm the correct delivery of a packet. In a scenario with transmission of ACK messages, the access mechanism of the non-persistent CSMA/CA MAC protocol is slightly modified. More precisely, after successful transmission of a data frame, a time interval, denoted as *s*hort *i*nter*f*rame *s*pacing (SIFS), is reserved. The duration of this interval is longer than the duration of an ACK message and shorter than the ACK window duration. Therefore, the receiving node can immediately send an ACK message back, avoiding any collision. If the SIFS is too short, this mechanism may incur some problems in the presence of a router. In this case, in fact, the sum of the transmission times of the two ACK messages may be longer than the SIFS: therefore, the second ACK message may collide with other on-going transmitted packets (from other nodes). This problematic behavior is exacerbated with two routers.

We remark that the medium access mechanism in Zigbee wireless networks makes use of a *back-off algorithm* to reduce the number of packet collisions. A node, before transmitting a new packet, waits for a period randomly chosen in an interval defined during the network start-up phase. After this period has elapsed, the node tries to send its packet: if it detects a collision, it doubles the previously chosen interval and waits; if, instead, the channel is free, it transmits its packet. This procedure is repeated five times, after which the waiting interval is maintained fixed to its maximum value. This back-off algorithm makes it likely, in the considered scenarios with low traffic loads, that a node will eventually succeed in transmitting its packet.

3. EXPERIMENTAL AND SIMULATION SETUP

Our experimental wireless sensor network is constituted by PICDEM Z nodes produced by Microchip [14]. All the experiments have been performed in an indoor scenario, and the PICDEM Z nodes have been configured in order to remain active all the time. Two network configurations have been considered for the experimental tests, with RFDs connected to the coordinator (i) directly, as shown in Figure 1a, or (ii) through a router, as shown in Figure 1b, respectively. In both these *experimental* scenarios, the number of packets transmitted by the RFDs is equal to 1000.

In order to validate and extend our experimental results, we make use of the Opnet Modeler 11.5 simulator [16] and a built-in Opnet model for a standard (without router) Zigbee sensor network created at the National Institute of Standards and Technology (NIST) [15]. Both in the case of simulation and in the case of experimental results, we consider a scenario where the sensors

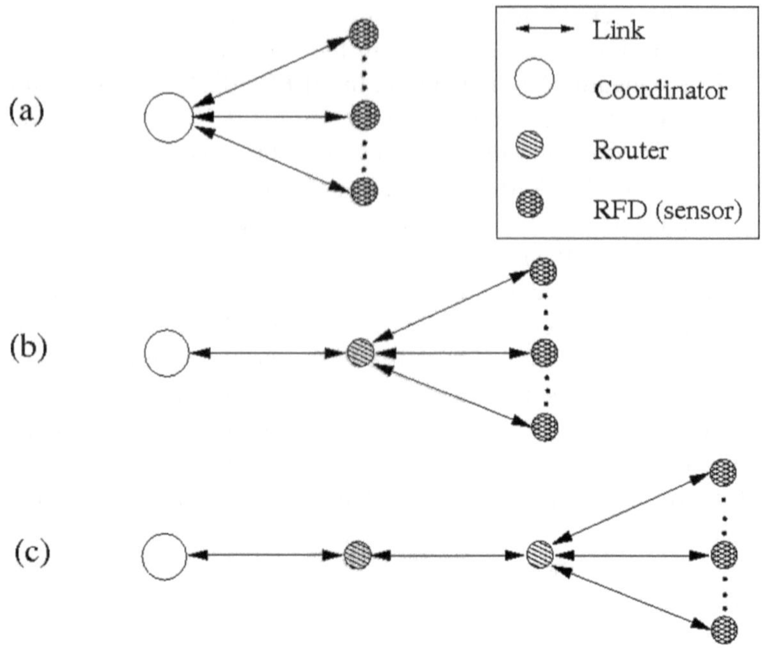

Figure 1. Possible sensor network topologies: three RFDs are connected to the coordinator (**a**) directly, (**b**) through one router, or (**c**) through two routers

transmit to the coordinator, which only collects the performance statistics and, if required, sends back the ACK message. The adopted simulation model does not consider power attenuation due to transmission in a non-ideal channel and, in addition, does not take into account multipath phenomena which, instead, influence the experimental results. In some cases, in fact, experiments had to be repeated because of the presence of people crossing the wireless communication links between nodes. Besides the topologies in Figure 1a and b (also analyzed experimentally), the developed simulator allows to evaluate a wider set of topologies, such as that shown in Figure 1c, corresponding to a scenario where the RFDs communicate to the coordinator through two routers. In Section 4.4, an even more complex network simulation setup will be used, as a generalization of the model shown in Figure 1b. More precisely, we will consider a *clustered* scenario, where RFDs are grouped into clusters and each cluster is associated with a relay, which forward the data received from its RFDs to the coordinator.

Since the Opnet model for a Zigbee network developed at NIST [15] does not make use of a router, we have implemented a router model which receives the packets (from the RFDs or another router) and properly changes some parameters (such as destination and source addresses) in order to allow the coordinator to send back ACK messages to the RFDs. More precisely, we assume an ideal "binding" between the source nodes (RFDs) and the destination node (coordinator). In other words, whenever the RFDs are not directly connected with the coordinator, they send data packets to the router which, in turn, substitutes the destination address embedded in the packet header with the new destination address (i.e., the coordinator address) and forward the packet. In the NIST Opnet simulator, the *channel* is modeled as an infinite buffer: when a node generates a packet, it inserts it in the channel buffer and schedules its transmission after a period of time associated with processing and transmission delays. After this interval, the simulator checks if there is any collision[1]: if so, the node reschedules its transmission; if not, the packet is sent and the destination node (the coordinator in the case of direct transmission or, in the case of transmission through routers, at first the routers and then the coordinator) starts processing the packet. Once the router has relayed the packet to the coordinator, it waits for the ACK message from the latter (for a time interval corresponding to the ACK window duration) and does not accept any other incoming packet: this is due to our implementation, where no transmission queue has been inserted in the router. A time description of this behavior is summarized in Figure 2. As will be shown later, this specific implementation of the router leads to a significant performance loss. We are currently working on the extension of the relay model, to make it more efficient and compliant with its experimental implementation in the PICDEM Z nodes.

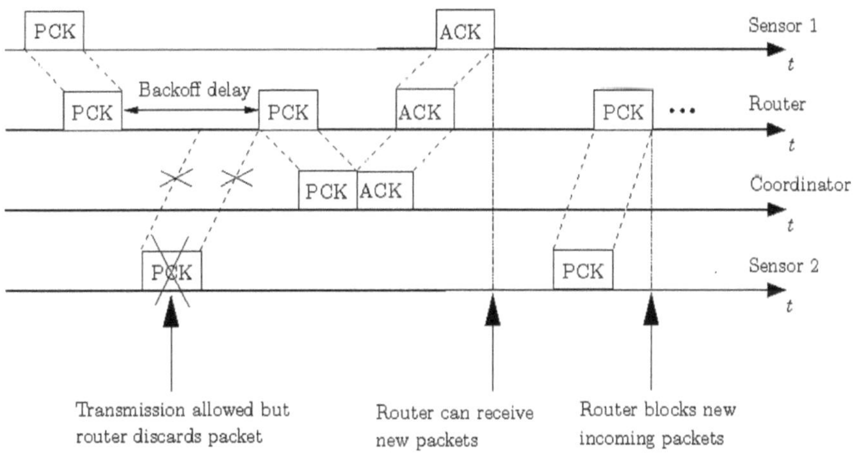

Figure 2. Temporal description of the behavior of a router in the presence of a new incoming packet

[1]The simulator just checks if there is any other transmission scheduled for the interval during which a node wants to transmit. If there is a transmission, it declares a collision. Otherwise, if no transmission is scheduled, the simulator declares the channel idle and notifies the destination node (either the coordinator or the router, depending on the scenario) that there is a transmission in progress.

We remark that both simulation and experimental tests have been performed in the *beaconless* mode. All *simulation* durations, except for those related to lifetime analysis, have been set equal to 1 h.

4. PERFORMANCE ANALYSIS

We now characterize the network performance, evaluating the impacts of (1) the number of sensors, (2) the traffic load, (3) the ACK window duration, and (4) the network lifetime.

4.1 Impact of the Number of Sensors

We distinguish between scenarios *with* and *without* the use of ACK messages, respectively.

4.1.1 *With* the Use of ACK Messages

The packet length is set to $L = 64$ byte/pck and a fixed packet generation rate, equal to $g_{\text{pck}} = 2$ pck/s, is considered. While the value of L has been chosen as intermediate between the minimum and maximum packet lengths admitted by the Zigbee standard, the value of g_{pck} has been selected to make the experiments feasible. The PICDEM Z nodes, in fact, use a single memory register for both transmitted and received packets. Therefore, if the transmission rate is too high, the router cannot relay the packets and may experience a buffer overflow.

The network transmission rate, defined as the average number of bits received per second, and the delay, defined as the time interval between packet generation and packet correct reception, are shown, as functions of the number of the nodes N, in Figure 3a and b, respectively. Two network configurations are considered: (i) direct transmission from the RFDs to the coordinator (see Figure 1a) and (ii) transmission through a router (see Figure 1b). We first comment on the network transmission rate, then on the delay.

As one can see from the simulation results in Figure 3a, in the presence of direct transmission from the sensors to the coordinator, the network transmission rate is an increasing function of the number of sensors and saturates at approximately $N = 128$ sensors. In the presence of a router, instead, there is a drastic reduction of the transmission rate for small values of N. The network transmission rate reaches its maximum in correspondence to $N = 3$ sensors and then starts decreasing with N. In Figure 3a, experimental results for the network transmission rate are also shown. In the case without the router, the experimental results are in excellent agreement with the simulations. In the case with the router, instead, the experimental results on the network

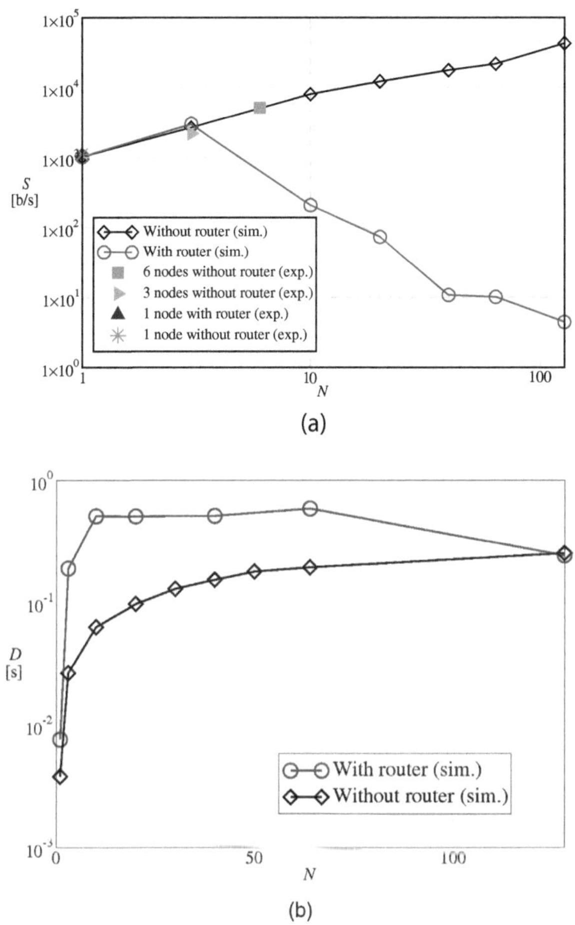

Figure 3. Network performance, as a function of the number of transmitting nodes, in terms of (**a**) network transmission rate (at the coordinator) and (**b**) delay, in the case with packet length $L = 64$ byte/pck and packet generation rate at the RFDs equal to $g_{\text{pck}} = 2$ pck/s. Two scenarios are considered: (i) direct transmission between remote nodes and coordinator (*lines with diamonds*) and (ii) relayed transmission through a router (*lines with circles*). The *symbols* (in **a**) are associated with experimental results, while the *curves* correspond to simulation results

transmission rate do not agree with the simulations. The reason for this mismatch will be explained in the following paragraph. In the case with a router, the slope change at $N = 3$, however, can be justified as follows. When the number of nodes is small, the transmission rate is high, since the relaying capabilities of the router can still compensate for the increasing number of collisions between the packets transmitted by the RFDs. For a number of nodes larger than 10, the decrease is due to the fact that the relaying capabilities of the router are reduced because the number of collisions is larger and it is more likely that the channel is occupied by another transmitting node.

In order to justify the simulation/experimental results, we propose a simple and intuitive approach to evaluate the saturation value of the network transmission rate in *ideal* conditions. In the case without the router, the transmission rate tends to saturate to an ideal maximum value *if*, after transmission of an ACK message from the coordinator, a new packet is immediately available from an RFD. Considering the experimental values for the data packet length ($L = L_{\text{payload}} = 64$ byte/pck), the packet header length ($L_{\text{header}} = 80$ b/pck), the data rate ($R_{\text{b}} = 250$ kbps), the ACK message length ($L_{\text{ack}} = 88$ b/pck), and neglecting processing and propagation delays, the saturation value of the network transmission rate is

$$S^{\text{sat}} = \frac{L \cdot R_{\text{b}}}{L + L_{\text{ack}} + L_{\text{header}}} = 188.2\ \text{kb/s} \tag{1}$$

In the presence of a router, instead, one has to consider the transmission not only from the sensors to the router, but also from the router to the coordinator: this doubles the number of bits to be transmitted. Therefore, the effective network transmission rate saturates at

$$S^{\text{sat}} = \frac{L \cdot R_{\text{b}}}{2 \cdot (L + L_{\text{ack}} + L_{\text{header}})} = 94.1\ \text{kb/s} \tag{2}$$

Comparing the analytical values in (1) and (2) with the simulation results in Figure 3a, the following comments can be made. In the absence of the router, the network transmission rate obtained with simulations (line with diamonds) approaches the saturation value in (1); at the opposite, in the presence of a router, the saturation value in (2) is not approached (in the considered range of values of N) by the simulation results (line with circles). Note that the theoretical limits do not take into account the impact of control messages and the particular router implementation in the simulator. This justifies why the maximum transmission rate predicted by the simulations is lower than the theoretical value in (2). In order to obtain more accurate analytical results, in the Appendix we adapt the analytical framework in [4], relative to IEEE 802.11 networks, to IEEE 802.15.4 networks.

Considering the delay performance shown in Figure 3b, in the *absence* of a router the delay between two subsequent received packets is an increasing function of N. As for the network transmission rate, the saturation value of the delay is approximately reached around $N = 128$ sensors. In the *presence* of a router, instead, the delay behavior can be characterized as follows. For small values of N, this delay is higher than the delay in the absence of the router, in agreement with the network transmission rate behavior in correspondence to small values of N. For large values of N, instead, the delay reduces and becomes basically equal to that in the absence of a router. This is due to the fact that the router (in the simulator) relays the received packets almost immediately (due to the limited number of packets which effectively are able to access the channel and reach the destination), so that the channel is likely to be free from other transmissions from the RFDs.

In Figure 4a and b, the network transmission rate and the delay are analyzed, as functions of the number N of sensors, in all topologies shown in Figure 1. Looking at Figure 4a, the curves related to the scenarios without a router and with a single router are the same as in Figure 3a. In the case with two routers,

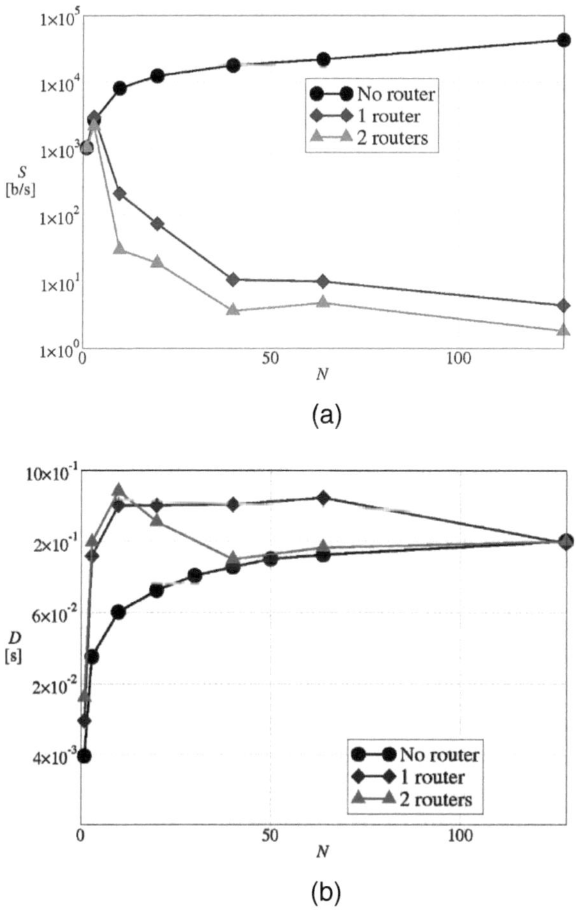

Figure 4. Network performance, as a function of the number of transmitting nodes, in terms of (**a**) network transmission rate (at the coordinator) and (**b**) delay, in the case with $L = 64$ byte/pck and packet generation rate $g_{\text{pck}} = 2$ pck/s. Three scenarios are considered: (i) direct transmission without a router (*curves with circles*), (ii) transmission through a router (*curves with diamonds*), and (iii) transmission through two routers (*curves with triangles*). An ACK message is sent back from the coordinator to the RFDs to confirm the correct data reception

the network transmission rate has a very similar trend with respect to the case with a single router. As explained in Section 3, if the router is waiting for an ACK message from the coordinator, it refuses any new incoming data packet until it can forward the ACK message received from the coordinator. In the case with two routers, when the traffic load is low, the network transmission rate is close to that related to the case with one router. When the offered traffic increases, the relaying capabilities of the routers drastically reduce, because of the larger number of nodes which try to access the channel. In particular, since each packet travels through a three-hop path, it is likely that each router can no longer receive new incoming packets. When the number of nodes is small, the number of collisions is small, so that each router is likely to deliver its packet almost immediately. On the opposite, if a router cannot relay a packet, data transmissions from other nodes fail.

In Figure 4b, the delay performance is evaluated in scenarios with no router, one router, and two routers, respectively. Note that while in the *absence of a router* the delay is an increasing function of N, in the presence of *one router* and in the presence of *two routers* the delay first increases significantly (it reaches its maximum for $N \simeq 10$) and decreases to the same asymptotic value observed in the absence of a router. In all cases, the delay saturates for increasing values of N. In particular, in the presence of one router, the delay is higher than in the case with two routers for large values of N. This fact can be explained considering that in the presence of two routers, the number of exchanged packets is small (and, consequently, the number of collisions is large). Therefore, a node, which can access the channel, transmits almost immediately.

4.1.2 *Without* the Use of ACK Messages

In this section, the network performance is analyzed in a scenario with *no* transmission of ACK messages. As in the case with the use of ACK messages, the network topologies shown in Figure 1 are considered. Unlike in Figure 2, in this case the ACK message transmission disappears: as soon as the router sends a data packet to the coordinator, the former is ready to receive a new incoming data packet from an RFD. Therefore, the probability of finding the router ready to process new packets increases. The simulation setup is, but for the absence of ACK messages, the same as in the case with ACK messages.

In Figure 5a, the network transmission rate is shown as a function of the number of RFDs. Various network configurations are considered: (i) absence of the router (as in Figure 1a), (ii) presence of one router (as in Figure 1b), and (iii) presence of two routers (as in Figure 1c). As one can see from the results in Figure 5a, the behavior of the network transmission rate is similar in all these network configurations. For small values of N, the network transmission rate rapidly increases until it reaches its maximum between $N = 10$ and $N = 50$

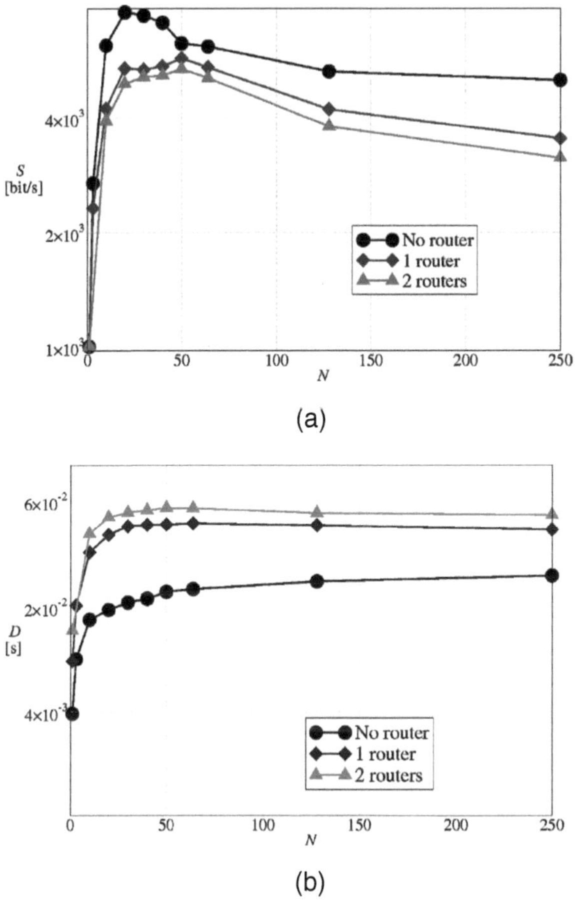

Figure 5. Network performance, as a function of the number of transmitting nodes, in terms of (**a**) network transmission rate (at the coordinator) and (**b**) delay, in the case with $L = 64$ byte/pck and $g_{\text{pck}} = 2$ pck/s. Three scenarios are considered: (i) direct transmission without a router (*curve with circles*), (ii) transmission through a router (*curve with diamonds*), and (iii) transmission through two routers (*curve with triangles*). ACK messages are not used

(depending on the network configuration). Beyond this maximum value, the network transmission rate decreases, with a trend typical of networks using the CSMA/CA MAC protocol [11]. Expression (1) for the saturation value of the network transmission rate in a scenario with no router can be rewritten as follows:

$$S^{\text{sat}} = \frac{L \cdot R_{\text{b}}}{L + L_{\text{header}}} = 216.2 \text{ kb/s} \tag{3}$$

Comparing (1) with (3), one can observe that, at the denominator, the term related to L_{ack} has been eliminated. Therefore, in the absence of ACK messages, the saturation value of the network transmission rate is expected to be higher than in the case with ACK messages.[2]

The delay is shown, as a function of the number of RFDs, in Figure 5b. For all considered network configurations, the delay increases rapidly for small values of N and then saturates at values of N between 10 and 20. Since no ACK message is sent, the delay is marginally influenced by the presence of a router, as already highlighted in Figure 4b. In this case, in fact, because the number of exchanged control messages per data packet is smaller (no ACK message is sent, for instance), it is more likely that the channel is found free and, therefore, the router can process new data packets.

4.2 Impact of Traffic Load

In Figures 6 and 7, the network transmission rate and the delay are shown, as functions of the packet length L, in two scenarios with (a) $N = 20$ RFDs and (b) $N = 100$ RFDs, respectively.

The packet generation rate is fixed to the value $g_{\text{pck}} = 2$ pck/s and ACK messages are used. Looking at Figure 6, both in the presence and in the absence of a router, the transmission rate increases rapidly and then tends to saturate (except for some fluctuations introduced by the limited number of delivered packets). As one can observe, the network transmission rate with direct transmission to the coordinator (no router) is basically equal to the total network traffic load, whereas it is approximately two orders of magnitude lower in the presence of a router. The network transmission rate in the presence of one or two routers is heavily affected by the limited relaying capabilities of the router in the presence of a high traffic load. The delay, on the other hand, is approximately constant, with respect to L, in the absence of the router and slightly varying in the case of transmission through one or two routers. Intuitively, the delay in the presence of the router is given by the sum of the delays (i) between the RFDs and the router and (ii) between the router and the coordinator. The former delay is shorter than the latter since, according to our Opnet model, the router relays a packet right after receiving it, so that it is likely to find the channel free. Processing and propagation delays are negligible—they are three orders of magnitude lower than the transmission delay.

In Figures 6a and 7a, the performance results (in terms of network transmission rate and delay) in a scenario with $N = 20$ RFDs are shown. The performance behaviors are basically the same as those shown in Figures 6b and 7b, relative to scenarios with $N = 100$ RFDs. This fact can be justified as

[2] We remark that the simple analytical approach used to derive (3) is approximated. Therefore, the value in (3) is optimistic with respect to the maximum network transmission rate in Figure 5a.

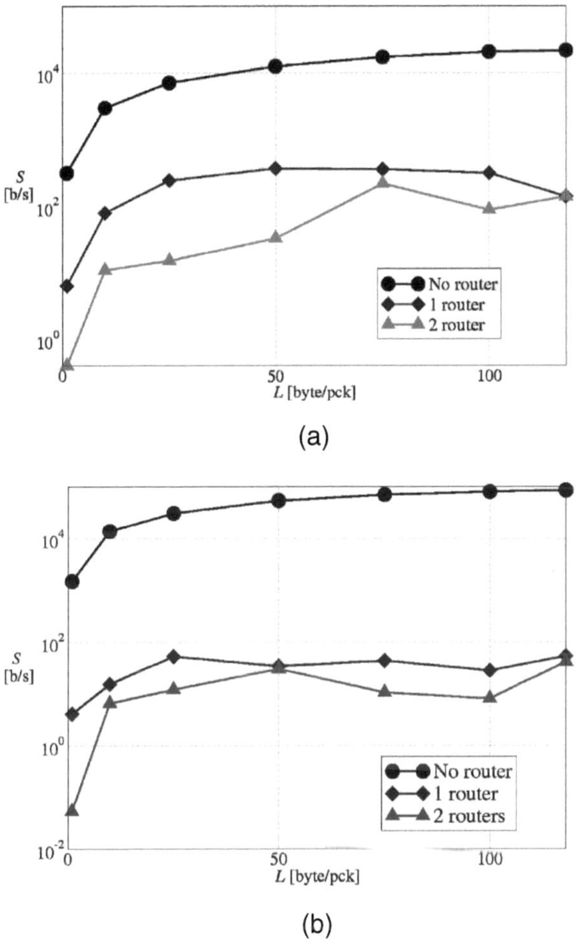

Figure 6. Network transmission rate, as a function of the packet length, in networks with (**a**) $N = 20$ RFDs and (**b**) $N = 100$ RFDs, respectively. The packet generation rate at the RFDs is $g_{\mathrm{pck}} = 2$ pck/s. Three scenarios are considered: (i) direct transmission between RFDs and coordinator (*lines with circles*), (ii) transmission through a router (*lines with diamonds*), and (iii) transmission through two routers (*line with triangles*). ACK messages are sent back from the coordinator to confirm correct data reception

follows. According to the chosen packet generation rate, the network traffic load is small and, therefore, the number of collisions is limited. In the presence of a router, instead, the network performance is heavily influenced by the packet loss due to the absence of a transmission queue at the relay node, which discards any new incoming data packet—while relaying another data packet—without taking any trace of it. Therefore, when the offered traffic load increases, the network performance improves.

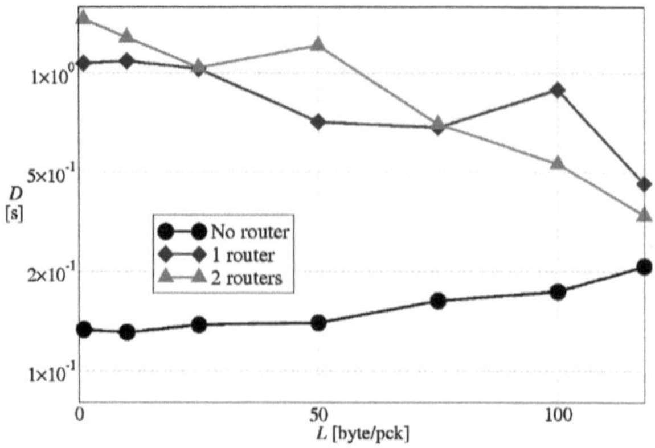

(a)

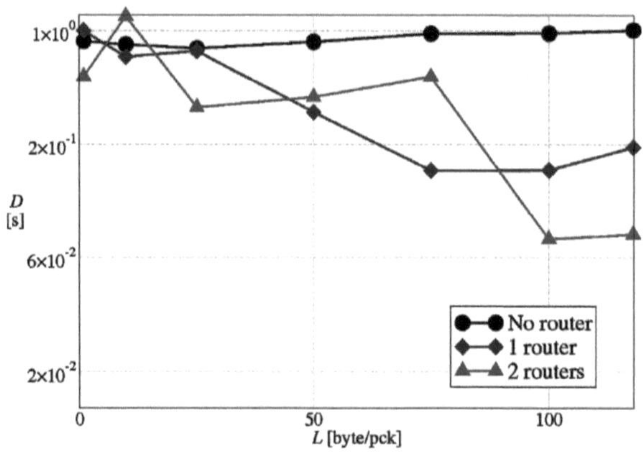

(b)

Figure 7. Delay, as a function of the packet length, in networks with (**a**) $N = 20$ RFDs and (**b**) $N = 100$ RFDs, respectively. The packet generation rate at the RFDs is $g_{\text{pck}} = 2$ pck/s. Three scenarios are considered: (i) direct transmission between RFDs and coordinator (*lines with circles*), (ii) transmission through a router (*lines with diamonds*), and (iii) transmission through two routers (*line with triangles*). ACK messages are sent back from the coordinator to confirm correct data reception

In Figure 8, the *throughput*, defined as the ratio between the number of data packets correctly delivered at the coordinator and the number of data packets sent by the RFDs, is shown as a function of the packet length. Two possible values for N are considered, namely 3 and 10. As one can see, the throughput in the case of *direct transmission* is basically equal to 1, regardless of the value of N, and decreases slightly for increasing values of L. This behavior is due to the fact that the total network traffic load for the considered values of L is too low to create congestion with the used CSMA/CA MAC proto-

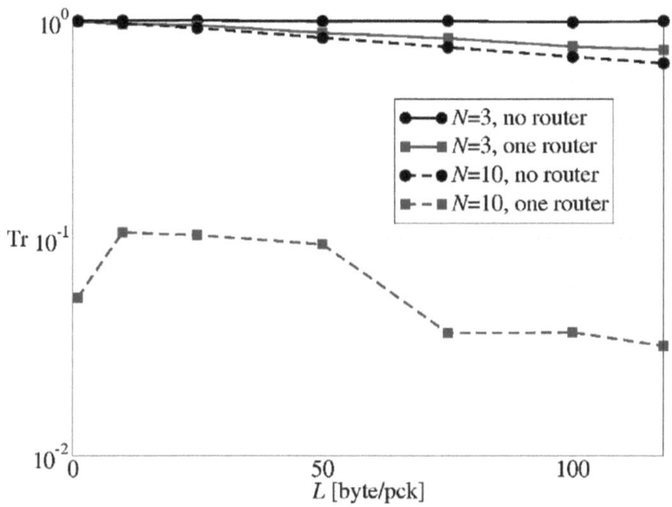

Figure 8. Throughput, as a function of the packet length L, in four scenarios: (i) 3 RFDs sending messages directly to a coordinator (*solid line with circles*) or (ii) through a router (*solid line with squares*), (iii) 10 RFDs sending messages directly to a coordinator (*dashed line with circles*) or (iv) through a router (*dashed line with squares*)

col. Both in the case with direct transmission and in the case of transmission through one router, the throughput reduces for increasing values of N. This is due to the used MAC protocol and the specific router implementation. In fact, since the router immediately relays a received packet, for increasing number of sensors the probability of a collision in an idle interval increases significantly. In the scenario with $N = 3$ RFDs and one router, the throughput is close to that in the case of direct transmission, due to the limited offered traffic and collisions. In the case with $N = 10$ RFDs and one router, instead, the throughput is heavily affected by the traffic load. In this case, in fact, due to the larger number of collisions, the network transmission rate first increases for small values of L and then decreases for larger values of L. In this case, it is possible to identify an "optimal" traffic load (very low) which maximizes the throughput. This maximum is around $L = 10$ byte/pck.

In Figures 9 and 10, *experimental results* relative to the impact of the traffic load on the network performance are presented. The considered network configurations are those shown in Figure 1a and b: the RFDs are connected, either directly or through a router, to the coordinator. As discussed in Section 3, the used experimental nodes belong to the PICDEM Z family. In order to make a meaningful comparison, in the experiments we have considered the same packet generation rate ($g_{\text{pck}} = 2$ pck/s) used in the simulator.

In Figure 9a, the network transmission rate is shown, as a function of the packet length L, in scenarios with (i) one, (ii) two, and (iii) three RFDs,

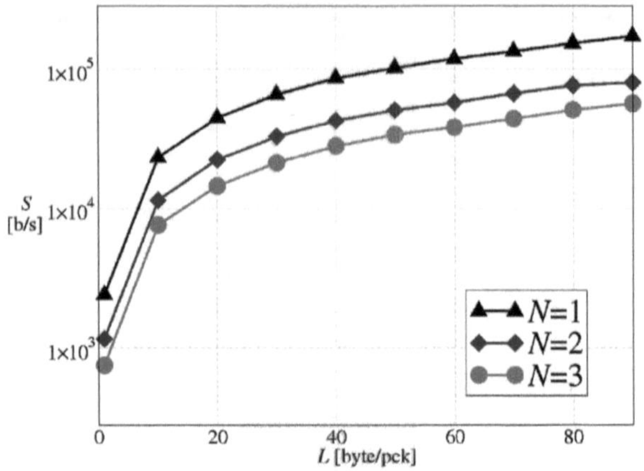

(a)

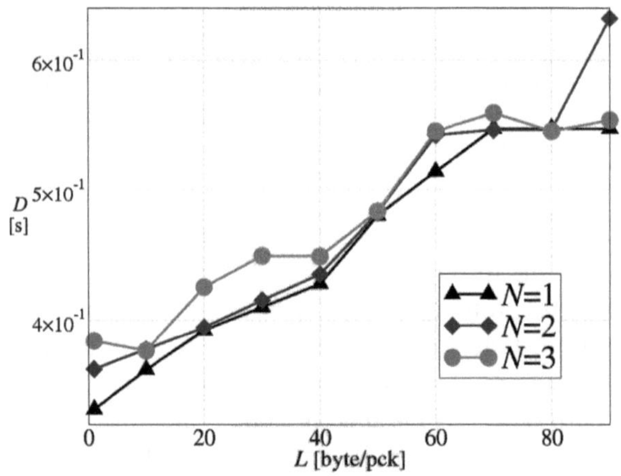

(b)

Figure 9. Experimental network performance, as a function of the packet length, in terms of (**a**) network transmission rate and (**b**) delay. Three scenarios with direct transmission to the coordinator are considered: (i) one RFD (*lines with triangles*), (ii) two RFDs (*lines with diamonds*) or (iii) three RFDs (*lines with circles*). The considered packet generation rate at the RFDs is $g_{\text{pck}} = 2$ pck/s

in the case with direct transmission to the coordinator. As intuitively expected, the scenario with the highest network transmission rate is the one with only one RFD transmitting to the coordinator. This is due to the fact that in the presence of more than one RFD, the RFDs trying to access the channel may incur into collisions. Therefore, retransmissions may be needed (according to the used back-off algorithm) and the network transmission rate reduces. In all cases, for short packet lengths, the throughput is low for all network configurations; for

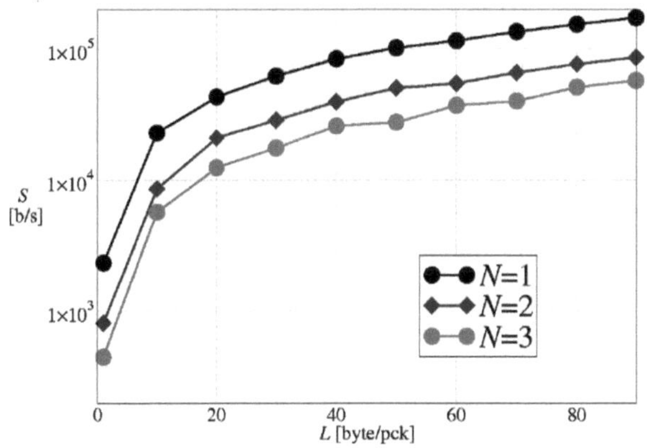

(a)

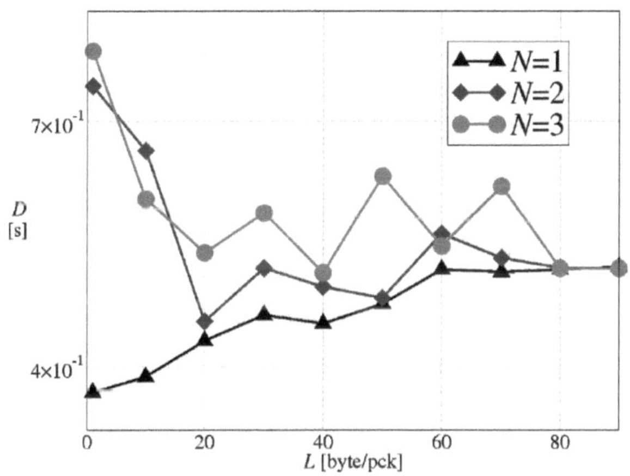

(b)

Figure 10. Experimental network performance, as a function of the packet length, in terms of (**a**) network transmission rate and (**b**) delay. Three scenarios with transmission to the coordinator in the presence of a router are considered: (i) one RFD (*lines with triangles*), (ii) two RFDs (*lines with diamonds*), and (iii) three RFDs (*lines with circles*). The considered packet generation rate at the RFDs is $g_{\mathrm{pck}} = 2$ pck/s

increasing values of L, it rapidly increases and tends to saturate for large values of L (i.e., high traffic loads).

In Figure 9b, the delay is shown as a function of the packet length. It can be noticed that the delay behavior is similar in the three considered scenarios, i.e., with (i) one RFD, (ii) two RFDs, and (iii) three RFDs. This is due to the fact that for the considered values of the network parameters, the traffic load is low and, therefore, an RFD is likely to send a data packet either at its first or second attempt. Therefore, the delay is not influenced by the network traffic.

In Figure 10a, the network transmission rate is shown, as a function of the packet length, in the presence of a router. In this case, unlike for the simulation results shown in Figure 4, the network transmission rate is not affected by the use of ACK messages in a scenario with a router. In Figure 10a, the network transmission rate rapidly increases, even at small values of L, and then saturates for large values of L. Comparing the results in Figure 9a with those in Figure 10a, it can be noted that the behavior of the network transmission rate is almost identical. This fact can be explained considering that the software implementation of the networking and communication protocols in the PICDEM Z nodes provides all the functionalities required by the Zigbee standard. The nodes employed in the simulations, at the opposite, have only the functionalities of the first two layers of the ISO/OSI stack [19]: therefore, they lack the upper-layer network capabilities that contribute to correct ACK message management. In addition, in the PICDEM Z nodes the router is equipped with a queue where received messages can be stored—we recall that our Opnet router model, instead, does not accept any new incoming packet until it has delivered the current one.

In Figure 10b, the delay is shown, as a function of the packet length, in a scenario with a router. In the case with $N = 1$, the delay is slightly increasing, since the RFD always finds the channel free. In the scenarios with $N = 2$ and $N = 3$, instead, the delay is high for low values of L and then decreases for large values of L. This behavior is quite surprising at first sight, since one intuitively expects a higher delay for large values of L. However, the observed behavior can be explained considering that, after a collision, a packet can be delayed for a certain amount of time because of the back-off algorithm. In the experiments with $N = 2$ RFDs, we have observed that while an RFD could not transmit for a certain time interval, the other RFD was sending its messages regularly. In other words, out of the two RFDs one starves and the other gets all the resources. In general, since the RFDs are in the same carrier sensing range, while an RFD is transmitting a packet, the other RFDs must wait (according to the back-off algorithm) before transmitting their packets. In unfortunate situations, it may happen that a node detects the channel as busy for a long time and, therefore, cannot transmit any message. This behavior might also be related to the *capture effect*, which affects IEEE 802.15.4 wireless networks [18].

4.3 Impact of the ACK Window Duration

Under the use of ACK messages, it is of interest to investigate the impact of the ACK window duration. While a data packet is transmitted according to the CSMA/CA protocol, an ACK message is transmitted using another protocol.

In fact, after successful transmission of a data packet, all RFDs must wait a SIFS interval before transmitting, as this interval is used by the receiving node to send back the ACK message. Therefore, an RFD can estimate the amount of time, denoted as *ACK window duration*, it will have to wait for the ACK message. If an RFD does not receive the ACK message within this estimated time, it stops generating new packets. The standard provides a maximum ACK window duration, after which the packet is declared lost. After a maximum number of attempts (which is pre-determined during the network setup), the data packet is discarded and a new data packet may be generated.

In order to investigate the impact of the ACK window duration, we resort to our Opnet simulator. More precisely, we consider a scenario where $N = 3$ RFDs are sending data messages to a coordinator (i) directly or (ii) through a router (topologies a and b in Figure 1, respectively). The major network parameters are fixed as in the previous simulations (in particular, $g_{pck} = 2$ pck/s and $L = 64$ byte/pck). In Figure 11, the delay is shown as a function of the ACK window duration, in the cases with no router (solid line) and with a router (dashed line). Both in the case of *direct transmission* and with *transmission through a router*, with low traffic, the delay between two subsequent packets is short and increases slowly for increasing offered traffic load. However, the delay curve, in the case with transmission through a router, presents some fluctuations, introduced by the presence of the router, which, in addition, introduces a further delay due to the retransmission of the packet from the router to the coordinator.

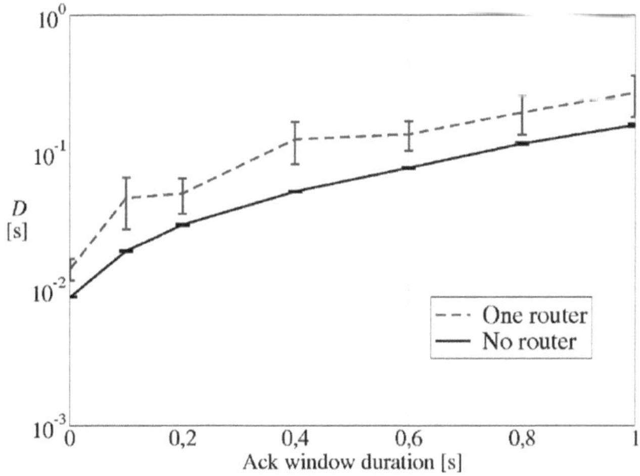

Figure 11. Transmission delay, as a function of the ACK window duration, in a network with $N = 3$ RFDs, packet generation interval at the remote nodes given by $g_{pck} = 2$ pck/s, and fixed packet length $L = 64$ byte/pck. Two scenarios are considered: (i) direct transmission (*solid line*) and (ii) transmission with router (*dashed line*)

In Figure 11, we also show the confidence intervals $(\eta - \sigma, \eta + \sigma)$ associated with each simulation point, with

$$\eta \triangleq \frac{\sum_{i=1}^{n_{\text{Trials}}} D_i}{n_{\text{Trials}}} \qquad \sigma \triangleq \sqrt{\frac{\sum_{i=1}^{n_{\text{Trials}}} (D_i - \eta)^2}{n_{\text{Trials}}}}$$

where n_{Trials} is the number of simulation runs and D_i is the delay value collected in each simulation run. As one can observe, the simulation results in the absence of a router are very reliable (i.e., the confidence interval is very small), whereas, in the presence of a router, the confidence interval is larger, according to the fluctuations introduced in the performance results. Similar considerations (especially in terms of accuracy) hold also for the other simulation results presented in this chapter. We did not explicitly indicate the confidence intervals in all figures for the sake of illustrative clarity.

4.4 Impact of Network Lifetime

A critical issue in wireless sensor networking is the network lifetime, since nodes are typically equipped with a limited-energy battery and may be subject to failures. In order to save as much energy as possible, the sensors may be grouped into *clusters*, i.e., they transmit their data to intermediate nodes (denoted as *cluster-heads*), which may properly modify these data and relay them to the coordinator. In the remainder of this section, we try to analyze the impact of the network lifetime on the network performance (evaluated in terms of network transmission rate), considering various clustering configurations.

First, one has to define until when the network has to be considered "alive." Several definitions have been proposed in the literature [6]. In general, the network can be considered alive until a proper QoS condition is satisfied. Obviously, the more stringent the QoS condition, the shorter the sensor network lifetime. We consider, as QoS condition, the *tolerable network death level* χ_{net} (adimensional), which is defined as the percentage of RFDs' deaths at which the overall network is assumed to be dead. In other words, if the QoS condition is stringent, the network is considered dead just after a small percentage of the RFD population dies. The lifetime of a single RFD is modeled as an exponential[3] random variable with mean value equal to 300 s. We remark that our results have been obtained in the absence of ACK messages to confirm the correct reception of the packet.

In Figure 12a, the network transmission rate is shown, as a function of the tolerable network death level, in a scenario with $N = 16$ RFDs. Various configurations are considered, including the case with no cluster (i.e., the RFDs

[3] We point out that an exponential distribution is often considered to characterize the lifetime of technological devices [20]. However, we did not investigate, from an experimental viewpoint, a more accurate lifetime distribution. Our approach, however, can be applied with any sensor lifetime distribution.

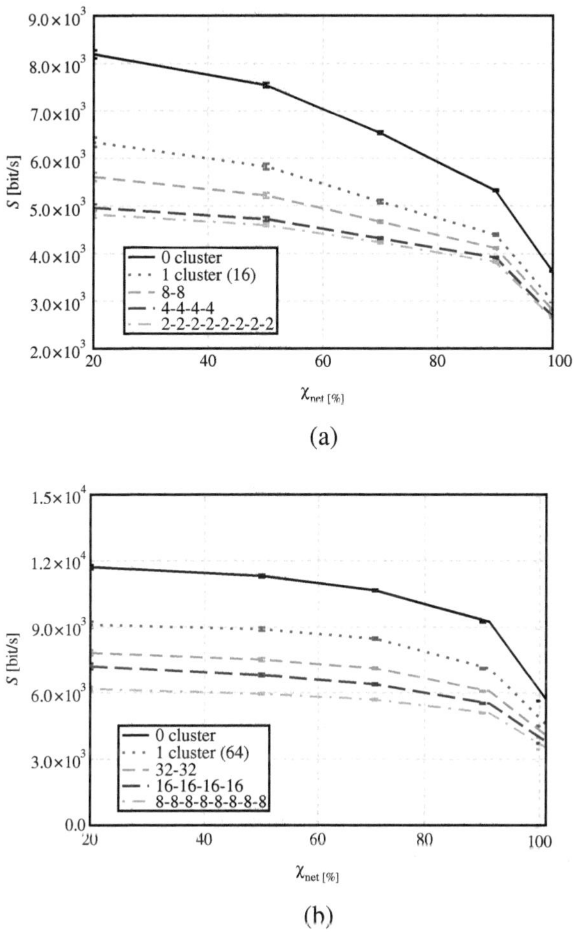

Figure 12. Network transmission rate, as a function of the tolerable network death level provided by the network (i.e., the number of RFDs' deaths required to declare the network dead), in various clustering configurations. Two scenarios are considered: (**a**) 16 RFDs transmitting to the coordinator and (**b**) 64 RFDs transmitting to the coordinator. The packet length is $L = 64$ bytes/pck and the packet generation rate at each RFD is $g_{\text{pck}} = 2$ pck/s

communicate directly to the coordinator) and a few clustered cases (the RFDs in each cluster communicate to an intermediate relay which, in turn, communicate to the coordinator). For a low tolerable network death level χ_{net}, i.e., when the network is considered dead after a small number of sensors' deaths, the network transmission rate slightly decreases. When the tolerable network death level becomes less stringent, i.e., when the number of RFDs' deaths necessary to make the network dead increases, the network transmission rate quickly decreases. This can be explained considering that when the average number of transmitting nodes is large (this happens for stringent tolerable death-level conditions), under the condition of low traffic load, the number of delivered packets is large. When the average number of transmitting nodes decreases (this is allowed for less stringent tolerable network death level), the number of delivered packets decreases as well. Moreover, the overall maximum network transmission rate is reached in a scenario with no clustering. Increasing the number of clusters has a negative impact on the network performance, according to the intuition that the relays (i.e., the cluster-heads) act as bottle-necks for data transmission.

In Figure 12b, the network transmission rate is shown, as a function of the tolerable network death level, in a scenario with $N = 64$ RFDs. The network performance trend is very similar to that shown in Figure 12a. In this case as well, the best performance is obtained when no clustering is used. Moreover, a higher number of cluster-heads causes a reduction of the network transmission rate. However, it is immediate to observe that the network performance trend is similar in the two considered scenarios. This fact suggests that it could be possible to create a general simulator and, consequently, an analytical model which allows to predict the performance of an IEEE 802.15.4 wireless network.

5. CONCLUSIONS

In this chapter, we have analyzed the impact of *relaying* on the performance of wireless sensor networks. In particular, we have presented simulation and experimental results, together with simple analytical considerations. Our simulator has been built with Opnet, and our experiments have been carried out using PICDEM Z nodes. Various topologies have been considered, including scenarios with direct transmission from the RFDs to the coordinator (absence of relaying) and the presence of one relay or two relays. In all scenarios, the system performance has been evaluated considering the impact of (1) the number of sensors, (2) the traffic load, (3) the ACK window duration, and (4) the tolerable network death level (a measure of the network lifetime). In scenarios with at least one router, our results show that the performance is highly degraded if ACK messages are used and can be improved either by limiting the number of contemporaneous transmitting nodes (maybe through synchronization between RFDs) or by eliminating completely the use of ACK messages—this improvement is, of course, possible because of the low traffic load typical of sensor networks. In a scenario with clustering, we have discovered that the best system configuration (in terms of number of clusters) does not depend on (i) the number of RFDs and, consequently, (ii) the tolerable network death level. In particular, the performance trend is similar for $N = 16$ and $N = 64$ RFDs.

APPENDIX. ANALYTICAL APPROXIMATION OF THE SIMULATION RESULTS

In IEEE 802.11 networks, the MAC protocol is given by the distributed coordination function (DCF), which corresponds to a *slotted 1-persistent* CSMA/CA MAC protocol. The MAC protocol of the IEEE 802.15.4 standard differs from the DCF in the sense that (i) the access is *unslotted* and (ii) the used CSMA/CA protocol is *non-persistent*. In order to apply the framework in [4] (valid for a slotted scenario) to a Zigbee network (which uses an unslotted communication scheme), we consider a "mini-slotted" time subdivision. The mini-slot dimension, denoted as σ, corresponds to the duration of a symbol in the Opnet simulations—this is also expedient for comparison purposes between analytical and simulation results. In fact, all time parameters available in the Opnet simulator are expressed as multiples of this value. In [4], the analysis is presented only for a scenario with many RFDs connected directly to the coordinator, as shown in Figure 1a. Therefore, a comparison between analysis and simulations is possible only in this case. A simple analytical framework applicable to a scenario with a router, in fact, does not lead to reasonable, yet accurate, results. Our conclusion is that a novel Markov chain-based approach, inspired by [4], should be developed in a two-router scenario, and we are currently pursuing this research direction.

In [4], the following expression for the network transmission rate is derived:

$$S = \frac{L}{T_s - T_c + \frac{\sigma(1-P_{tr})/P_{tr}+T_c}{P_s}} \tag{4}$$

where L is the data packet length, σ is the duration of a mini-slot, T_s is the average time interval during which the channel is sensed busy, T_c is the average time interval during which the channel is sensed busy by each station during a collision, P_{tr} is the probability that there is at least one transmission in the slot, and P_s is the probability of successful transmission through the channel. In particular, P_{tr} can be expressed as $P_{tr} = 1 - (1-\tau)^N$, where τ is the probability that an RFD transmits and N is the number of nodes. The probability P_s, instead, can be expressed as

$$P_s = \frac{N\tau(1-\tau)^{N-1}}{1-(1-\tau)^N}$$

Therefore, expression (4) for S can be rewritten as follows:

$$S = \frac{L}{T_s - T_c + \frac{\sigma(1-\tau)^N + T_c[1-(1-\tau)^N]}{N\tau(1-\tau)^{N-1}}} \tag{5}$$

In Figure 13, the network transmission rate is shown, as a function of the number of nodes N, in a scenario with direct transmission from the RFDs to the

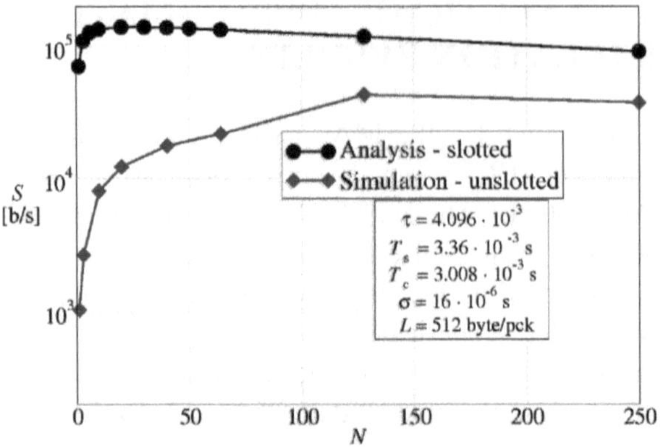

Figure 13. Network transmission rate, as a function of the number of RFDs, in a scenario with direct transmission to the coordinator. The results are obtained through simulation (*line with squares*) and analysis (*line with circles*). The parameters used in the simulation are shown in the figure

coordinator. The parameters employed in the analysis are summarized in the figure. The difference between the curve obtained with the analysis (line with circles) and the curve given by the simulations (line with diamonds) can be explained by observing that the analysis is based on a slotted network, whereas the simulation corresponds to an unslotted networking scenario—note that the simulation curve in Figure 13 is the same as that in Figure 4a relative to the case with no router. A direct comparison of the two curves shown in Figure 13 shows that asymptotically (for increasing values of N) the analytical results are in agreement with the simulation results. In fact, when the offered load is high, the main difference between slotted and unslotted networks is the presence of synchronization, which, in the former case, leads to a (slight) performance improvement. In order to provide a more detailed comparison between the analytical framework and the simulation setup, we are extending the Markov chain-based approach to the CSMA/CA-based medium access technique to the case with one or more routers.

REFERENCES

[1] R. Abileah and D. Lewis, "Monitoring high-seas fisheries with long-range passive acoustic sensors," in *Proc. Int. Conf.* OCEANS'96*: Prospects for the 21th century*, Fort Lauderdale, FL, USA, September 1996, vol. 1, pp. 378–382.

[2] I. F. Akyildiz, W. Su, Y. Sankarasubramaniam, and E. Cayirci, "A survey on sensor networks," *IEEE Commun. Mag.*, vol. 40, no. 8, pp. 102–114, August 2002.

[3] S. Barberis, E. Gaiani, B. Melis, and G. Romano, "Performance evaluation in a large environment for the awacs system," in *Proc. Int. Conf. on Universal Personal Communications* (ICUPC'98), Florence, Italy, October 1998, vol. 1, pp. 721–725.

[4] G. Bianchi, "Performance analysis of the IEEE 802.11 distributed coordination function," *IEEE J. Select. Areas Commun.*, vol. 18, no. 3, pp. 535–547, March 2000.

[5] C. Y. Chong and S. P. Kumar, "Sensor networks: evolution, challenges, and opportunities," *Proc. IEEE*, vol. 91, no. 8, pp. 1247–1256, August 2003.

[6] G. Ferrari and M. Martalò, "Extending the lifetime of sensor networks through adaptive reclustering," *EURASIP J. Wireless Commun. and Networking, Special issue on "Novel Techniques for Analysis and Design of Cross-Layer Optimized Wireless Sensor Networks"*, 2007, 20 pages. Doi: 10.1155/2007/31809.

[7] G. Ferrari, P. Medagliani, S. Di Piazza, and M. Martalò, "Wireless sensor networks: performance analysis in indoor scenarios," *EURASIP J. Wireless Commun. and Networking, Special Issue on "MobileMAN (Mobile Multi-hop Ad Hoc Networks): From Theory to Reality"*, 2007, 14 pages. Doi: 10.1155/2007/81864.

[8] O. Hyncica, P. Kacz, P. Fiedler, Z. Bradac, P. Kucera, and R. Vrba, "The Zigbee experience", in *Proc. Int. Symposium on Communications, Control and Signal Processing (ISCCSP'06)*, Marrakech, Morocco, March 2006.

[9] IEEE, "IEEE 802.11 Std: Wireless LAN Medium Access Control (MAC) a Physical Layer (PHY) specifications," IEEE Computer Society Press, pp. 1–459, June 1997.

[10] IEEE, "IEEE 802.15.4 Std: Wireless Medium Access Control (MAC) and Physical Layer (PHY) Specifications for Low-Rate Wireless Personal Area Networks (LR-WPANs)," IEEE Computer Society Press, pp. 1–679, October 2003.

[11] L. Kleinrock and F. A. Tobagi, "Packet switching in radio channels: part I—carrier sense multiple-access modes and their throughput-delay characteristics," *IEEE Trans. Commun.*, vol. 23, no. 12, pp. 1400–1416, December 1975.

[12] J. S. Lee, "An experiment on performance study of IEEE 802.15.4 wireless networks," in *Proc. 10th Conference on Emerging Technologies and Factory Automation 2005, (ETFA'05)*, Catania, Italy, September 2005, vol. 2, pp. 451–458.

[13] M. Madou, *Fundamentals of Microfabrication*, CRC Press, New York, NY, USA, 1997.

[14] Microchip Website, http://www.microchip.com.

[15] National Institute of Standards and Technology (NIST) Opnet Model Website, `http://w3.antd.nist.gov/Health.shtml`.

[16] Opnet Website, `http://www.opnet.com`.

[17] S. N. Simic and S. Sastry, "Distributed environmental monitoring using random sensor networks," in *Proc. 2nd Int. Work. on Inform. Processing in Sensor Networks*, Palo Alto, CA, USA, April 2003, pp. 582–592.

[18] N. Son, B. Krishnamachari, and J. Heidemann, "Experimental study of concurrent transmission in wireless sensor networks," in *Proc. 4th Int. Conf. on Embedded Networked Sensor Systems* (SenSys'06), Boulder, Colorado, USA, November 2006, pp. 237–250.

[19] A. S. Tanenbaum, *Computer Networks*, 4th Edition, Prentice-Hall, Upper Saddle River, NJ, USA, 2003.

[20] R. E. Ziemer, *Elements of Engineering Probability and Statistics*, Prentice Hall, Upper Saddle River, NJ, USA, 1996.

[21] Zigbee Alliance Website, http://www.zigbee.org.

IMPLEMENTATION OF AVERAGE CONSENSUS PROTOCOLS FOR COMMERCIAL SENSOR NETWORKS PLATFORMS

R. Pagliari[1] and A. Scaglione[2]

[1] *University of Genova, Genova, Italy*

pagliari@dibe.unige.it

[2] *School of Electrical and Computer Engineering, Cornell University, Ithaca, NY, USA*

anna@ece.cornell.edu

Abstract In sensor networks, average consensus and gossiping algorithms, featuring only near neighbor communications, present advantages over flooding and epidemic algorithms in a number of distributed signal processing applications. This chapter looks into the implementation of average consensus algorithms within the constraints of current sensor network technology. Our event-based protocols work in the real event-based environment provided by a common Mica2 platform and use its wireless CSMA packet-switched network interface. Within this architecture our chapter derives different protocols according to an event-based software architecture that are suitable for an environment like TinyOS, the most used operating system for low-power mote platforms. Theoretical and simulation results are presented, and the main advantage over traditional routing protocols is given by the fully distributed and scalable nature this approach follows.

1. INTRODUCTION

Wireless sensor networks have been widely studied in recent years. The most immediate applications are represented today by environmental monitoring, surveillance of seismic areas, intrusion detection and target tracking [4, 12]. In many of these applications the scale of the network renders impractical the direct transmission of data from the sensors to a fusion center, because of the energy consumption it entails; also, a centralized approach, in which a single unit processes all the data received from the sensors by direct or multi-hop transmissions, creates a communication bottleneck causing increased delay, non-uniform accuracy in the data collection and possibly a delay in the sought response mechanism, when the monitoring is associated to a feedback control mechanism. Yet, in many practical engineering systems this is the architecture of choice.

Alternatives, however, are not lacking. Fully distributed algorithms appeared first in the computer science literature as a tool to provide distributed means of querying the data and performing distributed computations: examples are the information dissemination/gossiping methods (see, e.g., [13]) and the epidemic protocols in [1, 7]. More recently, a class of algorithms that requires only near neighbors interactions has gained increasing interest: they are called consensus or agreement protocols [2, 18–20, 25].

In a nutshell, in all these protocols the nodes iteratively update and exchange a state variable, which converges to be identical and, in the average consensus case, it is the average of the initial states of every node. In several protocols the communication is described as analog and noiseless or affected by a constant independent additive white Gaussian noise added to it. To capture the presence of a wireless channel, several recent works have investigated the effect of a switching topology [22, 25] as well as completely asynchronous protocols [15].

1.1 Specific Contributions of the Paper on Communication Protocols for Average Consensus

Agreement protocols possess features that a good algorithm for sensor network applications should have, such as being: (1) decentralized; (2) robust to link failures, link asymmetries [11] and time delays; (3) adaptable to switching topologies.

The objective of our study is that of examining some of these options on a common Mica2motes platform, which is considered representative of the current state of the art. Our contribution is the design, analysis and comparison of different implementations for the agreement protocols over this network.

The chapter is organized as follows: we first review agreement protocols in Section 3. In Section 4, we describe different implementations on commercial platforms. Section 5 discusses theoretical and simulation results. Section 7 summarizes our main results and observations. Sections 2 and 6 describe in detail the wireless model we assumed for our simulations and possible applications of average consensus algorithms.

2. WIRELESS NETWORK MODEL

The Mica2Motes RF interface is compliant with the IEEE802.15.4 standard. Two bands are supported, one at the 2.4/2.4835 GHz band with data rates of 250 kb/s and one supporting 20/40 kb/s in the 868/915 MHz band. In both bands the physical layer adopts orthogonal modulation, enabling simple non-coherent detection. The channel can only operate in half-duplex. The channel access scheme is carrier sensing multiple access with collision avoidance (CSMA/CA). The so-called distributed coordination function (DCF) mechanism handles retransmissions of collided packets using a binary exponential back-off delay and requiring a two-way hand-shaking performed with ACK messages. The stable throughput of CSMA/CA was analyzed in [8].

In this context, links can be modeled as packet erasure channels, whose erasure probability depends on the fading state of all the links active including the one under examination. We can also assume that data transferred in the payload are represented with infinite precision, as other errors exist that will make the error due to finite representation negligible.

As a coarse approximation one can assume that packets are exempt from collisions, only if one transmitter transmits at a time, as assumed in [8]. Equivalently, one can imply that the collision avoidance mechanism is so effective that only nodes that are causing negligible interference to each other can ever transmit simultaneously. In this case the successful reception of the packet can be mapped approximately in the requirement of having a sufficient signal-to-noise ratio (SNR) over the link, while no one in a certain range r^* from the receiver.[1] This model is similar to the so-called *protocol model* in [9]. The SNR of the link from node j to node i can be modeled as

$$\mathrm{SNR}_{ij} = \kappa\beta_{ij}/r_{ij}^{\eta} \tag{1}$$

where β_{ij} is a random small-scale fading coefficient, r_{ij} is the link distance, η is a path loss exponent and κ is the transmit power-to-noise ratio. Fading as well as collisions can cause an erasure. According to this model the network is a random geometric graph $G2(n, r^*)$ [9] of packet erasure links, which allows to calculate a series of interesting statistics, including the distribution of each node degree. For example, for a uniform deployment of n nodes over an area A, the node degree d_i is a Bernoulli random variable $\mathscr{B}(n-1, p)$ with probability of success $p = \pi r^{*2}/A$.

An alternative model includes interfering signals and allows also to describe the so-called *capture* phenomena that occur in wireless random access. In it, it is assumed that the receiver experiences an SNR equal to

$$\mathrm{SINR}_{ij} = \frac{\beta_{ij} r_{ij}^{-\eta}}{\kappa^{-1} + \sum_{p \neq i,j} \beta_{ip} r_{ip}^{-\eta}} \tag{2}$$

where the sum of interfering signals at the denominator includes all nodes that are simultaneously active.

[1] Truly, the transmitting node can only preventively sense the area around itself and cannot monitor the area around the receiver which can have, as well known, hidden terminals.

Simulations are based on the common variant of Mica motes, composed of an Atmel ATMEGA103 microcontroller unit, 128 kB program memory, 4 KB RAM and a RFM TR1000 radio transceiver, with 50 kB of transmission rate at 916.5 MHz. The platform simulator is *event-based*, handling interrupts such as messages arrival and clock firing.

The MAC layer is a simplified CSMA-CA, in which every node, before transmitting a packet, waits, the first time, for a random time interval, given by the sum of a constant (200 bit times) plus a uniform random contribution of 128 bit times. After the first time, if the channel is sensed busy, the node begins the back-off mechanism. A generic node, instead, checks for the incoming signal power and, depending on the two interference models described above, is able to know if a message has arrived safely or collided. In fact, during the reception of a frame, other nodes far from a generic transmitter can send messages to the same receiving node causing, under some conditions, the collision of two or more packets.

3. PRELIMINARIES ON AVERAGE CONSENSUS

Consensus protocols are gossip-based algorithms and allow a network of nodes to reach an agreement on a common value that depends on the initial state of all the sensors.

The generic network, directed or undirected, is typically described as a graph $G = G(V,E)$, where V is the set of vertices, E is the set of edges that, as mentioned before, can be modeled as a packet erasure channel. We are going to describe discrete time consensus protocols, since computations in our event-driven network are naturally described using a discrete index that points to each iteration of the algorithm. Let

$$N_i = \left\{v_j \in G : (v_i, v_j) \in E\right\} \tag{3}$$

be the set of neighbors of node v_i. The $k+1$th computation update at node i is as follows:

$$x_i(k+1) = f(x_i(k), u_i(k)) \tag{4}$$

where $x_i(k)$ represents the state variable of node i, and $u_i(k) = g(x_1(k), \ldots, x_{N_i}(k))$ is a *coupling* or *update* variable, function of the external signals that convey state information of neighbor nodes, defined in (3). In its simplest form the state-updating rule is an integrator:

$$x_i(k+1) = x_i(k) + \varepsilon u_i(k) \tag{5}$$

where ε is a step-size parameter, and $u_i(k)$ is linear:

$$u_i(k) = \sum_{j \in N_i} a_{ij}(x_j(k) - x_i(k)) \tag{6}$$

It can be seen that the global state, at each time-step, is updated as follows:

$$x(k+1) = (I - \varepsilon L)x(k) \triangleq Wx(k) \tag{7}$$

where L is the graph Laplacian matrix, defined by

$$l_{ij} = \begin{cases} \sum_{j \in N_i} a_{ij} & j = i \\ -a_{ij} & j \neq i \text{ and } j \in N_i \\ 0 & \text{else} \end{cases} \tag{8}$$

and $W \triangleq (I - \varepsilon L)$. The goal of a generic f-consensus protocol is to reach an agreement on the state variable of the whole network driving all node states to become equal to a function of the initial state. Thus, if $\overline{x}$ is the fixed point of the system (7), $\overline{x} = f(x(0))$, $f(\cdot)$ the desired function of the initial state. In average consensus problems $f(x(0)) = \frac{\sum_i x_i(0)}{n}$.

Simple choices to ensure the convergence of (7) were derived in [25]. In particular, the convergence is guaranteed when $\varepsilon \in (0, 1/\text{deg}_{\max})$, where $\text{deg}_{\max}$ is the maximum degree of the network, or $\varepsilon_{ij} < 2/(\max(d_i, d_j))$. Even if other choices of the weight associated to the update rule are known, based on the optimization of the particular topology a network describes, the two possibilities presented above are the most suitable for a sensor network, in which every node is aware of its neighbors only. Section 4 presents two approaches to estimate the degree of a generic node and its neighbors, while in Section 4.2 a protocol with constant weights is described.

4. METROPOLIS-HASTINGS AVERAGE CONSENSUS

Each node updates its internal state variable according to the following rule:

$$x(k+1) = x(k) + \sum_{j=1}^{N_i} \frac{1}{\max(d_i, d_j)} (x_j(k) - x_i(k))$$

where d_i and d_j represent the degree of nodes i and j, respectively. The algorithm is so composed by two stages: (i) a discovery phase and (ii) an averaging phase. This method comes from the Metropolis-Hastings algorithm, and guarantees convergence, provided the graph is not bipartite [3].

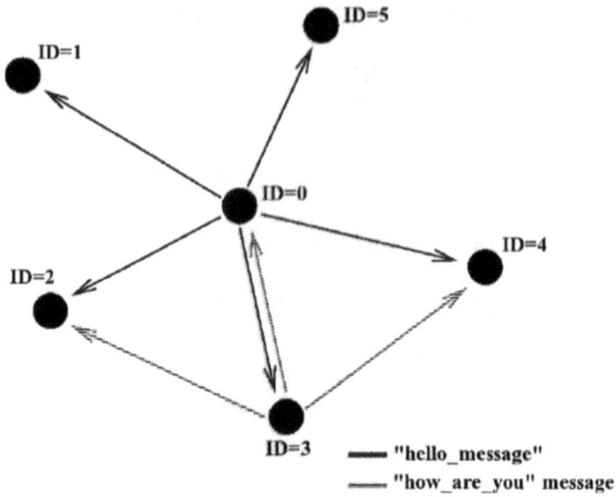

Figure 1. Discovery phase

Each node discovers its neighbors, calculates its own degree and the degree for each of the neighbors it can reach. As shown in Figure 1, every time an internal clock fires, node 0 sends a message, i.e., a "hello_message", inserting within the message the ID number and the actual local degree. Each receiving node updates the degree of node 0, d_0, and answers by a "how_are_you" message: the format is the same, but the session ID is different to avoid infinite loops. After a node i has answered, all the nodes that can hear it update their internal neighbor table with respect to i, if the degree is greater than the previous one (this is done to avoid delays due to the CSMA-based medium access).

The nodes can be, or not, synchronized. For this latter case the protocol takes into account the possibility that different nodes finish the discovery phase at different absolute clock times, and that was the scenario we simulated. This is a realistic assumption, since even if nodes are synchronized, an initialization phase, during which a starting signal has to be disseminated into the network, causes different starting times.

The main concepts of this algorithm can be described for each node:

- A fixed number of # hello_message are sent. Every time a node answers, upgrade the local degree.

- If a hello_message was received, answer with a howAreYou_msg; if howAreYou_msg are received from other nodes, do not answer (to avoid loop cycles), but update.

- When the discovery phase is finished start the averaging phase (divided in clock periods).

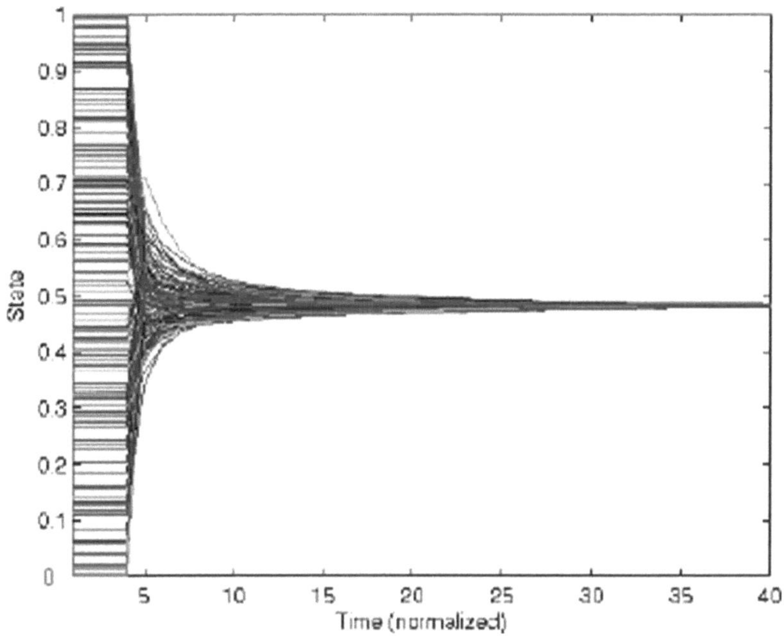

Figure 2. Evolution, over time, of a network with $N = 160$ nodes in a $500 \times 500\,\text{m}^2$ area and radius $R = 100\,\text{m}$

- If an avg_msg is received update a neighbors_table (if the value is not present from that node). If some hello_msg arrives, answer and update local degree.
- For every clock fire, within the average phase, update value and broadcast it.

As an example, the evolution of a scenario with $N = 160$ sensors deployed in a $500 \times 500\,\text{m}^2$ area, with a radius of $R = 100\,\text{m}$, is depicted in Figure 2. After the discovery phase, each node starts to update its internal variable and interact with the other. As we can observe, if all the sensors are connected, the internal values converge to an agreement that is, in this case, the average value.

In Algorithm 1 an event-based code implementation is derived. The node can be in "discovery" or "averaging phase". Initially, the node sets the initial value. The internal state is then set to "discovery phase", since that is the first stage of the algorithm, and it takes # number of clock fires to complete the algorithm. Within this stage the node sends "hello_message" every time the clock fires, and records all the incoming answers from its neighbors. After the last clock fire has expired, the state turns to "averaging", and it starts the local state update. During this period, characterized by a fixed number of iterations, if a "hello_message" arrives an update is performed. The averaging phase takes a fixed number of iterations, and the final value is considered the average of initial values.

Algorithm 1 Local degree with training

```
Initialization Phase
read value;
Set_Clock( next_time );
state ← discovery_phase;
num_hello_msg ← 0;
neighbor_table ← [];
Event Packet_Received:
if discovery_phase then
    if found(msg.ID, neighbor_table) then
        update( );
    else
        add_entry( );
    end if
    if msg.type == hello_msg then
        Send_HowAreYou_msg( local_degree );
    end if
else
    // we are in averaging_phase
    if hello_msg then
        update( neighbor_table, msg.degree );
        Send_HowAreYou_msg( local_degree );
    else
        if found(msg.ID, neighbor_table) then
            update( );
        else
            add_entry( );
        end if
    end if
end if
Event Clock_Fired:
if discovery_phase then
    Send_HowAreYou_msg( ID, value, local_degree );
    num_hello_msg = num_hello_msg +1;
    if num_hello_msg == max_num_hello_msg then
        state ← averaging_phase;
    end if
    Set_Clock( next_time );
else
    // we are in averaging_phase
    sum = 0;
    for i = 1 : size(neighbor_table) do
        if neighbor_table(i).received == 1 then
            sum ← sum + max(neighbor_table(i).degree, local_degree) · (neighbor_table(i).value - value);
        end if
    end for
    value ← value + sum;
    Send_Avg_msg( ID, value, local_degree );
    neighbors_table(:, received) ← 0;
    Set_Clock( next_time );
end if
```

4.1 Metropolis-Hastings Without Training

This algorithm differs from the previous one by the fact that the training (or discovery) phase is absent. During the iterations each node receiving messages from the neighborhood simply updates its local degree as messages from new nodes come, so that its nearby nodes learn the new values of that cardinality progressively.

Algorithm 2 Local degree without training

```
Initialization Phase
read value;
Set_Clock( next_time );
neighbor_table ← [];
Event Packet_Received:
if found(neighbors_table) then
    add_entry( );
else
    update( degree );
    if already received then
        update( value);
    end if
end if
Event Clock_Fired:
for i = 1 : size(neighbor_table) do
    if neighbor_table(i).received == 1 then
        sum ← sum + max(neighbor_table(i).degree, local_degree), · (neigh-
        bor_table(i).value - value);
    end if
end for
value ← value + sum;
neighbors_table(:, received) ← 0;
Send_Avg_msg( ID, value, local_degree );
Set_Clock( next_time );
```

4.2 Constant Weight

In this case every node uses the same weight to compute the updated value. Hence, the knowledge of the local degree is not required and every time an internal clock fires, the sensor simply has to check all messages arrived within the previous time window, calculate the new value and broadcast it to the neighborhood. The common weight must fall within the interval $0 < \alpha < \frac{2}{\lambda_1(L)}$, and the optimum value is given by $\alpha^* = \frac{2}{\lambda_1(L)+\lambda_{N-1}(L)}$. This value was found by averaging on Monte Carlo simulations and set to be the constant weight.

Algorithm 3 Constant weight

```
Initialization Phase
read value;
Set_Clock( next_time );
neighbor_table ← [];
Event Packet_Received:
if found(neighbors_table) then
   add_entry( );
end if
Clock_Fired:
for i = 1 : size(neighbor_table) do
   if neighbor_table(i).received == 1 then
      sum ← sum + const_weight · (neighbor_table(i).value - value);
   end if
end for
value ← value + sum;
neighbors_table = [];
Send_Avg_msg( ID, value, local_degree );
Set_Clock( next_time );
```

5. PERFORMANCE EVALUATION

In our analysis we will resort to a statistical characterization of the delay derived based on the following *outage probabilities* averaged over all random graphs G with identically distributed nodes as

$$P^{(k)}_{\text{out-node}}(\zeta) \triangleq E\{P(|\delta_i(k)|^2 > \zeta|G)\} \tag{9}$$

$$P^{(k)}_{\text{out}}(\zeta) \triangleq E\{P(||\delta(k)||^2 > \zeta|G)\} \tag{10}$$

If $x(0)$ is zero mean and jointly normal $\mathcal{N}(0, R_x)$, $\delta(k)$ is jointly normal with covariance $R_{\delta(k)} = (W^k - n^{-1}\mathbf{1}\mathbf{1}^T)R_x(W^k - n^{-1}\mathbf{1}\mathbf{1}^T)^T$. Hence, for $k > 0$, assuming all nonzero eigenvalues $\nu_i(k)$ of the auto-covariance of $\delta(k)$ are distinct, we have

$$\begin{aligned} P^{(k)}_{\text{out-node}}(\zeta) &= E\left\{ \mathrm{e}^{-\frac{\zeta}{[R_{\delta(k)}]_{ii}}} \right\} \\ P^{(k)}_{\text{out}}(\zeta) &= \sum_{i=1}^{n-1} E\left\{ \prod_{m \neq i} \frac{\nu_i(k)}{\nu_i(k) - \nu_m(k)} \, \mathrm{e}^{-\frac{\zeta}{\nu_i(k)}} \right\} \end{aligned}$$

Upper and lower bounds for this expression are

LEMMA 1 *The following bounds on the per node and total average outage probability hold:*

$$P^{(k)}_{\text{out-node}}(\zeta) \leq P^{(k)}_{\text{out}}(\zeta)$$

$$P^{(k)}_{\text{out}}(\zeta) \leq E\left\{P\left(\chi^{(n-1)} > \frac{\zeta}{\lambda_1(R_x)(1-\varepsilon\lambda_1(L))^k}\right)\right\} \tag{11}$$

$$P^{(k)}_{\text{out-node}}(\zeta) \geq E\left\{\mathrm{e}^{-\frac{\zeta}{\lambda_n(R_x)(1-\varepsilon\lambda_1(L))^k}}\right\} \tag{12}$$

where $\chi^{(n-1)}$ is a chi-square random variable with $n-1$ degrees of freedom.

Proof The first equation is because $||\delta(k)||^2 \geq |\delta_i(k)|^2$. The second bound is obtained considering that $||\delta(k)||^2 = \sum_{i=1}^{n-1} \nu_i(k) z_i^2$ where $z_i \sim \mathcal{N}(0,1)$ are the i.i.d. Gaussian random variables obtained whitening the vector $\delta(k)$. Assuming $\nu_i(k)$ are in descending order, this implies that $\nu_{n-1}(k)\chi^{(n-1)} \leq ||\delta(k)||^2 \leq \nu_1(k)\chi^{(n-1)}$, where $\chi^{(n-1)} = \sum_{i=1}^{n-1} z_i^2$ is a central chi-square distribution with $n-1$ degrees of freedom.

Hence, $P^{(k)}_{\text{out}}(\zeta|G) \leq P(\chi^{(n-1)} > \zeta/\nu_1(k))$. The inequality then follows from the fact that $\nu_1(k) \leq \lambda_1(R_x)\lambda_{n-1}(W) = \lambda_1(R_x)(1-\varepsilon\lambda_1(L))$. In Figure 3 an

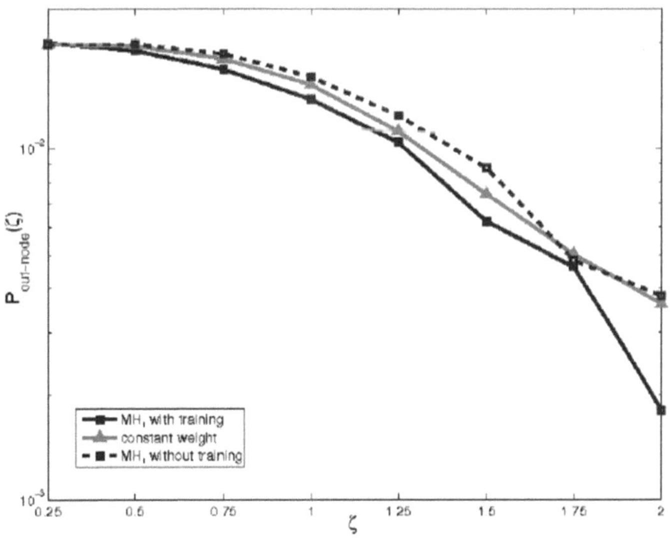

Figure 3. Outage probability of a network with $N = 50$ sensors over a $500 \times 500\,\text{m}^2$ area, with a radius of 100 m, and $k = 25$ iterations, with Metropolis-Hastings algorithms (*MH* with/out training) and constant weight algorithm. Simulations are based on SINR wireless channel model

example is reported with $N = 75$ nodes, a normalized radius of $r = 0.2$ and $k = 10$ iterations.

As an example, Figure 3 depicts the outage probability $P_{out}(\zeta)$, of the proposed protocols, obtained using network simulator. As expected, the "local degree without training" algorithm performs worse than the other ones. However, the "local degree with training" and "constant weight" have similar performance in a real network simulation.

6. APPLICATIONS FOR AVERAGE CONSENSUS ALGORITHMS

The most obvious application of average consensus algorithms is to synchronization problems, where these protocols address an old problem that has been studied since the first appearance of distributed systems [10, 16] and has been investigated in database theory as well [5]. There are two main families of algorithms that have been proposed for sensor networks synchronization: the main protocol falling in the first class is the reference broadcast synchronization (RBS) algorithm (RBS) [6], where a reference node signal sent to the neighborhood allows the nodes to exchange the time stamp of their reception time and then compute their relative clock difference (improvements of RBS are in [17, 21, 23]); the second class of algorithms is derived from the network time protocol (NTP), and it uses a global clock time scale combined with flooding algorithms, which disseminate the information on the timing throughout the network [14, 24]. Agreement protocols allow the nodes to calculate a common average clock phase iteratively, only using local communications, and they naturally support EBS-like protocols.

Synchronization is only one of many applications. Any query on sensor data that can be reduced to having a linear synopsis can be calculated through an agreement protocol. Let s be the vector of sensor data, with entries s_i. Agreement protocols permit to compute at sensor j any function that can be decomposed as follows:

$$g_j(s_1, \ldots, s_N) = p_j\left(N^{-1}\sum_{i=1}^{N} f_i(s_i)\right) \tag{13}$$

This can be easily accomplished by initializing the state variable at node i so that $x_i(0) = f_i(s_i)$ and, assuming that convergence is attained at time/iteration T, by computing the function $p_j(x_j(T))$.

There are several statistical signal processing functions that can also be calculated under certain constraints, beginning with a simple histogram of the data. For example, for identically distributed sensor data, to estimate the cumulative distribution function $F_s(\gamma)$ one can simply set $f_i(s_i) = 1$ if $s_i \leq \gamma$ and

$f_i(s_i) = 0$ otherwise. Certain statistical inference problems also have this structure: for example, the log-likelihood ratio $\Lambda_{ij}(s) = \ln(p(s|H_i)/p(s|H_j))$, in the case where the sensor data are conditionally independent, becomes a sum of likelihood ratios:

$$\Lambda_{ij}(s) = \sum_{k=1}^{N} \ln(p(s_k|H_i)/p(s_k|H_j)) = \sum_{k=1}^{N} \Lambda_{ij}(s_k)$$

In a parameter-estimation problem, with parameter vector θ, if the sensor data are conditionally independent, also $\Lambda_\theta(s) = \sum_{k=1}^{N} \Lambda_\theta(s_k)$.

Because of their decentralized nature, these methodologies are ideal to enforce coordinated actions and control mechanisms which are based on the network state.

Given the array of applications for consensus protocols discussed above, it is evident that the interplay with the physical aspects of communications, congestion problems, network delays, energy consumption, deserves close attention.

7. CONCLUSIONS

In this chapter we derived different implementations for average consensus protocols. In a fully distributed and decentralized scenario a generic node of the network is likely to be aware of its neighborhood and, consequently, the available information can be only the local degree, i.e., the cardinality, and the degree of neighbors. This leads to the design of cooperative protocols based on a few information about the local structure of a network. In such a way, we proposed and evaluated algorithms (i) based on the estimation of the local cardinality, and neighbors degree, and (ii) based on constant weights. In the latter case no additional training is required, nor additional updates during the averaging phase, since every node knows in advance the parameter to use within the updating rule. The performance results indicate that the "constant weight" protocols are advantageous over the alternatives because their potential performance difference compared to the other choices is negligible when the algorithm is run over a realistic network scenario, and the algorithms are much simpler to implement, because they do not require an initial network discovery phase.

REFERENCES

[1] D. Agrawal, A. El Abbadi, and R. C. Steinke, "Epidemic algorithms in replicated databases," in *PODS '97: Proceedings of the Sixteenth ACM SIGACT-SIGMOD-SIGART Symposium on Principles of Database Systems*, New York, NY, USA, 1997, ACM Press, New York, pp. 161–172.

[2] W. Ren and R. W. Beard, "Consensus seeking in multi-agent systems under dynamically changing interaction topologies," *IEEE Trans. Autom. Control, Technote*, vol. 50, no. 5, pp. 655–661, 2003.

[3] S. Boyd, P. Diaconis, and L. Xiao, "Fastest mixing Markov chain on a graph," *SIAM Rev.*, vol. 46, no. 4, pp. 667–689, 2003.

[4] C. Y. Chong and S. P. Kumar, "Sensor networks: evolution, opportunities, and challenges," in *Proceedings of IEEE*, San Antonio, TX, August 2003, pp. 1247–1256.

[5] A. Demers, D. Greene, C. Hauser, W. Irish, J. Larson, S. Shenker, H. Sturgis, D. Swinehart, and D. Terry, "Epidemic algorithms for replicated database maintenance," in *PODC '87: Proceedings of the Sixth Annual ACM Symposium on Principles of Distributed Computing*, New York, NY, USA, 1987, ACM Press, New York, pp. 1–12.

[6] J. Elson, L. Girod, and D. Estrin, "Fine-grained network time synchronization using reference broadcasts," *SIGOPS Oper. Syst. Rev.*, vol. 36, no. SI, pp. 147–163, 2002.

[7] D. Ganesan, B. Krishnamachari, A. Woo, D. Culler, D. Estrin, and S. Wicker, "An empirical study of epidemic algorithms in large scale multihop wireless networks," Tech. Rep., University of California, Los Angeles, 2002.

[8] P. Gupta and P. Kumar, "Performance analysis of the IEEE 802.11 distributed coordination function," *IEEE Trans. Inf. Theory*, vol. 18, no. 3, pp. 535–547, 2000.

[9] P. Gupta and P. Kumar, "Capacity of wireless networks," *IEEE Trans. Inf. Theory*, vol. 46, no. 2, pp. 388–404, 2000.

[10] L. Lamport, "Time, clocks, and the ordering of events in a distributed system," *Commun. ACM*, vol. 21, no. 7, pp. 558–565, 1978.

[11] Q. Lindsey, D. Lymberopoulos, and A. Savvides, "An empirical analysis of radio signal strength variability in IEEE 802.15.4 networks using monopole antennas," Tech. Rep., Yale University, ENALAB, 2006.

[12] A. Mainwaring, J. Polastre, R. Szewczyk, D. Culler, and J. Anderson, "Wireless sensor networks for habitat monitoring," in *ACM WSNA*, Atlanta, GA, 2002.

[13] M. Maroti, "Directed flood-routing framework for wireless sensor networks," in *Middleware '04: Proceedings of the 5th ACM/IFIP/USENIX International Conference on Middleware*, New York, NY, USA, 2004, Springer, New York, pp. 99–114.

[14] M. Maroti, B. Kusy, G. Simon, and A. Ledeczi, "The flooding time synchronization protocol," in *SenSys '04: Proceedings of the 2nd International Conference on Embedded Networked Sensor Systems*, New York, NY, USA, 2004, ACM Press, New York, pp. 39–49.

[15] M. Mehyar, D. Spanos, J. Pongsajapan, S. Low, and R. Murray, "Asynchronous distributed averaging on communication networks," *IEEE Trans. Netw.*, vol. 18, no. 3, pp. 535–547, 2007.

[16] D. L. Mills, "Internet time synchronization: the network time protocol," *IEEE Trans. Commun.*, vol. 39, no. 10, pp. 1482–1493, 1991.

[17] M. Mock, R. Frings, E. Nett, and S. Trikaliotis, "Continuous clock synchronization in wireless real-time applications," in *The 19th IEEE Symposium on Reliable Distributed Systems, SRDS'00*, Washington–Brussels–Tokyo, 2000, pp. 125–133.

[18] L. Murray, "Stability of multiagent systems with time-dependent communication links," *IEEE Trans. Autom. Control*, vol. 50, no. 2, pp. 169–182, 2005.

[19] R. Olfati Saber and R. M. Murray, "Consensus protocols for networks of dynamic agents," in *American Control Conference*, 2003.

[20] R. O. Saber and R. M. Murray, "Consensus problems in networks of agents with switching topology and time-delays," *IEEE Trans. Autom. Controls*, vol. 49, no. 9, pp. 1520–1533, 2004.

[21] K. Romer, "Time synchronization in ad hoc networks," in *MobiHoc 2001*, Long Beach, USA, 2001.

[22] R. O. Saber and R. M. Murray, "Consensus problems in networks of agents with switching topology and time-delays," *IEEE Trans. Autom. Controls*, vol. 49, no. 9, pp. 1520–1533, 2004.

[23] M. L. Sichitiu and C. Veerarittiphan, "Simple, accurate time synchronization for wireless sensor networks," in *Proceedings of the IEEE Wireless Communications and Networking Conference, WCNC'2003*, New Orleans, LA, USA, 2003.

[24] W. Su and I. F. Akyildiz, "Time-diffusion synchronization protocol for wireless sensor networks," *IEEE/ACM Trans. Netw.*, vol. 13, no. 2, pp. 384–397, 2005.

[25] L. Xiao and S. Boyd, "Fast linear iterations for distributed averaging," *Syst. Control Lett.*, vol. 53, pp. 65–78, 2004.

Part II

Networking and the Grid

BIOLOGICALLY INSPIRED APPROACHES TO NETWORK SYSTEMS

(Invited Contribution)

T. Suda[1], T. Nakano[1], M. Moore[1], and K. Fujii[1]

[1] *University of California, Irvine, Irvine, California 92697, USA*

Abstract This chapter describes two branches of biologically inspired approaches to networks: biologically inspired computer networks (i.e., the bio-networking architecture) and biologically inspired nanoscale biological networks (i.e., molecular communication). The first branch, biologically inspired computer networks, applies techniques and algorithms from biological systems to the design and development of computer networks. The second branch, biologically inspired nanoscale biological networks, applies techniques and algorithms from biological systems to the design and engineering of nanoscale biological networks.

1. INTRODUCTION

Information processing systems today are composed of a larger number of devices gathering, exchanging, and processing information (e.g., grid computing). In addition, system components (e.g., computational and sensing devices) are becoming smaller, potentially less reliable, and integrated directly into the environment (e.g., sensor networks for environmental monitoring). As a result, information processing systems today face research challenges of scaling to a larger number of system components and adapting to changes in the environment and failures that may occur in system components.

Observation of the biological world shows that a biological system is composed of a massive number of nano/micro-scale biological components that adapt to changes in the environment and failures of biological components. In biological systems, biological components self-organize into complex systems in a hierarchical manner, where a higher level structure is composed of a number of lower level components. For example, an organism is composed of multiple organs, each of which is composed of multiple cells, and so on. Components at each level communicate and coordinate through a variety of mechanisms (e.g., electric signals, molecule diffusion), allowing biological systems to adapt to environmental changes and component failures.

Design principles from biological systems may thus be applicable to designing of today's information processing systems that need to integrate large number of components and be adaptive to environmental changes and component failures. This chapter describes two branches of biologically inspired approaches to network systems: (1) *biologically inspired computer networks* and (2) *biologically inspired nanoscale biological networks*. The first branch, biologically inspired computer networks, applies design principles found in biological systems to improve computer networks design. The second branch, biologically inspired nanoscale biological networks, applies the principles to engineering nano to microscale biological networks.

The remainder of this chapter is organized as follows. Section 2 describes an example of biologically inspired computer networks (i.e., the bio-networking architecture), and Section 3 describes an example of biologically inspired nanoscale networks (i.e., molecular communication). Section 4 concludes this chapter.

2. BIOLOGICALLY INSPIRED COMPUTER NETWORKS

Future computer networks are expected to be autonomous, scalable, and adaptive. They autonomously operate with minimal human intervention. They scale to and support billions of nodes and users. They adapt to diverse user demands and remain available to users. They also adapt to dynamically changing network conditions such as network failures.

Key features of future networks mentioned above, i.e., autonomy, scalability, and adaptability, have already been achieved in biological systems. For example, scalability is observed in large-scale social insect colonies (e.g., ant colonies) that contain millions of entities (e.g., worker ants), yet exhibit highly sophisticated and coordinated behaviors (e.g., division of labor in foraging and nest building). Insect colonies also adapt to dynamically changing conditions (e.g., change in the amount of food in the ant colony) through local interaction of entities that adjust their behavior based on environmental conditions (e.g., more ants go out of a colony and gather food when the food level in the colony becomes low.). Based on the observation that biological systems exhibit key features required for future networks, an increasing number of researchers are currently investigating the feasibility of applying biological concepts and principles to computer networks design [1, 3, 4, 7, 9, 10, 16–18, 20].

We also believe if a network is modeled after biological principles, network applications are able to satisfy key requirements such as autonomy, scalability, and adaptability. We have therefore designed a biologically inspired network, called the bio-networking architecture [26, 28] where key biological principles and concepts (e.g., self-organization, emergence, redundancy, natural selection, and diversity) are applied to the design of computer networks and network applications.

In this section, we first describe biological principles applied to the bio-networking architecture and then describe our design of a middleware framework for this architecture.

2.1 Biological Principles Applied

A key concept in biological systems is emergent behavior. In biological systems, useful behavior often emerges through the collective, simple, and autonomous behaviors of individual biological entities. For example, when a bee colony needs more food, a large number of bees will leave the hive and go to the flower patches in the area to gather nectar. When the bee colony is near its food storage capacity, only a few bees will leave the hive to gather nectar. This adaptive food gathering function emerges from the relatively simple and local interactions among individual bees. If a returning food-gathering bee can quickly unload its nectar to a food-storing bee, it means that the food-storing bees are not busy and that there is little food in the hive. This food-gathering bee then encourages other nearby bees to leave the hive and collect nectar by doing the well-known "waggle dance." If a returning food-gathering bee must wait a long time to unload its nectar, it means that the food-storing bees are busy and that there is plenty of food in the hive. This food-gathering bee then remains in the hive and rests. Since the food-gathering bee performs a localized interaction only between itself and the food-storing bee, the interaction can scale to support the large or growing nutritional needs of a bee colony. The bee colony also exhibits other types of emergent behavior, such as self-organization, evolution, and survivability. Thus, emergent behavior is the formation of complex behaviors or characteristics through the collective, simple, and autonomous behaviors or characteristics of individual entities.

In the bio-networking architecture, we apply the concept of emergent behavior by implementing network applications as a group of autonomous entities called the cyber-entities (see Figure 1). This is analogous to a bee colony (an application) consisting of multiple bees (cyber-entities). Each cyber-entity implements a functional component related to the application and also follows simple behavior rules similar to biological entities (such as reproduction, death, migration, relationship establishment with other cyber-entities). In the bio-networking architecture, useful application functionality emerges from

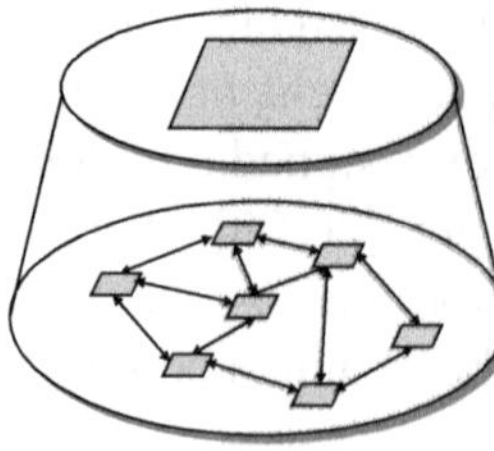

Figure 1. Cyber-entities and network applications

the collaborative execution of application components (or services) carried by cyber-entities, and useful system behaviors and characteristics (e.g., self-organization, adaptation, evolution, security, survivability) arise from simple behaviors and interaction of individual cyber-entities.

Another key concept that we borrow from biological systems is evolution. Evolution in biological systems occurs as a result of natural selection from diverse behavioral characteristics of individual biological entities. Through many successive generations, beneficial features are retained while detrimental behaviors become dormant or extinct, and the biological system specializes and improves itself according to long-term environmental changes.

Within a biological system, diversity of behaviors is necessary for the system to adapt and evolve to suit a wide variety of environmental conditions. In the bio-networking architecture, we design multiple cyber-entities to perform the same task with different behaviors to ensure a sufficient degree of diversity. For instance, we design multiple cyber-entities to implement the same service with different behavior policies (e.g., one cyber-entity with a migration policy of moving toward a user and another cyber-entity with a migration policy of moving toward a cheap resource cost node). In the bio-networking architecture, we design cyber-entity reproduction to include mutation and crossover mechanisms to produce new behaviors and new behavior policies.

Within a biological system, food (or energy) serves as a natural selection mechanism. Biological entities naturally strive to gain energy by seeking and consuming food. Similar to an entity in the biological world, each cyber-entity in the bio-networking architecture stores and expends energy for living. Cyber-entities gain energy in exchange for performing a service, and they pay energy to use network and computing resources. The abundance or scarcity of stored energy affects various behaviors and contributes to the natural selection process in evolution. For example, an abundance of stored energy is an indication of higher demand for the cyber-entity; thus the cyber-entity may be designed to favor reproduction in response to higher levels of stored energy. A scarcity of stored energy (an indication of lack of demand or ineffective behaviors) may eventually cause the cyber-entity's death.

Reproduction to create diverse cyber-entity behaviors along with a mechanism for natural selection will result in the emergence of evolution to allow applications to adapt to long-term environmental change in the bio-networking architecture.

2.2 Middleware Framework

The bio-networking architecture is a middleware framework that supports design and development of network applications. It consists of interconnected bio-networking nodes as shown in Figure 2. A virtual machine capable of resource access control (such as the Java virtual machine) runs atop the native operating system. The bio-networking platform software, which provides an execution environment and supporting facilities for cyber-entities, runs using the virtual machine. The platform software also manages underlying resources such as CPU and memory.

Cyber-entities that provide network applications run atop the platform software. A cyber-entity consists of three main parts: attributes, body, and behaviors. Attributes carry information regarding the cyber-entity (e.g., cyber-entity ID, description of a service it provides, cyber-entity type, stored energy level, and age). The body implements a service that a cyber-entity provides and contains materials relevant to the service (such as data, application code, or user profiles). For instance, a cyber-entity's body may implement control software for a device, while another cyber-entity's body may implement a hotel reservation service. A cyber-entity's body that implements a web service may contain a web page data. Cyber-entity behaviors implement non-service-related actions that are inherent to all cyber-entities. Major behaviors of cyber-entities are listed below:

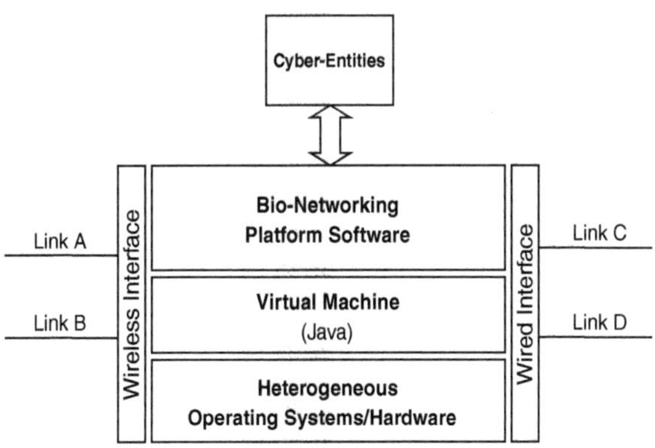

Figure 2. Bio-networking node

- Energy exchange and storage: Cyber-entities may gain/expend and store energy as described above.
- Communication: Cyber-entities communicate with others for requesting a service, fulfilling a service, or routing messages (e.g., discovery messages) for other cyber-entities.
- Migration: Cyber-entities may migrate from bio-net platform to platform.
- Replication and reproduction: Cyber-entities may make copies of themselves (replication). Two parent cyber-entities may create a child cyber-entity (reproduction), possibly with mutation and crossover.
- State change: Each cyber-entity may have autonomous, active, and inactive (sleeping) states during its lifetime. Cyber-entities in different states consume resources differently; therefore they expend energy at different rates.
- Death: Cyber-entities may die because of old age or energy starvation.
- Relationship establishment: Cyber-entities may have their own relationships with others.
- Discovery: Cyber-entities may discover other cyber-entities by sending a discovery message through their relationships.
- Resource sensing: Cyber-entities may sense the type, amount, and unit cost of resources (e.g., CPU cycles and memory space) available on both local and neighboring platforms. Each platform determines the unit cost of resources provided on that platform based on their availability.

2.3 Current Status

As an application of the bio-networking architecture, we have designed Aphid, a web content distribution application [27]. Aphid cyber-entities accept requests for web pages and deliver them using the HTTP protocol. Simulation studies have been conducted to demonstrate that the Aphid exhibits unique and desirable features such as scalability, adaptability, and survivability/ availability. Some other network applications have also been implemented using the bio-networking architecture in [15]. In addition, the bio-networking architecture has been implemented and empirically evaluated in [26]. Extensive simulation studies on evolvability of the bio-networking architecture have been performed in [22].

3. BIOLOGICALLY INSPIRED NANOSCALE NETWORKS

Current bionanotechnology has advanced to the point where engineering of biological systems (e.g., receptors, nano-scale reactions) is feasible, as demonstrated through modification of DNA to produce new cell functionality. Techniques of engineering biological systems can now be used to design nano-scale or (at least small-scale) biological network systems.

Along this line, we have proposed molecular communication [14, 21] as one solution for nano-scale communication between nanomachines. Nanomachines (e.g., biological molecules, artificial devices) represent small devices or components that perform computation, sensing, or actuation. Molecular communication provides a mechanism for one nanomachine to encode or decode information into molecules and to send information to another nanomachine.

Molecular communication represents a more direct and precise mechanism for interacting with biological cells and it may be used, for instance, in implant devices [8] to directly interact with human cells for human health monitoring. Molecular communication may also be useful in molecular computing. In the molecular computing area, researchers are currently creating molecular logic gates (e.g., an inverter and a NAND gate) and memory [5, 11, 12, 18, 19, 24] using components from biological systems. Those molecular components may be interfaced through molecular communication to perform more complex computing functionality.

In Section 3.1, we first introduce biological communications with two examples of how communication is performed within/between cells. In Section 3.2, we describe a molecular communication architecture. In Section 3.3, we present a design of two molecular communication systems based on specific biological communications.

3.1 Biological Communications

Biological nanomachines (e.g., cells) exhibit a wide variety of mechanisms for exchanging information at the nano and micron scales. Some specific biological mechanisms include intracellular communication and intercellular communication [2].

In intracellular communication, communication occurs between components within a single biological cell. For example, acetylcholine is transported within a neuron along the axon and released, causing a response in the adjacent neurons that have a receptor for acetylcholine. A component of the cell sends molecules (e.g., vesicles containing acetylcholine) carried by molecular motors. A molecular motor (e.g., kinesins and dyneins) is a protein or protein complex that transforms chemical energy (e.g., ATP hydrolysis) to mechanical movement of the molecular motor along a cytoskeletal track (e.g., microtubules).

In intercellular communication, communication can occur through cell–cell channels called gap junction channels. Gap junctions allow connected cells to share small molecules such as Ca^{2+} (calcium ions) and inositol 1,4,5- trisphosphate (IP_3) and, therefore, enable coordinated actions among adjacent cells. For example, ciliated airway epithelial cells communicate with each other through the diffusion of IP_3 via gap junctions. A stimulated cell first increases the intracellular IP_3 concentration that results in the release of Ca^{2+} from the intracellular calcium store. IP_3 diffuses through gap junctions to adjacent cells, resulting in the release of Ca^{2+} from intracellular stores in each of the adjacent cells. Diffusion of IP_3 continues and thus propagates calcium waves over a number of cells.

3.2 Molecular Communication Architecture

Molecular communication can be described relative to a generic communication architecture that consists of system components and communication processes of a molecular communication: encoding, sending, propagation, receiving, and decoding (Figure 3).

3.2.1 Molecular Communication System Components

The generic molecular communication system consists of *information molecules* (i.e., molecules such as proteins or ions) that carry the information to be transmitted, *sender nanomachines* that transmit the carrier molecules, *receiver nanomachines* that receive carrier molecules, and the environment in which the carrier molecules propagate from the sender to the receiver.

The sender and receiver nanomachines may be biological cells that use peptides, ions, or phosphates such as inositol-triphosphate as carrier molecules

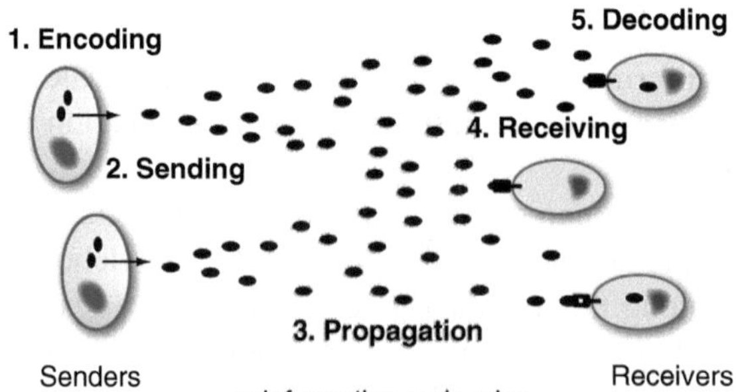

Figure 3. Molecular communication architecture

(e.g., cells that communicate through calcium ions). The sender stores carrier molecules and releases them according to some event (e.g., high concentration of an external signal molecule). For instance, calcium ions may be pumped into a sender cell from the external environment and stored in the endoplasmic reticulum of a sender cell. A sender may then release the stored calcium when a ligand binds to a receptor of a sender cell. The receiver includes calcium-sensitive components to detect increased calcium concentration and reacts to it. For example, calcium ions bind to receptors of the receiver, resulting in neuron action potential or a cellular immune response such as inflammation.

The information molecules propagate from the sender to the receiver in the environment. The environment of molecular communication may be an aqueous medium with various ions and molecules dissolved in solution.

3.2.2 Molecular Communication Processes

The communication processes of molecular communication include encoding, sending, propagation, receiving, and decoding (Figure 4).

Encoding is the process by which a sender translates information into carrier molecules that the receiver can capture or detect. Information may be encoded

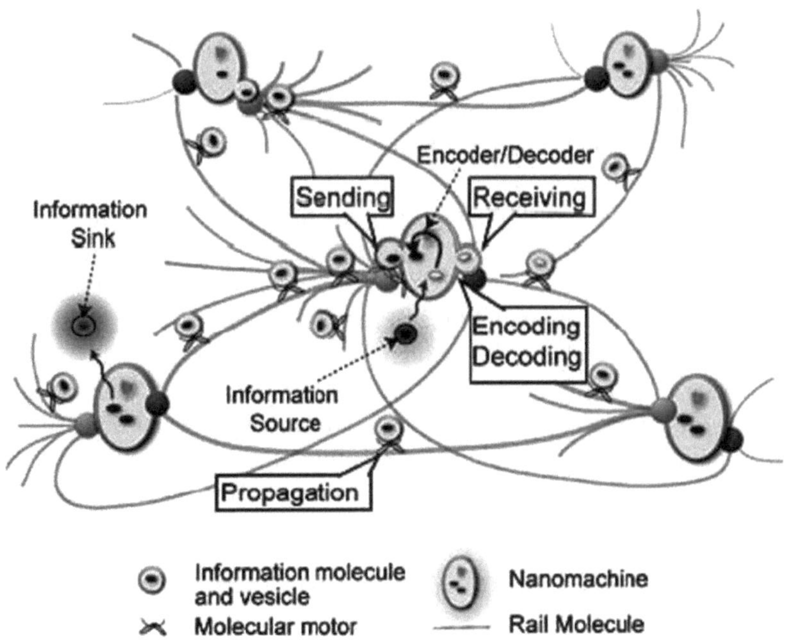

Figure 4. Molecular communication using intracellular communication (molecular motors and rail molecules)

in the specific molecules used, in a subcomponent of the molecule (e.g., subsequence of a DNA sequence), or in characteristics of the molecules. Information may also be encoded in the environment by, for example, the sender emitting molecules that modify the environment, and a receiver that detects the changes in the environment.

Sending is the process by which the sender emits the carrier molecule into the environment. For example, a sender may emit ligands toward membrane receptors of nearby cells, which results in the generation of calcium waves that propagate from cell to cell. Another example of sending is the sender emitting carrier molecules using peptide translation machinery. In this case, the carrier molecule may be a peptide sequence that is encapsulated into a vesicle, transported by molecular motor machinery of the cell from the endoplasmic reticulum (site of vesicle encapsulation) to the cellular membrane (site of vesicle exocytosis), and emitted outside of the sender cell using vesicle exocytosis.

Propagation is the process by which carrier molecules move through the environment from a sender to a receiver. Propagation may occur through simple passive propagation (e.g., Brownian motion) in which the carrier molecules do not actively use energy to move through the environment. Propagation may also be controlled by constraining the volume of the environment in which carrier molecules can move. For example, in propagation through gap junctions, propagation is limited to inside the cell and the gap junction and thus molecules do not propagate in all directions. Another example of controlled propagation is molecular motors that walk over rail molecules to transport carrier molecules.

Receiving is the process by which the receiver captures carrier molecules propagating in the environment. The receiver may contain a selective receptor (e.g., sensitive to calcium ions or specific peptides) to capture the carrier molecule. The receiver may contain gap junctions that allow molecules (e.g., calcium ions) to flow into the cell without using receptors. Another option for receiving is to use fusion of vesicles (observed in vesicle transport) containing carrier molecules into the membrane of receivers.

Decoding is the process by which the receiver, after receiving carrier molecules, decodes the received molecules into a reaction. The design of a reaction is dependent on the application. If biological cells are used as receivers, potential reactions include enzyme-mediated reactions or protein synthesis. For instance, to report a detected molecule, the receiver may express GFP (green fluorescent protein) in response to the transmitted signal.

3.3 Molecular Communication Systems

We have designed two molecular communication systems, one using intracellular communication and the other using intercellular communication. The following illustrates their design relative to the generic molecular communication processes described in Section 3.2.2.

3.3.1 Molecular Communication Using Intracellular Communication

In the molecular communication system using intracellular communication (Figure 5) [6, 20], the sender nanomachine sends information molecules that are contained in an interface molecule. The interface molecule is a container that stores various information molecules and isolates information molecules from inside the propagation environment and, thus, reduces interference from environmental noise. Transport molecules (e.g., molecular motors such as dynein or kinesin) propagate along guide molecules (e.g., microtubule) and carry interface molecules (containing information molecules) from the sender to the receiver nanomachines. The network topology of guide molecules determines the direction that molecular motors will move. Communication processes in this molecular communication system are described as follows.

Encoding: Sender nanomachines encode information on information molecules (e.g., DNA molecules, proteins, peptides). For example, nanomachines encode information on sequences of peptides and inject the peptides into vesicles. Vesicles can be loaded on molecular motors, and thus a variety of encoded molecules can be sent.

Sending: Sender nanomachines then emit information molecules. The information molecules are then attached to molecular motors that move along rail molecules.

Propagation: Propagation is performed through molecular motors that move along rail molecules from sender nanomachines to receiver nanomachines in a directed manner.

Receiving: Receiver nanomachines are assumed to retrieve interface molecules from molecular motors using protein tags. When molecular motors approach receiver nanomachines, interface molecules such as vesicles may be fused into receiver nanomachines and thus release information molecules into the receiver.

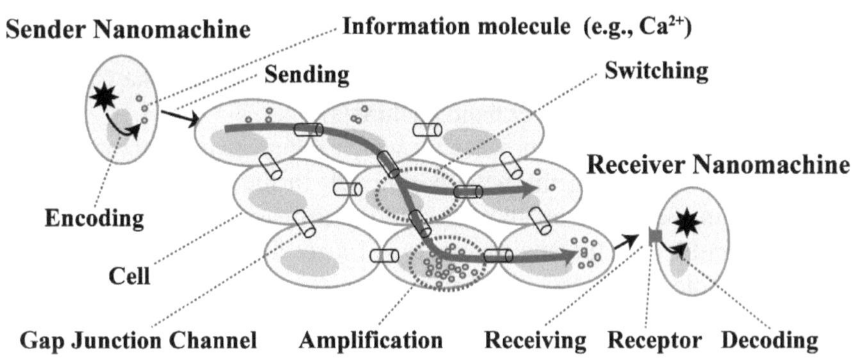

Figure 5. Molecular communication using intercellular communication

Decoding: In decoding, receiver nanomachines invoke reactions in response to information molecules. For example, peptides (e.g., neurotransmitters) transported through molecular motors in a neuron cause receiver neurons to generate an action potential.

3.3.2 Molecular Communication Using Intercellular Communication

In the molecular communication system using intercellular communication, a sender nanomachine (i.e., a cell) communicates with a receiver nanomachine (i.e., a cell) through a network of cells that passively propagate information molecules from a sender nanomachine to a receiver nanomachine (Figure 5) [23]. The particular intercellular communication mechanism used in this system is cell–cell calcium signaling that is described in Section 3.1. Communication processes in this molecular communication system are described as follows:

Encoding: Ca^{2+} waves are used to encode various cellular information such as muscle contraction, chemical secretion in biological systems. Similarly, in this molecular communication, a sender nanomachine encodes various information onto Ca^{2+} waves by varying the properties of Ca^{2+} waves (e.g., frequency, amplitude, and duration of Ca^{2+} waves). Encoding in this system is thus the process of selecting the properties of Ca^{2+} waves (e.g., release of agonistic substances in certain amounts) that represent different information at the receiver nanomachine(s).

Sending: A sender nanomachine releases chemical substances (e.g., Ca^{2+} and/or Ca^{2+}-mobilizing molecules such as IP_3) in the manner decided in the encoding process. The released chemical substances stimulate a nearby cell to initiate the following propagation process.

Propagation: A stimulated cell increases its Ca^{2+} level, and the increased Ca^{2+} level propagates from cell to cell through gap junction channels. The network of cells may perform amplification of the Ca^{2+} level to increase the distance that the Ca^{2+} wave propagates. The network of cells may also perform switching by opening and closing gap junctions so that the Ca^{2+} wave propagates to a specific receiver nanomachine(s).

Receiving: Receiving is a process by which the receiver nanomachine detects cellular responses of the neighboring cell in the environment. For example, cells in the environment receiving the propagating Ca^{2+} waves may release molecules into the environment, and a receiver nanomachine detects the released molecules in the environment.

Decoding: The receiver nanomachine invokes an application-specific response corresponding to the information molecule it receives. Possible decoding at a receiver nanomachine (e.g., a nanomachine based on a cell) includes differential gene expression, secretion of molecules, and generation of movement.

3.4 Current Status

We are currently designing and investigating the feasibility of the two molecular communication systems described in this section. In addition, researchers in molecular communication are designing and testing different systems [13,22,25].

Research of molecular communication has just started and there are many research issues that need to be addressed, including (1) representing information such that nanomachines can understand and biochemically react (encoding), (2) selectively addressing destination receiver nanomachines at a sender nanomachine (sending), (3) controlling propagation of information molecules (propagation), (4) selectively receiving information molecules at a receiver nanomachine (receiving), and (5) decoding of information molecules to cause desired chemical reaction and status change at a receiver (decoding).

4. CONCLUSIONS

This chapter described biologically inspired approaches to two different classes of networks: computer networks and nanoscale biological networks. As an example of each biologically inspired network, we described our work of the bio-networking architecture and molecular communication. In the bio-networking architecture, biological principles such as emergence and evolution are applied to design a broad range of network applications that need to scale to a large number of network components and be adaptive to environmental changes and various failures. In the molecular communication, components and mechanisms from biological communications are used to design and implement nano- and cellular-scale biological networks. In both classes of networks, biological inspiration is explored to construct scalable and adaptive networking systems.

REFERENCES

[1] Aickelin, U., J. Greensmith, and J. Twycross, "Immune System Approaches to Intrusion Detection – A Review", *Proceedings of the 3rd International Conference on Artificial Immune Systems*, LNCS 3239, 316–329, 2004.

[2] Alberts, B., A. Johnson, J. Lewis, M. Raff, K. Roberts, and P. Walter, *Molecular Biology of the Cell*, 4th Bk&Cdr edition, Garland Science, London, 2002.

[3] Di Caro, G., and M. Dorigo, "AntNet: Distributed Stigmergetic Control for Communications Networks", *Journal of Artificial Intelligence Research (JAIR)*, 9, 317–365, 1998.

[4] Dorigo, M., G. Di Caro, and L. M. Gambardella, "Ant Algorithms for Discrete Optimization", *Artificial Life*, 5(3), 137–172, 1999.

[5] Elowitz, M. B., and S. Leibler, "A Synthetic Oscillatory Network of Transcriptional Regulators", *Nature*, 403, 335–338, 2000.

[6] Enomoto, A., M. Moore, T. Nakano, R. Egashira, T. Suda, A. Kayasuga, H. Kojima, H. Sakibara, and K. Oiwa, "A Molecular Communication System Using a Network of Cytoskeletal Filaments Communication", *2006 NSTI Nanotechnology Conference*, May 2006.

[7] Forrest, S., S. Hofmeyr, and A. Somayaji, "Computer Immunology", *Communications of the ACM*, 40(10), 88–96, 1997.

[8] Freitas Jr, R. A., *Nanomedicine, Vol. I: Basic Capabilities*, Landes Bioscience, Austin, TX 1999.

[9] George, S., D. Evans, and S. Marchette, "A Biological Programming Model for Self-Healing", *Proceedings of the 2003 ACM Workshop on Survivable and Self-Regenerative Systems*, 72–81, 2002.

[10] Hariri, S., B. Khargharia, H. Chen, J. Yang, Y. Zhang, M. Parashar, and H. Liu, "The Autonomic Computing Paradigm", *Cluster Computing: The Journal of Networks, Software Tools, and Applications*, 9(1), 5–17, 2006.

[11] Head, T., M. Yamamura, and S. Gal, "Aqueous Computing – Writing on Molecules", *Proceedings of CEC'99*, 1006–1010, 1999.

[12] Hiyama, S., Y. Isogawa, T. Suda, Y. Moritani, and K. Suto, "A Design of an Autonomous Molecule Loading/Transporting/Unloading System Using DNA Hybridization and Biomolecular Linear Motors in Molecular Communication", *European Nano Systems*, December 2005.

[13] Hiyama, S., Y. Moritani, T. Suda, R. Egashira, A. Enomoto, M. Moore, and T. Nakano, "Molecular Communication", *Proceedings of the 2005 NSTI Nanotechnology Conference*, 2005.

[14] Hofmeyr, S. A., and S. Forrest, "Architecture for an Artificial Immune System", *Evolutionary Computation*, 8(4), 443–473, 2000.

[15] Itao, T., S. Tanaka, T. Suda, and T. Aoyama, "A Framework for Adaptive UbiComp Applications Based on the Jack-in-the-Net Architecture", *Kluwer/ACM Wireless Network Journal*, 10(3), 287–299, 2004.

[16] Kephart, J. O., "A Biologically Inspired Immune System for Computers", *Proceedings of the Fourth International Workshop on the Synthesis and Simulation of Living Systems*, 130–139, 1994.

[17] Kephart, J. O., and D. M. Chess, "The Vision of Autonomic Computing", *IEEE Computer*, 36(1), 41–50, 2003.

[18] Mao, C., T. H. Labean, J. H. Reif, and N. C. Seeman, "Logical Computation Using Algorithmic Self-Assembly of DNA Triple-Crossover Molecules", *Nature*, 407, 493–496, 2000.

[19] Montresor, A., "Anthill: a Framework for the Design and Analysis of Peer-to-Peer Systems", *Proceedings of the 4th European Research Seminar on Advances in Distributed Systems*, 2001.

[20] Moore., M., A. Enomoto, T. Nakano, R. Egashira, ,T. Suda, A. Kayasuga, H. Kojima, H. Sakakibara, and K. Oiwa, "A Design of a Molecular Communication System for

Nanomachines Using Molecular Motors", *Fourth Annual IEEE Conference on Pervasive Computing and Communications and Workshops*, March 2006.

[21] Moritani, Y., S. Hiyama, and T. Suda, "Molecular Communication Among Nanomachines Using Vesicles", *2006 NSTI Nanotechnology Conference*, May 2006.

[22] Nakano, T., and T. Suda, "Self-Organizing Network Services with Evolutionary Adaptation", *IEEE Transactions on Neural Networks*, 16(5), 1269–1278, 2005.

[23] Nakano, T., T. Suda, M. Moore, R. Egashira, A. Enomoto, and K. Arima, "Molecular Communication for Nanomachines Using Intercellular Calcium Signaling", *IEEE NANO 2005*, June 2005.

[24] Sasaki, Y., M. Hashizume, K. Maruo, N. Yamasaki, J. Kikuchi, Y. Moritani, S. Hiyama, and T. Suda, "Controlled Propagation in Molecular Communication Using Tagged Liposome Containers", *BIONETICS (Bio-Inspired mOdels of NEtwork, Information and Computing Systems) 2006*, December 2006.

[25] Suda, T., T. Itao, and M Matsuo, "The Bio-Networking Architecture: The Biologically Inspired Approach to the Design of Scalable, Adaptive, and Survivable/Available Network Applications", In K. Park (ed.), *The Internet as a Large-Scale Complex System*, The Santafe Institute Book Series, Oxford University Press, Oxford, 2005.

[26] Suzuki, J., and T. Suda, "A Middleware Platform for a Biologically-Inspired Network Architecture Supporting Autonomous and Adaptive Applications", *IEEE Journal on Selected Areas in Communications (JSAC)*, Special Issue on Intelligent Services and Applications in Next Generation Networks, 23(2), 249–260, 2005.

[27] Wang, M., and T. Suda, "The Bio-Networking Architecture: A Biologically Inspired Approach to the Design of Scalable, Adaptive, and Survivable/Available Network Applications", *Proceedings of the 1st IEEE Symposium on Applications and the Internet*, 2001.

[28] Weiss, R., S. Basu, S. Hooshangi, A. Kalmbach, D. Karig, R. Mehreja, and I. Netravali, "Genetic Circuit Building Blocks for Cellular Computation, Communications, and Signal Processing", *Natural Computing*, 2, 47–84, 2003.

DYNAMIC NETWORK RESOURCES ALLOCATION IN GRIDS THROUGH A GRID NETWORK RESOURCE BROKER

D. Adami[1], S. Giordano[2], and M. Pagano[2]

[1]*CNIT Research Unit – Dept. of Information Engineering, University of Pisa, Via Caruso 16, 56122 PISA (ITALY),*

[2]*Dept. of Information Engineering, University of Pisa, Via Caruso 16, 56122 PISA (ITALY)*

Abstract The design and deployment of grid environments for remote instrumentation services represent a key challenge for high-speed next-generation networks: not only is it necessary to implement a (G)-MPLS backbone offering Quality of Service at IP level, but it is also essential to design and develop new control plane mechanisms for dynamic resources allocation in explicit and optimized way. To this goal, we introduced the Grid Network Resource Broker (GNRB), a new architectural entity, which acts as an intermediate between the network and the grid environment. This chapter focuses on the design of the Path Computation Element, a key building block of the GNRB, which performs LSP path computation with QoS constraints. Several tests have been carried out to verify its applicability, in terms of LSP set-up time, to a grid environment scenario.

1. INTRODUCTION

Nowadays, the possibility of using unique and expensive laboratory equipment plays a key role for the success of many research and industrial projects. Remote instrumentation services and technologies, which allow virtualized, remote and shared access to industrial or scientific equipment, are therefore a key factor to open up new relevant opportunities for industry, science and business.

Grid technologies offer effective strategies to achieve such ambitious goals. Originally conceived to solve large-scale computation problems, according to [4, 6, 7], a grid may be defined as "a system which coordinates resources that are not subject to a centralized control, using standard, open, general-purpose protocols and interfaces, to deliver non-trivial qualities of service." Moreover, in most cases, grid applications heavily stress the network used to interconnect the grid resources, so the performance of the applications depends on the QoS (Quality of Service) provided by the network. Emerging e-Science applications include high-energy particle physics experiments, such as the Large Hadron Collider (LHC) at CERN [16], that produce data sets in the order of petabytes per year. Data are then transferred from the LHC to research centers, distributed all over the globe, to be processed and analyzed. End-to-end connections at 10 Gbps among these centers are necessary. These connections are usually scheduled in advance and are sometimes set up simultaneously.

Supernova [17] is another application, which allows scientists to monitor the behaviour of physical phenomena arising in space. Parallel simulations running on a supercomputer or high-performance grids generate terabytes of data, which are transferred to a cluster for visualization.

Furthermore, NASA has developed several applications to predict hurricanes and tsunamis. Several hundred terabytes of data are collected from observations at different locations across the globe and then transferred to different NASA sites for weather and climate forecasts.

Electronic Very-Long Baseline Interferometry (e-VLBI) [18] allows radio astronomers to study objects in the universe at ultra-high resolution and ultra-high accuracy of earth motion measurements. Data (several hundreds of gigabytes) simultaneously acquired from a global array of up to about 20 radio telescopes are collected and have to be processed at a central location.

In summary, emerging eScience applications have the following networking requirements [8]:

- Cost effectiveness.
- High degree of flexibility for users communications needs.
- Inter-domain and internetworking dynamic provisioning.
- Short/long-term, as well as low/high-capacity connections should be scheduled.
- Distributed fault management mechanisms.

The (G)-MPLS [10–12] distributed control plane has three main capabilities that allow it to meet most of the above requirements:

- Distributed routing.
- Near-real-term connection set-up using signaling.
- Network services and topology auto-discovery mechanisms.

Dynamic provisioning of end-to-end deterministic bandwidth pipes is an essential capability to enable grid applications.

Depending on the type of application, requests for grid or networking resources could be immediate or book-ahead, the duration of connection short (seconds to a few minutes) or long (tens of minutes to hours) and the bandwidth granularity fine (tens to few hundred Mbps) or large (1–10 Gbps).

In previous works [1,3], some of the authors of this chapter proposed a centralized approach, based on a Grid Network Resource Broker (GNRB) to dynamically provision Bandwidth-on-Demand (BoD) services for grid applications in DiffServ/MPLS networks. This chapter focuses on the PCE (Path Computation Element), a key building block of the GNRB, introducing a new algorithm, which allows to perform LSP path computation with QoS constraints.

The chapter is organized as follows. Section 2 describes the architecture of the Grid Network Resource Broker (GNRB). Section 3 proposes a new path computation algorithm, to be implemented within the GNRB, for the dynamic provisioning of end-to-end connections with bandwidth and delay constraints. In Section 4, we show the results of some simulations performed to validate our algorithm. Section 5 sums up the work with some final remarks.

2. GRID NETWORK-AWARE RESOURCE BROKER ARCHITECTURE

The cooperative working of a grid infrastructure with the network (Figure 1) requires the development of a new Grid Middleware Service, called Network Resources Manager (NRM), which can be invoked by the Grid Application Manager (GAM) to ask for different types of network services.

To achieve a Grid Network-aware environment, we adopted a centralized architecture where a new entity, the Grid Network Resource Broker (GNRB),

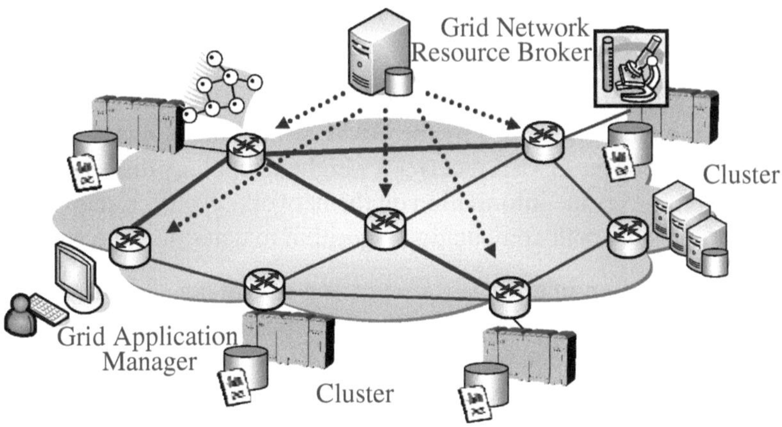

Figure 1. Grid-enabled network infrastructure

provides two basic functionalities: network resource management and network information retrieval.

In particular, the NRM is equipped with an interface to request a pre-defined network service and may include a Resources Management System (RMS) to optimize the utilization of network resources.

A Network Information Database (NID) is also necessary to collect information about the network status and to control service agreements.

In the following, before analyzing the architectural components, we briefly describe the services offered by the GNRB.

2.1 GNRB Network Services

Grid network services may be divided in two groups: path oriented and knowledge based [5]. The former includes a number of network services with specific connectivity requirements (e.g., visualization session, large data streaming coordinated with job execution, networked supercomputing, and file replica management), while the latter includes information-oriented services, related to network monitoring and usage scenarios of performance data (e.g., passive monitored data, administrative setup of measurement campaigns).

Most of remote instrumentation and distributed measurements services may be classified as a hybrid use case between path-oriented and knowledge-based grid network services.

In our GNRB-based architecture, the network services that the Grid Resource Broker can request are the following:

- **Network Topology Discovery:** this service provides information about the topology of the network and QoS metrics (available bandwidth, delay, etc.) associated to the links. In this case, the GAM may choose the grid resources taking into account also the network status. Anyway, the network only provides a Best Effort service.

- **Weighted Topology Discovery**: this service provides the best paths, according to a metric specified by the GAM, which interconnect a set of grid resources. When the GNRB receives a weighted topology discovery request, the NRM retrieves the information on the network resources status from the NID and runs the path computation algorithm to determine the paths.

- **QoS Provisioning**: in this case, a set of end-to-end connections among the grid resources, which satisfy specific QoS constraints (Bandwidth, Delay, Packet Loss), is provided for the whole running time of the application. Different types of services may be assured based on application requirements:

    1. Premium service, defined in terms of Peak Rate, Burst Size, Maximum Latency

2. Better than best-effort service, where Mean Rate, Burst Size and Mean Latency are specified.

Generally, it is necessary to reserve network resources on-demand to satisfy the application requirements or to reject the request if there are not enough available resources. In our architecture, an ad hoc protocol is used by the GNRB and the GAM to exchange query/response messages, formatted by using XML-based schemes. Figure 2 shows an example of two QoS Provisioning requests for Premiun (Peak Rate and Burst Size) and Better than best-effort services, respectively.

Notice that each mapping request consists of a service type and one or more LSP set-up requests. Every LSP has to connect the Ingress-LER and Egress-LER that allow the source and destination LANs, where grid resources are located, to access the MPLS backbone.


```
<mappingrequest>
<id>1</id>
    <service>PREMIUM</service>
        <componentPath>
            <srcLan entryRouter="10.0.0.1">100.2.250.23/24</srcLan>
            <dstLan entryRouter="10.0.0.2">200.243.23.11/30</dstLan>
                <Bandwidth>
                    <peak>10</peak>
                    <burst>10</burst>
                </Bandwidth>
        </componentPath>
        <componentPath>
            <srcLan entryRouter="10.0.0.2">200.243.23.11/30</srcLan>
            <dstLan entryRouter="10.0.0.5">193.168.1.2/24</dstLan>
                <Bandwidth>
                    <peak>40</peak>
                    <burst>10</burst>
                </Bandwidth>
        </componentPath>
</mappingrequest>
<mappingrequest>
<id>2</id>
    <service>BEST\_than\_BE</service>
        <componentPath>
            <srcLan entryRouter="10.0.0.5">193.168.1.2/24</srcLan>
            <dstLan entryRouter="10.0.0.1">100.2.250.23/24</dstLan>
            <Bandwidth>
                <mean>100</mean>
                <burst>10</burst>
            </Bandwidth>
        </componentPath>
</mappingrequest>
```


Figure 2. XML query messages

2.2 GNRB Architecture

The GNRB (Figure 3) consists of a Network Resource Manager (NRM), which interacts with the Network Monitoring System (NMS) and a Network Element Configuration Manager (NECM).

2.2.1 Network Resource Manager

The NRM performs the following tasks [3]:

- Policy-based provisioning.
- Path computation.
- Network resource scheduling.

When a network service request is delivered to the GNRB, the Policy-based module checks if the user is allowed to ask for the service by querying a database, which associates policies to users (or group of users).

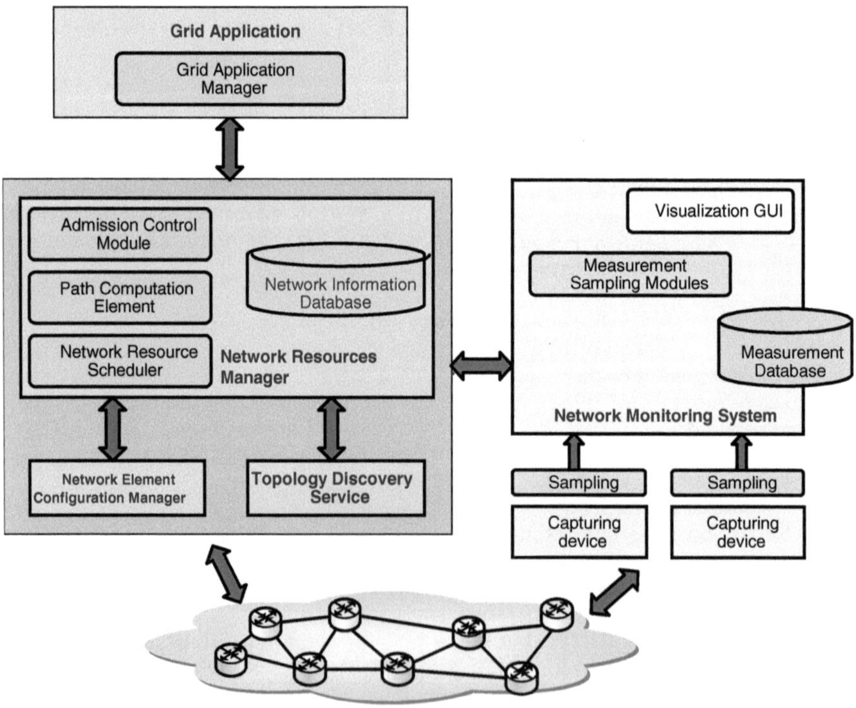

Figure 3. GNRB architecture and Network Monitoring System

At a lower-level, policies are also associated to network elements, so that the compliance of the request with the Service Level Agreement (SLA) established between the Network Provider and the Virtual Organization may be verified. Moreover, policy-based provisioning allows incorporating AAA (Authentication, Authorization, Accounting) policy and scheduling information into path computation, resource allocation and signalling functions.

The GNRB collects topology information about its domain by listening to the local OSPF-TE protocol and exchanges inter-domain routing information with external GNRB peers, so as to enable inter-domain path computations and traffic-engineered LSP (Label Switched Path) provisioning.

On behalf of network nodes, the Path Computation Element (PCE) performs route computation and determines if a network service request with QoS constraints may be accepted, must be rejected or reclassified. In more detail, the PCE executes advanced algorithms, which allow path computation with multiple constraints (e.g., CSPF, Wang-Crowcroft, etc.). The network topology is represented as a graph whose edges are weighted according to QoS metrics (available bandwidth, packet loss, delay, jitter) stored within the NID. Such database contains statistics concerning the status of network resource utilization at path, link and node level, as well as information on configured MPLS recovery strategies and performance metrics.

Finally, the Network Resource Scheduler allows planning resources utilization (type of resource, start time, end time).

2.2.2 Network Element Configuration Manager

This module is responsible for service provisioning, that is, for the actual configuration of the network elements.

Many tasks are performed by the NECM, such as switches and routers interfaces configuration, LSP set up and tear down, LSP protection or restoration at path/span level, etc.

The NRM sends to the NECM information concerning the configuration to apply to the network devices. Then the NECM applies the configuration statements to each network device by using, for each of them, specific configuration commands and languages.

2.2.3 GNRB and Network Monitoring System

The GNRB is interfaced with a Network Monitoring System, which allows to monitor the network status and to collect statistics, such as link utilization and network performance parameters (packet loss, delay jitter), at different architectural layers. Network status data, both static and dynamic, have to be reported to the GNRB through an Inter-Working API. Subsequently, network performance parameters must be evaluated and stored into the NID in a preliminary phase when the network infrastructure is set up. The knowledge of the physical and IP configurations enables the topology discovery. Moreover, to deliver real-time metrics reports to the GNRB, ad hoc measurement systems have to be deployed. For example, a network bandwidth probe might rely upon SNMP-enabled network devices, such as routers or switches.

At the end, network performance may be evaluated through experiments in the real network scenario and simulations carried out to achieve the bounds of the network performance metrics.

2.2.4 Inter-Domain Communication

When the network nodes or the Grid extension increase, scalability problems, due to the GNRB centralized approach, may arise.

In addition to this, if multiple autonomous domains are interconnected, administrative policies may limit the visibility of the information concerning the status and the topology of the network.

To solve these issues, however, GNRBs may be organized in a hierarchical manner (Figure 4). In this new scenario, each GNRB is responsible for the administration and management of network resources within its own network domain. Grid consumers or Grid manager applications may request a network service to the GNRB belonging to their own administrative domain.

Besides, Inter-Domain GNRB communication allows control and coordination of the assignment of the network resources across different domains.

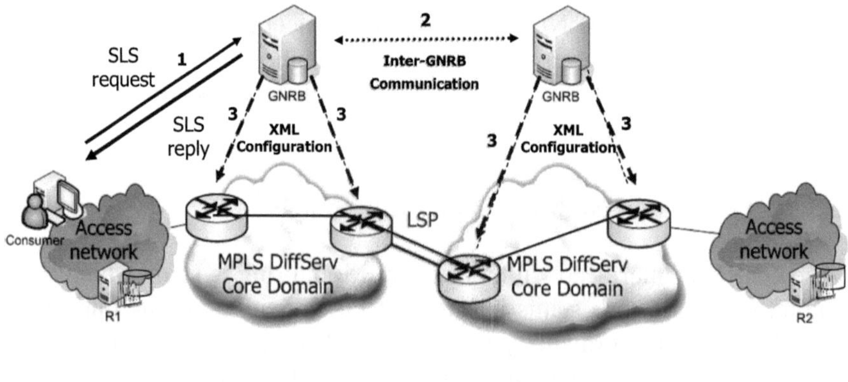

Figure 4. Grid Network Resource Broker Inter-domain communication

3. PATH COMPUTATION ALGORITHM

Given a set of N LSP set-up requests, the basic function of the Path Computation System is to find N network paths that satisfy the QoS constraints. Considering the QoS metrics, the network status and the QoS requirements, the problem is to determine a set of paths that maximizes the chances of each request to be successful with the minimum negative impact on the network ability to accept the remaining requests.

3.1 QoS Metrics

Routing metrics are links' features used to represent a network. Given the metric *d(i,j)*, associated to the link *(i,j)* and the path *p=(i,j,k, ... , l,m)*, the metric *d* is:

- Additive if d(p)=d(i,j)+d(j,k)+ ... +d(l,m).
- Multiplicative if d(p)=d(i,j)*d(j,k)* ... *d(l,m).
- Concave if d(p)=min[d(i,j);d(j,k); ... ;d(l,m)].

Routing protocols usually characterize a network through a single metric (the cost of the link). In the case of QoS routing protocols and constrained-based path computation algorithms, instead, the network is described by means of multiple metrics. The most common ones are the following:

- *Cost*: it is an additive metric because the cost of a path is the sum of the costs of the links.
- *Bandwidth (or residual bandwidth)*: it is a concave metric. Indeed, we define the bandwidth of a path as the minimum of the residual bandwidth of all links on the path (*bottleneck bandwidth*).
- *Delay*: it is an additive metric, since the delay of a path is the sum of the delays of its links. It consists of three components: *propagation* delay, *transmission* delay and *queuing* delay.

While the computation of the first two metrics is quite straightforward, it is not easy to properly evaluate the end-to-end delay because one of its components, the queuing delay, is not a deterministic quantity. By using the *Network Calculus* theory [9], we can compute an upper bound of the queuing delay for flow *i* as follows:

$$D_i = \frac{M}{r_i} + \sum_{j=1}^{K} S_i^j \qquad (1)$$

where M is the maximum burst size, r_i is the guaranteed rate, S_i^j is the delay due to node j and its value is strictly dependent on the service discipline adopted in the node. For instance, supposing that all nodes handle their output queues according to a WFQ (Weighted Fair Queuing)-scheduling discipline, S_i^j may be computed as:

$$S_i^j = \frac{L_{MAX}}{R_j} + \frac{L_i}{r_i} \tag{2}$$

where L_{MAX} is the maximum packet size, L_i is the maximum size of a packet belonging to flow i and R_j is the output link bandwidth.

3.2 Wang–Crowcroft with Bandwidth and Delay Sorting (WCBDS)

When a centralized NRM has to setup several LSPs with bandwidth and delay constraints, the number of requests that can be allocated may depend on the order in which they are processed by the Path Computation System. Therefore, we introduced a new algorithm, based on a heuristic approach (see Figure 5 for details), which tries to improve the performance of the Wang-Crowcroft (WC) algorithm [15], when all the path computation requests cannot be satisfied.

The Wang–Crowcroft path computation algorithm (WC) aims at finding a path that satisfies multiple QoS constraints, given in terms of bandwidth (B_{MIN}) and delay (D_{MAX}). Every link *(i,j)* of the network is characterized by two parameters: b_{ij} (residual bandwidth) and d_{ij} (propagation delay). Unlike the original version of the algorithm, which takes into account only the propagation delay, our implementation considers all the components of the delay, as described in the previous section.

The algorithm consists of the following steps:

1. set $d_{ij} = \infty$ if $b_{ij} < B_{MIN}$
2. compute the path P with the minimum delay D^* (applying Dijkstra algorithm)
3. compare the delay with D_{MAX}. If $D^* < D_{MAX}$ select the path, otherwise the request cannot be satisfied.

The basic idea behind the WCBDS algorithm is to analyze the path computation requests after re-ordering them based on their bandwidth and delay requirements.

In particular, given a random sequence of N path computation requests, our algorithm checks whether they can be satisfied as a whole by applying the WC

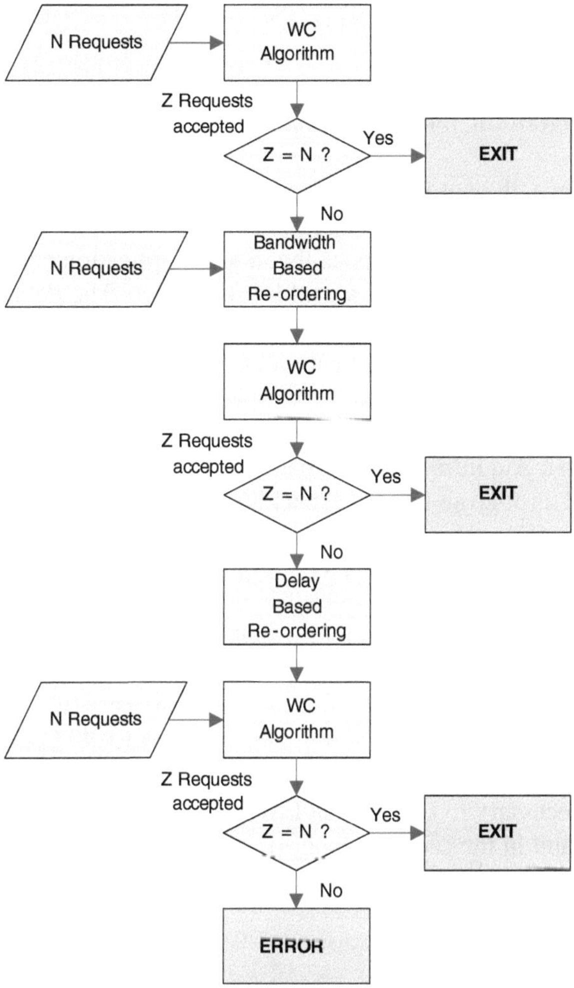

Figure 5. WCBDS Algorithm

algorithm. If some of them are rejected ($Z < N$), requests are sorted top-down according to the bandwidth, and the WC algorithm is applied again.

If $Z = N$, the WCBDS ends, since a solution has been found; otherwise, the requests are sorted bottom-up according to the delay D_{MAX} requirements and the number of requests that can be accepted is once again computed. If $Z = N$, a solution has been found, otherwise an ERROR message is generated.

Therefore, the WCBDS algorithm always performs no worse than the original WC algorithm.

4. SIMULATION RESULTS

This section shows the results of some tests performed to validate our path computation algorithm, through simulations [2, 13, 14], in a complex network scenario.

Figure 6 shows the network topology chosen for the tests and Table 1 lists the links metrics <B(Mbps) – D(ms)>.

In the following, we will suppose that a grid application has been clustered and mapped on the resources of a grid infrastructure. The details of the mapping algorithm are outside the scope of this chapter. Basically, the algorithm tries to map the nodes of the application on the Grid resources, while guaranteeing a pre-defined QoS level to the whole application. When multiple solutions are available, the one which minimizes the intra-LAN communications is selected. When a grid infrastructure spans over a wide area, LANs are typically interconnected through an IP/MPLS backbone. Inter-LAN communications requirements are affected by the way grid resources are distributed and clustered. In our network scenario, three LANs are interconnected by a DiffServ/MPLS backbone.

Several simulations have been performed to investigate the efficiency of the WCBDS algorithm in terms of number of requests satisfied and to evaluate the latency to set-up the LSPs.

Tables 2–5 report the LSP requirements (Ingress LER, Egress LER, bandwidth and delay constraints), the paths computed by the WCBDS algorithm and the time necessary to set-up each LSP, for four different scenarios.

In the first and in the second scenarios, the network was unloaded, while a background traffic was introduced in the remaining two. The main difference

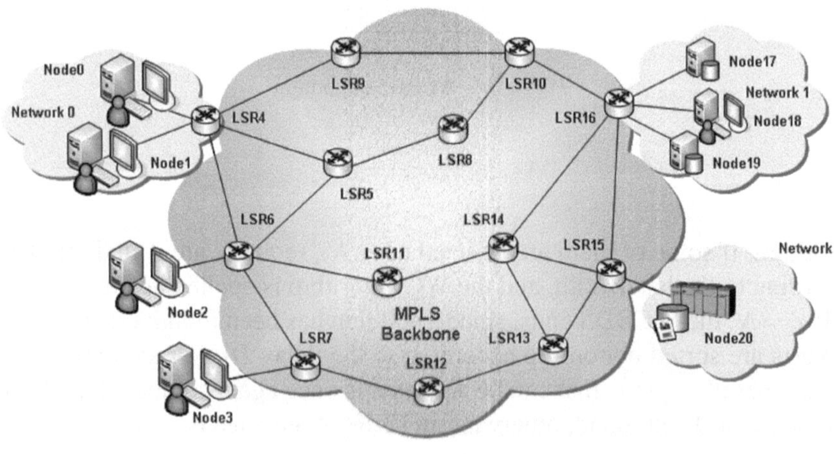

Figure 6. Network topology

Table 1. Link metrics

Link		Metrics	
Upstream node	Downstream node	Bandwidth (Mbps)	Delay (ms)
4	9	0.3	100
1	4	2.5	10
15	20	2.5	10
4	6	2.5	10
6	7	2.5	10
7	12	2.5	10
12	13	2.5	10
13	14	2.5	10
13	15	2.5	10
14	15	2.5	10
4	5	2	20
5	8	2	10
6	5	1	30
6	11	2.5	15
11	14	2	20
14	16	2	100
9	10	1	50
10	16	2	10
8	10	2	10
Other	Other	1	10

among our tests consists of the bandwidth requirements for the third LSP: 1,800 Kbps in the first and third scenarios, 2,400 Kbps in the others.

These two factors affect the choice of the path, as well as the latency time for the set-up of the third LSP only.

Note that in spite of high traffic load, the LSPs' set-up time is relatively low if compared with the typical holding time of the connections. Hence, dynamic resources allocation strategies may be used in IP/MPLS networks to satisfy QoS constraints of, for instance, remote instrumentation services or distributed environment measurements.

Table 2. First scenario

Ingress LER	Egress LER	Bandwidth (Kbps)	Delay (ms)	Path	Time (ms)
4	16	600	100	4-5-8-10-16	84
16	15	100	20	16-15	11
15	4	1800	200	15-14-11-16-4	110

Table 3. Second scenario

Ingress LER	Egress LER	Bandwidth (Kbps)	Delay (ms)	Path	Time (ms)
4	16	600	100	4-5-8-10-16	84
16	15	100	20	16-15	11
15	4	2400	200	15-13-12-7-6-4	95

Table 4. Third scenario (traffic load 50% on the path 16-14-11-6-7)

Ingress LER	Egress LER	Bandwidth (Kbps)	Delay (ms)	Path	Time (ms)
4	16	600	100	4-5-8-10-16	84
16	15	100	20	16-15	11
15	4	1800	200	15-14-11-6-4	1287

Table 5. Fourth scenario (traffic load 75% on the path 16-14-11-6-7)

Ingress LER	Egress LER	Bandwidth (Kbps)	Delay (ms)	Path	Time (ms)
4	16	600	100	4-5-8-10-16	84
16	15	100	20	16-15	11
15	4	2400	200	15-13-12-7-6-4	1287

5. CONCLUSIONS

The design and deployment of grid environments for remote instrumentation services requires the introduction of new control plane mechanisms for dynamic resources allocation in high-speed (G)-MPLS networks.

To this goal, we propose a centralized architecture built around the GNRB, a new entity, which acts as an intermediate between the network and the grid infrastructure. This chapter focuses on the Path Computation Element, a key building block of the GNRB, which performs LSP path computation with QoS constraints.

A new algorithm for the computation of paths with bandwidth and delay constraints has been proposed, and several tests have been performed to verify its applicability to a grid environment scenario.

Since the simulations results are really promising, the algorithm will be implemented within our GNRB and, the tests will also be performed in a real grid environment.

ACKNOWLEDGMENTS

This work was supported by the Italian Ministry of Education and University (MIUR) under FIRB project "Enabling platforms for high-performance computational grids oriented to scalable virtual organizations (GRID.IT)" and by RINGrid, co-funded by the European Commission under the 6th Framework Programme (contract no. 031891).

REFERENCES

[1] Adami, D., Baraglia, R., Ferrini, R., Giordano, S., Laforenza, D., Tonellotto, N., Yahyapour, R., "A Study on Network Resources Management in Grids", *COREGRID Workshop*, Warsaw, Poland, October 2006.

[2] Adami, D., Callegari, C., Giordano, S., Pagano, M., "A New Path Computation Algorithm and Its Implementation in NS2" to appear in *Proc. of ICC 2007*.

[3] Adami, D., Carlotti, N., Giordano, S., Repeti, M., "Design and Development of a GNRB for the Coordinated Use of Network Resources in a High Performance Grid Environment", in Davoli, F., Palazzo, S., Zappatore, S., Distributed Cooperative Laboratories: Networking, Instrumentation, and Measurements, Springer, New York, NY, 2006, pp. 295-307.

[4] Baker, M., Buyya, R., Laforenza D., "Grids and Grid technologies for wide-area distributed computing", *Software-Practice and Experience Journal*, 2002.

[5] Ferrari, T., Ed, "Grid Network Services Use Cases" v. 2.10, *GGF Grid High Performance Research Group*, July 2006.

[6] Foster, I., Kesselman, C., Nick, J., Tuecke, S., "The physiology of the Grid: An open Grid services architecture for distributed systems integration", http://www.globus.org/research/papers/ogsa.pdf, June 2002.

[7] Foster, I., Kesselman, C., Tuecke S., "The Anatomy of the Grid: Enabling scalable virtual organization", *International Journal of Supercomputer Application* http://www.globus.org/research/papers/anatomy.pdf, 2001.

[8] Habib, I.W., Song, Q., Li, Z., Rao, N.S.V., "Deployment of the GMPLS Control Plane for Grid Applications in Experimental High-Performance Networks", *IEEE Communications Magazine*, March 2003.

[9] Le Boudec, J.Y., Thiran, P., "Network Calculus – A Theory of Deterministic Queuing Systems for the Internet", *Springer Lecture Notes in Computer Science*, Vol. LCNS 2050.

[10] Le Faucheur, F., Wu, L., Davie, B., Davari, S., Vaananen, P., Krishnan, R., Cheval, P., Heinanen, J., "Multi-Protocol Label Switching (MPLS) Support of Differentiated Services", *RFC 3270*, May 2002.

[11] Mannie, E., Ed "Generalized Multi-Protocol Label Switching (GMPLS) Architecture", *IETF RFC 3945*, April 2004.

[12] Rosen, E., Viswanathan, A., Callon, R., "Multiprotocol Label Switching Architecture", *RFC 3031*, January 2001.

[13] The OSPF-TE Network Simulator Home Page: http://netgroup-serv.iet.unipi.it/ospf-te_ns/

[14] The RSVP-TE Network Simulator Home Page: http://netgroup-serv.iet.unipi.it/rsvp-te_ns/

[15] Wang, Z., Crowcroft, J., "Quality of Service Routing for Supporting Multimedia. Applications", *IEEE Journal on Selected Areas in Communication*, September 1996.

[16] http://public.web.cern.ch/Public/Welcome.html

[17] http://www.phy.orn.gov/tsi

[18] http://www.elvbi.org

EXPLOITING OVERLAY NETWORKS FEATURES TO ENHANCE THE PERFORMANCE OF THE FILE MOVER

C. Anglano[1] and M. Canonico[1]

[1]*Dipartimento di Informatica, Università del Piemonte Orientale, Alessandria (Italy)*

{cosimo.anglano,massimo.canonico}@unipmn.it

Abstract The *File Mover* is a file transfer infrastructure, based on the overlay networks paradigm, specifically conceived to provide high-performance file transfers for Data Grids. In its current implementation, the File Mover exploits only in part the potential benefits typical of overlay networks. In this chapter we consider three possible extensions of the File Mover, aimed at increasing its performance, that better exploit the characteristics of overlay networks. The performance improvements obtained with these extensions are demonstrated by means of a set of experiments, in which the performance obtained by an extended version of the File Mover has been compared with that attained by its standard version.

1. INTRODUCTION

A rather large set of scientific disciplines, such as high-energy Physics, Climate Modeling, and Life Sciences, requires the archival, retrieval, and analysis of massive data collections, whose size may well be in the order of tens to hundreds of megabytes (and sometimes even Petabytes) [17]. *Data Grids* [28], providing infrastructure and services for distributed data-intensive applications, are at the moment regarded as the best solution to the processing needs arising in the above application domains. As observed in [5], one of the crucial components of Data Grids is its file transfer infrastructure, that must be conceived in such a way to reduce as much as possible the time taken to transfer a file among the resources in the Grid. The need for high-performance data transfers arises from the observation that the completion time of typical data-intensive applications is given by the sum of their execution time and of the time taken to transfer the data they need [22] and is often dominated by the data transfer time.

The *File Mover* [3] is a file transfer infrastructure specifically conceived for Data Grids that use production networks (e.g., the Internet) to connect their resources. The main characteristics of such networks are (a) lack of control, that is, it is not possible to enforce any application-specific control decision (e.g., choosing the network path used to perform a transfer) on the network, and (b) presence of cross traffic, originated by other applications, that competes for the bandwidth with the traffic generated by the Data Grid applications and components of its infrastructure.

The File Mover is based on the *overlay network* [1, 8] paradigm, in which a set of machines (called *File Relays*) connected to the edges of the network are used to build a virtual topology layered on top of the physical one. File Relays communicate among them by using direct transport-layer connections and use application-level routing to overcome the inherent limitations of IP routing, namely (a) its inability to exploit alternative network paths that often exhibit better performance than those chosen to route traffic [23, 24], and (b) the slow convergence time [13] of routing protocols that results in very long time to recover routing tables after a network failure [7, 18, 19]. More precisely, the transfer of a file from a source to a destination machine exploits a sequence of File Relays (that we call *virtual paths*) to force the transfer to use a network path better than the one chosen by IP routing.

As shown in [3], the File Mover is able to achieve significantly better performance than state-of-the-art transfer solutions based on IP routing. However, its performance improvements have been obtained by exploiting only in part the potential benefits typical of overlay networks. In particular, the first version of the File Mover exploits only a single virtual path between the source and the destination machines, although several virtual paths may exist among them. In this chapter, we study a set of possible optimizations, that may be introduced in the File Mover design in order to further increase its performance, which are based on the idea of properly exploiting all the virtual paths existing between the source and the destination machines, and the storage resources available on the various machines of the overlay. In particular, we consider *multipath file transfers*, usage of *file caches* as transfer sources, and *striped file transfers* as possible means to increase file transfer performance. In order to evaluate the performance benefits yielded by these techniques, we have implemented a prototype version of the File Mover that includes them and performed a set of experiments aimed at comparing its performance against those attained by its "plain" implementation. The results of these experiments, performed on a network testbed built to accurately reproduce the operational conditions of realistic production networks (like the Internet) in terms of both path capacities and traffic flows, clearly indicate that the proposed techniques have the potential of greatly enhancing the performance attainable by the File Mover.

The remainder of this chapter is organized as follows. In Section 2, we briefly review the architecture of the File Mover, its behavior, and operations. In Section 3, we describe the various extensions we have introduced in the File Mover. Section 4 reports the results obtained in our experiments. Finally, Section 5 concludes the chapter and outlines future research work.

2. THE FILE MOVER

Let us briefly describe the main features of the File Mover, whose complete description, as well as a thorough evaluation of its performance, may be found in [3]. The File Mover is a system providing a high-performance transfer service for (potentially very) large files. Its architecture, depicted in Figure 1, features an overlay network – comprising a set of *File Relays* – used to carry out the transfer between a *File Server* and a *Client*. The overlay network is fully connected, that is, each File Relay may communicate directly with any other File Relay via transport-layer connections (henceforth denoted as *virtual links*). *File Proxies* represent the interface between the clients and the File

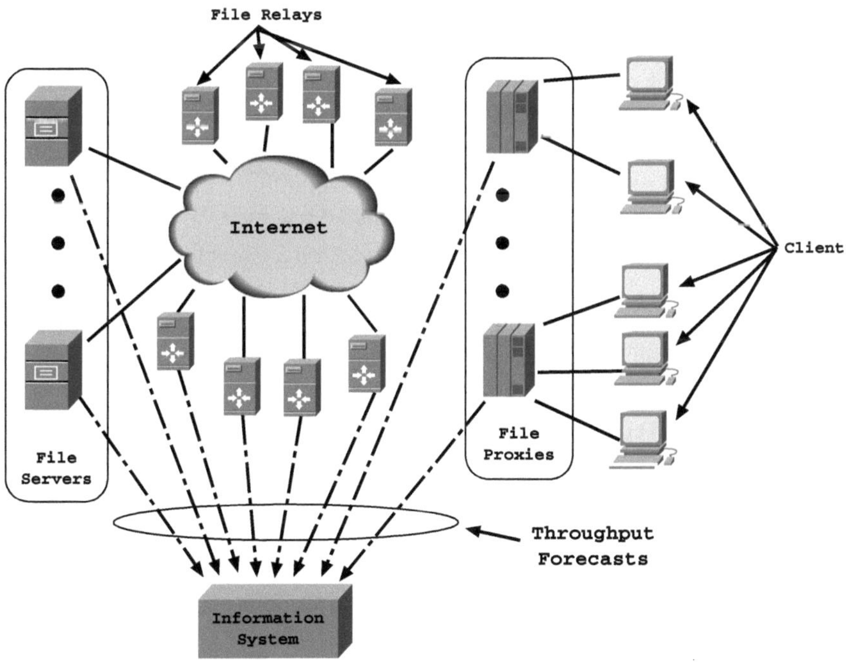

Figure 1. The File Mover architecture

Mover overlay network: when a client needs a file, it contacts one of the File Proxies, that in turn requests the file to a server holding one of its copies and notifies the requesting client when the transfer has been completed.[1]

Upon receiving a request, a Server computes the best *virtual path*, consisting in the sequence of virtual links (and of the corresponding File Relays) that connect itself with the requesting Proxy that provides the highest achievable throughput among all the possible virtual paths connecting them. For more details concerning the algorithm to compute the best path, the interested reader may refer to [3].

In order to maximize the throughput on individual virtual links (and, hence, on the virtual path they compose), the File Mover uses the *UDP-based Data Transfer* (*UDT*) protocol [11] to move data among pairs of overlay nodes. Furthermore, *cut-through* is used to transfer files, that is, the transferred file is divided into fixed-size blocks that are forwarded independently of each other.

The File Mover has been conceived to be used in production networks, where the presence of cross traffic due to other network applications can introduce potentially very large variations in the bandwidth available on individual virtual links, that result in proportional variations in the achievable throughput. This implies that a virtual path that at a given time provides the highest throughput may become less effective than another one that exhibits a lower throughput at the moment of virtual path computation while the transfer is still going on. In order to minimize the impact of these changes in cross traffic, the File Mover computes mid-term forecasts of the throughput achievable on individual virtual links and uses them to compute the virtual path characterized by the highest expected achievable throughput. These forecasts are computed by measuring (by means of *iperf* [12]) at regular time instants, the achievable throughput of each virtual link, and by feeding these measurements to the statistical forecasting algorithms of the *Network Weather Service* (NWS) [30]. These forecasts are then stored into the *Information System* for future use. Measurements are carried out on an individual basis, that is, every Server and Relay periodically measures the available bandwidth for all its outgoing virtual link (special care is taken to avoid interference among concurrent measurements).

3. EXTENDING THE FILE MOVER

As mentioned in the Section 1, although the File Mover architecture discussed in the previous section results in significant performance improvements over source-to-destination direct transfers [3], further performance improvements can be obtained by exploiting several virtual paths to

[1] We assume that the identity of the Server holding a copy of the requested file is determined by the Client by means of a data location service external to the File Mover, as for instance the *Replica Location Service* [4].

transfer a single file. As a matter of fact, there are situations in which the number of available (i.e., not engaged in any file transfer) File Relays allows the set up of multiple virtual paths directed toward the same Proxy that can be used to transfer different parts of the same file. Furthermore, the possibility of storing files (or part of them) at individual Relays enables one to transfer a file either from a location different from the Server holding the file, or from multiple locations simultaneously. In the remainder of this section, we discuss three optimizations that result from the above observations, namely *multipath transfers*, *transfer from cached copies*, and *striped transfers*.

3.1 Multipath File Transfers

The first optimization we consider consists in performing *multipath transfers* of the requested file, where the server splits the file into independent chunks that are sent along different virtual paths connecting the Server to the requesting Proxy. The benefits of this approach, also known as multipath streaming, are well documented in the scientific literature, as discussed in [2, 6, 9, 10, 16]. In our scenario, it is often the case that, after the best virtual path has been chosen, the remaining set of File Proxies can be used to set up other virtual paths characterized by lower values of the expected throughput. The idea behind multipath transfers is to use these lower-performance virtual path – in addition to the best one – so that the total transfer time can be further reduced. In order to balance the work that has to be done by the various virtual paths, the size of the chunk sent on each of them is proportional to its achievable throughput. More specifically, the size $S(C_i)$ of the chunk sent along the ith virtual path VP_i is computed as:

$$S(C_i) = file\_size * \frac{AT(VP_i)}{\sum_{j=1}^{n} AT(VP_j)} \qquad (1)$$

where n denotes the number of different virtual paths available at the moment of the request, $file\_size$ denotes the size (in bytes) of the requested file, and $AT(VP_i)$ is the expected value of the achievable throughput for virtual path VP_i (the achievable throughput is defined in [3]). For example, to transfer a 1 GB file over two distinct virtual paths VP_1 and VP_2 such that $AT(VP_1)$ $= 40$ Mbps and $AT(VP_2) = 60$ Mbps, $S(C_1) = 1GB * 40/(40+60) = 0.4GB$ and $S(C_2) = 1GB * 60/(40+60) = 0.6GB$.

To perform a multipath transfer, the Server computes first the best virtual path, and then proceeds with the computation of the next one that does not include the Relays already used for the first one, and so on until no more available Relays exist. Note that the Server–Proxy direct virtual link is a particular case of virtual path consisting in a single link and no intermediate Relays, and as such may be used for multipath transfers.

3.2 Transfer from Cached Copies

The next optimization we consider is based on the observation that, being File Relays standard computers equipped with hard disks, temporary copies of the file being transferred may be stored locally on all the Relays belonging to the chosen virtual path. Subsequent transfers for the same file may be then performed using the Relays that results in the best virtual path toward the requesting Proxy, rather than from the origin server, if this exhibits a lower achievable throughput.

The introduction of cached replicas requires a change in the way transfers are carried out. As a matter of fact, the original File Mover design assumes that at any given point in time only a single instance of each file is present, so that only the corresponding Server has to be contacted in order to request the transfer. However, the introduction of cached copies implies that the number of potential sources to which a transfer may be requested is larger than one, so the following three problems arise: (a) the various existing replicas must be located, (b) consistency among copies must be maintained if and when the original file is modified, and (c) the replica to be used as the source must be selected. The first two problems are solved by introducing into the File Mover architecture a *Replica Location and Management* (*RLM*) subsystem, in charge of keeping track of the various existing replicas, and of maintaining consistency among them by means of a simple write-invalidate protocol that – upon writing of the original file – invalidates all the existing copies. It is worth to point out that systems of this type have been already designed and developed for Grid infrastructures [4], so one of them may be exploited for the File Mover. The third problem, namely replica selection, may be solved by having either the Server or the requesting Proxy (a) query the RLM subsystem to find all the available replicas, (b) run the virtual path computation algorithm for each of them, and (c) choose the replica that results in the virtual path exhibiting the highest expected achievable throughput. Although this selection might be performed by either the requesting Proxy or the Server, our choice is to assign this functionality to the requesting Proxy because this allows a simpler integration into the File Mover architecture of the optimization discussed in the next subsection.

As a final consideration, it should be noted that caching may be combined with multipath transfers, that is, the Relay holding the replica that will be transferred may set up multiple virtual paths toward the requesting Proxy and transfer different chunks of the file across them, as discussed in the previous section.

In this chapter, however, we do not consider this option, that is left as part of our future work.

3.3 Striped File Transfers

The presence of multiple replicas of the same file enables another optimization, namely, the usage of *striped transfers* in which different chunks of the same file are simultaneously transmitted *from different replicas*. The size of the chunk sent along a given virtual path is computed according to (1). One important potential advantage of striped transfers with respect to multipath ones (as defined in Section 3.1) is that, by using separate sources for the various file chunks, the transmission of each chunk may exploit the full bandwidth available on the link connecting the source to the Internet, so potential bottlenecks that may arise when a single source is used are removed. As a final consideration, note that although it is possible (and probably profitable) to combine multipath and striped transfers, in which the file chunk sent by a given cache can be further split across several virtual paths connecting the same source–Proxy pair, in this chapter we do not consider this option, that is left as part of our future work.

4. EXPERIMENTAL EVALUATION

In order to evaluate the performance benefits of the three optimization techniques described before, we developed a prototype implementation of the File Moreover that included them and used it to carry out a set of experiments aimed at comparing its performance with those attained by the standard File Mover implementation described in Section 2 for different background traffic scenarios.

In order to obtain reproducible and at the same time realistic results, we ran our experiments on a local testbed built in such a way to mimic as realistically as possible the characteristics and dynamics of real networks. On this testbed, in addition to the Information System (a machine running a MySQL DBMS), we deployed five Relays, one Server, one Proxy, and one Client (the motivations for using a small overlay configuration are discussed in Section 4.2). We decided to use our own, locally controlled testbed rather than publicly available ones (e.g., PlanetLab [27] or Emulab [29]) in order to maintain control over the workload run on the composing machines and the traffic load injected into the network. In this way, we could ensure experiment repeatability, tool comparison in the same operational conditions, and accuracy of result interpretation, while at the same time maintaining the realism of the environment in which experiments were executed.

In the remainder of this section, we describe (a) the characteristics of our network testbed (Section 4.1), (b) the scenarios in which the comparisons were performed (Section 4.2), and (c) the results we obtained (Section 4.3).

4.1 The Network Testbed

For our testbed, we used a cluster of PCs running the Linux operating system, each one equipped with an AMD 1.6 GHz Athlon, 724 MB of main memory, a 60 GB hard disk, and two Fast Ethernet network interfaces. The machines were dedicated to our testbed (i.e., no other applications were executed during the experiments) and did not have any external network connections, so no exogenous traffic could enter the testbed network.

In order to reproduce the behavior of a real network, we equipped our testbed with suitable mechanisms for the specification and enforcement of the capacities of virtual links, as well as for the injection of background traffic streams (i.e., traffic streams due to other applications using the network) on them.

Virtual link capacity is set and enforced by means of the *Traffic Control* (TC) mechanisms available in the Linux kernel [14]. TC works by associating with each network interface a set of queues, each one representing a virtual (unidirectional) link toward a specific host, and by enforcing a user-defined transmission rate for each of them. In our testbed, on each machine a queue was created for each virtual link directed toward the other File Mover components. We chose TC rather than alternative possibilities (e.g., NistNET [25]) since this resulted in maximal stability and minimum overhead.

Injection of background traffic is accomplished by means of the *Distributed Internet Traffic Generator* (D-ITG) [20], as discussed in [3].

4.2 Experimental Scenarios

For our experiments, we considered three scenarios, corresponding to injecting three different background traffic workloads on an overlay configuration obtained by setting to specific values the capacities of individual virtual links.[2]

Overlay Configuration. The overlay configuration used for our experiments includes one *Server*, one *Proxy*, and five *Relays*. The choice of a relatively small configuration depends on the fact that the amount of CPU capacity required by D-ITG to generate the background traffic is directly proportional

[2]In this chapter, we define the capacity of a network path as the maximum throughput that the path can provide to a flow when there is no cross traffic [21].

to the number of streams departing from the node (i.e., the node *fan-out*). In our case, being the overlay fully connected, the fan-out of each node is $N-1$, where N is the total number of nodes in the overlay. The machines used in our testbed, being relatively old, could sustain the generation of at most six D-ITG streams, thus limiting the size of the overlay we could experiment with. However, in spite of its limited dimension, this topology allows the emergence of interesting performance differences between the plain File Mover version and the extended ones.

The capacities of individual virtual links have been chosen in such a way to closely reproduce the characteristics of a realistic network setting by using the results presented in [15] (summarized in Table 1), where the capacity distribution of the network paths among PlanetLab nodes has been reported (the specific values of the capacities of each virtual link are reported in [3]).

Background Traffic Load. As already discussed, in order to mimic the behavior of a realistic production network, we injected background traffic on all the virtual links of our overlay testbed. This traffic consists in a set of data streams transmitted between pairs of testbed nodes, each one characterized by a set of parameters related to its *intensity* (i.e., how much resources it uses) and *dynamism* (i.e., how it uses these resources). The parameters of a stream related to its dynamism are its lifetime, the protocol used to transport data, and the distribution of packet size, while intensity is expressed in terms of the time elapsing between the generation of two consecutive packets (the *inter-packet time*).

For the sake of realism, for our experiments the characteristics of these streams have been set according to the latest research results in the field of Internet traffic characterization. Although an analysis of the relevant literature reveals that no single traffic model exists that takes into consideration all the parameters mentioned above, there are studies that address each of the above characteristics separately. We have therefore developed our own traffic model, whose parameters are summarized in Table 2, that puts together the results of these studies by setting the various parameters, as discussed in [3]. In order to perform our comparison in various operational conditions, we defined three different background traffic workloads corresponding to situations of low,

Table 1. Distribution of virtual link capacities

Capacity *(Mbps)*	Fraction *(%)*
Uniform(0,20)	30
Uniform(20,50)	9
Uniform(50,80)	6
Uniform(80,100)	55

Table 2. Common background traffic parameters

Stream parameter	Value
Lifetime distribution	45% less than 2 sec 53% less than 15 minutes 2% up to 2 h
Protocol type distribution	80% TCP 20% UDP
Packet size distribution	25% Uniform(40,44) 50% Uniform(45,576) 25% Uniform(577,1500)
Inter-packet time distribution	Exponential (parameter set according to (2)

intermediate, and high intensity. The intensity of background traffic is defined as the fraction of the *average overlay capacity* (*AvgCap*) that it consumes, that is, in turn defined as the arithmetic average of the capacities of the individual virtual links. The three background traffic workloads considered in our study, denoted in the rest of the chapter as *Low*, *Medium*, and *High*, correspond to intensities of 20, 50, and 80%, respectively. The parameter *PPS* of the inter-packet time distribution for the background traffic streams transmitted over the various virtual links is computed as

$$PPS = \lfloor \frac{AvgCap * BgTrafIntensity}{MaxPacketSize} \rfloor \tag{2}$$

where $BgTrafIntensity$ is the intensity of the background traffic and it takes value in the set $\{20, 50, 80\%\}$ as mentioned before, while $MaxPacketSize$ is the maximum packet length. For the overlay configuration used in our experiments, we have that $AvgCap = 63$ Mbps, so the value of *PPS* are $1{,}050, 2{,}625$, and 4,200 for the *Low, Medium*, and *High* background traffic intensity, respectively.

4.3 Experimental Results

To assess whether the extensions proposed in this chapter actually result in performance improvements, we performed a set of experiments consisting in the repeated transfers of a 2 GB file for all the scenarios discussed above. Our experiments were aimed at evaluating the following two features of the proposed extensions:

- the *performance gain* that they provide with respect to the plain File Mover version. The performance metric used to carry out this comparison was the *average effective throughput*, computed as the arithmetic average of the throughput obtained in a set of five transfers of the same file, which has been in turn computed as the ratio of the file size over the time taken to transfer it;

- its *adaptability* to variations of the network conditions due to background traffic measured by the number of different virtual paths used by the File Mover to carry out all the transfers of the same file (two paths are considered different if they differ in at least one Relay).

In order to perform the comparison in the same operational conditions, we ran first all the experiments involving the plain version of the File Mover, and then we restarted the testbed and performed the experiments involving the extended version exactly at the same time instant from the beginning of the injection of background traffic.

The results obtained in our experiments, discussed in detail in the following subsections, demonstrate that the File Mover extensions we propose result in significantly improved performance.

Evaluation of Transfer from Cached Copies. In order to evaluate the effectiveness of using cached copies as sources for file transfers, in our experiments we placed a replica of the transferred file on the various *Relays*, and then performed a set of experiments where, for each background traffic intensity, we requested the transfer of the file to the *RLM* that selected the replica resulting in the best expected performance. The results of our experiments, reported in Table 3, show that the use of cached copies (*Caching FM* column) greatly enhances performance for all the background traffic intensities we considered with respect to the plain File Mover version (*plain FM* column). More specifically, the gain decreased for increasing values of the background traffic intensity and ranged from 18.12% (*Low* intensity) to 7.81% (*High* intensity). This demonstrates the effectiveness of cached file transfers, especially in situations where the amount of background traffic limits mostly the virtual links departing from the *Server*, as in the case of our experiments.

The ability of the File Mover to adapt to variations in the background traffic intensity on individual virtual links is demonstrated by the results reported in column *Num. paths* (reporting the different number of virtual paths used for

Table 3. Comparison of cached versus standard File Mover transfers

Background	Average throughput (Mbps)			Number
Traffic Load	Caching FM	Plain FM	Gain	paths
Low	82.69	70.00	18.12%	1
Medium	76.99	67.43	14.18%	2
High	70.58	65.47	7.81%	3

the experiments): while only one path has been used to carry out all the five transfers for the *Low* intensity, two and three paths have been selected for the *Medium* and *High* intensities.

Evaluation of Multipath Transfers. The second set of experiments was aimed at evaluating the possible performance benefits resulting from multipath transfers. For these experiments, all the transfers were performed by using the *Server* as the source. For each transfer request, the *Server* computed the virtual paths to be used to transfer the different chunks of the file and determined the size of each chunk as indicated by (1). In order to force the *Server* to use several virtual paths, we set *a priori* to two the number of virtual paths used simultaneously to carry out the transfer. Without this restriction, the *Server* could have used all the available *Relays* to set up a single virtual path, an event relatively likely given the small number of *Relays* of our topology, thus preventing us of performing experiments using multipath transfers.

The results obtained in our experiments are reported in Table 4, where we can observe that, again, the introduction of multipath transfers (*Multipath FM* column) results in significantly better performance with respect to the plain File Moreover version for all the background traffic intensities we considered. More specifically we note that the performance gains are substantial (they range from 79.8 to 42.06%), and they decrease for increasing background load intensities. This is a consequence of the fact that a multipath transfer terminates when *all* the transfers of all the chunks have been completed, so a reduction of the achievable throughput *on any* virtual path degrades the performance of the whole transfer. The probability of such a degradation increases with the number of used virtual paths, and is more likely to occur for higher intensities of the background load.

Evaluation of Striped Transfers. The last set of experiments we carried out was aimed at assessing the performance of striped transfers. We modified the overlay topology by replacing one of the *Relays* (Relay 5) with a *Server*, on which we created another copy of the file to be transferred in the experiments. This modification was required by the need of keeping constant the size of the overlay, for the reason discussed in Section 4.2. For each transfer request, the

Table 4. Comparison of multipath versus standard File Mover transfers

Background Traffic Load	Average throughput (Mbps)			Number paths
	Multipath FM	Plain FM	Gain	
Low	125.86	70	79.80%	2
Medium	100.46	67.43	48.99%	2
High	93.01	65.47	42.06%	3

Table 5. Comparison of striped versus standard File Mover transfers

Background Traffic Load	Average throughput (Mbps) Striped FM	Plain FM	Gain	Number paths
Low	82.32	70.00	17.60%	2
Medium	71.48	67.43	6.01 %	3
High	68.18	65.47	4.14%	3

Proxy contacted the original Server first, that computed its best virtual path, and then the new one that used the remaining Proxies to compute its best virtual path. Each Server then independently computed the size of the chunk it sent by using (1), that was transferred to the Proxy. The transfer of the file was considered concluded when both chunk transfers were completed. The results of our experiments, shown in Table 5, once again show that the proposed extension (*Striped FM* column) results in performance improvements over the plain File Mover version. We note, however, that the gain strongly decreases for increasing values of the background load intensity. This is explained by the fact that the choice of computing the first virtual path first, and then the second one by using the Relays that had been discarded by the first Server, resulted in virtual paths characterized by very different throughput values. Consequently, variations in the throughput exhibited by the slower virtual path may have a negative impact on transfer performance even if the faster one is working at full speed. While this is unlikely to occur for relatively low intensities of the background traffic, thanks to the way the size of the individual chunks is computed, its negative impact tends to become higher for increasing intensities, as indicated by the results reported in the *Medium* and *High* rows of Table 5. This phenomenon calls for better strategies for the computation of the virtual paths used by the various sources used to carry out a striped transfer that we plan to include in our future work. In any case, in spite of this suboptimal choice, striped transfers always resulted in performance improvement, thus demonstrating the potentials of this technique.

5. CONCLUSIONS

In this chapter, we have considered various file transfer policies (namely, *multipath* and *striped*, as well as transfers using *cached copies* as source) for possible inclusion into the File Mover, a software infrastructure providing a high-performance file transfer service for Grid platforms. We have implemented a prototype version of the File Mover that included these extensions and used it to carry out an experimental study aimed at comparing its performance versus that attained by the plain File Mover implementation. As expected, the techniques we considered result in significant performance improvements but, given the limited size of the overlay we considered, these results should be considered only preliminary. A more thorough experimentation with larger configurations is required in order to make definitive conclusions about them. Nevertheless, although preliminary, our results can be considered encouraging.

There are several avenues of research we plan to explore in the future. First of all, we plan to study and implement transfer strategies obtained by combining the various extensions described in this chapter. More precisely, in addition to the rather obvious combination of multipath and cached transfers, in which multipath transfers are used also by the Relays selected as source for the transfer, we plan to study the combination of striped and multipath transfers, in which each chunk sent by a separate source is transferred by using a multipath approach.

A second avenue of research is concerned with the investigation of better algorithms for the selection of virtual paths when more sources are used for the same transfer (as in the case of striped transfers), or when independent file transfers are performed by different sources.

Third, we plan to improve the current method used to collect measurements of the throughput achievable on individual virtual links by replacing the application-level probes used in the current version of the File Mover (that, as already mentioned, uses *iperf*) with a more scalable and accurate approach, working at the network level, like *Plab* [26]. Finally, on the experimental side, we plan to perform experiments by using larger, and more realistic, overlay configurations thanks to the use of more powerful machines for our network testbed.

REFERENCES

[1] D.G. Andersen, H. Balakrishnan, M.F. Kaashoek, and R.Morris. "Resilient overlay networks". In *Proc. of 18th ACM Symp. on Operating Systems Principles*, Banff, Canada, Oct. 2001.

[2] D.G. Andersen, A.C. Snoeren, and H. Balakrishnan. "Best-path vs. multi-path overlay routing". In *IMC '03: Proc. of the 3rd ACM SIGCOMM conference on Internet measurement*, pages 91–100. ACM Press, New York, NY, USA, 2003.

[3] C. Anglano and M. Canonico. "The file mover: High-performance file transfer for the grid". *Concurrency and Computation: Practice and Experience*, 2007. Submitted for publication. Available from http://dcs.di.unipmn.it/.

[4] M. Cai, A. Chervenak, and M. Frank. "A peer-to-peer replica location service based on a distributed hash table". In *Proc. of Supercomputing 2004*, Pittsburgh, PA, USA. IEEE Press, 2004.

[5] A. Chervenak, I. Foster, C. Kesselman, C. Salisbury, and S. Tuecke. "The data grid: Towards an architecture for the distributed management and analysis of large scientific datasets". *Journal of Network and Computer Applications*, 23:187–200, 2001.

[6] A.L.H. Chow, L. Golubchik, J.C.S. Lui, and W.-J. Lee. "Multi-path streaming: optimization of load distribution". *Performance Evaluation*, 62(1-4):417–438, 2005.

[7] M. Dahlin, B. Baddepudi V. Chandra, L. Gao, and A. Nayate. "End-to-end wan service availability". *IEEE/ACM Transactions on Networking*, 11(2):300–313, 2003.

[8] D.D. Doval and D. O'Mahony. "Overlay networks: A scalable alternative for p2p". *IEEE Internet Computing*, pages 79–82, July–August 2003.

[9] L. Golubchik and J.C.S. Lui. "Multi-path streaming: is it worth the trouble?". *ACM SIGMETRICS Performance Evaluation Review*, 30(3):12–14, 2002.

[10] L. Golubchik, J.C.S. Lui, T.F. Tung, A.L.H. Chow, W.-J. Lee, G. Franceschinis, and C. Anglano. "Multi-path continuous media streaming: What are the benefits?". *Performance Evaluation*, 49(1-4):429–449, 2002.

[11] Y. Gu and R.L. Grossman. "Optimizing udp-based protocol implementation". In *Proc. of the Third International Workshop on Protocols for Fast Long-Distance Networks PFLDnet*, 2005.

[12] Iperf: The TCP/UDP Bandwidth Measurement Tool. http://dast.nlanr.net/Projects/Iperf, 2007.

[13] C. Labovitz, A. Ahuja, A. Bose, and F. Jahanian. "Delayed internet routing convergence". In *Proc. of ACM SIGCOMM*, pages 175–187, Stockolm, Sweden, September 2000.

[14] Linux Advanced Routing & Traffic Control. http://lartc.org/, 2007.

[15] S.-J. Lee, P. Sharma, S. Banerjee, S. Basu, and R. Fonseca. "Measuring bandwidth between planetlab nodes". In *Proc. of the Passive and Active Measurement Workshop (PAM 2005)*, pages 292–305, Springer, Berlin, 2005.

[16] Z. Ling and I. Lee. "Adaptive multi-path video streaming". In *ISM '06: Proc. of the Eighth IEEE International Symposium on Multimedia*, pages 399–406. IEEE Computer Society, Washington, DC, USA, 2006.

[17] H.B. Newman, M.H. Ellisman, and J.H. Orcutt. "Data-intensive e-Science frontier research". *Communications of the ACM*, 46(11), Nov. 2003.

[18] V. Paxson. "End-to-end routing behavior in the Internet". In *Proc. of ACM SIGCOMM*, pages 25–38, Stanford, CA, August 1996.

[19] V. Paxson. "End-to-end internet packet dynamics". In *Proc. ACM SIGCOMM*, pages 139–152, Cannes, France, September 1997.

[20] A. Pescape, S. Avallone, and G. Ventre. "Analysis and experimentation of internet traffic generator". In *Proc. of the New2an*, pages 70–75, 2004.

[21] R.S. Prasad, M. Murray, C. Dovrolis, and K. Claffy, "Bandwidth estimation: metrics, measurements techniques, and tools. In *IEEE Network*, Dec. 2003

[22] K. Ranganathan and I. Foster. "Simulation studies of computation and data scheduling algorithms for data grids". *Journal of Grid Computing*, 1(1):53–62, 2003.

[23] S. Savage, A. Collins, E. Hoffman, J. Snell, and T. Andersonr. "The end-to-end effects of internet path selection". In *SIGCOMM '99: Proc. of the conference on Applications, technologies, architectures, and protocols for computer communicationr*, pages 289–299. ACM Press, New York, NY, USA, 1999.

[24] H. Tangmunarunkit, R. Govindan, S. Shenker, and D. Estrin. "The impact of routing policy on Internet paths". In *Proc. IEEE Infocom '01*, Anchorage, Alaska, Anchorage, AK, USA, April 2001.

[25] The nistnet project. http://snad.ncsl.nist.gov/nistnet/, 2007.

[26] The plab project. http://www.grid.unina.it/software/Plab, 2007.

[27] The planetlab project. http://www.planet-lab.org, 2007.

[28] S. Venugopal, R. Buyya, and K. Ramamohanarao. "A taxonomy of data grids for distributed data sharing, management, and processing". *ACM Computing Surveys*, 38(1), June 2006.

[29] B. White, J. Lepreau, L. Stoller, R. Ricci, S. Guruprasad, M. Newbold, M. Hibler, C. Barb, and A. Joglekar. "An integrated experimental environment for distributed systems and networks". *ACM/SIGOPS/Operating Systems Review*, 36(SI):255–270, 2002.

[30] R. Wolski. "Dynamically forecasting network performance using the network weather service". *Cluster Computing*, 1(1):119–132, Jan. 1998.

ADVANCED SATELLITE INFRASTRUCTURES IN FUTURE GLOBAL GRID COMPUTING: NETWORK SOLUTIONS TO COMPENSATE DELIVERY DELAY*

B. Bonito[1], A. Gotta[1], and R. Secchi[1]
[1]*Information Science and Technology Institute ISTI - CNR Italian National Research Council, Via G. Moruzzi 1, San Cataldo, 56124 Pisa (Italy)*
{blasco.bonito, alberto.gotta, raffaello.secchi} @isti.cnr.it

Abstract Scientific equipment in remote locations may be integrated in a grid network by means of satellite links. When the grid is an Internet overlay network, TCP-like congestion control protocols tend to introduce large delays and to achieve poor bandwidth utilization during the start-up phase due to the large bandwidth-delay product of the satellite link. Recent studies demonstrate that the loss of performance is high when a demand assignment multiple access (DAMA) scheme is adopted to manage satellite resources. To overcome this problem, we propose a new mechanism based on Quick Start, a recent IETF recommendation, to enhance the startup performance of data and streaming flows. We show, by means of extensive simulations, that our modifications to enable a QS-based DAMA algorithm allow the initial end-to-end delay to be significantly reduced and the queue length in transmitting nodes stabilized. Moreover, our results point out that the start-up delay is quite independent of the flow rate, which makes this algorithm suitable to the high bitrate required by grid computing applications.

1. INTRODUCTION

Science practicing has embraced a number of new paradigms in the 20th century promoted by the worldwide diffusion of the Internet, including the growth of large geographically dispersed teams and the use of distributed

*Work funded by the European Commission in the framework of the RINGrid project (contract no. 31891) and partially in the framework of the "SatNEx" NoE (contract no. 507052).

computational platforms. Essential to this new branch of e-Science applications is the capability of transferring immense amounts of data, ranging from hundreds of TeraBytes to PetaBytes. During this decade, crossing the "Peta" line is even expected: Petabytes in data storage, Petaflop in CPU processing, and Petabit/s in network bandwidth will be of common use in scientific applications. Grid computing will be the fundamental tool to deal with this new challenge.

An obstacle to the deployment of high-speed grid networks is the difficulty of retrieving data from equipment and sensors in remote locations. In these contexts, the satellite connections could be the only practical way to provide high-speed connectivity. However, the large bandwidth-delay product that characterizes these networks, which is actually comparable with the one of very high-speed optical networks, determines inefficient link utilization when TCP (or a TCP-friendly protocol) is used for congestion controlled transport of data. Indeed, during the Slow Start phase, TCP requires several rounds of transmission to achieve efficient bandwidth utilization. In addition, it has been shown [5] that the demand assignment multiple access (DAMA) allocation scheme, nowadays increasingly adopted to satisfy bandwidth requirements, negatively impacts TCP performance.

In such grid computing scenario with a mix-up of multimedia and data contents, network over-provisioning is a must but nonetheless congestion control [8] is mandatory for granting overall network stability. Consequently, it is likely that employment of protocols like datagram congestion control protocol (DCCP) [3, 7], which provide congestion control to slowly variable flow rates, will be necessary. Since these algorithms emulate TCP behavior, they also exhibit bad performance on DAMA-controlled links during the start-up phase. The authors of [11] indicate Quick Start (QS) as a valid technique to solve the startup problems of TCP-like applications over satellite links, though their analysis is limited to the case of constant bandwidth allocation only.

In this chapter, we present a preliminary study concerning the integration of QS and DAMA. The basic idea is simple: the DAMA algorithm, made aware of QS signalling at layer three, reacts quickly to the entry of a new flow by asking the amount of bandwidth indicated by QS. The congestion control protocol assumed in this paper is the TCP-friendly rate control (TFRC), the one that is currently most likely to be used for streaming on the Internet. In order to maintain compatibility with current TFRC implementations, we do not assume any explicit cross-layer coordination between TFRC and DAMA.

Since the system tuning up and algorithm parameters may depend on the particular system implementation, we chose Skyplex^{TM}(produced by Eutelsat) as reference system for configuration. However, the methodology to select the parameters in general is exemplified in the chapter.

The rest of the chapter has the following structure. In Section 2, we give the description of the technologies involved in our satellite scenario. In Section 3, we discuss our preliminary results in a simulated test-bed. Finally, Conclusions (Section 4) highlight the performance improvements and trace the guideline for future developments.

2. TECHNOLOGY OVERVIEW

2.1 Satellite Architecture

The IP/DVB architecture defines the basic set of functionalities required for interactivity of users across a satellite distribution system. Basically, IP packets are first encapsulated into an MPE (multi-protocol encapsulation) structure, which contains the source/destination MAC addresses of the traffic stations involved in the data transfer, then delivered using a transport stream made up of 188-bytes MPEG-2 cells.

Figure 1 shows the main elements of a mesh satellite network, which are possibly elements of a larger grid network. The network is divided into regions, corresponding to the geographic areas, each covered by a satellite spot beam. The group of traffic terminals (TTs) that belong to a specific region is monitored and controlled by a centralized authority, called regional network control center (RNCC), which is connected through a leased line with the higher level network management system. We distinguish the forward link and the return link in a satellite communication, since two different access methods are used

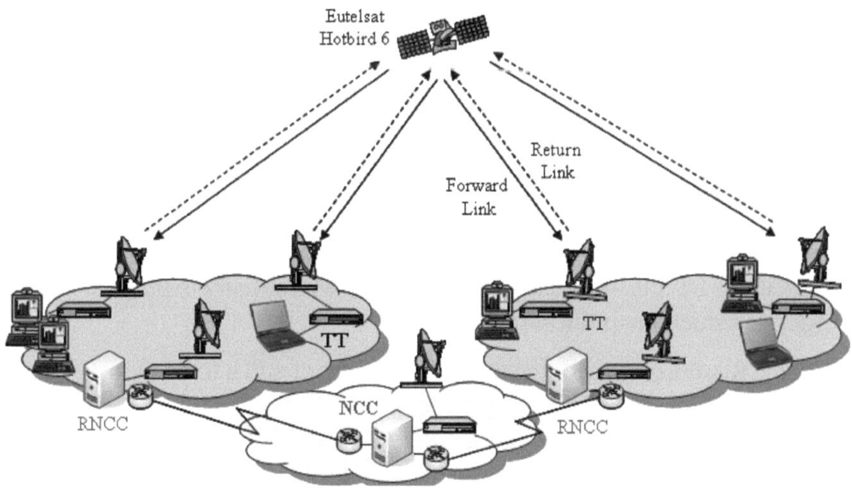

Figure 1. Network Architecture

for each link. The information content is typically transmitted to the satellite with a time division multiple access (TDMA) scheme on the return link, as for example the DVB-RCS standard [1] does. Then, these signals are demodulated, regenerated, and multiplexed by the on-board unit and the satellite forward them over the forward link in time division multiplexing (TDM) mode, by adopting the DVB-S standard. The application of this technology imposes both receiver and transmitter capabilities at the ground station, thus permitting the implementation of a network in which all terminals can communicate with each other over bi-directional links with a single satellite hop.

In DVB-RCS, the time–frequency domain is organized into a hierarchical structure, which consists of super-frames (the largest containers), multi-frames, and frames. In order to establish the time-slot assignments for the traffic terminals, the RNCC broadcasts periodically a burst time plan (BTP) message to its group of traffic terminals. The DVB-RCS standard describes several methods in order to demand an assignment of bandwidth. These methods are

- **Constant rate assignment** (CRA). The link resources are negotiated at the beginning of the transmission and are maintained for all the duration of the connection. Thus, this scheme does not consume bandwidth for signalling from TTs to RNCC but allocating resources, even when not used, usually leads to waste precious bandwidth.
- **Rate-based dynamic capacity** (RBDC). A TT periodically submits a capacity request message to the RNCC on the basis of the local measurement of the incoming traffic rate. Every explicit request overrides the previous one and new requests are submitted only if needed.
- **Volume-based dynamic capacity** (VBDC). A TT dynamically signals the data volume required to empty its buffer. The scheduler assigns the capacity according to the requests that terminals send when more traffic is queued at their input. Since the DAMA controller cannot guarantee the requested bandwidth to the TTs, this access scheme is more suitable for best effort traffic flows.
- **Free capacity assignment** (FCA) is an extension that allows the unassigned capacity to be redistributed among the TTs without any explicit request, in order to reduce the jitter in traffic flows.

When CRA is concerned, the static case is trivial and a fixed number of time slots is assigned to each TT.

In the dynamic case, instead, each terminal sends a request to the RNCC on the basis of its own instantaneous need, and the time slots are assigned in a best-effort mode. The assignment does not guarantee any priority between terminals and quality of service. In such last case, the assignment algorithm involves principally two processes:

1. Each TT computes a Bandwidth request, which contains the information about the bandwidth required on the basis of the local traffic analysis every T_r period and sends it to the RNCC.

2. The RNCC computes the bandwidth to be assigned on the basis of the information received with the BWreq, by assigning a portion of bandwidth that is proportional to each request. If the sum of the requests overcomes the availability, it is normalized with the respect to the maximum capacity.

Figure 2 shows the sequence of messages exchanged between a TT and the RNCC for the bandwidth negotiation.

- The BWreq is the first message transmitted by the TT and includes the results of the local traffic monitoring.
- The RNCC receives the message, it computes the value of the bandwidth to be allocated and, after an expiration period, delivers the BTP in multicast mode. The expiration period is the sum of the propagation delay in up- and down-link and the processing time in order to compute the allocations.
- Once the TT has received the reply from the RNCC, it waits the safety frame period after which the BTP can be applied. The safety period accounts for delivery time drifts among remote locations.

Thus, the allocation delay is the delay between a request and the availability of the desired bandwidth. During this very long period – more than 500 ms with GEO satellites – the TT can only serve the MAC queue at the previous

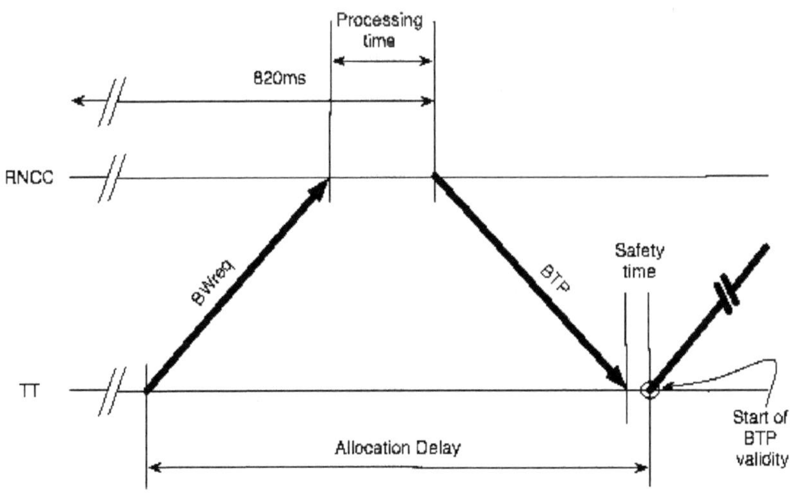

Figure 2. Messages for bandwidth request and assignment

assigned rate. Hence the exceeding incoming traffic is queued in the MAC buffer, waiting for delivery.

2.2 TCP-Friendly Rate Control Algorithm

To alleviate the problem of UDP's inability to provide congestion control, TFRC was designed to allow multimedia applications to provide congestion control when using UDP with RTP [12]. TFRC is a rate-paced equation-based congestion control transport layer protocol that requires both the sender and the receiver to participate in determining the allowed sending rate. In TFRC, the receiver periodically sends a feedback report informing about the recent loss event rate that has been witnessed by the connection. The sender uses this to calculate the allowed sending rate for the next RTT period by using an equation that models the equivalent throughput that would have been obtained by a TCP flow. This equation, given below, provides the throughput that a TCP connection would receive under steady state conditions given the loss event rate and RTT of the connection.

$$T = \frac{s}{RTT \cdot \sqrt{\frac{2bp}{3}} + 3\,T_{RTO} \cdot \sqrt{\frac{2bp}{3}} \cdot p(1+32p^2)} \tag{1}$$

where T is the transmit or sending rate in bytes/s, s is the packet size in bytes, RTT is the round-trip time, p is the steady state loss event rate, T_{RTO} is the TCP retransmission timeout, and b is the number of packets acknowledged by a single TCP acknowledgment. In this case b is set to 1.

Using this formula, TFRC exhibits a much lower variation of throughput over time compared to TCP, making it more suitable for multimedia applications such as TVoIP and VoIP since it allows the sending rate to vary more smoothly by decreasing and increasing the sending rate gradually, while ensuring that it competes fairly with TCP. However, this makes TFRC to respond slower to changes in available bandwidth compared to TCP. TFRC can be used for both unicast and multicast traffic.

TFRC, which was originally intended for applications that use a fixed packet size, was designed to be reasonably fair when competing for bandwidth with TCP connections using the same packet size. TFRC's conventional way of calculating the sending rate in packets per round-trip time, using the loss event ratio, has consequences: a low bandwidth TFRC flow using small packets sharing a bottleneck with high bandwidth.

TCP flows using large packets may be forced to slow down, even though the nominal rate in bytes per second of the TFRC flow is less than the rate achieved by the TCP flows. To solve this problem, TFRC-SP [4], a Small Packet (SP) variant of TFRC was designed for applications such as VoIP that send small packets. TFRC-SP seeks to achieve the same bandwidth in bps as a TCP flow using packets of up to 1,500 bytes. TFRC-SP also enforces a minimum interval of 10 ms between packets, to prevent a single flow from sending small packets arbitrarily frequently.

Regarding TFRC on satellite links, the Round Trip Time (RTT) delay is an important factor in determining the performance of a TFRC-SP flow because the sending rate is inversely proportional to the RTT delay. Satellite networks typically have high RTT delays. Therefore, particular attention needs to be directed to the performance and protocol behavior over such links. Such experience is essential prior to deployment over operational satellite links. Some problems of using TFRC or TFRC-SP on satellite networks are

- Initial Slow Start: TFRC starts with four packets per RTT. This is not sufficient enough for applications with large encoding rates. If the RTT is small, then this initial rate should be fine. For satellite delays, the sending rate to reach the encoding rate is going to be too long.
- Idle period: Similarly to slow start, during an idle period, the sending rate may get reduced to a minimum of two packets per RTT. Slow starting from a rate of two packets per RTT to the encoding rate is going to be too long while operating over satellite links.
- Sending rate limit: TFRC sending rate can be at most twice the current receiver rate. This growth rate is not sufficient to keep up with the encoding rate when the application oscillates between silence and talk periods.

2.3 QS Protocol

Quickstart (QS) [10] is a new IETF experimental protocol that has been designed to provide lightweight signaling of the level of congestion (specifically available capacity) between routers and a pair of communicating end hosts. QS was originally conceived to improve the performance of TCP bulk transfers over lightly loaded network paths. QS may also be useful for multimedia flows. In this case, it can alleviate the effect of slow starting on the encoding rate, and after periods of silence.

The sender sends a QS request for its required sending rate (measured in bytes per second) using a Quick-Start option placed in the IP header. Each router along the path should, in turn, approve the requested rate, reduce the requested rate, or indicate that the Quick-Start request is not approved (possibly by ignoring this new option). In approving a Quick-Start request, a router does not give preferential treatment to subsequent packets from that connection; the router is only asserting that it is currently underutilized and believes there is sufficient available capacity to accommodate the sender's requested rate. The Quick-Start mechanism can determine if there are routers along the path that do not understand the Quick-Start option or have not agreed to the Quick-Start rate request. On receipt of the QS request, the receiver communicates the final rate request value (if the requested rate is approved) that it receives to the sender in a transport-level Quick-Start Response. If the Quick-Start request is approved by all routers along the path, then the sender can send at up to the approved rate for a window of data. Subsequent transmissions will be governed by the default transport protocol's congestion control mechanisms of that connection. If the Quick-Start request was not approved, then the sender would use the default congestion control mechanisms.

Even though QS was proposed with TCP in mind, QS could be used with any congestion control protocol that would prefer to inflate their sending rates without effectively slow starting from a small initial rate. Using QS, the standard Internet protocols can effectively and efficiently work over a wide range of links – including those with satellite delay. QS may also be useful for multimedia flows. In this case it can alleviate the effect of slow start on the encoding rate, and after periods of silence.

In this paper, we also explore the use of QS with TFRC to alleviate the effects of slow start over a DAMA-enabled satellite link. The satellite terminal takes the role of a QS-enabled router. This paper proposes QS as a mechanism to inform the satellite sub network of the desired capacity, and for this to then allocate resources (or prepare itself to do so at a later time). This implies an element of cross-layer communication.

3. NUMERICAL RESULTS

In this section, we show by means of a set of detailed simulations the drawbacks of using the TFRC protocol over a DAMA driven link. Since the round-trip delay of a geostationary satellite link is around half a second, the typical values of the bandwidth-delay product range, depending on the available bandwidth, from hundreds to thousands of kilobytes. In the case of multimedia streaming, this corresponds to hundreds of small size multimedia packets. Since the TFRC protocol starts doubling its sending rate from one packet per RTT, several RTTs are required to reach the steady state throughput. During this period, the workload generated by the sending application is backlogged into the TFRC transmission buffer, and the packet delivery is delayed. When the satellite link is controlled by DAMA, the end-to-end delay may be significant due to the latency induced by the demand-assignment mechanism.

Simulations were carried out with the ns-2 simulator [9] implementing the basic bandwidth allocation mechanisms [6] of a real satellite platform (Skyplex [2]). The system we consider has a single carrier with a capacity of 6.3 Mb/s and implements an RBDC access scheme. The TDMA frame consists of 144 slots and has a duration of 273 ms.

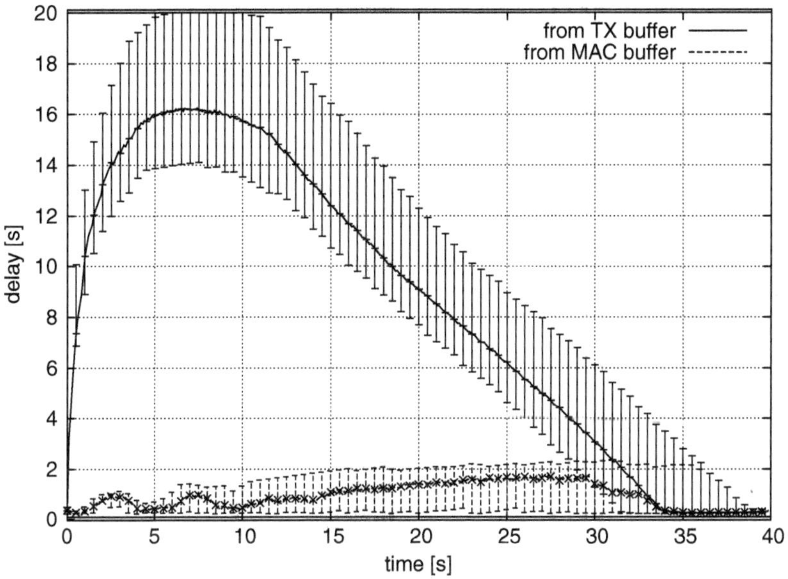

Figure 3. OWD and E2ED of a single TFRC connection over Skyplex link

Packet delays of the first 40 sec of transmission of a CBR video stream (1,024 kb/s) using TFRC-SP are illustrated in Figure 3. The graph reports the one-way delay (OWD), that is, the delay from wire-line transmission of a packet to its wire-line reception, and the end-to-end delay (E2ED), the delay between the time the packet is generated by the transmitting application to the time the packet arrives to the receiving application. The difference between the two delays represents the time spent in TRFC buffers and reflects the mismatch between the application sending rate and the TFRC rate. In order to evaluate the dispersion of delays, together with the median value (solid line), we reported the minimum and maximum delays (vertical bars) at each time obtained starting the TFRC connection with different offsets with respect to the beginning of the TDMA frame.

Figure 3 shows OWD and E2ED of a single TFRC connection over the Skyplex link. This figure points out how DAMA and TFRC result in a bad interaction. The high end-to-end delay (up to ten seconds) at the very beginning of transmission has some unwanted consequences. First, it enforces the use of large de-jitter buffers to avoid loss of data, which leads to unpleasant long delays to get the video play-out. Second, the long transient phase leads to considerable drop of performance when the source generates bursty traffic, or the network is subjected to variable congestion. In addition, the delay exhibits a large variability, which indicates the potential instability of TFRC during the startup phase. Actually, this phenomenon is related to the way the receiver estimates the rate. Since the flow rate is estimated as the number of packets seen by the TFRC receiver in a RTT, the estimator suffers from a quantization effect when the number of packets transmitted is small.

In order to assess the benefits of QS, we repeated the simulations using a TFRC source with QS enabled and placing a QS agent into the satellite terminal. The QS agent follows a simple rule: when a new QS request arrives to the terminal asking for satellite bandwidth, it checks the current bandwidth utilization and approves the new request if enough bandwidth is available. To measure the available bandwidth, various strategies can be adopted, including an explicit signaling from the master station. Since in Skyplex the bandwidth is allocated only when requested, the available bandwidth can be derived by counting the unused slots in the BTP.

Figure 4 shows the beneficial effects of QS in terms of speeding up the connection during the slow start phase. In this case, indeed, the E2ED is reduced by nearly 40% and the delay required in order to complete the startup phase passes from 35 to 25 sec. Though QS is undoubtedly useful, by analyzing carefully the trace during the first 5 sec, we can note that, starting from time 2.5 sec, the E2ED is subjected to a second increasing phase. As explained in the following, this behavior is a consequence of the lack of coordination between QS and DAMA protocols.

When the TFRC source receives a QS-response that notifies the availability of bandwidth resources (about one RTT after the beginning of transmission), it starts sending packets at the application encoding rate (1,024 kb/s), which

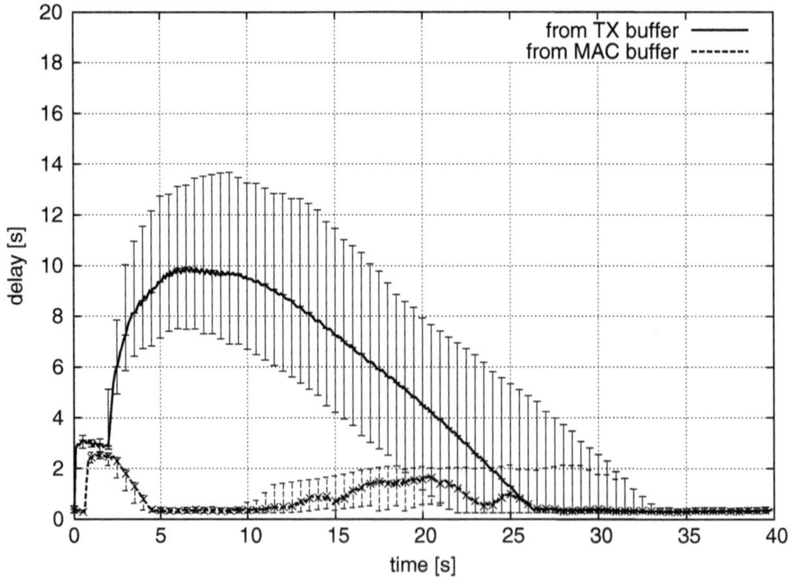

Figure 4. Delay measures of a single TFRC connection with QS

allows the TFRC transmission buffer to be emptied. However, since the bandwidth is not available at that time, the packets are queued into the MAC buffer of the satellite terminal, which causes a sudden increase of the OWD. During the subsequent RTTs, only few packets are delivered to the TFRC receiver through the satellite link and the TFRC receiver measures a low incoming rate. Thus, it generates feedback reports with low, throughput indications, which eventually induce the TFRC sender to decrease its rate.

To overcome this problem preserving the TFRC end-to-end semantic, we can choose an alternative DAMA scheme aware of QS (QSD). Our idea is to modify the DAMA request to take into account the QS requests approved during a prefixed observation period. For a rate-base allocation scheme, this corresponds simply to increment the DAMA request with the sum of QS requests received in the last allocation period. Thus, the necessary bandwidth is demanded in advance and there is room for the new flow when it begins its fast transmission.

The implementation presents a number of issues. Since QS and DAMA may have independent dynamics, the DAMA assignments might be too early or too late with respect to the time the TFRC receives the QS approval, which might invalidate the benefits of QS. Indeed, if the QS response is too early, the bandwidth is not available when TFRC quick-starts, eventually causing the same problems observed in the QS-only case. Conversely, if the QS response is too late or the QS request rejected, the bandwidth allocated is wasted. The lack of time alignment is compensated by forwarding the QS request with a delay of T_q seconds from the satellite terminal when satellite bandwidth demand/grant is slower than QS transaction, or the DAMA request when it is faster. In Skyplex scenario, the long allocation delay (about 1,600 ms) determines a DAMA response always late with respect to the QS response. Thus, in our simulations we hold the QS request packet at the satellite terminal for T_q seconds.

Figure 5 compares the maximum E2ED measured in QS and QSD cases (for different the T_q settings) as a function of application encoding rate. While the improvements achieved by QSD with $T_q = 0$ sec are marginal with respect to QS, setting $T_q = 1.23$ sec allows not only to reduce considerably the initial end-to-end delay but it allows drastically decrease the delay variability, which indicates a more stable flow control. Moreover, the delay is almost independent of the flow rate, which is an appealing feature for grid-computing applications that require high flow bitrates.

Finally, Figure 6 shows the initial maximum E2ED as a function of the T_q. A delay margin of just 0.3 sec is sufficient to trigger Quick Start into the TFRC source. However, the correct selection of T_q parameter is strongly dependent of satellite frame parameters and delaying the request might be unnecessary in a different system (e.g., when the allocation delay is smaller).

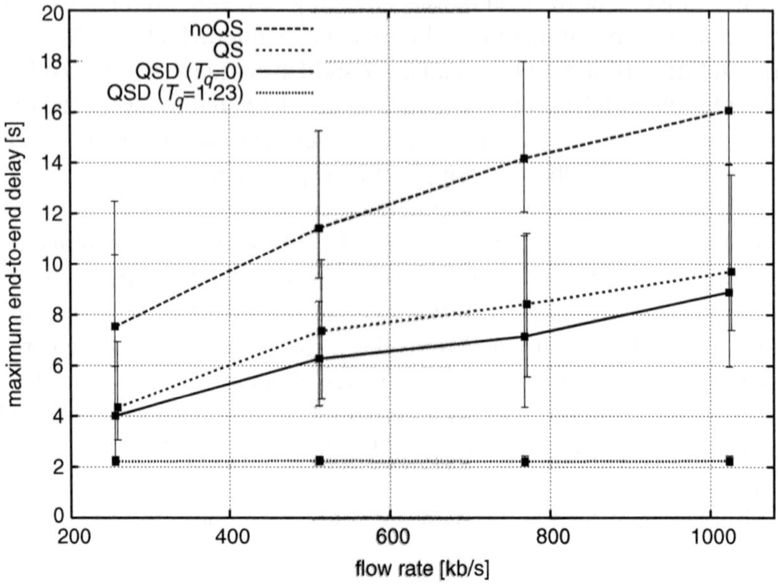

Figure 5. Maximum end to end delay for a single TFRC connection with QS

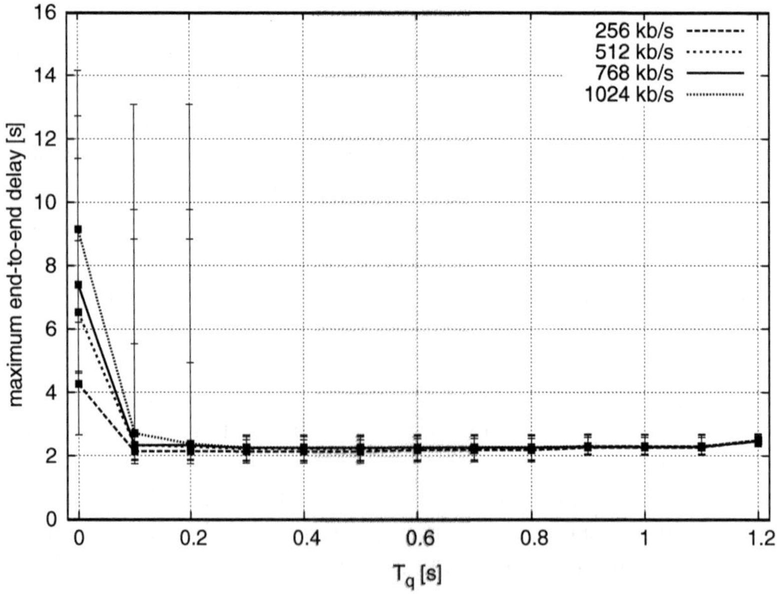

Figure 6. Maximum end-to-end delay as a function of T_q for different flow rates

4. CONCLUSIONS

When a high bit rate stream possibly carrying scientific data is transmitted from a remote site over a grid-network, the satellite links could be the only cost-effective solution. However, if a TCP-friendly protocol is used as congestion control mechanism for such streams, the connection startup phase suffers significant performance degradation in terms of end-to-end delay and delay jitter, especially over DAMA-controlled satellite links.

In this paper, we presented a novel mechanism based on quick-start, a recent IETF proposal, to enhance the performance of multimedia streaming applications at the beginning of connection or after long idle periods. Our simulations reveal that, though QS reduces the initial end-to-end delay of TFRC packets, the performance improvements are marginal, compared with analogous tests over terrestrial links because of the lack of coordination between transport layer and satellite MAC layer. Hence, we introduced a modified DAMA scheme (QSD) to manage the QS request upon its arrival. The new (implicit) cross-layer approach not only allows us to speed up the connection startup but also is able to regulate the size of the transmission buffer. More important, the startup delay is independent of the flow rate, which permits to support the bulk flows required by grid computing applications.

As a future work, we plan to define a criterion to properly select the parameters involved in the QSD algorithm and investigate QSD performance in more complex traffic scenarios.

ACKNOWLEDGMENTS

The authors wish to thank Dr. Francesco Potortí for his valuable contribution to this work.

REFERENCES

[1] Digital Video Broadcasting (DVB); Interaction Channel for Satellite Distribution Systems, ETSI EN 301 790 V1.4.1 - (2005–09).

[2] E. Feltrin, E. Weller, E. Martin, and K. Zamani, "Design, Implementation and Performances of an on Board Processor-Based Satellite Network", in *Proceedings of ICC 2004*, pp. 3321–3325 vol.6.

[3] S. Floyd, M. Handley, and E. Kohler, "Problem Statement for the Datagram Congestion Control Protocol", *RFC 4336*, March 2006.

[4] S. Floyd and E. Kohler, "TCP Friendly Rate Control: the Small-Packet Variant", *draft-ietf-dccp-tfrc-voip-05*, 2006.

[5] A. Gotta, F. Potortì, and R. Secchi, "An Analysis of TCP Startup over an Experimental DVB-RCS Platform", in *Proceedings of IEEE IWSSC 2006*, pp. 176–180, Madrid (Spain), Sept.14–15, 2006.

[6] A. Gotta, R. Secchi, and F. Potortì, "Simulating Dynamic Bandwidth Allocation on Satellite Links", in *Proceedings of Valuetools 2006*, ACM, Pisa (IT), October 2006.

[7] M. Handley, S. Floyd, J. Padhye, and J. Widmer, " TCP Friendly Rate Control (TFRC): Protocol Specification", *RFC 3448*, 2003.

[8] V. Jacobson, "Congestion Avoidance and Control", in *Proceedings of ACM Sigcomm*, pp. 273–288, 1988.

[9] S. McCanne and S. Floyd, Network Simulator. http://www.isi.edu/nsnam/ns/

[10] P. Sarolahti, M. Allman, and S. Floyd, "Determining an Appropriate Sending Rate over an Underutilized Network Path", *Computer Networks*, November 2006.

[11] A. Sathiaseelan and G. Fairhurst, "Using Quickstart to Improve the Performance of TFRC-SP over Satellite Networks" in *proceedings of IEEE IWSSC 2006*, pp. 46–50, Madrid (Spain), Sept.14–15, 2006.

[12] H. Schulzrinne, S. Casner, R. Frederick, and V. Jacobson, *RTP: A Transport Protocol for Real-Time Applications*, RFC 1889, January 1996.

MONITORING SOLUTION FOR OPTICAL GRID ARCHITECTURES

A. Binczewski[1], L. Grzesiak[1], E. Kenny[2], M. Stroinski[1], R. Szuman[1], S. Trocha[1], and J. Weglarz[1]

[1] *Poznan Supercomputing and Networking Center, Noskowskiego 12/14, 61-704 Poznan, Poland,*
[2] *HEAnet, 5 George's Dock, IFSC, Dublin 1, Ireland*

Abstract Today there are very few open source optical network monitoring solutions available for grid users and even network operators. Most of the available optical monitoring solutions are proprietary and single vendor solutions. Even the existing software is often limited in functionality and extremely difficult to apply in a distributed grid environment. grid users can only successfully use optical grid infrastructures spanning over multiple domains if they have efficient monitoring and fault detection tools available to them. In this chapter we present a tool for monitoring wavelengths. Our approach recognizes and facilitates the ability of independent networks to set policies while facilitating the status and performance monitoring by users interested in optical paths connecting distributed resources.

1. INTRODUCTION

Grid [1] is an infrastructure that involves integrated and collaborative use of computers, networks, databases and scientific instruments owned and managed by multiple organizations. grid applications often involve large amounts of data and/or computing resources that require secure resource sharing across organizational boundaries. This makes grid application management and deployment a complex undertaking.

The development of computer networks and the increase in the number of grid environments over the last decade has driven the requirement for new grid monitoring solutions (to guarantee resource availability). The monitoring of network resources dedicated to grid environments is now an essential requirement for grid users, but as the importance and size of grid environments have increased so too have the importance and complexity of monitoring solutions.

Currently there are remarkably few monitoring tools available for monitoring grid networks. In order to overcome the lack of suitable available monitoring tools, the grid community has created some tools which perform some level of monitoring. The following are three grid-related monitoring tools.

The most commonly used grid monitoring solution is the R-GMA (relational grid monitoring architecture) [9] which was developed within the European Data Grid Projects to bring the power of SQL to an information and monitoring system for the grid. It provides producer and consumer services to both publish and retrieve information from anywhere within a grid environment. R-GMA is used for the information system as well as for making details about grid resources available for use by other middleware components. R-GMA monitors grid jobs by members of CMS and D0 (virtual organizations of grid users) collaborations where information about jobs is published from within a job wrapper, transported across the grid by R-GMA and made available to the users.

The next example of grid monitoring solutions is MonALISA [6] (monitoring agents using a large integrated services architecture) which has been developed by Caltech and its partners. The framework is based on dynamic distributed service architecture and is able to provide monitoring, control and global optimization services for grid architectures. Focusing on all the aspects of grid monitoring, MonALISA can provide the following information:

- System information for parts of grid architecture (i.e. computer and clusters)
- IP layer network information (traffic, flows, connectivity, topology) for grid architecture
- Monitoring the performance of applications, jobs or services.

Another commonly used solution for network resources monitoring in a grid environment is designing and developing custom-made tools dedicated to a particular project. This is a very popular solution especially in various R&D projects, e.g. EGEE-II SA2 Activity [3], GridLab – Adaptive Components System (Delphoi) [4] and CLUSTERIX (National CLUSTER of LInuX Systems) [2] but it is also not a very often publicly available software and limited only to the grid architecture chosen in that project.

In the Section 2 we describe the biggest limitations and disadvantages of existing solutions for grid architectures and network resources monitoring.

In Section 3 we present requirements for optical networks monitoring system.

Section 4 includes detailed description of our implementation of the "Lambda Monitor" software which solves the most important disadvantages and limitations of existing grid monitoring solutions.

In Section 5 we describe a set of propositions for further development of the "Lambda Monitor".

The last section (6) includes our summary and conclusions

2. PROBLEMS WITH EXISTING APPROACHES AND TOOLS

The biggest limitation and disadvantage concerning all the above-mentioned solutions for grid architecture monitoring is the lack of publicly available and free (open source) software with the functionality for monitoring of the physical layer (L1) of the grid network resources (especially for optical networks).

All previous solutions monitor only higher layers. The R-GMA monitors mainly grid jobs, the MonALISA system monitors IP network layer for information like traffic, flows, connectivity and logical topology, even the solutions created to meet only the selected project requirements are limited to IP layer monitoring.

Traditionally only large proprietary applications and vendor-specific tools were available for monitoring DWDM networks. Vendor and platform-specific tools are unsuitable for monitoring in a grid environment as multiple vendors and equipment platforms need to be integrated into a monitoring solution. Typically grid monitoring solutions have ignored monitoring of the physical layer of the grid's network resources. As wavelengths become key parts of grid network resources, the requirement to monitor them will become a crucial task. This is a non-trivial problem for optical grids based on dedicated lambdas in the next-generation DWDM network.

In this chapter we propose a solution that overcomes many of these issues associated with the existing tools.

3. REQUIREMENTS FOR OPTICAL NETWORKS MONITORING SYSTEM

Our idea was to create a monitoring system dedicated to optical grid architecture which would be characterized by the following features:

- Capable of monitoring the physical layer of the grid optical network resources.
- A user must be able to view all lambdas that are part of a particular grid network.

- Easy to identify and localize any alarms and faults. Including identifying which part of the optical layer of the grid network is faulty.
- Provide detailed information on the configuration of the grid lambdas and the physical optical characteristics and parameters of each wavelength.
- Speed up and facilitate the work of the grid NOC.
- Open source software tool which could be developed and extended by independent grid/network communities or R&D projects for their own purposes and special requirements.

The next section describes our implementation of this idea which solves the most important disadvantages and limitations mentioned before.

4. IMPLEMENTATION OF THE "LAMBDA MONITOR" SOFTWARE

Our software implementation called "Lambda Monitor" is a monitoring solution which can be used for optical grid architectures. It is this feature that distinguishes Lambda Monitor from other currently available grid monitoring tools. This application makes it possible to monitor many lambda circuits simultaneously and to notify, in real time, of current problems occurring in the network physical layer of the optical grid. By means of this application we are capable of affirming in a quick and exact manner which part of the lambda circuit (exactly which device) causes problems.

This system also provides users with much detailed information about the managed circuits on both lambda and device levels. The Lambda Monitor architecture and operation scheme is presented in Figure 1.

The Lambda Monitor has a special alarms checker module which gathers information about the states of the optical layer of the grid architecture every 5 minutes, and this interval time can be easily changed by the user. This module makes use of the snmpwalk function for checking standard as well as vendor-specific MIB subtrees for determining of alarms in each of the monitored devices which form the grid architecture. If any alarm occurs, the alarms checker module saves this information into the MySQL database. The Lambda Monitor software also uses some secure mechanisms (SSL) for user's authentication which ensures that only registered users can log into the system.

The Lambda Monitor has three levels of user interface (called layers) and additionally current and historical alarms browsers. The first part of the user interface called "Lambdas layer" shows the current status of the monitored lambda circuits (end-to-end). All three possible states of the lambdas along with a short description are presented below:

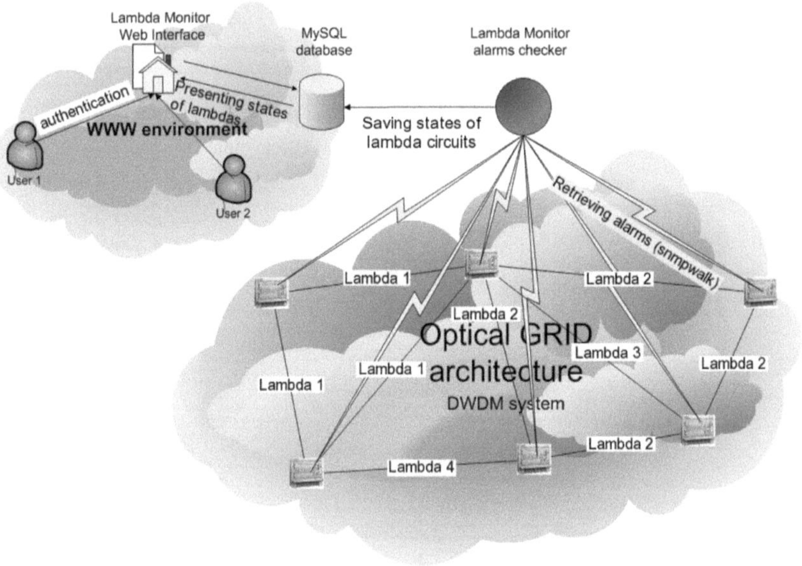

Figure 1. Lambda Monitor architecture and operation scheme

- OK state – lambda circuit works OK without any alarms (Figure 2)
- Warning state – occurs when some warning type alarms are present in the lambda circuit (Figure 3)
- Critical state – occurs when any of critical alarms appeared in lambda circuit (Figure 4)

Figure 5 illustrates an example view of different lambda states in "Lambdas layer" interface.

The second level of the user interface is the "Devices layer" where the lambdas are presented in the device-oriented view. This interface's level shows the

Figure 2. Lambdas layer – OK state

Figure 3. Lambdas layer – warning state

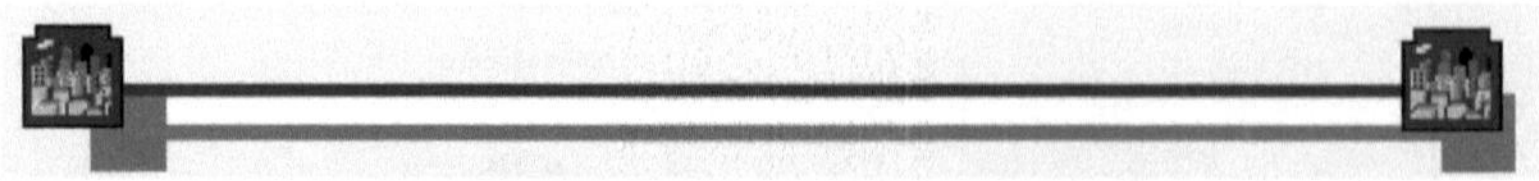

Figure 4. Lambdas layer – critical state

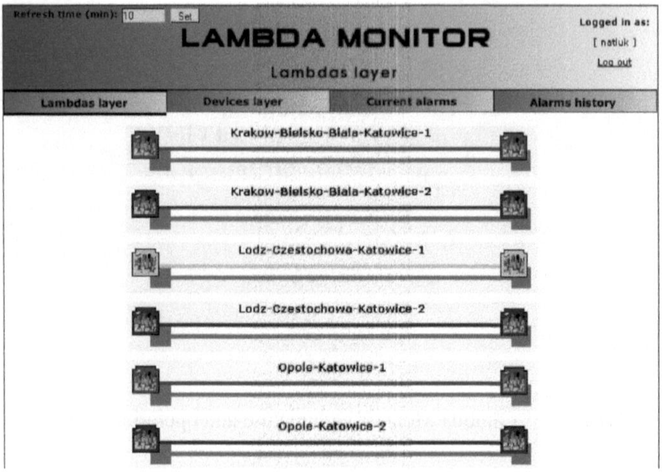

Figure 5. Lambdas layer interface in Lambda Monitor

current status of the DWDM devices terminating selected lambda. The possible states of the devices and their interfaces are shown in Figure 6.

Figure 7 depicts an example view of device interfaces' states in the "Devices layer" interface.

The last and most detailed level of user interface is called the "Parameters layer". This layer presents the current values of various monitored parameters on selected device's interface in the form of graphical tables. In these tables

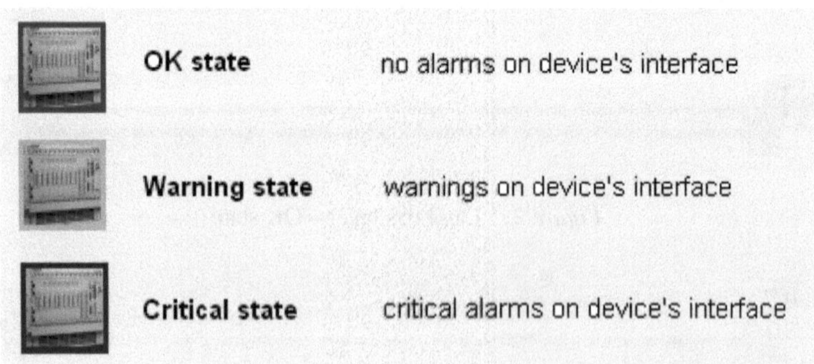

Figure 6. Possible states of the DWDM devices' interfaces

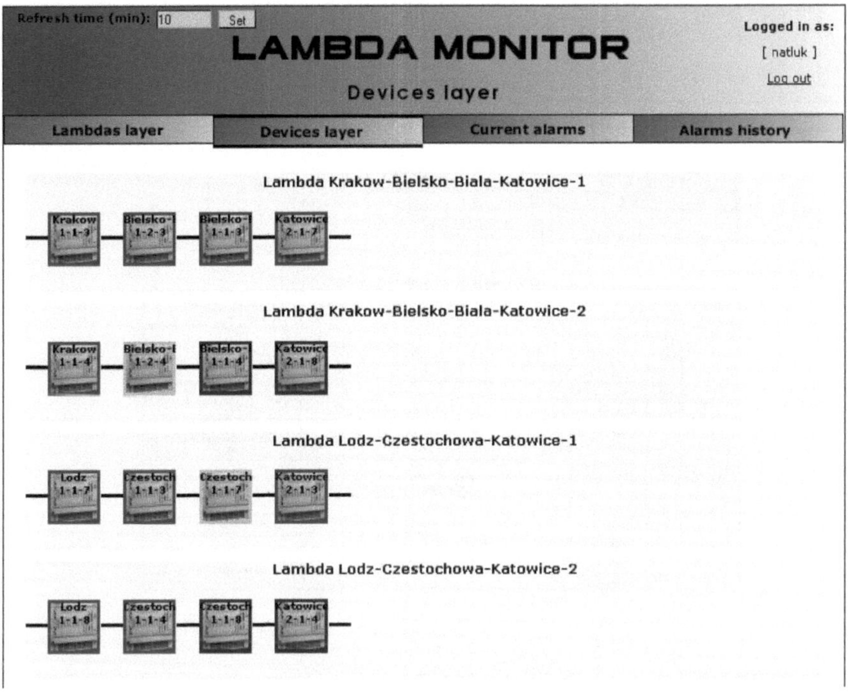

Figure 7. The states of the DWDM devices' interfaces in the "Devices layer"

users can find very detailed information about many parameters which determine and influence the operation of lambdas used by, e.g. the grid infrastructure (Figure 8).

The "Parameters layer" is divided into the following tables:

- Trail Termination – contains parameters for optical interfaces that terminate an optical trail
- TTPLP15min – The Interface Trail Termination (TT) Physical Layer Performance (PLP) 15 Min Interval Table contains statistics collected by an interface over the last 16 intervals (15 minutes each)
- TTPLP24h – The Interface Trail Termination (TT) Physical Layer Performance (PLP) 24 h Interval Table contains statistics collected by an interface over the last 2 intervals (24 hours each one)
- FecCurrent15min – The otuFec Current 15 Minute Table – values from the FEC mechanism
- FeCurrent24h – The otuFec Current 24 Hour Table – values from the FEC mechanism

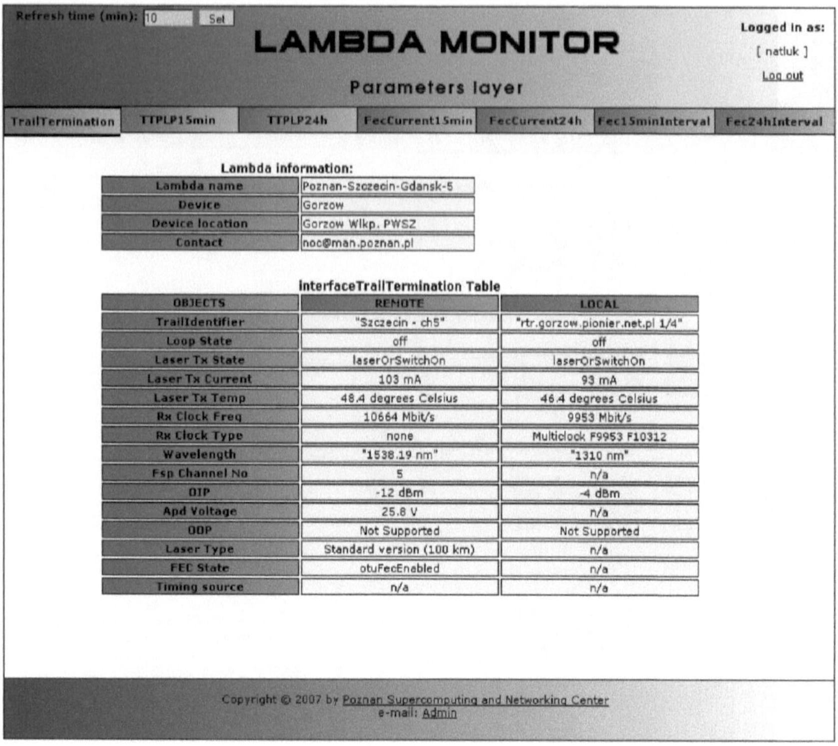

Figure 8. Trail Termination table in the "Parameters layer"

- Fec15minInterval – The otuFec 15 min Interval Table contains statistics collected by the FEC mechanism for an interface over the last 16 intervals

- Fec24hInterval – The OtuFec 24h Interval Table contains statistics collected by the FEC mechanism for an interface over a maximum of the previous 24 hours of operation.

Lambda Monitor also has the "Current alarms" module which enables viewing of information about alarms that occur currently in the monitored lambda circuits (Figure 9).

The "Alarms history" module presents all the alarms that occurred formerly (Figure 10).

The Lambda Monitor system has been successfully deployed in the National Research and Education Network in Poland – Pionier (Polish Optical Internet [8]) as well as in the HEAnet network in Ireland [5]. The architecture of the Pionier network embraces 21 ADVA FSP 3000 devices which form the core of the DWDM system, and HEAnet has a network consisting of 13 ADVA

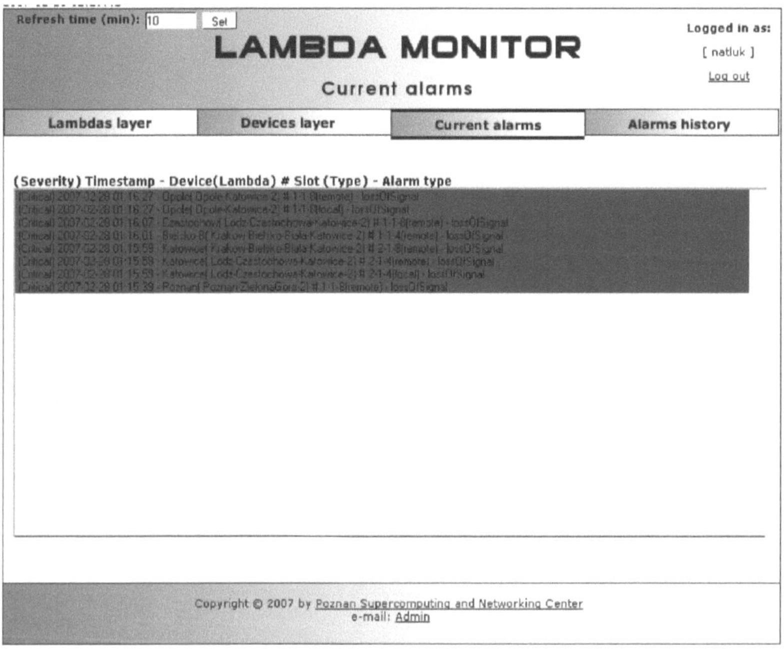

Figure 9. Current alarms browser

FSP 3000 devices. In these two large networks deployments Lambda Monitor proved to be a reliable and suitable physical layer monitoring system for such next-generation networks which can also be used and dedicated to any optical grid architectures.

HEAnet as Ireland national education and research network is closely involved in providing networking resources and services to the Irish grid community as a whole. As part of developing new services, HEAnet is investigating ways to provide grid users with direct access to wavelengths. In order to achieve this, new methods of monitoring optical resources must also be provided to the grid community in order for the grid community to manage their own optical resources. As part of this evaluation, HEAnet has been testing the Lambda Monitor (LM) from PSNC. The Lambda Monitor has been set up to monitor 9 of the 10 Gbit/s point-to-point wavelengths used in the HEAnet national backbone. The Lambda Monitor has a user-friendly web-based front end that provides a visual representation of each wavelength in the network. Each wavelength is colour coded to represent the state of the wavelength: Red (Fault), Yellow (Warning) or Green (Ok). The main advantage of this visual representation of each wavelength is that it provides a simple visual indication to the viewer of the current status of a wavelength without having to understand large numbers of complex alarm messages from numerous DWDM network

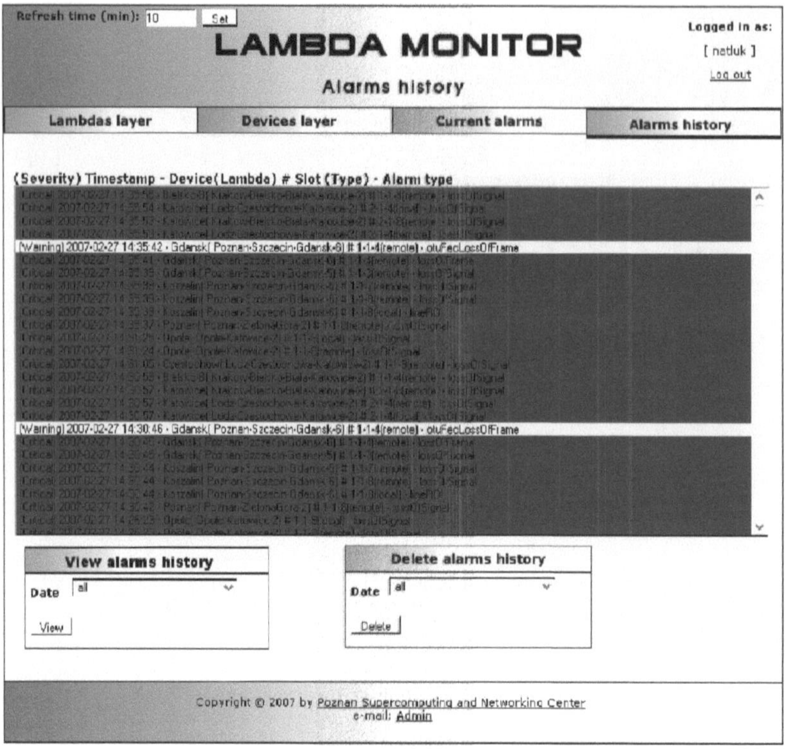

Figure 10. Alarms history browser

elements and line cards. Hence LM is excellent at hiding the complexity of the network from the viewer. In addition LM provides more detailed information on a per wavelength basis if required, by selecting the individual network element in the device layer tab.

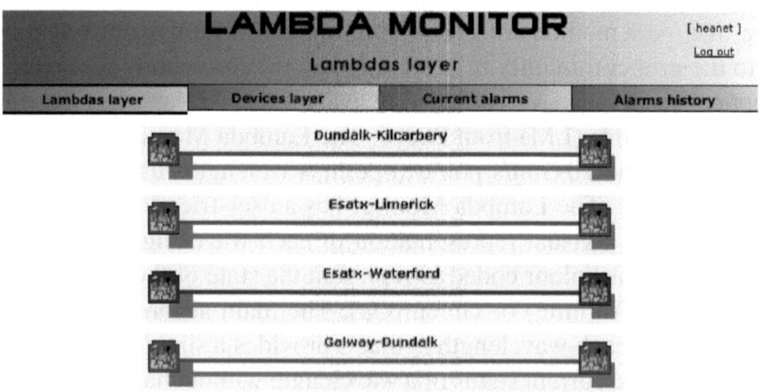

Figure 11. Lambda Monitor interface (HEAnet)

An example of what Lambda Monitor looks like for a subset of the HEAnet-monitored wavelengths is shown in Figure 11.

The green-coloured wavelengths indicate that the operational status of these wavelengths is ok, while the red wavelength indicates that there is a fault with the wavelength Esatx to Waterford.

5. FURTHER DEVELOPMENT OF "LAMBDA MONITOR"

One of the main advantages of Lambda Monitor is its open source license which will enable individual grid users to use and modify Lambda Monitor for their own grid applications. Future developments of Lambda Monitor may include the ability to partition the monitored wavelengths into individual grid user groups that would enable grid users to see only their wavelengths.

Additionally web services interface may be developed for grid applications to directly query the status of grid wavelengths. This web services interface could also be used for integration into the GEANT2 joint research activity perfSONAR. perfSONAR [7] is currently working on an inter-domain end-to-end monitoring solution.

Currently the Lambda Monitor supports only the ADVA DWDM system, but we plan to extend support to other DWDM systems in the near future.

6. CONCLUSIONS

To summarize, we have shown that monitoring of the optical grid architecture is a very important and essential activity which requires the use of the dedicated monitoring solutions. We have covered different types of grid monitoring solutions available on the market along with their short characteristics. The available software for grid monitoring focuses its attention on monitoring of upper layers only (IP and application layers). Unfortunately there is a significant lack of solutions which can monitor the physical (especially optical) layer of the grid network resources, what surely is a considerable limitation and disadvantage. Next, we have outlined the main features of the optical grid monitoring solution of a system which should fill the existing gap. Then we have described in detail the "Lambda Monitor" – our own software implementation consistent with our idea of solving the mentioned problems and tested in various real WAN environment. Our system dedicated to monitor the optical grid architectures exemplifies the grid monitoring solutions needed today as well as in the nearest future. Last but not least, there is an additional paragraph which describes different suggestions for further development of Lambda Monitor software and various visions of its extensions for future needs and requirements which can be done even within independent projects and by communities thanks to its open source nature.

REFERENCES

[1] R. Buyya, S. Venugopal, "A Gentle Introduction to Grid Computing and Technologies", *CSI Communications*, pp. 9–19, Vol. 29, No. 1, ISSN 0970-647X, Computer Society of India (CSI), Mumbai, India, July 2005.

[2] CLUSTERIX website, http://clusterix.pcz.pl

[3] EGEE-II: SA2 Activity, http://egee-sa2.web.cern.ch

[4] GridLab – Adaptive Components System (Delphoi), http://www.gridlab.org/WorkPackages/wp-7/

[5] HEAnet website, http://www.heanet.ie/

[6] MonALISA website, http://monalisa.cacr.caltech.edu

[7] PerfSONAR activity website, http://www.perfsonar.net/

[8] PIONIER website, http://www.pionier.gov.pl/eindex.html

[9] A.J. Wilson, et al., CHEP 2004, Intarlaken, Switzerland, September 2004.

GRIDS AND NETWORKS MONITORING - PRACTICAL APPROACH

J. Jajor[1], C. Mazurek[1], and W. Procyk[1]
Poznan Supercomputing and Networking Center, Noskowskiego 12/14, 61-704 Poznan, Poland

Abstract The article describes a practical approach to the process of designing a monitoring system. Several available tools are discussed with their features, pros, and cons, giving a rough overview of the possibilities and limitations. Then the authors point out a few requirements that are necessary to provide a unified interface tool, which could combine all the tools in a single, consistent web portal, with the support for user authentication and authorization. Several design considerations are discussed in-depth, along with the implementation examples of the tool written at PSNC.

1. INTRODUCTION

In today's network and computer installations, monitoring and accounting play a very important role, helping administrators manage and track system performance and resource usage as well as foresee possible problems.

Poznan Supercomputing and Networking Center (PSNC) is the operator of the National Research and Education Network PIONIER [9] and of the Metropolitan Area Network POZMAN. PSNC is also a High Performance Computing Center and hosts and maintains many different network services, including web portals, audio/video streaming, mail, DNS, and computational services in grid environments.

Heterogeneous service domains to which monitoring is applied in PSNC are corresponding to heterogeneous tools used to monitor particular parameters as well as general systems performance. This diversity is a source of problems for administrators at both application/services and infrastructure level. The problems that we summarize further on became the main reason for working on the interoperability of monitoring domains to build a unified presentation of different monitoring data, e.g., statistics, network weather maps, values that need to be correlated between different sources as well as aggregated on the basis of abstract parameters.

Administrators use different monitoring tools to accomplish their daily tasks. Most of the tools used in PSNC to gather statistics are based on GPL license and are presented further in the next section. All mentioned tools store their data in the common data format RRD (Round Robin Database); however, each tool has its own specific configuration format. Many tools also include some scripts for the presentation of collected data, but presentations concern only data specific to the tool.

RRD data can be presented not only as line charts but also as bar graphs or color arrows on the map (network weather maps). Some more sophisticated tools allow viewing the chart after pointing to a connection of interest represented as an arrow on the map. The problem is that these tools have their own additional configuration, which has to be kept in sync with the configuration of presentation scripts for the creation of the charts and configuration of the monitoring tool.

For quick and accurate diagnosis there is a need for reports that correlate data from different sources and present the results in a readable, easy to understand form. Network administrators, after receiving signals regarding, for instance, problems in network performance, have to check the level of errors, collisions, bandwidth usage (MRTG) and RTT/packet loss (SmokePing). Lost packets can confirm the problem; however, it is not clear whether the loss is caused by errors or by (D)DoS – (Distributed) Denial of Service – protection. Similarly, network services administrators need access not only to the system performance parameters (CPU load, memory and disks usage, I/O stats) but also to the parameters of the underlying network, which is used to connect with the service. Access to correlated reports via a single unified interface reduces the time needed to accurately diagnose the problem.

Another common problem is the presentation of aggregated values, e.g., in the case of trunked (teamed) connections using many interfaces or light wavelengths (lambdas) in parallel.

A very important issue is the authorization of the access to data and reports. Some institutions connected to national or metropolitan networks are interested in reports of their connections. These reports should be available for the institution only, and users of different institutions should not be granted access to these data.

For the above-mentioned reasons and to allow higher level administrators (i.e., application and service level) to eliminate problems at the infrastructure level in their diagnosis tasks, the unified access to monitoring tools became mandatory, to provide a robust and up-to-date interface to data about performance of all infrastructure components.

In the following Sections, the main design issues of the proposed system are discussed (Section 2) followed by a general description of monitoring tools that are being used in PSNC (Section 3). Then, the proposed unified interface solution is presented (Section 4) followed by the description of future work and the conclusions (Section 5).

2. DESIGN ISSUES

2.1 Methodology

The first step in designing the monitoring system is determining the scope and requirements of the project. The answers to the following questions should be collected:

- What is the purpose of monitoring the systems?
- Which parameters should be measured, what is the source of the data, and how can the data be obtained?
- How frequently should the data be collected and how should the collection process be scheduled?
- How will the performance data be analyzed?
- How should the data be visualized and reported, and who will be viewing the data?
- How long should the data record be kept?
- What resources are available?

A real-time monitoring system should be built using several components: data collection utilities, a data storage system, and a data presentation system. To reduce the cost of creating the system before writing one's own tools to capture and analyze data, existing tools and packages should be considered. In all cases, in order to use the tools efficiently, it is important to know how data is collected, what types of data are collected, and how to use the data to keep the system at its best. Naturally, the method used to monitor certain parameters or collect data should be as non-invasive as possible.

Because of the general aim of statistics gathering, we can single out three domains of monitoring, presented below. These three domains have much in common; for example, accessibility can be seen as response time below a certain level, covered by the Service Level Agreement. Finding how these domains cover our needs plays a key role in the selection of monitoring tools.

2.2 Performance Monitoring

Monitoring performance is a necessary part of preventive maintenance for the computer system. Through monitoring it is possible to obtain performance data that are useful in diagnosing system problems and in planning for the growth in demand for system resources. Routine performance monitoring starts with establishing a default set of counters to track. By using the performance monitoring tools for real-time observation and periodic logging, it is possible to acquire important information about the health of the system. In addition, while troubleshooting system problems, performance data give some information about the behavior of system resources at the time the problem occurs, which is useful in pinpointing the cause. Finally, the analysis of the trends in system performance provides data to plan the future growth and to estimate how changes in the system configurations might affect future operation.

Generally, performance monitoring concentrates on the operating system resource usage (i.e., system load, processor and memory usage, disk throughput, network bandwidth), as well as any applications or services that are capable of performance data collection (for example, the number of web server users, number of streaming server listeners, etc.).

The data source for performance monitoring is the monitored system or application itself. Data can be fetched through some well-defined interfaces. Most commonly used interfaces are the SNMP protocol (for both network devices and operating systems monitoring), SE Toolkit (for Solaris operating system), ProcFS file system (for Linux operating systems), Performance Counters (for some Microsoft family operating systems), or other system utilities (for example, vmstat for Unix-family operating systems).

2.3 Monitoring Versus Accounting

While monitoring as previously described is generally perceived as real-time and on-line data collection, viewing, and reporting, system accounting is the methodology and set of tools that provide methods of collecting per-process or per-user resource usage data and charging fees to specified users or organizations. Even if no real fee charging is used, it is important to know which users are consuming which resources for both cost allocation and capacity planning purposes. Accounting concentrates on presenting accumulated or average resource usage (depending on the resource type). For "consumable" resources, like power, processor time, an accumulated value over a specified period of time can be presented, while for "non-consumable" resources like disk space, accounting, and fee charging can be based on average resource usage level or on assigned resource quota. Usage accounting is even more challenging in the case of large systems because of the larger quantities of usage data that are captured and must be managed. Accounting data collection is tightly related to the monitored application or in case of system resource usage, to the system kernel.

2.4 Availability Monitoring

Another type of monitoring methodology is availability monitoring. In this case tools are oriented toward checking whether the host and its services are working correctly and are capable of servicing user requests. Different approaches can be adopted based on the monitoring targets. The most common solution for host monitoring is using the ICMP Echo Request (so-called ping) packets. Another solution mostly applied to network devices is querying with SNMP protocol requests. This solution can also provide means for monitoring applications and other system components with the use of SNMP agents. However, a typical solution for application testing is to write custom tools, which imitate the application protocol (for example, send HTTP and interpret the response). Such tools serve as a data collection source, returning arbitrary constants as service status.

The topic of availability monitoring slightly overlaps performance monitoring in areas where apart from the sole application response the response time should also be measured. In such case the response time can also be treated as data source and recorded along with the service state.

3. COMMONLY USED TOOLS

3.1 RRDTool

The most popular format for storing data is the Round Robin Database (RRD) format [10]. RRD provides a flexible binary format for the storage of numerical data measured over time. RRD files consist of a few data series, archives, and aggregation methods. Every data series accept real values as an input. Using aggregation and built-in archives it can cover quite a long period in parallel with a small file size, and automatic multi-level aggregation (e.g., automatic computing of daily, monthly, yearly archives). The size of the file remains the same during the data collection process, so there is no risk concerning storage overfill. Because of the database internal structure, access to the data stored in RRD files is extremely fast.

The tools provided to manage RRD archives also offer very good support for charts creation; however, there is no way to access parameters of a created chart. It is easy to embed the created chart on a webpage, but any attempt to use the *onMouseOver* event over the chart or legend area to display additional information is really a hard task because the coordinates of the image areas where the legend or plot is placed are unknown. It is possible to combine data from different files, compare them, create mathematical formulas, and make nice-looking charts with a colorful legend.

Despite all its features, the RRD format also has some limitations. RRD databases are binary files, so they cannot be shared between systems with different data architectures. In some cases, it means the data have to be stored and visualized on the same machine. Moreover, RRD files do not provide any mechanism to store metadata. Thus it is not possible to store any description of the data series or information on events happening during data collection.

Another problem, which needs to be very carefully considered, is the usage of different units and scales for different data series. An example of such situation is RTT and *PacketLoss* presentation. It can be done using a certain tricky way (like the *SmokePing* does).

3.2 MRTG

MRTG [6] stands for Multi Router Traffic Grapher. It is one of the oldest examples of GPL software to monitor traffic on network devices. MRTG is devoted to log and visualize traffic, but it can be customized to collect any pair of values with a common unit and scale. Similarly, in basic configuration it is just enough to point the interface by its *ifIndex* number, description, name, IP address or MAC address, and MRTG polls for *ifInOctets* and *ifOutOctets*. It can also query an arbitrary oID, combine it with another oID using the supplied mathematical formula, or even fetch data from an external script.

Data can be stored in two exclusive formats: human readable log or binary RRD format. The human readable log is a native data format. Assuming that the data are viewed on the same station it is collected, no web servers are necessary to present the data, and even if it is necessary to publish the results no support for dynamic technologies like php or cgi is required. Among others, it provides better security. By its simplicity, this format is laden with some disadvantages. Every time a new data sample is collected, the whole log file has to be rewritten and each webpage and image have to be generated. It can cause huge I/O load. On the contrary, the RRD format is much more efficient and concise. It has great support in different visualization scripts.

Besides the choice of data format, there are also other decisions to be taken. As an example from the network monitoring – what to do when a network node is reconnected from one network device to another? Should the historical volume traffic data be moved along with the node to its new attachment point or should the data be duplicated on both interfaces: the old one and the new one? In other words: which relation should be really maintained:

- data file to network device interface,
- data file to IP address,
- data file to function/destination.

Nowadays MRTG has some successors like Cricket or Cacti but still remains very popular software.

3.3 Smokeping

Smokeping [11] is a tool for Round Trip Time and packet loss monitoring. It supports RRD and has a built-in script for data visualization.

The design of the tool allows to configure different probes, so it can be used not only for RTT timings and packet loss but it can also monitor HTTP, DNS, and SSH availability. It is also possible to define parameters of probing and patterns for alerts, making *Smokeping* a very flexible tool.

Two series of data collected by *Smokeping* are visualized on the same chart with a clever use of line color and line level to represent two data domains on a single graph. The Y-axis is the response time and the color of the line indicates host or service availability. In fact this is done in a very tricky way. Every section of the line is a part of the bar, and the chart creation is a two-phase process. In the first phase the script gets the resolution of the whole image and min/max values to calculate the thickness of the line; on the second phase the real job of drawing grayscale bars (representing jitter) and color lines (representing packet loss) is done.

3.4 Netflow Tools

Netflow is a proprietary protocol developed by Cisco Systems for collecting IP-layer traffic information. However proprietary, the protocol specification has been published in an open form and, starting from version 9, has been published as RFC3954. The protocol is based on a definition of the flow. Cisco proposed to use the common 5-attribute tuple definition (see [4]). Routers should output a record when they determine that the flow is finished. Flow expiration is based on flow aging counters and TCP session termination. Routers can also be configured to output a record at a specified time interval, regardless of the fact that the flow is still ongoing. Further versions of NetFlow added more fields to the record definition.

Flow accounting can be computationally expensive and, as such, unacceptable for many applications. To avoid problems concerning router CPU exhaustion, Cisco proposed a "*Sampled NetFlow*" solution. Rather than checking each packed processed by the router, only every n-th packet is accounted. The n value can be a configured constant (in case of a Deterministic NetFlow) or be randomly selected (in case of Random Sampled NetFlow).

While there are numerous projects developed to collect and present data provided by a NetFlow-enabled device, nfdump [1] and NfSen [2] proved to be the most powerful and flexible.

The nfdump project is a set of tools to collect and process netflow data. The goal of the design is to be able to analyze netflow data from the past, as well as to track interesting traffic patterns continuously. All data are stored to a disk before analyzing. This separates the process of storing and analyzing the data. The amount of time back in the past is limited only by the disk space available for all the netflow data.

NfSen provides a graphical web-based front end for nfdump tools. It allows display netflow data organized as flows, packets, or bytes using RRDs. It is possible to easily navigate through netflow data and define filter profiles and process the netflow data within specified time range. One of the most powerful features of NfSen is the ability to write and use additional plug-ins to process and store collected data (backend plug-ins) and display various views of the data (frontend plug-ins).

3.5 Orca

Orca [8] is a Perl script designed to plot arbitrary data from text files to a directory on a Web server. It can plot the same type of data from different files into different or the same images, group multiple columns into a single plot using regular expressions on the column titles, plot the results of arbitrary Perl expressions, including mathematical formulas.

To build a complete monitoring system one needs to separately run a tool that collects the data. Several such tools have been developed for Orca, especially for monitoring server performance. For Solaris there is, for example, a SEToolkit with an appropriate *orcallator.se* program, while for Linux one can use another Perl script known as *procallator.pl*. With some changes, Orca can be also used to process MS Windows Performance Monitor logs.

Configuration for Orca is stored in a file. The directories specified in the configuration are being watched for new files, and all found files are monitored for updates. Source data are read from white space separated files. The last modification time for each file is stored so they do not have to be reread continuously. The timestamp of each data record is also parsed from the data file. The data read for the files are passed through an arbitrary number of consolidation functions (as defined in Orca configuration) and are permanently stored in RRD databases.

The Web interface for Orca is a collection of HTML files and PNG plots created by Orca. Among these HTML pages there is also an index of URL links listing all available targets, as well as an index of URL links listing all different plot types (allowing to easily compare the same parameter of different monitoring targets). In this type of interface no separate CGI set up is required. However, one of the drawbacks is the need to recreate all plot images in addition to updating all the RRDs after each data read.

3.6 Nagios

Nagios [7] is an open source host, service, and network monitoring program. It watches hosts and services that are defined in the configuration, alerting the administrator of arising problems. Some of Nagios include monitoring of hosts and network services, monitoring host resources, network topology awareness, simple plug-in APIs, and monitoring redundancy.

All hosts and services data, as well as network topology, are defined in text files. The main process – Nagios daemon – is responsible for reading and parsing configuration data. The work of Nagios is based on the scheduled execution of the so-called plug-ins, which can be any binary executables or script files. The scheduling queue is maintained by the Nagios daemon, while the timing between subsequent plug-in executions is taken from the configuration file. The value returned by the plug-in determines the actual service or host state as seen by Nagios. As defined in the plug-in API, the returned value defines the service (OK, WARNING, CRITICAL, UNKNOWN) or host state (DOWN, UP, UNREACHABLE). Host or service administrators can be notified about any state changes, and some actions (including restarting the service) can be automatically taken by Nagios itself. Another interesting feature of Nagios is the possibility to easily create a redundant and distributed monitoring setup.

All features presented (which are only a fragment of full Nagios functionality) make it a very powerful monitoring tool; yet some steps could be taken to slightly improve the Nagios web interface.

4. UNIFIED MONITORING INTERFACE

4.1 Project Motivation

The tools described in the previous chapter are very well designed and do their job very well, but to work with them all efficiently one needs to combine them under a single consistent user interface. Each of the tools provides its own set of HTML pages or CGI scripts to present the content; in many cases there is usually no support for user authentication and authorization with the possibility to limit the contents to some subset of pages or graphs making it suitable only for internal purposes. A service provider has no possibility to easily provide access to distinct subsets of graphs for each of the clients. These problems created the need to design another tool, which could integrate the tools described and would be flexible enough to be able to join other existing and future tools.

The early unification scenario, which the project started with, was to give the administrators the systems' overview, for example to build a map to present the bandwidth usage on each of the links of the Polish Research and Education Network PIONIER. Searching for existing tools did not give any immediate answers. First of all there is a very limited GPL licensed number of so-called weathermaps. Among them there are only few supporting RRD and all of them in a very limited manner – for example, only MRTG's RRDs. Even accepting all the limitations it is required to maintain at least two configurations and keep them in sync: the configuration for monitoring/chart creation scripts (e.g., MRTG) and the configuration for the weathermap. After a few iterations of corrections and additions to the chosen tool it became obvious that the system should be redesigned and rewritten to provide the required functionality and flexibility.

The two main issues, which had to be considered while designing a unified monitoring interface are data flow model and security and access control. Both issues are presented below.

4.2 Data Flow

To integrate all mentioned tools within a single monitoring system the data flow model presented in Figure 1 was proposed.

Each tool requires its own storage for the most recent monitoring data. Data format is at this point arbitrary for a specific tool – it can be an RRD database (e.g., for MRTG or Smokeping) or a plain text file (e.g., for Orcallator, PerfMon). As mentioned before, the main data format for the monitoring system is the Round Robin Database. Because RRD files are binary data, they are not directly accessible between different architectures. While designing this model, these architectural concerns had to be taken into account. Therefore an intermediate data storage in plain text or XML files is used at some stage of data collection to support integration between 32-bit and 64-bit, as well as big-endian and little-endian architectures. This is the reason for placing appropriate parsers, which translate the intermediate format and populate local Round Robin Databases with collected data.

If the data on the server participating in the monitoring infrastructure are in compatible RRD format, then it can be accessed directly by means of a

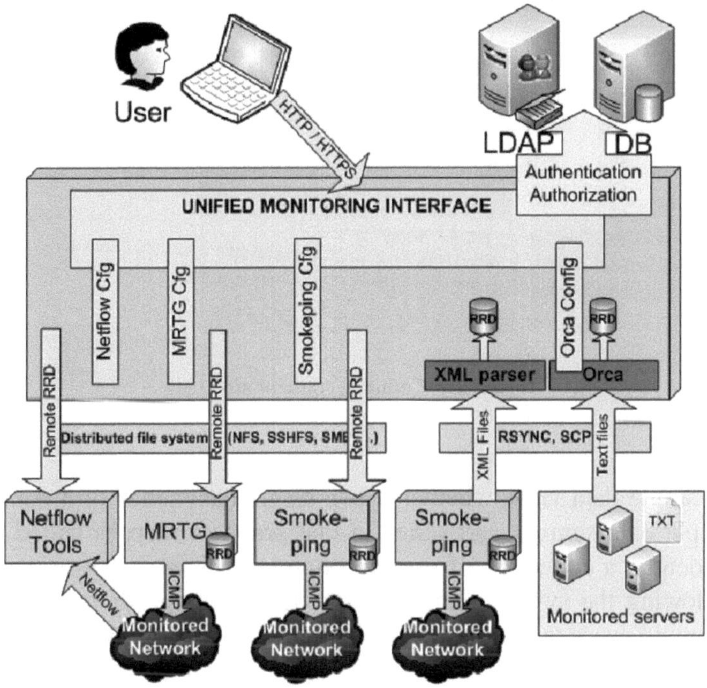

Figure 1. System data flow

distributed file system (such as NFS, CIFS, SSHFS, or similar). In case of RRD files created on binary incompatible architectures and text files, the data is replicated to the monitoring server by copying and synchronization with SCP, RCP or rsync, and additionally processed by an appropriate parser (e.g., XML Import, Orca) and imported to local RRD files.

4.3 Security and Access Control

Another important concern was security and access control. The system was designed to provide fine-grained access control to all presented data. This required a unified configuration view layer over all tools combined by the system. A common configuration access is accomplished by providing specific plug-ins for each format used by monitoring tools (see Figure 2).

Besides the access control, a common configuration layer over all integrated tools is required to combine data from different sources into a single data view, e.g., one map or one graph. This gives the possibility to show, for example, web server response time together with allocated network bandwidth or packet loss.

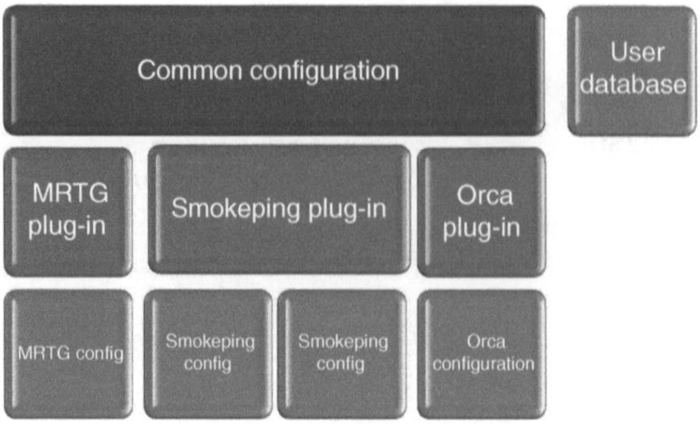

Figure 2. Common configuration idea

Each single data view, whether it is a single data series or a combination of multiple data sources and series in one graph or map, is called a target. Independently, a collection of user accounts and hierarchy of user groups is used, allowing the system's administrator to assign specific permissions for each group to every target. Also some system roles can be assigned to groups, allowing users to perform certain tasks – e.g., create new objects or update data in the configuration, administer user accounts and so on.

For all configuration data (monitoring and access control) a separate database is used. User accounts can be also stored in the database; however, the system integrates with any LDAP compatible directory server (e.g., Sun Directory Server, or Microsoft Active Directory) to provide enterprise-wide unified user account database.

4.4 Implementation and Deployment

The created software has a very wide functionality and seems to remain flexible and customizable. While the present paper is being written, two graphical modules are implemented: charts and a map. The charts are capable of displaying MRTG graphs, Smokeping RTT and Packet Loss graphs, Netflow packet/flow/traffic graphs, Netflow graphs aggregated by the AS number, and graphs from users' custom RRDs. The map module can display a map with a configuration-supplied background, nodes, arrows, lines, pie-charts, whose appearance can be based on values computed from the RRDs or taken from the Nagios status files. A single map also supports multiple legends for specified groups of objects. Each object on the map supports properties to provide actions performed on mouse over and on mouse click, allowing the implementation of multi-level submaps and pop-up windows that display the details of the selected object.

For the implementation of the system, the PHP5 programming language was chosen because of its popularity on many web servers, as well as its multi-functionality. The interface is designed and implemented in the object model, providing clear dependencies and well-defined interfaces. It can be easily expanded by users with knowledge of the PHP5 programming language, to support other monitoring tools, different data stores, and several authentication and authorization providers.

The unified monitoring interface is deployed at Poznan Supercomputing and Networking Center, where it is used with MRTG, Smokeping, Orca and Nagios to monitor the National Research and Education Network PIONIER [3], Poznan Metropolitan Area Network POZMAN, PSNC's local area network, and servers hosted at PSNC. It provides a front-end for a system, which monitors bandwidth usage on about 50 network devices with the total of 1,500 network interfaces/subinterfaces in PIONIER, 80 devices with the total of 2,000 network interfaces/subinterfaces in POZMAN, and 1,000 network interfaces in the LAN. Smokeping is used to monitor about 400 network nodes in PIONIER and POZMAN from three different locations. There are about 30 performance parameters monitored on each of the 20 SUN/Solaris servers, and 20 Intel/Linux servers providing structure services, database access and hosting environment for PSNC, educational society in Poland, and other users. The system monitors also the infrastructure of the Content Delivery System for Polish Interactive Television (iTVP) [5], where it gathers network statistics from several network devices with the total of about 500 network interfaces, as well as performance data from 50 multimedia streaming and servers. Additionally, the system interfaces with a netflow probe, which gathers data from about 80% of PIONIER's core routers. The statistics are gathered on six monitoring hosts, with MRTG and smokeping running on three hosts, orca running on two hosts and one host collecting netflow data. The distribution is caused by different accessibility of different parts of the network and gets a wider view of systems' performance.

5. CONCLUSIONS

The architecture of the presented tool is extensible, allowing many future additions to cope with many tasks. The software is used in the administrators' community of Poznan Supercomputing and Networking Center. The work is still in progress, and the tool is under constant development. Current implementation tasks aim to provide authentication and authorization with the use of text files, extend the map component to be able to draw multiple parallel links between two nodes of a grid or a network in a more readable way.

Monitoring is a must in today's computer systems and networks, but before starting there are aims to be established and questions to be answered.

The mentioned tools are flexible enough to support a wide variety of tasks and cover different environments of computer systems. The created common user interface allows easily integrate data from different monitoring tools and preserve their flexibility.

REFERENCES

[1] http://nfdump.sourceforge.net/
[2] http://nfsen.sourceforge.net/
[3] http://weathermap.pionier.gov.pl/
[4] http://www.faqs.org/rfcs/rfc3954.html
[5] http://www.itvp.pl/
[6] http://www.mrtg.org/
[7] http://www.nagios.org/
[8] http://www.orcaware.com/orca/
[9] http://www.pionier.gov.pl/
[10] http://www.rrdtool.org/
[11] http://www.smokeping.org/

ANALYZING AND OPTIMIZING THE LINUX NETWORKING STACK

R. Bolla[1], R. Bruschi[1], A. Ranieri[1], and G. Traverso[2]

[1]*Department of Communications, Computer and Systems Science (DIST), Via Opera Pia 13, 16145 Genova, Italy*

[2]*InfoCom Genova, Piazza Alessi 2, 16128 Genova, Italy*

Abstract It is well known that the Linux Operating System provides a complete and effective support to networking functionalities, to be widely used in many advanced fields, like software routers and Grid architectures, among others. Notwithstanding the rising popularity and the continuous development of the Linux Operating System, there is no clear indication about the maximum networking performance level that it can achieve and no information about the computational weight required by packet processing operations. Our main aim in this work is to analyze and to optimize the Linux networking stack performance, trying to evaluate the impact of networking functionalities on the whole system. To this purpose, to better characterize the performance bottlenecks, we have performed both external and internal measurements, benchmarking both a standard Linux kernel and an optimized version.

1. INTRODUCTION

Today, many open source operating systems (OSs), like Linux and FreeBSD, include networking capabilities so sophisticated and complete that they are often employed in all such cases where an intensive and critical network usage is expected. In fact, owing to the flexibility of the open source approach, these OSs are becoming more and more widespread in many application fields, as, for example, in network firewalls, web servers, software routers and Grid architectures.

Notwithstanding their arising popularity and their continuous development, there is no clear indication about the maximum performance level that these architectures can achieve and about the computational weight that they introduce. In fact, the knowledge of both maximum packet transmission and reception rates, as well as of the related computational load, can help to better dimension all such systems with intensive network usage or to design new task allocators for Grid clusters that take into account the performance limits and the resources needed by networking operations. In more detail, networking operations generally have a high CPU-intensive nature, and they need to be carefully taken into account, especially in a Grid environment. In fact, in such case, if one decides to move some processing tasks to remote machines, one should take into account how many networking and CPU resources are spent to transfer the input data. In other words, when moving one or more tasks to remote computers, a non-negligible computational overhead is introduced.

With this idea in mind, we have decided to analyze and to optimize the Linux networking stack performance, trying to evaluate the impact of networking functionalities on the system. Linux can boast of one of the most complete networking stacks (i.e., networking source code is about 40% of the whole kernel) and it offers support to a wide range of hardware components. Besides, we have to underline that Linux is still deeply inclined to the networking usage, but it is a general purpose architecture, and so it presents a large number of aspects that can be tuned or optimized to enhance its networking performance. Moreover, to simplify the working environment, we have decided to take only layer 2 and 3 functionalities into account, since in every respect they are the backbone of the overall software architecture. Thus, all the benchmarking tests reported here have been carried out by forwarding IPv4 datagrams (i.e., reception and transmission) through a single Gigabit Ethernet interface. To this purpose, in addition to the "classical" external performance measurements (i.e., latency and throughput), we have decided to perform also internal measurements, with the objective of better understanding the software structures and dynamics included in the Linux networking code. These internal measurements have been carried out with both an advanced tool called Oprofile and some kernel patches, proposed by us, to dump the values of some critical Linux internal variables.

The chapter is organized as follows. The state of art is briefly surveyed in the next section, while Section 3 describes the Linux networking architecture, and some parameter tuning and kernel patches that we have used to obtain the maximum performance. Section 4 describes the external and internal performance evaluation tools used in the tests, while Section 5 reports the numerical results of all the most relevant experiments. The conclusions and future activities are presented in Section 6.

2. RELATED WORK

Some interesting works regarding the Linux networking stack can be found in the scientific literature. Reference [1] reports some performance results (in packet transmission and reception) obtained with a PC Linux-based test bed. Some evaluations have been realized also on the network boards, see for example [8]. Other interesting works regarding Linux-based routers can be found in [2, 3], where Bianco et al. report some interesting performance results. This work tries to give a contribution to the investigation by reporting the results of a large activity of optimization and testing realized on a router architecture based on Linux software.

In our previous works [4–6], carried out inside the BORABORA project [7], we have focused our attention on testing, optimizing and evaluating the performance of the basic packet forwarding functionalities. To this purpose, besides classical external (throughput and latency) measurements, we have also adopted profiling tools to analyze in depth and to optimize the internal behavior of Linux. In [5], we have presented a survey on how the presence of control plane functionalities in software routers can impact the forwarding performance.

3. THE LINUX NETWORKING ARCHITECTURE

As outlined in [4], while all the packet processing functions are realized inside the Linux kernel, the large part of daemons/applications requiring network functionalities run in user mode. Thus, we have to outline that, unlike most of the high-end commercial network equipment (e.g., IP routers), the packet processing functionalities and the control ones have to share the CPUs in the system. Reference [5] reports a detailed description of how the resource sharing between the applications' plane and the packet process can have effect on the overall performance in different kernel configurations (e.g., SMP kernel, single processor kernel).

The critical element for the network functionalities is the kernel, where all the link, network and transport layer operations are realized. During the last years, the networking support integrated in the Linux kernel has experienced many structural and refining developments. For these reasons, we have chosen to use a last generation Linux kernel, more specifically a 2.6 version.

Since the older kernel versions, the Linux networking architecture is fundamentally based on an interrupt mechanism: network boards signal the kernel upon packet reception or transmission, through HW interrupts. Each HW interrupt is served as soon as possible by a handling routine, which suspends the operations currently processed by the CPU. Until completed, the runtime cannot be interrupted by anything, even by other interrupt handlers. Thus, with the clear purpose to make the system reactive, the interrupt handlers are designed to be very short, while all the time-consuming tasks are performed by the so-called "Software Interrupts" (SoftIRQ) for a second time. This is the well-known "top half–bottom half" IRQ routine division implemented in the Linux kernel [13].

SoftIRQs are actually a form of kernel activity that can be scheduled for later execution rather than real interrupts. They differ from HW IRQs mainly in that a SoftIRQ is scheduled for execution by an activity of the kernel, like for example a HW IRQ routine, and has to wait until it is called by the scheduler. SoftIRQs can be interrupted only by HW IRQ routines.

The "NET_TX_SOFTIRQ" and the "NET_RX_SOFTIRQ" are two of the most important SoftIRQs in the Linux kernel and the backbone of the whole networking architecture, since they are designed to manage the packet transmission and reception operations, respectively. In details, the forwarding process is triggered by a HW IRQ generated by a network device, which signals the reception or the transmission of packets. Then the corresponding routine makes some fast checks, and schedules the correct SoftIRQ, which is activated by the kernel scheduler as soon as possible. When the SoftIRQ is finally executed, it performs all the packet layer 2 and 3 forwarding operations.

Thus, the packet forwarding process is fundamentally composed by a chain of three different modules: a "reception API", which handles the packet reception (NAPI), a module that carries out the IP layer processing (both in reception and in transmission) and, finally, a "transmission API" that manages the forwarding operations to egress network interfaces. In particular, the reception and the transmission APIs are the lowest level modules, and are composed by both HW IRQ routines and SoftIRQs. They work by managing the network interfaces and performing some layer 2 functionalities.

More in details, the NAPI [10] was introduced in the 2.4.27 kernel version, and it has been explicitly created to increase the packet reception process scalability. It handles network interface requests with an interrupt moderation mechanism, which allows to adaptively switch from a classical interrupt management of the network interfaces to a polling one. This is done by inserting, during the HW IRQ routine, the identifier of the board generating the IRQ into a special list, called "poll list", by scheduling a reception SoftIRQ and by disabling the HW IRQs for that device. When the SoftIRQ is activated, the kernel polls all the devices, whose identifier is included in the poll list, and a maximum of quota packets are served per device. If the buffer (RxRing) of a device is emptied, then its identifier is removed from the poll list and its HW IRQs re-enabled; otherwise, its HW IRQs are left disabled, the identifier kept in the poll list and a further SoftIRQ scheduled.

While this mechanism behaves like a pure interrupt mechanism in the presence of low ingress rate (i.e., we have more or less a HW IRQ per packet), when traffic raises, the probability to empty RxRings, and so to re-enable HW IRQs, decreases more and more, and the NAPI starts working like a polling mechanism.

For each packet received during the NAPI processing, a descriptor, called skbuff, is immediately allocated and used for all the layer 2 and 3 operations.

A packet is processed in the same NET_RX SoftIRQ, till it is enqueued in an egress device buffer, called Qdisc. Each time a NET_TX SoftIRQ is activated or a new packet is enqueued, the Qdisc buffer is served. When a packet is dequeued from the Qdisc buffer, it is placed on the Tx Ring of the egress device. After the board transmits one or more packets successfully, it generates a HW IRQ. Then, during a SoftIRQ, all the descriptors of transmitted packets are deallocated and the Tx Ring is refilled by new packets coming from the Qdisc buffer.

3.1 Performance Tuning

The whole networking kernel architecture is quite complex, and it has many aspects and parameters that should be refined or tuned for a system optimization. For what concerns the parameter tuning, we have used the same configuration adopted in [4]. The 2.6.16.13 kernel images include two structural patches that we have produced with the aim of testing and/or optimizing the kernel functionalities. In particular, the used patches are described in the following.

3.1.1 Skbuff Recycling Patch

We studied and realized a new version of the skbuff recycling patch, originally proposed by R. Olsson [12] for the "e1000" driver. In particular, the new version is stabilized for 2.6.16.13 kernel version and extended to the "sundance" driver.

This patch allows to intercept the skbuff descriptors of sent packets before their deallocation and to re-use them for new incoming packets. As shown in [6], this architectural change heavily reduces the computation weight of the memory management operations, and it allows to achieve a very high-performance level (i.e., about 150–175% of the maximum throughput of standard kernels).

3.1.2 Performance Counter Patch

To study the Linux internal behavior in depth, we have decided to introduce a set of counters in the kernel source code, with the aim of understanding how many times a certain procedure is called, or how many packets are kept per time. In detail, we have introduced the following counters:

- IRQ: number of interrupt handlers generated by a network card
- tx/rx IRQ: number of tx/rx IRQ routines per device
- tx/rx SoftIRQ: number of tx/rx software IRQ routines

- Qdiscrun and Qdiscpkt: number of times when the output buffer (Qdisc) is served and number of served packets per time.
- Pollrun and Pollpkt: number of times when the RxRing of a device is served and the number of served packets per time.
- tx/rx clean: number of times when the tx/rx procedures of the driver are activated.

The values of all these parameters have been mapped in the Linux "proc" file system.

4. TESTBED AND MEASUREMENT TOOLS

For what concerns the traffic generation and measurement, we have used a professional equipment, namely the Agilent N2X Router Tester [11], which allows obtaining throughput and latency measurements with very high availability and accuracy levels (i.e., the minimum guaranteed timestamp resolution is 10 ns).

The internal measurements have been carried out by using, besides the counter patch described in the previous section, a specific SW tool (called profiler) placed inside the Linux box. This software tool can trace the percentage of CPU utilization for each SW module running on the node. The problem is that many of these profilers require a relevant computational effort that perturbs the system performance. We have experimentally verified that one of the best is Oprofile [9], an open source tool that realizes a continuous monitoring of system dynamics with a frequent and quite regular sampling of CPU HW registers. Oprofile allows the effective evaluation of the CPU utilization of both each SW application and each single kernel function with a very low computational overhead.

5. NUMERICAL RESULTS

To benchmark the performance of the Linux networking stack, we have selected a new generation Linux kernel, a 2.6.16.13 version, and used it with two hardware architectures. The presence of more hardware architectures gives us the possibility to better understand which performance bottlenecks can depend on the selected hardware and how, and to evaluate how networking performance scales according to hardware capabilities.

With this idea we have selected two hardware architectures, identified as Gelso and Magnolia, respectively, which include fast I/O buses (i.e., PCI-X and PCI Express) and fast CPUs:

- Gelso: based on SuperMicro X5DL8-GG mainboard, equipped with a PCI-X bus and with a 32-bit 2.4 GHz Intel Xeon
- Magnolia: based on a SuperMicro X7DBE mainboard, equipped with both the PCI Express and PCI-X buses, and with a 5050 Intel Xeon, which is a dual core 64-bit processor.

In particular, Magnolia can be considered as the "evolution" of Gelso, which does not include 64-bit CPUs and PCI Express buses.

About the network interface cards (NIC), we have used the Intel PRO 1000 XT server adapters for the PCI-X bus and the Intel PRO 1000 PT dual port adapters for the PCI Express.

With such hardware architectures, we have benchmarked the maximum performance of the Linux networking stack in a very simple environment: a single data flow that is forwarded at IPv4 layer. The data flow has a constant bit rate, with variable packet size, and it is received and transmitted by a single Gigabit Ethernet interface (a test bed with two interfaces showed comparable results), similar to a server with very high network usage. The absence of "user-level" processing is clearly justified by the fact that kernel space operations (i.e., HW and SoftIRQ handlers) are the highest priority activities in a Linux system, and, as shown in [5], all this can cause heavy performance deteriorations to "user plane" processes.

Observing Figure 1, which shows the maximum forwarding rate according to different packet sizes, we can note how all the software/hardware architectures can achieve almost the maximum theoretical rate only for packet sizes equal to or larger than 256 bytes. This kind of bottleneck can be better

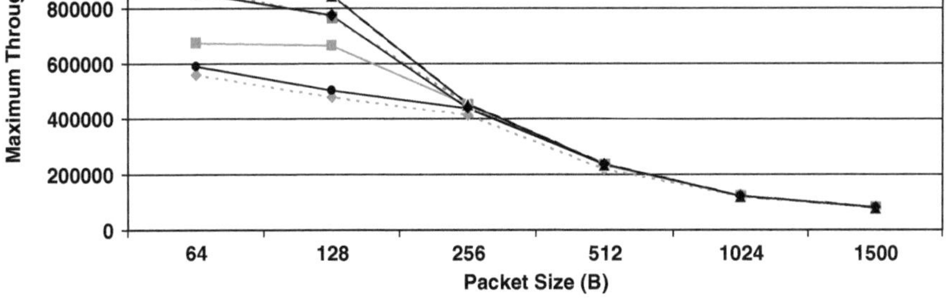

Figure 1. Maximum forwarding rates according to different packet sizes (layer 2 sizes)

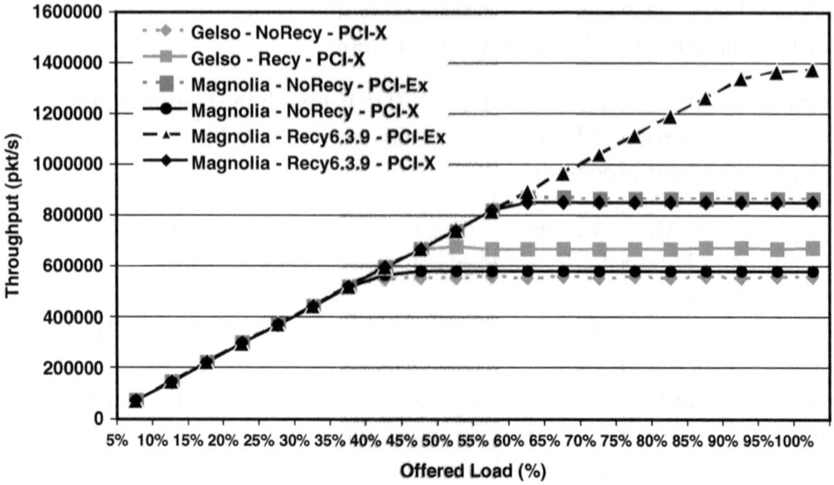

Figure 2. Forwarding rate according to different traffic load values with 64-byte-sized packets. The traffic load percentage is calculated on the Ethernet Gigabit speed

appreciated in Figure 2, where the throughputs versus packet ingress rates are reported in most critical case of 64-byte packet size, while Figure 3 shows the corresponding latency values.

From all these figures, we can note how the Magnolia hardware architecture provides higher performance with respect to the Gelso one. Under similar

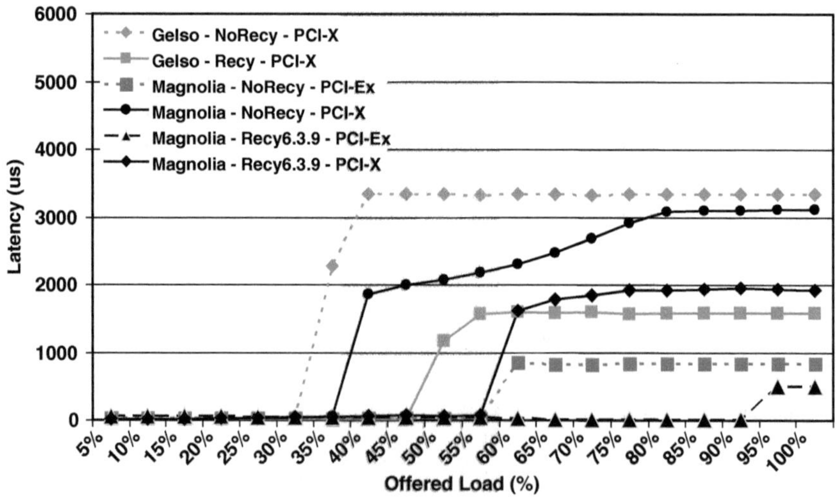

Figure 3. End-to-end latency values according to different traffic loads with 64-byte-sized packets. The traffic load percentage is calculated on the Ethernet Gigabit speed

conditions, if the NIC is placed on the PCI-X bus and the same standard kernel version is used, Magnolia achieves only nearly 20 kpkt/s more than Gelso. But, when the skb recycling patch is applied, the performance difference becomes much greater: while Magnolia can forward up to 830 kpkt/s, Gelso reaches a bit less than 650 kpkt/s.

The use of the PCI Express bus enhances considerably the performance of both the standard and the optimized kernels to 840 and 1390 kpkt/s, respectively.

This rise is due to the efficiency of the PCI Express bus. In fact, it allows DMA transfers with very low latencies and with a very low control overhead, which probably lead to less heavy accesses to the RAM, with a consequent benefit for the CPU memory accesses. Thus, in other words, this performance enhancement is substantially due to a more effective memory access of the CPU caused by the features of PCI Express DMA.

This is also evident by observing the latency values shown in Figure 3, where the curves exhibit the classical "step" shape (i.e., the step position coincides with the saturation point, after which the buffers fill up): forwarding through a PCI Express bus allows to halve the latency at saturation with respect to the PCI-X tests.

Moreover, since packet latency at saturation mostly depends on the average packet processing time, a lower latency is somehow an index of a lower computational weight of the forwarding process. In fact, the skb recycling patch permits to further lower the packet latency with respect to the standard kernel, too.

Figures 4 and 5 show the internal measurements obtained with Oprofile, with Gelso and with both standard and optimized kernels. All profiled kernel

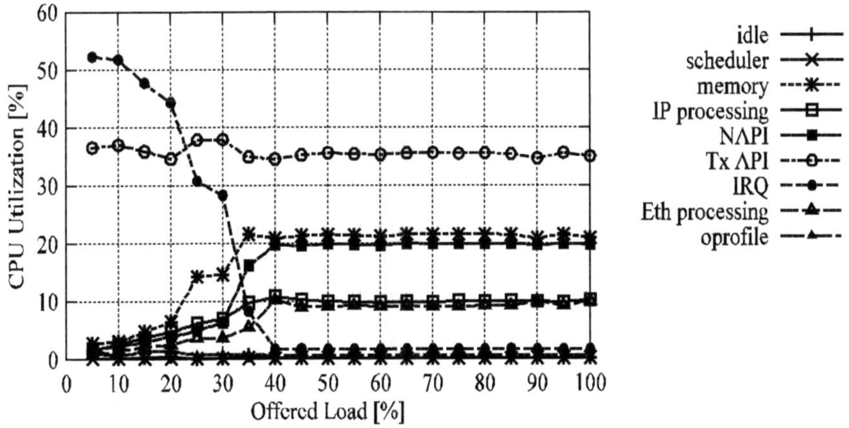

Figure 4. CPU utilization of the kernel networking functionalities according to different traffic offered load (64-byte sized packets) for Gelso with the standard kernel

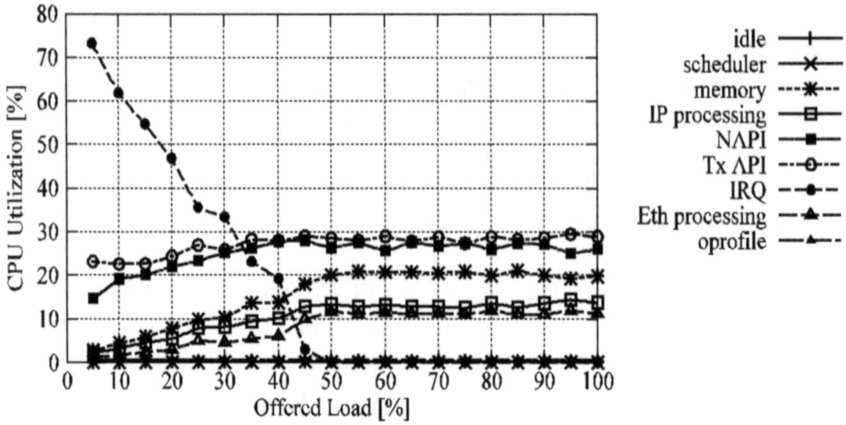

Figure 5. CPU utilization of the kernel networking functionalities according to different traffic offered load (64-byte sized packets) for Gelso with the skbuff recycling patch

functions have been grouped into different sets that represent the most important activities. The sets that we have selected are "idle", (OS) "scheduler", "memory", "IP processing", "NAPI", "TxAPI", "IRQ", "Ethernet processing" and "oprofile", respectively.

Comparing Figures 4 and 5, we can note how memory management functions rise more slowly in the optimized kernel case, till they achieve saturation (35% of the full Gigabit load for the standard kernel and about 50% for the patched one).

For what concerns the other functionalities, their behavior is clearly related to the number of forwarded packets: the weight of almost all classes raises linearly up to the saturation point, and after it remains more or less constant.

On the contrary, the IRQ handlers show a particular behavior, caused by the NAPI paradigm: when the traffic load is low, their computational weight is high, since the Rx API works like an interrupt mechanism; for higher loads, it starts to lower more and more. When the IRQ weight becomes zero, the kernel reaches the saturation point and works according to a polling mechanism.

Figure 6 reports the Oprofile results in the 1500-byte packet case. With respect to the 64 byte packets, the CPU utilization of IRQ handlers remains quite high for a wide range of traffic offered load (up to 80–90%). This is because in the presence of such packet size, the kernel has to process a lower number of packets per second, and NAPI begins to behave as a polling mechanism only for higher rates. Moreover, when NAPI works as an interrupt mechanism (e.g., one packet at each IRQ), the CPU idle time is maintained over 10%; when the packet rate rises and the forwarding process requires more computational resources, CPU idle time lowers to 0%. All the operations performed per

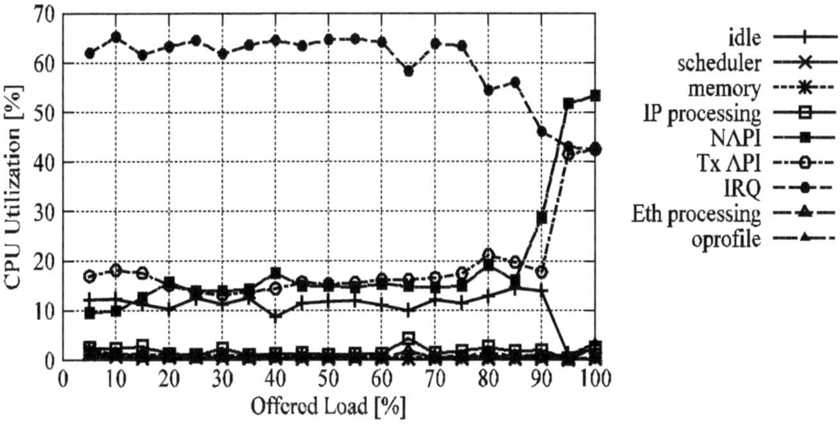

Figure 6. CPU utilization of the kernel networking functionalities according to different traffic offered load (1500-byte sized packets) for Gelso with the skbuff recycling patch

packet (IP and Ethernet Processing) are obviously lower than in the 64 byte packet case.

This is also confirmed by the performance indexes reported in Figures 6 and 7, where we can view both the tx and rx boards reducing their IRQ generation rates, while the kernel passes from polling the rx ring twice per received packet, to about 0.22 times. Also the number of Rx SoftIRQ per received packet decreases as offered traffic load raises. Moreover, as shown in [5], such a high IRQ rate at low traffic offered loads can negatively impact on user plane processes. Since IRQ handlers are originated by PCI boards, and they can interrupt any other kernel/user activities without being interrupted by anything,

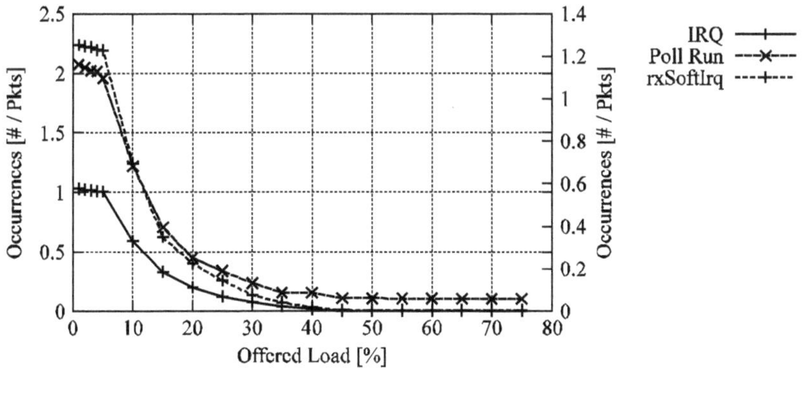

Figure 7. Number of IRQ routines, of polls and of Rx SoftIRQ (*second y*-axis) for the RX board for the skbuff recycling patched kernel, in the presence of an incoming traffic flow with 64-byte-sized packets

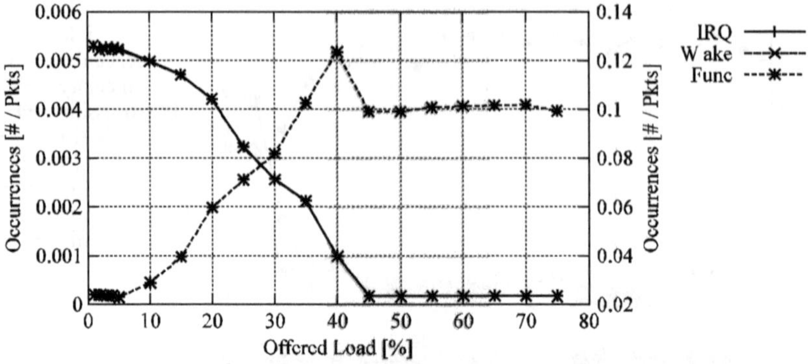

Figure 8. Number of IRQ routines for the TX board, for Tx Ring clearing by TxSoftIRQ ("func") and by RxSoftIRQ ("wake") for the skbuff recycling patched kernel, in the presence of an incoming traffic flow with 64-byte-sized packets. The *second* y-axis refers to "wake"

a high number of interrupts can result in a computational overhead not controlled by the OS, with a consequent performance deterioration of any other software activities.

For what concerns the transmission dynamics, Figure 8 shows very low function occurrences: the Tx IRQ routines lower their occurrences up to the saturation, while the "wake" function, which represents the number of times that the Tx Ring is cleared and the Qdisc buffer is served during a Rx SoftIRQ, shows a mirror behavior: this is because when the kernel reaches the saturation, all the tx functionalities are activated when the Rx SoftIRQ starts.

6. CONCLUSIONS

The main contribution of this work has been reporting the results of a deep activity of optimization and testing realized on the Linux networking stack, and, more in particular, on layers 2 and 3. The main objective has been the performance evaluation (with respect to IPv4 packet forwarding) of both a standard kernel and its optimized version on two different hardware architectures. The benchmarking has been carried out with both external (i.e., throughput and latency) and internal (i.e., kernel profiling and internal counters) measurements. The benchmark results show that, while for large packet sizes almost all the hardware architectures (with high bandwidth I/O busses) can achieve the full Gigabit speed, for small-sized packets the maximum performance level generally decays, since there is a bottleneck on the packet processing rate. However, using some kernel enhancements (i.e., skb recycling patch) and PCI Express bus together, the maximum throughput with minimum-sized packets can achieve about 1390 kpkt/s (nearly 95% of full Gigabit speed). Besides this considerable performance rise, the profiling results show that packet receiving and transmitting processes generally consume a non-negligible share of CPU time.

REFERENCES

[1] A. Barczyk, A. Carbone, J.P. Dufey, D. Galli, B. Jost, U. Marconi, N. Neufeld, G. Peco, V. Vagnoni, "Reliability of datagram transmission on Gigabit Ethernet at full link load", *LHCb Technical Note*, LHCB 2004-030 DAQ, Mar. 2004.

[2] A. Bianco, R. Birke, D. Bolognesi, J. M. Finochietto, G. Galante, M. Mellia, M.L.N.P.P. Prashant, F. Neri, "Click vs. Linux: two efficient open-source IP network stacks for software routers", *HPSR 2005 (IEEE Workshop on High Performance Switching and Routing)*, Hong Kong, May 12–14, 2005

[3] A. Bianco, J. M. Finochietto, G. Galante, M. Mellia, F. Neri, "Open-source PC-based software routers: a viable approach to high-performance packet switching", *Third International Workshop on QoS in Multiservice IP Networks*, Catania, Italy, Feb. 2005

[4] R. Bolla, R. Bruschi, "A high-end Linux based open router for IP QoS networks: tuning and performance analysis with internal (profiling) and external measurement tools of the packet forwarding capabilities", *Proc. of the 3rd International Workshop on Internet Performance, Simulation, Monitoring and Measurements (IPS MoMe 2005)*, Warsaw, Poland, Mar. 2005, pp. 203–214.

[5] R. Bolla, R. Bruschi, "IP forwarding performance analysis in presence of control plane functionalities in a PC-based open router", *Proc. of the 2005 Tyrrhenian International Workshop on Digital Communications (TIWDC 2005)*, Sorrento, Italy, Jun. 2005; in F. Davoli, S. Palazzo, S. Zappatore, Eds., "*Distributed Cooperative Laboratories: Networking, Instrumentation, and Measurements*", Springer, New York, NY, 2006, pp. 143–158.

[6] R. Bolla, R. Bruschi, "RFC 2544 performance evaluation and internal measurements for a Linux based open router", *Proc. of IEEE 2006 Workshop on High Performance Switching and Routing (HPSR 2006)*, Poznan, Poland, Jun. 2006, pp. 9–14.

[7] Building Open Router Architectures based on Router Aggregation (BORA BORA) project, website at http://www.telematica.polito.it/projects/borabora/

[8] P. Gray, A. Betz, "Performance evaluation of copper-based Gigabit Ethernet interfaces", *27th Annual IEEE Conference on Local Computer Networks (LCN'02)*, Tampa, FL, Nov. 2002, pp. 679–690.

[9] Oprofile, website at http://oprofile.sourceforge.net/news/

[10] J. H. Salim, R. Olsson, A. Kuznetsov, "Beyond Softnet", *Proc. of the 5th annual Linux Showcase & Conference*, Oakland, CA, USA, Nov. 2001.

[11] The Agilent N2X Router Tester, http://advanced.comms.agilent.com/n2x/products/ index.htm

[12] Skb recycling patch, ftp://robur.slu.se/pub/Linux/net-development/skb_recycling/

[13] K. Wehrle, F. Pählke, H. Ritter, D. Müller, M. Bechler, "*The Linux Networking Architecture: Design and Implementation of Network Protocols in the Linux Kernel*", Pearson Prentice Hall, Upper Saddle River, NJ , USA, 2004.

Part III

Instrumentation and Measurement

A GRID ARCHITECTURE FOR EFFICIENT MEASUREMENT OF ELECTRICAL DISTURBANCES

A. Carullo[1], M. Parvis[1] and A. Vallan[1]

[1] *Dipartimento di Elettronica, Politecnico di Torino, corso Duca degli Abruzzi, 24 - 10129 Torino (Italy)*

alessio.carullo@polito.it

Abstract This chapter deals with the extension of the grid paradigm to distributed measuring systems. The main problems that arise when distributed measuring systems are configured as grid are discussed and a solution is also proposed in order to integrate measuring nodes with low processing capabilities in a grid architecture. The problem that is mainly discussed in the chapter is that related to the configuration of the grid that allows good performance to be obtained in terms of processing time. An example application is developed, consisting of a system for measuring conducted disturbances of electrical equipment, which is based on a parallel processing algorithm. A grid architecture has been deployed in order to implement the proposed algorithm and a specific service has been developed, whose task is the choice of the processing strategy that minimizes the total processing time. Experimental results are also provided that refer to a real scenario.

1. INTRODUCTION

The extension of the grid paradigm to distributed measuring systems has enhanced both performance and usability of such systems.

At a very basic level, a distributed measuring system can be organized as a grid architecture, where the measuring units act as the grid nodes, thus allowing a cooperative measuring process. Unfortunately, such a simple idea not always can be employed due to the specific nature of the available measuring units.

In fact, to be able to act as a grid node, a measuring unit has to exploit a non-negligible processing capability, which in several cases is not available.

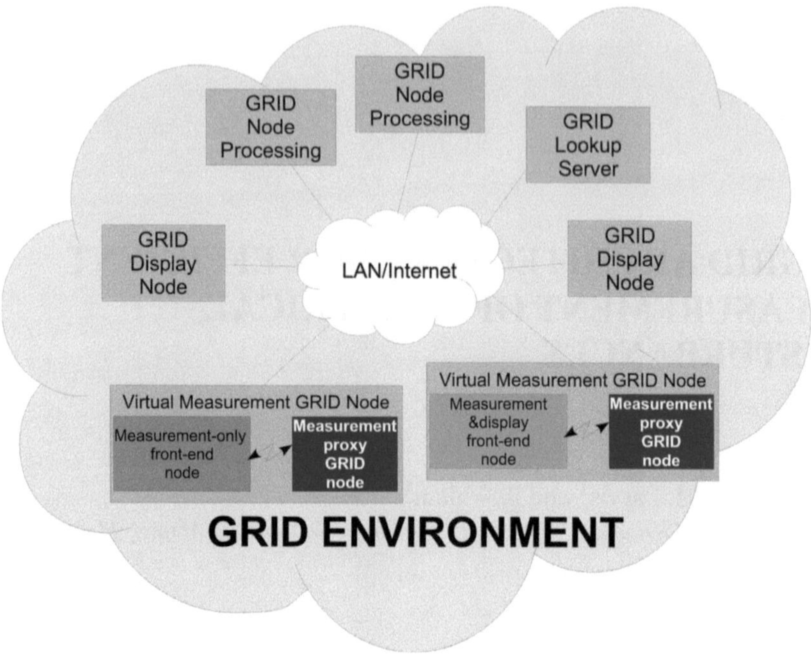

Figure 1. The extension of the grid paradigm to a distributed measuring system

However, in such cases, a proxy-based solution can be adopted by employing a conventional computer connected to the measuring device in order to implement the full grid requirements. A node with such characteristics will be referred to as Virtual Measurement grid Node. Figure 1 shows an example of mixed grid with normal and virtual nodes and with processing-only nodes without measuring capabilities, as well as with display nodes, which are intended only for output tasks.

The application of the grid paradigm to distributed measuring processes opens new problems that are not faced in a common measuring system. Among the different problems, three can be recalled that are of primary importance for grid measurement systems:

- The system metrological management
- The effects on the system performance of the data transport among different nodes
- The difficulty in configuring the grid to efficiently distribute the acquisition and processing load among the nodes

The authors have already investigated the first two problems and some solutions have been proposed: the metrological management of a grid measuring system can be performed by employing a grid of traveling standards [2], while the data transport among the nodes can be improved if suitable algorithms are employed to compress the data. From this point of view, the authors have shown that commercial compression algorithms are not suitable for measurement data, since high compression rates can only be obtained if a considerable data degradation can be accepted, that results in a high added uncertainty. For this reason, a specifically designed compression algorithm has been proposed that gives good results in terms of compression rate while assuring a specified uncertainty threshold not to be exceeded [1].

The last problem is still open and this chapter explores different solutions to achieve an efficient implementation of a measurement grid using, as a clarifying example, the setup that is required for the measurement of conducted disturbances. This example has been chosen because it refers to an application that can experience great advantages from a grid implementation, since the measuring process is intrinsically distributed, the required data processing is rather heavy and the numerical algorithm can be parallelized. In addition, since the processing employs a huge amount of data, the example may lead to different architectures, depending on the speed of the network that is used to connect the nodes, and thus the example allows one to explore different grid aspects. In the following sections, after a brief discussion of the measurement problem, three different solutions are presented that are suitable in different scenarios.

2. MEASUREMENT PROCESS

The considered problem is the measurement of conducted disturbances that an electrical equipment sends toward the mains through its power cord. This is one of the test that is required to assess the electromagnetic compatibility of an electrical equipment. Specific international standards dictates how these tests have to be performed and the maximum emission levels. The CISPR-16 standard [6] is the reference document for the measurement of conducted disturbances, which recommends the arrangement of the setup that is shown in Figure 2: a line impedance stabilization network (LISN) is employed to separate the mains signal from the wide-band disturbance signal produced by the equipment under test (EUT), while the receiver, which is a specially designed spectrum analyzer, gives the frequency contents of such a disturbance.
The CISPR-16 standard establishes

- The maximum emission levels in the B band (0.15–30 MHz)
- The receiver detector characteristics (quasi-peak detector with charge and discharge time equal respectively to 1 and 160 ms);

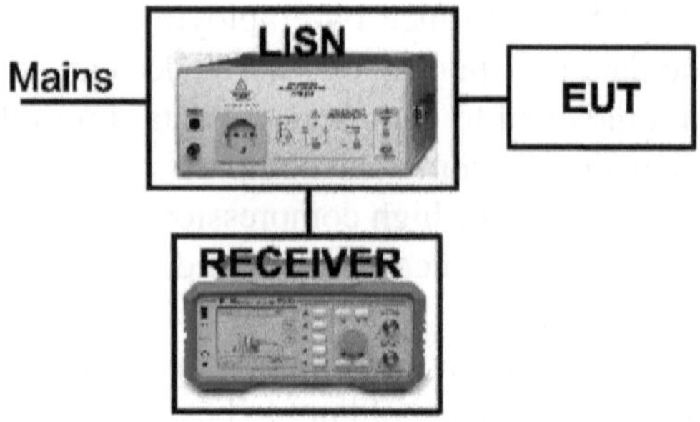

Figure 2. The measurement setup recommended in the CISPR-16 standard for conducted disturbance measurements

- The receiver intermediate filter characteristics (9 kHz bandwidth at the −6 dB points)

With the aim to avoid the use of very expensive spectrum analyzers, alternative solutions have been investigated that allow pre-compliance tests to be performed. Such solutions are commonly based on a digitizer and on numerical data processing [3, 5], which employs a numerical algorithm that simulates the behavior of both the receiver intermediate filter and the peak detector.

These numerical approaches are effective, but they require a large amount of data to be processed. Actually, the disturbance spectrum has to be analyzed up to 30 MHz, thus requiring a sampling rate greater than 100 MHz and the digitizer has to acquire more than one period of the main voltage signal. As an example, if the disturbance signal is acquired for five periods of the European mains frequency (50 Hz), about 100 ms have to be acquired at a sampling rate of 100 MHz, thus resulting in 10 MSamples that have to be stored and processed. The processing of such a large amount of data with numerical filters is a heavy process that could require intolerable times when a single personal computer (PC) is employed.

2.1 Parallel Version of the Measurement Process

The numerical algorithm that has to be implemented to simulate the receiver can be executed in a parallel way, so that several PCs can be employed to process the data: this allows the required processing time to be drastically reduced.

In order to fully parallelize the numerical algorithm that simulates the setup recommended in the CISPR-16 standard, about 6000 band-pass filters have to be implemented to estimate the spectrum amplitude of the disturbance signal in the range of 150 kHz to 30 MHz with the required frequency resolution. Such an analysis is time consuming, because of the large amount of samples that have to be processed by means of the 6000 selective filters. Fortunately, the

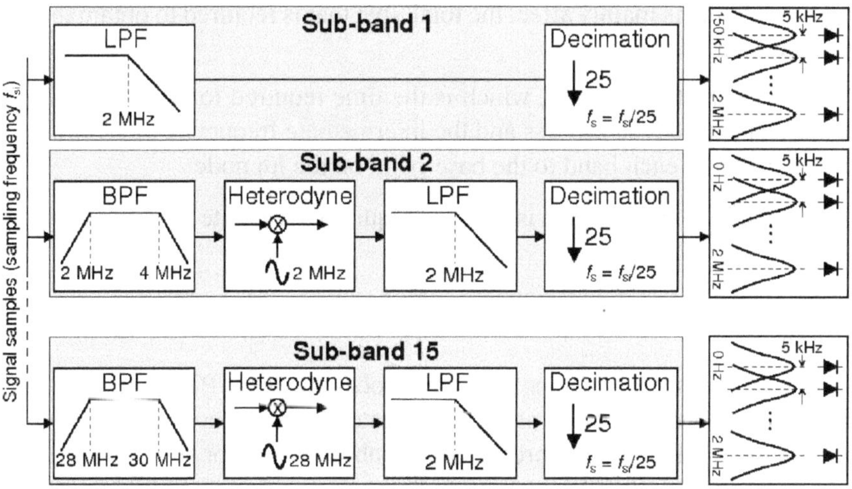

Figure 3. The proposed algorithm that allows the processing to be executed in a parallel way by reducing the number of samples to be processed

processing can be parallelized in a flexible way, as summarized in Figure 3. The acquired signal, sampled at the sampling frequency f_{si}, is first divided into 15 sub-bands 2 MHz wide, then each sub-band is translated in the base-band. The signal related to each sub-band is decimated by a factor 25, i.e., it is re-sampled at the frequency $f_s = f_{si}/25$: this way, each band-pass filter bank has to process a number of samples that is reduced by a factor 25 with respect to the initial number. Furthermore, this solution allows identical filter banks for each sub-band to be employed.

The proposed algorithm allows one to explore at least three different processing strategies:

- **Sequential**: the whole algorithm is sequentially implemented on a single PC
- **Partially distributed**: the processing that allows the signal to be divided into sub-bands is performed on a PC, then the obtained signals, which are decimated by a factor of 25, are transferred to other PCs where the band-pass filters are implemented in parallel
- **Fully distributed**: the raw signal is sent to different PCs, each one extracting the sub-band and then computing a set of filters

Apart from the first strategy, which neither requires nor takes advantage from a grid environment, the choice of which of the other two strategies allows a faster processing to be obtained depends on the number of available processing nodes and on the data transfer rate between the nodes.

Three elements mainly affect the total time that is required to obtain the final measurement:

- The sub-band time SBT_j, which is the time required to implement the numerical heterodyne process and the intermediate frequency filter, which are used to bring each band to the base-band on the *j*th node
- The filter time FT_j, which is the time required to execute the band-pass filter bank on the *j*th node
- The average transfer rate TR: transfer rate among the grid nodes expressed in byte/s

By using these parameters, the total processing time PT corresponding to each strategy can be approximatively estimated for a scenario where the virtual measurement grid node has processing capabilities and for similar processing-node performance. In this situation, that is very common in real grid scenarios, the total processing time for the sequential strategy is estimated as

$$\mathrm{PT} \approx 15 \cdot (\mathrm{SBT} + \mathrm{FT}) \tag{1}$$

since 15 is the number of sub-bands to be processed.

For the partially distributed strategy, the processing time is obtained as

$$\mathrm{PT} \approx 15 \cdot \left(\mathrm{SBT} + \frac{\mathrm{DS}/25}{\mathrm{TR}} \right) + \mathrm{FT} \tag{2}$$

where DS is the size of the sampled data.

Eventually, for the distributed strategy the processing time is obtained as

$$\mathrm{PT} \approx \frac{\mathrm{DS}}{\mathrm{TR}} \cdot N + \left\lceil \frac{15}{N} \right\rceil \cdot (\mathrm{SBT} + \mathrm{FT}) \tag{3}$$

where N is the number of employed processing nodes.

In the case of very slow bit-rate networks, the time that is required to transfer the data on the network is too high; therefore, the sequential processing becomes the best solution. For medium bit-rate networks and low number of available processing nodes, it is usually more convenient to employ the partially distributed solution. The third strategy that corresponds to the distributed processing is useful with networks that assure higher transfer rates.

3. GRID IMPLEMENTATION

The grid measurement architecture has been deployed by means of the freely available Globus Toolkit [4], which provides a programming environment where it is easy to integrate the measurement nodes with the processing and display nodes.

A specific service has been developed that is dedicated to the management of the acquisition and processing operations. Such a service, which will be referred to as the manager service, is responsible for the distribution of the processing load among the available processing nodes on the basis of the estimated processing time (PT).

The proposed implementation of a measurement system for conducted disturbance in form of a grid system is shown in Figure 4. The figure shows one virtual measurement node and several processing and display nodes, but a general implementation can contain several measurement nodes. This way, the complete measurement system has two important advantages over the conventional solution: the digitizer node, which could be required to be installed in an unmovable recessed site, can be physically separated from the display node site where qualified technicians operate the assessment; the structure enables the assessment at several nodes without replicating the processing structure.

The virtual measurement grid node is composed of a digital storage oscilloscope (DSO), which is set to acquire 8 MSamples with a sampling frequency of 250 MHz and of a PC. Since the employed DSO does not exhibit computational power, the local PC is used both to acquire and pre-process the signal samples and to act as a proxy grid node by offering common services, such as the grid security infrastructure (GSI) and the monitoring and discovery system (DMS).

The samples acquired by the DSO are transferred to the local PC, where a pre-processing of the data is implemented that consists in a low-pass filter and a decimation, in order to reduce the sampling frequency to 125 MHz. In such a way, a reduction of both aliasing error and quantization noise is obtained.

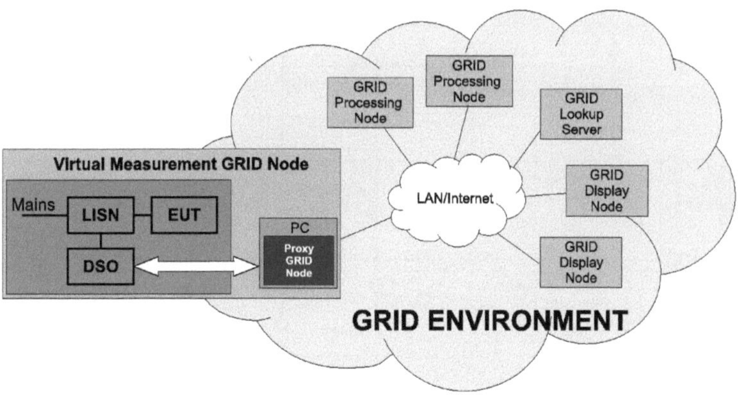

Figure 4. The grid implementation of the process for conducted disturbance measurements

At the grid startup and when a new node joins the grid, the manager service is designed to ask the nodes to perform dummy processing to estimate SBT_j, FT_j and TR for the different nodes.

Each time a user located at a display node requests a measurement execution for a specific virtual measurement node, the manager service estimates the PT times for the different strategies and starts the measurement process, distributing the computation load according to the best identified strategy.

4. EXAMPLES OF MEASUREMENT RESULTS

The architecture has been tested on a grid composed of 14 PCs based on Pentium IV-2.8GHz processors connected by means of a local 10 Mbit/s LAN. The tests have been performed by adding one PC at a time up to 14 PCs. During the tests, the process that chooses the processing strategy inside the service manager selection has been disabled to force all the processing combinations to occur.

The obtained results are summarized in Figure 5. The black dot on the very left represents the processing time on a single PC, which is about 4 minutes. The triangles refer to the partially distributed processing, which can be implemented when at least two nodes are available. The required time is about 3 minutes and remain constant even though more processing nodes are added.

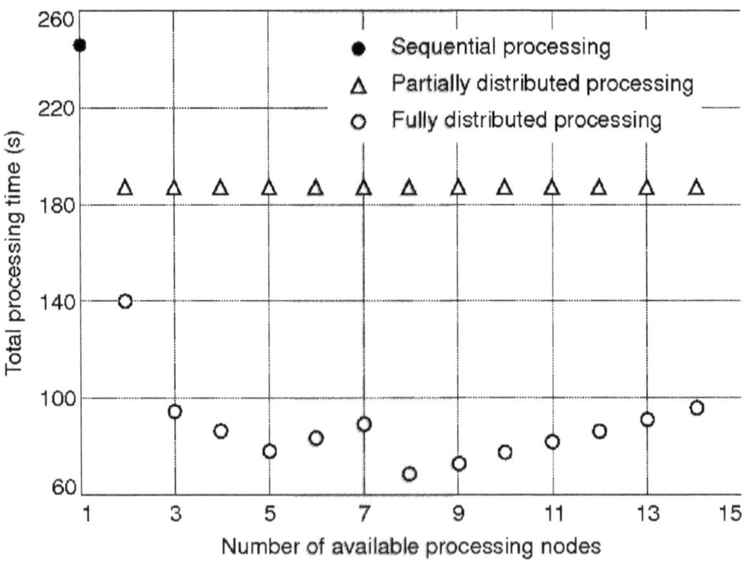

Figure 5. Comparison of the total processing time for three different scenarios

This behavior is expected, since the heaviest process is the slicing into sub-bands that is performed entirely on the measurement node.

The hollow dots represent the total processing time as a function of the number of available nodes for the fully distributed processing. Figure 5 shows that the total processing time decreases as the number of available processing nodes increases and reaches a local minimum for five nodes. Then, the processing time increases till seven nodes and reaches the minimum for eight nodes, where its value is reduced to about 1 minute.

This behavior depends on the distribution of the processing load among the nodes: when five nodes are available, the maximum processing time $(\mathrm{SBT} + \mathrm{FT})$ in (3) is the one required to extract three sub-bands and to implement the band-pass filters related to the three bands. With six or seven available nodes, the time required to transfer the signal samples to the nodes increases, but the maximum processing time does not change, since there is at least one node that has to process three sub-bands. When the available processing nodes reach eight, the maximum processing time corresponds to the time required to process two sub-bands. Increasing the number of available processing nodes up to 14 does not introduce further benefits, since the processing time does not change, while the data transfer time increases.

In order to evaluate the effectiveness of the proposed system from a measurement point of view, the amplitude spectrum provided by the proposed algorithm has been compared with that obtained by means of a commercial spectrum analyzer that is usually employed for pre-compliance tests. This comparison has shown that the algorithm results are in good agreement with the output of the spectrum analyzer up to 15 MHz (amplitude differences within ± 2 dBμV), while for frequencies greater than 15 MHz the internal noise of the oscilloscope becomes predominant, however, not compromising the compliance test.

5. CONCLUSIONS

This paper has described a grid implementation of a measurement system for disturbance measurement. The grid implementation allows a reduction of the processing time with respect to a single computer of a factor four down to about 1 minute, thus making it comparable with conventional measurement approaches. The measuring grid nodes, which are based on a DSO without processing capabilities, are integrated into the grid by using a PC that acts as a proxy grid node, thus creating a virtual measurement grid node. A specific grid service has been designed that is intended for managing the distribution of the processing load among the available nodes.

A processing time reduction of a factor four has been obtained by employing a fast inter-node connection at 10Mbit/s and the tests have shown that when the number of nodes increases, the network transfer time becomes a limiting factor. For this reason, the authors are now working on the application of the compression algorithm proposed in [1] to the signal acquired by the DSO. The use of such a compression algorithm should allow benefits in terms of processing time and should allow the proposed architecture to be employed also in the presence of slower connection networks.

REFERENCES

[1] Cacciatore V., Carullo A., Parvis M. and Vallan A. "Uncertainty Effects of Data Compression in Measurement Applications," in *Proceedings of the IEEE Instrumentation and Measurement Technology Conference*, pp. 1462–1467, April 2006, Sorrento (Italy).

[2] Carullo A., Corbellini S. and Parvis M. "Traceability Assurance of Tele-Measurements," in *Distributed Cooperative Laboratories – Networking, Instrumentation and Measurements*, pp. 363–372, Springer, 2006.

[3] De Capua C. and Landi C. "A Digital Measurement Station for RF-Conducted Emissions Monitoring," in *IEEE Transaction on Instrumentation and Measurement*, vol. 51, n. 6, pp. 1363–1366, December 2002.

[4] Foster I. "Globus Toolkit Version 4: Software for Service-Oriented Systems," in *IFIP International Conference on Network and Parallel Computing*, pp. 2–13, Springer-Verlag LNCS 3779, 2005.

[5] Parvis M., Perrone G. and Vallan A. "A Precompliance EMC Test-Set Based on a Sampling Oscilloscope," in *IEEE Transaction on Instrumentation and Measurement*, vol. 52, n. 4, pp. 1220–1223, August 2003.

[6] CISPR 16-2-1:2003-11 "Specification for radio disturbance and immunity measuring apparatus and methods," Part 2-1: Methods of measurement of disturbance and immunity-Conducted disturbance measurements. *CISPR16*, 1999.

SYSTEMATIZATION AND ANALYSIS OF REMOTELY ACCESSED INSTRUMENTS AND USER COMMUNITY REQUIREMENTS

Y. Kalvachev[1], V. Ganev[1], L. Macheva[1], J. Harizanova[2], and B. Harizanov[3]
[1]*Central Laboratory of Mineralogy and Crystallography, Bulgarian Academy of Sciences, 1113 Sofia, Bulgaria,*
[2]*Central Laboratory of Optical Storage and Processing of Information, Bulgarian Academy of Sciences, 1113 Sofia, Bulgaria,*
[3]*University of Sofia St. Kliment Ohridski, 1504 Sofia, Bulgaria*

Abstract This chapter describes a procedure for identifying high-level scientific equipment having the potential for remote instrumentation access, as well as the capacities of this equipment for integration into next-generation grids. The requirements of the user community for such kind of access are evaluated and analyzed. The information is implemented in a database that should be used as an interface between instrument OWNERS and USERS groups. This product could be considered as a prototype of future large-scale databases that will contain wide-structured information concerning remotely accessed instruments and services integrated into the next-generation grids.

1. INTRODUCTION

The need to use unique and expensive equipment, which is often locally unavailable, and the broad international cooperation are key issues for the success of the great number of scientific or industry ventures. This is especially important for scientific communities from countries with relatively low financial support for their research infrastructure. The possibility of accessing expensive and sophisticated equipment has a significant role in the scientific communities. Performing research through access to grid-embedded remote instrumentation, with the utilization of next-generation high-speed networks, will boost chances for scientific progress. The goals of many infrastructure projects (as RINGrid) are focused on opening new possibilities for remote-instrumentation-based cooperation among scientists from different countries, working in the same or closely related scientific domains. One of the main objectives is to provide the critical initial infrastructure necessary to facilitate the next generation of interdisciplinary science at both national and international levels, as well as in support of multi-institutional collaborative efforts. The RINGrid project responds to the pressing needs in different research fields to interlink and share expensive high-level scientific instruments located distantly.

The first and very important step to reach this goal is to collect and systematize detailed information about expensive, rare and high-level scientific instruments suitable for realizing the idea of remote access, as well as sharing the instruments with other groups of potential users in order to achieve optimization of equipment utilization. Some valuable experience in this aspect has been gained by many other GRID-oriented projects – EUROGRID, EXPReS, EGEE, GEON, GRIDCC, SEE-GRID, EELA, etc., some of which are currently active and include participation from RINGrid members. In this work we present results from the collection of information and the creation of a database connected with the RINGrid project.

2. IDENTIFICATION OF SCIENTIFIC INSTRUMENTS AND USER COMMUNITIES

The identification of user communities and scientific instruments involves various activities. An initial task was to increase the awareness about major RINGrid project goals, and therefore to create an opportunity for different communities to participate in any possible way in this research. Collected systematic information concerned instrument specifications, services and access requirements.

Technical instrument specifications and services were collected using the Instrument Information Form (IIF) tool, designed during the initial stage of the planned work. The information on user access requirements to remote instruments and services was collected by means of the User Information Form (UIF) questionnaire.

The performed activities, which served as a necessary input for the next steps of the instruments' identification procedure, were

- Preparation of the IIF as a basic tool for collection of information about equipment with the potential for remote instrumentation access
- Distribution and collection of the completed IIFs
- Systematization and analysis of the collected information

- Collection of information for specific user workgroups, mailing lists and other kinds of discussion environments

The summarized information from IIFs on instrument specifications and services should be easily accessed by users interested in performing experiments and measurements on this equipment. The fact that instrument owners should conventionally use the instruments, which will be accessible for remote access, should be also taken into account. This will benefit both users and owners.

On the other hand, information collected by UIFs concerning user needs and requirements for remote access has to be conveniently used by instrument owners in the search of proper user communities. Therefore, the creation of a database for storing the collected information was expected to make it easily available. These tasks are time consuming and their results require frequent updating. As an initial activity in the process of database creation, the information preliminarily collected by IIFs and UIFs was converted into a proper database format.

The process of identifying remotely accessed instruments and their potential user communities related to the aim of the RINGrid project covers a large variety of scientific fields: material science, astronomy and astrophysics, ecology, engineering and industry. The main goal is the integration of a large variety of remote instrumentation implicating expensive and widely used techniques and equipment. In all the specific areas the network environment needs, owner-specific facilities, user communities and access requirements should be investigated. The work tasks are fulfilled through collaboration with scientists, involved in the design, construction and operation of this equipment, and the groups of present and potential users of such equipment. The list of instruments and their owners was preliminarily established as a base for task initiation. It was enlarged with additional instruments through dissemination activities.

The identification of scientific instruments and user communities was carried out by the preparation and dissemination of IIFs and UIFs.

2.1 Preparation and Dissemination of the Instrument Information Form (IIF)

The Instrument Information Form (IIF) was adopted as the most appropriate tool for collecting appropriate information about the instruments. It consists of two major sections: I. Equipment configuration and II. Description of remote accessibility.

The first section contains details, connected with

- Basic information about the instrument – type, manufacturer, model, year of manufacture, approximate price, owner, location, contact person

- Main technical parameters of the instrument concerning user access and experimental measurements – the way of using the equipment, fields of application, current and potential new groups of users of the instrument as well as a description of the owner's access policies for external users, details concerning the sample preparation (where appropriate), measurements and data obtaining procedures, the time required for a single measurement, presentation of measurement results

The second section deals with the current and potential possibilities of remote access to the instruments – remote access interface features and requirements, the need for processing raw data before archiving, service and infrastructure requirements in terms of computational and storage requirements, requirements for streaming media, for visualization renderings, offline data processing, network requirements, etc.

The following reasons were taken into consideration in order to select the above-mentioned approach:

- Unification of the collected information
- Exhaustiveness of the data
- Flexibility of information processing
- Possibility for statistical processing of the collected data

All the RINGrid partners are involved in filling in the IIFs for their own instruments, as well as in the systematization of instrument specifications concerning potentially shared business and/or scientific use, and the establishment of working contacts for shared research use of the instruments belonging to the project consortium. The additional information regarding other equipment that comes from other sources, where the partners participate or have working contacts, was also collected.

The IIFs was disseminated among all participants in RINGrid as well as to other EU and non-EU institutions which are owners of expensive scientific equipment.

2.2 Analysis of the IIF Data

The returned completed IIFs were analyzed and the data collected summarized. They were classified into several domains, as follows: material science; satellite communications, telecommunications and networking; astronomy and high-voltage engineering. The distribution of the instruments by application domain is presented in Figure 1.

As an initial task, instruments are checked to satisfy the requirement of being expensive and/or rare equipment. As the remote access to the instruments

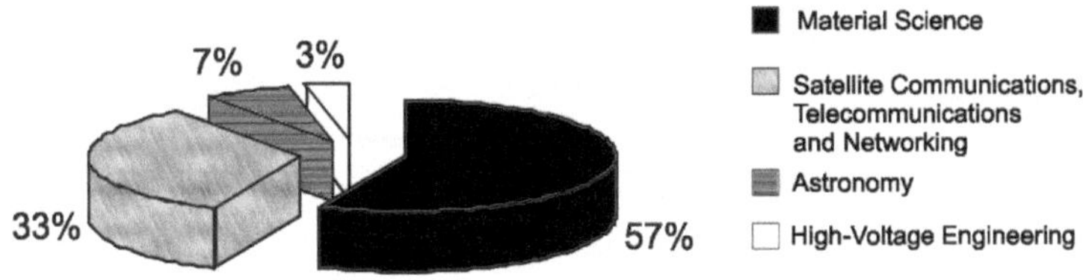

Figure 1. Distribution of instruments by application area

is one of the major factors for its range of application at the next step, the capabilities of network interfaces and instrument remote services were evaluated.

In the IIF, the owners indicated the current field of application of the instrument. In fact, some of the equipments can be utilized in many other areas and fields, so it was found out that the actual application domain could be enlarged significantly. Therefore the analyzed instruments were rearranged in a new manner (Table 1), based on their augmented field of application. The grouping by type–criteria (local, regional, global) is discussed below from the significance point of view. Additionally from this initial group, several new instruments were identified and included in the study.

Table 1. The RINGrid consortium's instruments and instruments belonging to external owners

Country/application area (domain)	Equipment	Type*
Brazil		
	Synchrotron light source with 11 beamlines	G
	JEOL high-resolution transmission electron Microscope (HR-TEM), model JEM 3010 UPR	R
	GEOL field emission scanning electron microscope (FEG-SEM), model JSM-6330F	R
	JEOL low-vacuum scanning electron microscope (LV-SEM), model JSM-5900LV	R
Bulgaria		
Material Science	BRUKER Tensor 37 FTIR	L
	VARIAN Cary 100 UV-Vis	L
Chile		
Material science	BRUKER AC-250 P (manufactured 1991)	R
	Laser scan microscope Zeiss LSM 410 Confocal microscope	R
	Diffractometer Siemens D-5000	R
	Analytical electronic microscope SEM-PROBE Cameca Camebax SU-30	R
Astronomy	4.1 m optical telescope at Southern Astrophysical Research Telescope (SOAR), Cerro Pachon, Chile	G
Italy		
Satellite communications, telecommunications and networking	Satellite network (mesh topology); 24 earth stations; audio and video multicasting	R

Table 1. (continued)

Country/application area (domain)	Equipment	Type*
	Vector Signal Generator Agilent ESG E4438C (250 kHz–6 GHz, IEEE 802.11b option)	L
	WLAN Spectrum Analyzer Agilent ESA E4404B (250 KHz–6 GHz)	L
	Wideband Radio Channel Simulator Electrobit Propsim C2 (2 channel 350 MHz–6 GHz with AWGN/Shadowing options)	R
	2 Multimeter/Swich Systems Keithley 2750 with 6 modules 7711 (up to 2.5 GHz)	L
	Router tester Agilent 900 series including various interface cards and software packages	R
	Switching Matrix Keithley Matrix 707E + 2750 DMM/DATA	R
	Vector Signal Analyzer Anritsu Signature MS2781A	L
	Telecommunication measurement test bed Agilent 33250 A, Agilent E4438C, Agilent 8563EC, Agilent Infinium 54820A	R
	Digital Oscilloscope Le Croy Wave Pro 7100	L
Mexico		
Material science	Gas Chromatograph VARIAN 38000	R
	Atomic Absorption VARIAN AA 800	L
	VARIAN Cary 1E UV-Vis Spectrophotometer	L
	ICP- Mass Spectrometer	R
Poland		
Material science	VARIAN UNITY 300 MHz NMR Spectrometer	L
	NMR Spectrometer Bruker Biospin GmbH Avance 600	L
	BRUKER AVANCE 600 MHz NMR Spectrometer	L
Astronomy	32 m Radio Telescope in Piwnice, Poland	G
Romania		
Material Science	Dielectric Spectrometer ALPHA-Novocontrol (1 MHz–8 GHz)	L

The common owner access policies do not impose restrictions on external users and several different user communities can use most of the instruments from the list. The technical information, taken from the received IIFs, shows that practically all the described instruments have hardware interfaces allowing some kind of remote access. Great interest of science communities in remote access to these instruments is expected, due to their relatively high price (in the range of 10 k–20 MEuro) and their wide geographical scattering.

Commonly, the remote operation of the identified instruments can be realized by

- Multimedia communication with an operator at the instrument site
- Direct commands sent to the instrument interfaces

- Both of these alternatives

There are several possible levels of operative control:

- Through sending and receiving data offline
- Through interactive real-time data control and data transfer
- Through dynamic instrument control

In most cases, interactive real-time data control and data transfer is sufficient, except when an online-controlled data transfer of high-quality images is required (in the field of electron microscopy and astronomy). For some of the equipments analyzed, several types of remote access could be implemented.

The modern instruments suitable for remote accesses could be classified first of all by the cost parameter. This is closely related to instrument distribution and research possibilities. Following this principle, three main groups of instruments could be distinguished, in accordance with the pattern of remote access involved:

I. **Local** (or conventional), which could be accessed by one or few geographically close organizations, as a rule with already well-established collaboration. Usually, the price of each of these instruments is approximately in the range of 10–100 kEuro and they are shared among these institutions. This kind of equipment is widespread and it has a relatively low cost. We expect a limited interest in the remote access of this group of devices.

II. **Regional**, which could be accessed by a great number of research institutions, located usually at national level. The price of these instruments is estimated between 100 k and 1 MEuro and could be classified as moderate. In some cases, international access to this kind of equipment based on regional and/or bilateral collaborations associated with specific scientific projects could be expected as well. Increased interest in remote access usage of this group of instruments could be supposed.

III. **Global**, which are extremely scarce and expensive pieces of equipments. They are used by a great number of researchers all over the world. The cost of these instruments can vary on a large scale, starting from around 1 MEuro. Usually, they are acquired, maintained and utilized by large international scientific groups. Great interest in remote access to this group of instruments is expected.

As is shown in Figure 2, the information collected through the IIFs is in good agreement with the classification described above.

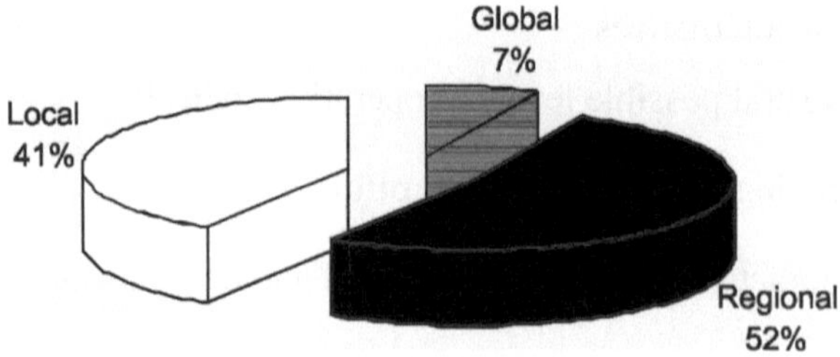

Figure 2. Distribution of instruments by their significance

The examples of particular instruments could be given as follow:

Local group of instruments: UV-Vis and IR spectrometers that are used in the field of material science (CLMC)

Regional group of instruments: TEM and SEM microscopes are also used in the material science field (CLARA)

Global group of instruments: optical and radio telescopes (CLARA, PSNC) used in astronomy

According to the classification presented above, it could be expected that **L** > **R**. In fact the relation was found to be the opposite (i.e., **L** < **R**, see Figure 2). This is due to the fact that a large number of instruments belonging to the category defined as **Local** have no suitable hardware interfaces for networking and, consequently, they were excluded from the more detailed future investigation and statistics.

2.3 User Community Identification and Requirements

Possible users of remote instrumentation systems are the most important source of information about the potential requirements and suggestions concerning remote access to various types of scientific instrumentation. The identification of user communities is accomplished in several ways:

- By filling the UIFs
- By extensive search for institutions that meet the requirements for remote instrumentation and will benefit from the access to scientific instruments

Information regarding user communities and potential new users of the remotely accessed equipment has been received, partially through IIFs and work contacts among RINGrid partners. More detailed information regarding user requirements for the remote access to scientific instruments in specific research domains was evaluated by means of the UIF tool. The received data show that user interest for remote-accessed instruments is directed mainly in two scientific domains – material science and astronomy.

In material science, technical requirements concerning adequate sample manipulation from remote site are leading, while in astronomy the possibility to transfer large amounts of visual information in real time is preferred. Fast, reliable and secure network connections are required from users of both scientific domains. Preliminary remote training possibilities are also of high demand. The users prefer "SCIENTI C collaboration" as an appropriate form for remote access usage.

3. TECHNOLOGY FOR THE CREATION OF THE DATABASE

The database was planned and engineered as information interface between both groups: instrument **OWNERS** and instrument **USERS**. It could be considered as a prototype of future large databases that should contain wide-structured information on remotely accessed instruments and services integrated in the next-generation grids.

Due to the requirements for an easier and more flexible data access, we decided to build up the above-mentioned database prototype by using web technology. In addition, we accepted to adopt free, commonly used, multiplatform-compatible programming tools to create the database.

Regarding that, we chose **MySQL** [5] as **Database Server**, **Apache Tomcat** [1] as **Application Server**, as well as **JavaServer Pages** [2] and **Java Servlet** [3,4] technologies. In this connection, the following stages are accomplished:

- Creation of MySQL Database
- Importing data from XML files of the collected IIFs into MySQL database
- Creation of the Java Web Application
- Creation of JavaServer Pages and Servlets for database search, which dynamically generate HTML for presentation of search results (text and images)

3.1 Overview of MySQL

The MySQL® database has become the world's most popular open source database because of its consistent fast performance, high reliability and simple usage. It is used in more than 10 million installations ranging from large corporations to specialized embedded applications [5]. MySQL is not only the world's most popular open source database, but it becomes also the database of choice for a new generation of applications built on the LAMP stack (Linux, Apache, MySQL, PHP/Perl/Python). MySQL runs on more than 20 platforms, including Linux, Windows, OS/X, HP-UX, AIX and Netware, giving one the kind of flexibility that puts one in control. Whether the MySQL user is new to database technology or an experienced developer or DBA, MySQL offers a comprehensive range of certified software, support, training and consulting [5].

3.2 JavaServer Pages Overview

JavaServer Pages (JSP) technology [2] enables web developers and designers to perform rapid development and easy maintenance of information-rich, dynamic web pages that leverage existing business systems. As a part of the Java technology family, JSP technology enables rapid development of Web-based platform-independent applications. JSP technology separates the user interface from content generation, enabling designers to change the overall page layout without altering the underlying dynamic content.

3.3 The Java Servlet Technology

Servlets are the Java platform technology [3, 4] of choice for extending and enhancing web servers. Servlets provide a component-based, platform-independent method for building web-based applications, without performance limitations of CGI programs. Unlike the proprietary server extension mechanisms (such as the Netscape Server API or Apache modules), the servlets are server and platform independent. This leaves one free to select a "best of breed" strategy for one's servers, platforms and tools.

The database structure diagram is shown in Figure 3.

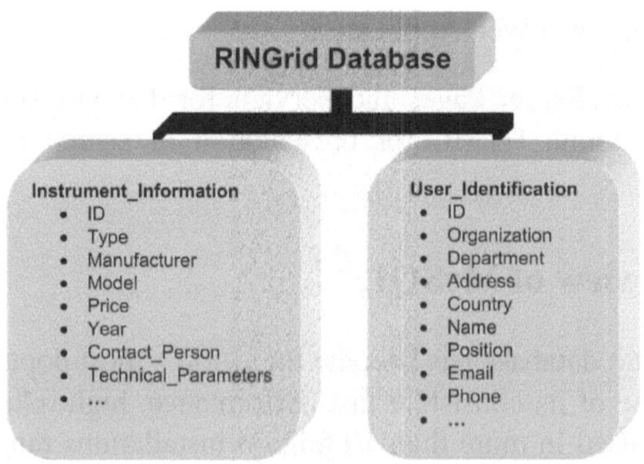

Figure 3. The database structure

4. DATABASE FUNCTIONAL DESCRIPTION

The database contains two main sections – Instrument Information and User Identification. Following the above-described instruments' and users' information, a suitable way for gathering, keeping and processing all mentioned

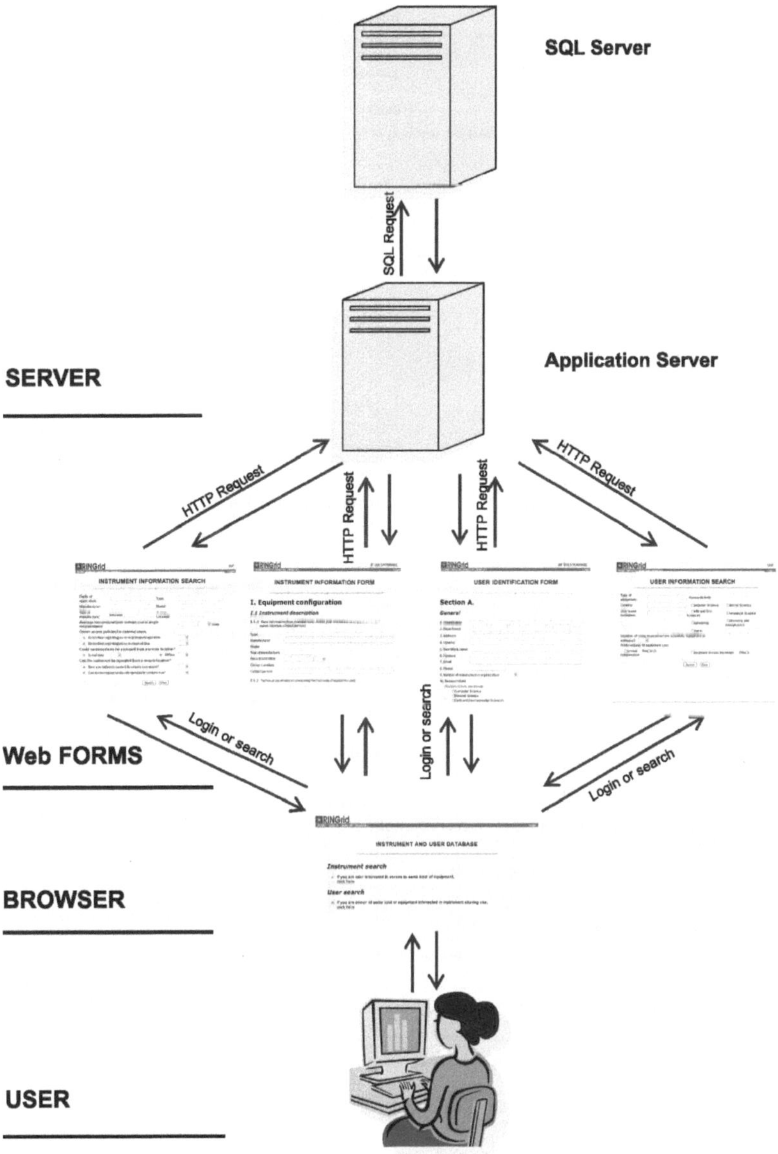

Figure 4. Web application and DB-functional diagram

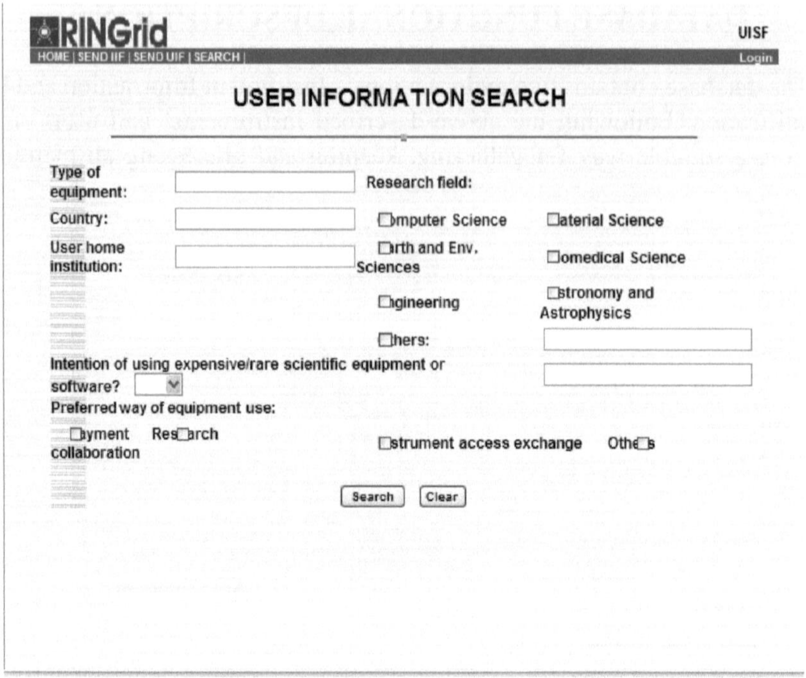

Figure 5. The main user search form

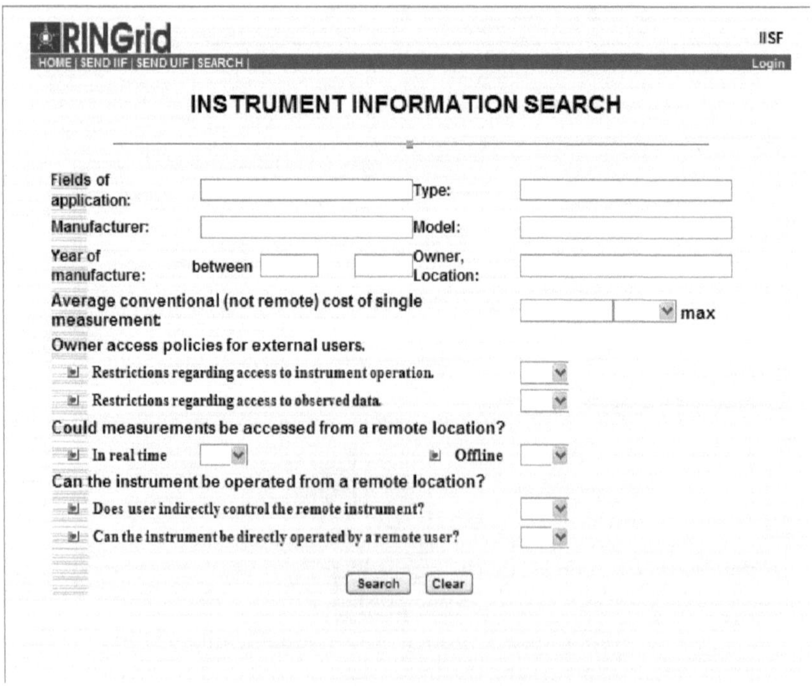

Figure 6. The main instrument search form

information is needed. With respect to these requirements and using the tools mentioned above, a working prototype of the database was developed. It consists of all IIF and UIF information collected so far, and it should be used as an interface between instrumentation OWNERS and USER communities, as well as in analytical tasks in the next RINGrid activities.

This product could be considered as a prototype of the future large databases that should work on the remote access information gates/gateways (RAIG) in the Internet, containing large amounts of structured information regarding remotely accessed scientific instruments and their user communities (Figure 4).

The database has the following advantages, concerning instruments and users information: unification of the collected information; exhaustiveness of the data; flexibility of information analysis; possibility for statistical data processing.

The group **(OWNERS)** should use it in order to search for proper users and user communities with potential interest in accessing their equipment according to defined and described preliminary conditions (Figure 5).

The second group **(USERS)** can access the database in their search for suitable remotely accessed instruments that meet their set criteria (Figure 6).

5. CONCLUSIONS

The systematized and analyzed information about collecting procedures, technical specifications of instruments, services and possibilities for remote access to scientific equipment has been presented. The equipment is divided into the following generalized domains: material science, astronomy, networking and engineering. The modern instruments suitable for remote access are classified by their significance into three groups – local, regional and global. The analysis of users requirements is based on the information collected from IIFs and UIFs. The criteria thus outlined could be used on a wider scale to classify a greater number of instruments. The development of the instruments and users database gives new and greater opportunities for collecting, systematizing and analyzing information regarding remotely accessed equipment. The database works as an interface between **OWNER** and **USER** groups with respect to their activities in planning and searching instruments available for remote access usage.

ACKNOWLEDGMENTS

This work was supported by the RINGrid project (contract No 031891 – FP6 program of the EC).

REFERENCES

[1] Apache Tomcat 6.0 Documentation, http://tomcat.apache.org/tomcat-6.0-doc/index.html

[2] JavaServer Pages 2.0 Specification, http://jcp.org/aboutJava/communityprocess/final/jsr152

[3] Java Servlet 2.4 Specification, http://jcp.org/aboutJava/communityprocess/final/jsr154

[4] Java Platform, Standard Edition 6 API Specification, http://java.sun.com/javase/6/docs/api

[5] MySQL 5.0 Reference Manual, http://dev.mysql.com/doc/refman/5.0/en/index.html

SERVICE-ORIENTED ARCHITECTURES FOR DISTRIBUTED COOPERATIVE INSTRUMENTATION GRIDS

A. Bagnasco[1], D. Cipolla[1], A. Poggi[1], and A. M. Scapolla[1]
[1] *DIBE – University of Genoa, Genova, Italy*

Abstract A large number of areas, including scientific research, phenomena monitoring, environment sensing and distance education, are looking for new infrastructures to set up distributed measurement systems. They need combining measurement instrumentation, networks of cooperating low-cost nodes, control processes, and computing facilities. This chapter shows how this goal can be achieved by extending the concept of distributed cooperating resources to deal with all these components in a similar way and proposes a Grid-based middleware to set up the distributed measurement system.

1. INTRODUCTION

A large number of areas, including scientific research, phenomena monitoring, environment sensing and distance education, are looking for new infrastructures to set up distributed measurement systems. They need combining measurement instrumentation, networks of cooperating low-cost nodes, control processes and computing facilities. This goal can be achieved by extending the concept of distributed cooperating resources to deal with all these components in a similar way. We need to detach their specificities and offer flexible and secure mechanisms to access and to aggregate different resources.

The web service technology is a promising candidate for the development of such a middleware. Web services use standard Internet protocols for inter-process communication and adopt a loosely coupled model, characterized by clients that connect to and use functions from servers without the specific knowledge of the servers themselves.

A further improvement to this technology can be found in Grid computing architectures. They emerged as middleware for sharing computing and storage resources, but nowadays a relevant number of projects document the interest in extending the approach to measurement instrumentation and large-scale data acquisition platforms.

The next section presents the main issues and advantages of the service-oriented approach for distributed measurement systems and analyzes the feasibility of porting web services on small devices, which are, more and more, important components of the next-generation measurement systems. The third section looks at the improvements coming from the emerging Grid infrastructures and evaluates their application to instrumentation control. The last section describes the Grid-based middleware that is proposed to set up a distributed measurement system. Specific services are designed to deal with instruments as Grid resources and to compose and to execute on-line measurements. Currently, the proposed architecture has been adopted to set up an educational laboratory on electronics.

2. SERVICE-ORIENTED ARCHITECTURES

A generic measurement process requires accessing and tuning instruments and data acquisition devices in a well-defined sequence. If we assume that these are distributed, the measurement system must access them remotely by efficient and secure remote procedure call mechanisms. To improve the capability of aggregating heterogeneous devices, it is important to adopt an approach based on loosely coupled components. Already existing distributed technologies, such as CORBA, RMI and EJB, result in highly coupled distributed systems, where the client and the server are very dependent on one another. These technologies work well in intranet contexts, but tend to have problems on a wider scale. For instance, they have a lot of troubles in passing firewalls, and clients require a deep knowledge about how servers work.

On the contrary, a loosely coupled model is characterized by clients that can connect to and use functions from a server without having specific knowledge about the server itself. This is the model implemented by the web services that use standard Internet protocols for inter-process communication.

The most typical and widely used implementation of web services is based on the SOAP protocol [3]. SOAP is an open, standard protocol developed by the W3C and supported by the main IT players (Sun, Microsoft, HP, IBM, Oracle, BEA). SOAP adds a further application level to the TCP/IP stack, so that SOAP messages can be transported by any Internet application protocol (HTTP, SMTP, FTP, etc.).

The service-oriented approach, applied to instrumentation control [8], lets encapsulate the real instruments behind general interfaces and allows creating measurement programs as composition of services. Such mechanisms can promote interoperability, both inside the local networks of small devices from different vendors and toward external entities. The development and the management are simplified, and, at the same time, scalability and lifetime are preserved. We can imagine scenarios where we select devices scattered over the network together with elaboration processes to compose the measurement, and some of these resources can be duplicated and requested on the fly according to the needs.

However, the implementation and the use of web services on limited-resources devices are not trivial.

2.1 Web Services for Embedded Systems

The adoption of SOAP on embedded devices is limited by the verbosity of the communication and the overhead necessary to parse messages. Nevertheless, some attempts to port SOAP on embedded devices exist in the literature. Worth mentioning are gSOAP [14], eSOAP [9, 13] and kSOAP [18]. The ASP.NET family includes a web service framework for embedded devices, but being based on Windows CE, it can be considered heavily platform dependent. It should be noted, however, that all the presented solutions are still demanding in terms of resources. For example, kSOAP requires the presence of a Java Virtual Machine on the embedded system.

The heaviness and complexity coming from the usage of SOAP are experienced not only in the case of embedded devices, but in computer networks, too. A consequence is the diffusion of simpler and lighter web service-based solutions, designed according to the REST criteria [5]. The acronym "REST" stands for "REpresentational State Transfer," an architectural style for the design of large-scale distributed software, inspired on a subset of World Wide Web functionality. REST cannot be considered a real standard, because it does not refer explicitly to technologies, protocols or formats; in fact, it states the roles for the entities making up a distributed system and the guidelines to manage the communications among them. REST uses stateless client–server architecture, based upon a minimum set of universally defined remote methods for complete resource management. The key abstraction of REST is the concept of resource: resources are the actual information, exchanged in the system under different representations. The representation of a resource consists of data and metadata, which describe data nature and format. It is possible to access the representation of a resource by means of an identifier, which addresses the resource regardless of its representation. In a so-called RESTful system, the client accesses a server-hosted resource using its identifier and calling the required handling methods, which return a representation suited to the nature of data, to the type of request, to its interpretation abilities.

The RESTful implementation of a service, compared to the corresponding SOAP protocol solution, has an important difference: the design and implementation of an additional application layer is not required, because the HTTP protocol already provides the primitives for service access. URIs, natively supported by HTTP, provide a unique, uniform and universal addressing mechanism; at the same time, the system components are tied by loose coupling, granting a high degree of scalability. The architecture of a RESTful web service is simple because, at most, it involves the extension of the client-side and server-side functionalities for the management of the XML responses.

The application of the REST style to small devices is reported in [4, 10]. In [1], the authors propose a two-layer service-oriented architecture that merges small devices performing basic tasks, such as environmental sensing and conditioning, and nodes that offer more complex services on the basis of the collected results.

2.2 From Service-Oriented Architectures to Grid Frameworks

Grid computing offers a step forward toward the implementation of service-oriented systems, based on efficient and secure middleware. It emerged as a key infrastructure for "flexible, secure, coordinated resource sharing among dynamic collections of individuals, institutions, and resources" [7], finalized to the aggregation of resources into virtual organizations.

Scientific communities embraced Grid technologies to run projects involving distributed resources and using massive data sets, in the context of the so-called e-Science. Main goals were the improvement of the collaboration across distributed research groups, sharing experimental results that must be submitted to compute intensive analysis and optimizing the use of computational and storage resources. So far, the sources of data, such as measurement instrumentation and data acquisition devices, have not yet been considered as full Grid components. The interaction with the instruments is unmanaged and the instruments run off-line with respect to the Grid environment, the attention is mostly focused on data rather than on the role of the instrumentation.

Nonetheless, there is the interest in extending this approach to the measurement and control field and it is documented by a relevant number of international projects [11, 12, 17, 18]. The main motivation is the growing importance of distributed measurement systems that merge specialized devices, sensor networks and processing units. Using Grid frameworks, facilities for sharing and integrating instruments with computing and storage are expected.

3. GRID AND INSTRUMENTS

Dealing with instruments as Grid services makes easier their management and orchestration with other resources. We can imagine scenarios where we select devices scattered over the Grid to compose a measurement process, based on the cooperation of a specific device under test or a network of sensors, instrumentation and processing units.

The integration of instruments with other Grid resources requires mechanism for registering, locating and interacting with instruments. Grid services can be used to abstract the views of instruments, to create interfaces that expose their functionality and to allow a communication model based on standard protocols across all instruments and sensors. In addition, it is possible to define interfaces that can be shared among sets of instruments that offer similar functionalities, with a minimal rewriting of acquisition and processing codes when the underlying instrument and controller change.

In [2], the authors report the development of two prototypes of Grid services for the control of measurement instrumentation. Some performance results are included. This work proves that the Grid approach is feasible, despite it requires an initial effort to apply the service-oriented approach and to be acquainted with the Globus environment used to set up the Grid. Concerning the performance issues, Grid protocols add latency, but the overhead can be considered acceptable for many scenarios and it resulted insignificant for the educational laboratory application.

Instrument services are only the first step to build Grid-based measurement systems. Workflows allow to schedule and to coordinate the actions of different Grid resources.

4. A GRID-BASED DISTRIBUTED MEASUREMENT SYSTEM

This section presents the architecture of a Grid-based comprehensive measurement system.

The architecture has been conceived to manage real distributed resources (instrumentation, devices under test, data acquisition nodes) by a secure middleware that allows combining their functionalities with storage and computing facilities. Potential end users are able to run measurement processes, which require controlling instruments or acquiring data from networks of low-cost nodes, and to set up new measurement processes on the base of the services exposed by the controlled resources.

The middleware manages concurrent accesses to the real devices and interactive work sessions. In this case, the system must maintain the status of each

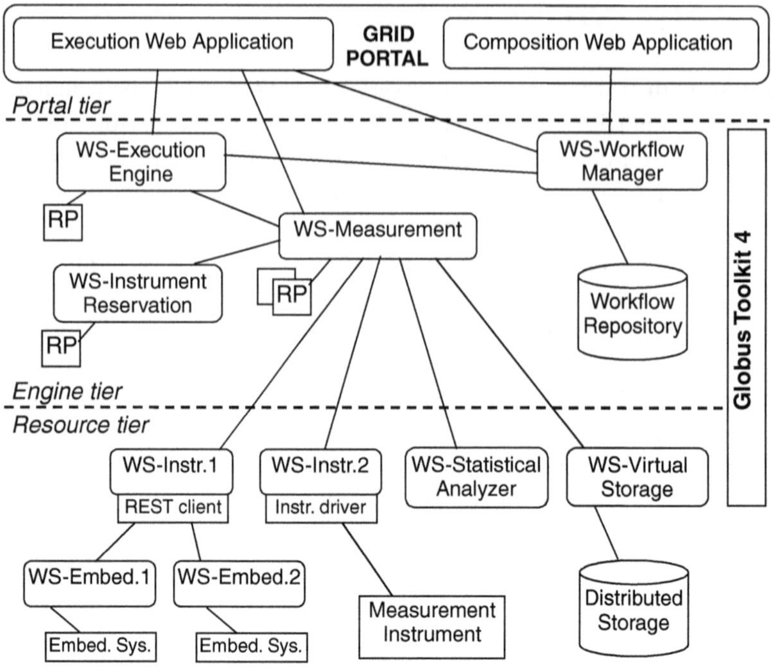

Figure 1. The system architecture

work session, as multiple phases of instrument configuration and data acquisition may be necessary to complete the measurement process.

The system is structured into three tiers (see Figure 1): the Resource, the Engine and the Portal tier. The design is based on the Globus Toolkit v.4 [6], which offers tools and useful features for building service-oriented applications, supporting secure user access mechanisms, reliable file transfer, job submission and indexing services. In Figure 1, the tiers that use Globus facilities are highlighted by the side bar. The services (WS-...) have been designed to face the complexity of the measurement process and to control real devices; they are deployed in the Globus container.

The Resource tier is the lowest layer and consists of the instrumentation services and the Storage Service. The Storage Service is invoked to archive the data acquired during the measurement processes, or the results of their elaboration. In a Grid environment, the Storage Service benefits of the Grid Distributed Storage system.

The Instrument Services (WS-Instrument) are in charge of communicating with the real devices. Basically, there are two kinds of WS-Instrument services: the former simply wraps the instrument drivers, the latter acts as a bridge toward a network of embedded devices. In this case, both REST and

SOAP protocols are used to merge different classes of devices and their functionalities. The result is an additional sub-layer: the WS-Instrument service is installed on a node that controls, at the lower layer, a set of embedded devices characterized by scarce computing power. The node exchanges SOAP messages with the other Grid services and REST messages with the embedded devices services (WS-Embedded). The web service paradigm is common to both WS-Instrument and WS-Embedded services; but the WS-Instrument services may be both service consumers for the embedded devices and service providers for the upper-layer WS-Measurement service.

The services provided by the embedded devices are available only inside the private network they belong to. This configuration allows centralizing the most demanding functionalities and maintains the control on the whole system. In particular, it is possible to centralize the control on concurrent accesses to the resources and to implement a unique security policy for the whole system.

The Engine tier is the core of the architecture and it supports the applications available at the portal level. Basically, the web services of this tier perform the business logic of the whole system: user session management, workflows management, measurement execution and dynamic invocation of the instrument services. The management of concurrent accesses to the measurement processes requires centralized booking mechanisms and instrument status tracking facilities. This is obtained using stateful services that have the capability of keeping state information across different invocations. A separate entity, called resource property (RP in Figure 1), manages the persistence of the state information. Each resource has a unique key (resource ID), and web services are called with references to resources.

The WS-Execution service is in charge to manage the work sessions and to obtain the measurement workflows by the WS-Workflow Manager, which reads the repository. The adoption of XML workflows makes it possible to add new processes and to modify the existing ones, whenever required.

The WS-Measurement is a generic service able to run any measurement. It works on the base of the workflow assigned to it at invocation time.

The WS-Instrument services are independent entities. They are invoked by the workflows of the measurement processes. When a new request arrives, the WS-Execution service creates a new instance of the WS-Measurement service. This calls the WS-InstrumentsReservation service, a stateful service that maintains the state of each device of the Resource tier, to check the availability of the devices present in the measurement workflow. If available, they are reserved and the measurement process can be run. The devices will be released once the process finishes and the status of the measurement session will be saved for further requests.

The Portal tier provides the user interfaces for the measurement execution and composition. The Execution Application allows the user to select the measure workflow he/she wants to execute, and then, to interact with the laboratory test bench, sending input data, i.e., the instruments configuration parameters, and receiving results, through the web services of the Engine tier. Similarly, the Composition Application allows the administrators to maintain the system and to publish new measurement workflows.

5. CONCLUSIONS

The architectural model and the services, which have been proposed in the previous section, have been validated to set up an educational laboratory on electronics. ISILab (Internet Shared Instrumentation Laboratory) [15] is the name of the laboratory, which is currently in use to support a few courses of the curricula in information engineering at the University of Genoa. The service-oriented approach, presented in this paper, makes available a flexible, modular and scalable system.

This activity is partially supported by the European project IDENTITY (Individualized Learning Enhanced By Virtual Reality) [16]. It aims at producing a high-level quality learning environment in an academic European network, ensuring an open access to improved educational resources, with a particular emphasis on practical laboratory activity, particularly in the electrical domain. The expected project results, in short and long terms, cover the reduction of the financial effort through multiplexed access to the remote resources, supporting European high-tech industry in the worldwide competition, establishing the co-operation of academia, secondary high schools with the industry and sharing of resources.

ACKNOWLEDGMENTS

The authors would like to thank Alberto Boccardo and Paolo Buschiazzo for the precious contribution to the prototype development.

This chapter is supported by the EU Education Audiovisual and Culture Executive Agency under contract number 229930-CP-1-2006 -1-RO-Minerva-M.

REFERENCES

[1] A. Bagnasco, D. Cipolla, D. Occhipinti, A. Preziosi, A.M. Scapolla, "Application of web services to heterogeneous networks of small devices", *WSEAS Transactions on Information Science and Application*, Volume 3, Issue 5, May 2006, ISSN 1790-0832.

[2] A. Bagnasco, A. Poggi, G. Parodi, A.M. Scapolla, "Exposing Measurement Instruments as Grid Services", *Distributed Cooperative Laboratories: Networking Instrumentation, and Measurements*, edited by Franco Davoli, Sergio Palazzo, Sandro Zappatore, Springer Science+Business Media, Inc., April 2006, ISBN: 0387298118.

[3] F. Curbera, M. Duftler, R. Khalaf, W. Nagy, N. Mukhi, S. Weerawarana, "Unraveling the Web services web: an introduction to SOAP, WSDL, and UDDI", *IEEE Internet Computing*, 2002.

[4] W. Drytkiewicz, I. Radusch, S. Arbanowski, R. Popescu-Zeletin, "pREST: A REST-based protocol for pervasive systems", *The 1st IEEE International Conference on Mobile Adhoc and Sensor Systems*, 2004.

[5] R.T. Fielding, "Architectural Styles and the Design of Network-based Software Architectures", Ph.D. dissertation, University of California, 2002.

[6] I. Foster, "Globus Toolkit Version 4: Software for Service-Oriented Systems", *Proceedings of the IFIP International Conference on Network and Parallel Computing*, Springer-Verlag LNCS 3779, pp 2-13, 2005.

[7] I. Foster, C. Kesselman, "The grid: blueprint for a new computing infrastructure", *Morgan Kaufmann*, San Francisco, 1999.

[8] F. Jammes, H. Smit, "Service-oriented paradigms in industrial automation", *IEEE Transactions on Industrial Informatics*, 2005.

[9] N. Kakanakov, G. Spasov, "Adaptation of Web service architecture in distributed embedded systems", *International Conference on Computer Systems and Technologies – CompSysTech*, 2005.

[10] T. Luckenbach, P. Gober, S. Arbanowski, A. Kotsopoulos, K. Kim, "TinyREST – a Protocol for Integrating Sensor Networks into the Internet", *Workshop on Real-World Wireless Sensor Networks* – REALWSN, 2005.

[11] T.E. Molina, G. Yang, A.W. Lin, S.T. Peltier, M.H. Ellisman, "A Generalized Service-Oriented Architecture for Remote Control of Scientific Imaging Instruments", *Proceedings of the First International Conference on e- Science and Grid Computing (e-Science'05)*, 2005.

[12] R. Bramley, K. Chiu, J.C. Huffman, K. Huffman, D.F. McMullen, "Instruments and sensors as network services: making instruments first class members of the grid", *Indiana University CS Department Technical Report 588*, December 2003.

[13] R. van Engelen, "Code generation techniques for developing LightWeight XML Web Services for embedded devices", *ACM Symposium on Applied Computing*, 2004.

[14] R.A. van Engelen, K.A. Gallivan, "The gSOAP toolkit for web services and peer-to-peer computing networks", *Proceedings of the 2nd IEEE/ACM International Symposium on Cluster Computing and the Grid*, 2002.

[15] http://isilab.dibe.unige.it

[16] http://iesc.unitbv.ro/identity

[17] http://www.gridcc.org

[18] http://sourceforge.net/projects/ksoap2

A WEB-BASED TOOL FOR COLLABORATIVE ACCESS TO SCIENTIFIC INSTRUMENTS IN CYBERINFRASTRUCTURES

R. Ranon[1], L. De Marco[1], A. Senerchia[1], S. Gabrielli[1], L. Chittaro[1], R. Pugliese[2], L. Del Cano[2], F. Asnicar[2], and M. Prica[2]

[1]*Human-Computer Interaction Lab, University of Udine, via delle Scienze, 206 - 33100 Udine, Italy,*

[2]*Sincrotrone Trieste S.C.p.A., Strada Statale per Basovizza 14 km 163.5, 34012 Trieste, Italy*

Abstract Cyberinfrastructures that are being built in the United States and in Europe have the goal of enabling large-scale distributed science and engineering collaborations. However, there is still the need for effective groupware tools that will enable distributed and heterogeneous teams of people to effectively collaborate within these global scenarios. This chapter presents a Web-based groupware tool that is being developed in the context of the GRIDCC European project, whose purpose is to extend current cyberinfrastructures to enable shared access to and control of distributed instrumentation. We discuss the challenges in developing groupware solutions for different social and organizational contexts, and present the main features of a prototype we have designed.

1. INTRODUCTION

In recent years, we have assisted to a growing interest for funding large-scale distributed science and engineering collaborations that are supported by the so called cyberinfrastructures. The main reason for their development consists in the increasing complexity of scientific problems, which nowadays require access to a range of resources (people, scientific instruments, and computing equipment) that cannot be afforded by a single investigator or institution [1]. As an example, the design of the next-generation linear particle accelerator (an electrical device for the acceleration of subatomic particles used in high-energy physics experiments) is a worldwide effort which, in Europe, involves a consortium of 27 institutes [5].

While in the past decade cyberinfrastructures for e-science have targeted access to computing power and data storage, more recently research projects have started to focus on other kinds of resources, such as scientific instruments. The Grid Enabled Remote Instrumentation With Distributed Control And Computation (GRIDCC) is a 3-year European co-funded project [9] started in 2004, whose goal is to extend current GRID infrastructures to enable shared access to and control of distributed instrumentation. Some examples of instruments and scenarios addressed by the project are geophysical sensors in geo-hazards monitoring applications, power generators in power grids, and detectors in high energy physics experiments.

One of the activities within the GRIDCC project has the goal of developing groupware tools that support distributed and heterogeneous teams of people (scientists, engineers, technicians, students, ...) in the collaborative access to a cyberinfrastructure of distributed computing, storage, and instruments resources.

As already pointed out in [1], an interesting synergy can be envisaged between the fields of Computer Supported Cooperative Work (CSCW) and cyberinfrastructures. While CSCW could benefit from a broader understanding of large-scale distributed science and engineering collaborations, the design of Cyberinfrastructures can capitalize on both social factors' analyses and technical solutions developed in CSCW to support various forms of cooperation. However, this kind of synergy does not seem to have occurred so far, main reasons being that [1]

- The projects to build Cyberinfrastructures have mostly tried to address the technical issues that arise in the intercommunication and sharing of heterogeneous resources, while focusing less on the social, organizational, and collaboration issues.
- Cyberinfrastructures are built to support a wide range of different scenarios and applications, but this makes it difficult to carefully analyze CSCW factors, as well as to design effective groupware tools for the different contexts involved. For example, the GRIDCC project includes seven pilot applications (real-world applications where the outcomes of the project will be tested and finally evaluated) operating in different domains, such as high-energy physics, geohazards prediction, power grids, and meteorology.

This chapter has two main goals. First, we describe the design process of groupware tools in the GRIDCC project, focusing on how we are approaching the problem of designing for a set of different contexts and requirements. Second, we present the outcome of our design efforts, i.e., a Web-based groupware tool, called virtual control room (hereinafter, VCR), whose purpose is

- to support the creation and management of (dynamic) Virtual Organizations [8] (hereinafter, VOs) of people, instruments, computing, and storage resources;
- to support distributed and heterogeneous teams of people in the collaborative monitoring and control of instruments, including support for distance communication and experimental activity planning.

The chapter is structured as follows. Section 2 describes related work; Section 3 presents the design approach as well as the main requirements that have been uncovered. Section 4 describes in detail the virtual control room, while Section 5 discusses some implementation issues and solutions. Section 6 concludes the chapter and mentions future work.

2. RELATED WORK

Many groupware tools specifically devoted to support scientific experimental activities are known as collaboratories [15]. The term collaboratory (a hybrid of collaborate and laboratory) was coined by William Wulf to denote technologies that enable researchers to remotely control and operate instrumentation, access data sets and digital libraries, and easily interact with colleagues, like research centers without walls. The main motivation for building collaboratories is to overcome typical problems in scientific research. For example, providing remote access to expensive and hard-to-duplicate equipment allows one to increase the availability of these instruments and reduce traveling by research groups. Allowing distributed teams to effectively collaborate allows one to increase the effectiveness of the experimental activity, since more experts can participate, give useful hints and solve problems. Finally, there is the need to facilitate multi-institutional consortia collaborations on large-scale projects.

According to Finholt and Olson [7], the core capabilities that constitute a collaboratory are technologies to link

- People to people (e.g., email and video conferencing), to make remote scientists more visible to one another, and allow them to recognize common interests that can form the basis for future collaborations.
- People with information (e.g., digital libraries), to provide faster and wider availability of data and results.
- People to facilities (e.g., viewers that display status of remote instruments), to enhance and expand utilization of scarce scientific resources.

Early collaboratories focused mostly on the sharing of large, expensive instruments, such as astronomical telescopes, particle accelerators, oceanographic instruments, atmospheric observatories, and space research applications [12]. For example, the Upper Atmospheric Research Collaboratory [13] provides access to instruments in Greenland for solar wind observation, and collaborators can exchange and archive multimedia information from the instruments and the measurements analysis.

More recently, collaboratories have focused also on communication and collaboration support through the integration of various groupware tools. For example, the Biological Collaborative Research Environment (BioCore) includes specific tools for biologists (such as visualization programs), asynchronous and real-time communication tools, as well as collaborative writing tools [2]. Other collaboratories have abandoned the tool-centric approach in favor of a more data-sharing approach. For example, the Biological Sciences Collaboratory enables the sharing of biological data through electronic notebooks, metadata capture, data organization views, and other tools [4].

3. DESIGN APPROACH

In early phases of our project, we conducted a series of interviews involving end users and stakeholders from the main pilot applications considered by the project. The goal was to uncover their fundamental needs for a collaborative tool supporting their practices in the remote monitoring and control of distributed instruments.

We also carried out in-site observations at one of the main pilot applications, i.e., the remote control of the ELETTRA accelerator. More specifically, we analyzed activities in the accelerator control room to better understand the kinds of employees involved, their roles, as well as communication and collaboration practices. While similar detailed analyzes for other pilot applications are planned in the future, this approach was impossible to follow for some pilot applications, simply because they do not exist yet. For example, one of the pilot applications considers the collaborative remote control of an instrument (the Compact Muon Solenoid, which will be part of the Large Hadron Collider at CERN) whose design and construction is being carried out by other research projects.

As a workaround, we thus asked all pilot applications to provide us with written stories describing or hypothesizing realistic working scenarios involving groups of collaborators and the remote operation of distributed instruments.

Results of this investigation fed our design of a number of paper sketches, like the one shown in Figure 1, based on the following ideas

- to partition the application window in a set of workspaces (organized in tabs), where each workspace can be dedicated to one of the VOs to which the user registers to, or to different tasks in the context of the same VO;

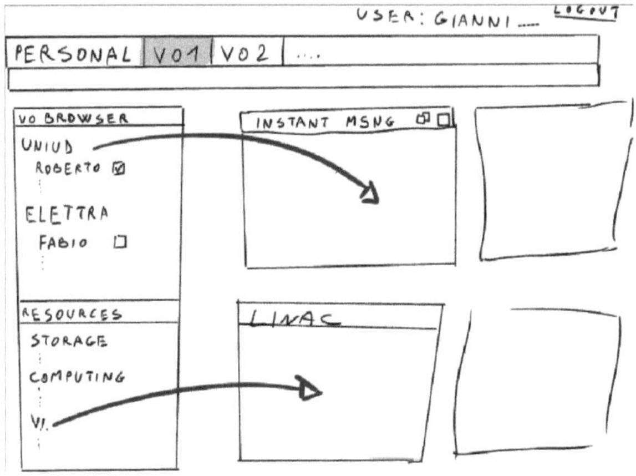

Figure 1. Paper sketch of VCR interface

- to organize each workspace into a set of panels, each one devoted to a specific purpose (e.g., monitoring of an instrument);
- to design the user activity with the virtual control room around a set of tools to browse, inspect, and access the various resources of the VO, including people, instruments, and other resources, such as computing and storage ones.

Moreover, a main outcome from the analysis of the feedback received was that the specific requirements pointed out by each pilot application differed too much from those of the others to make it possible the design of a unique VCR matching all needs. As a solution, we eventually decided to adopt the following design approach for the VCR:

- to develop a general architecture and set of features supporting the common requirements of all pilot applications (i.e., creation and management of VOs, basic communication and collaboration support, generic tools to browse resources, and access remote instrumentation);
- in a subsequent phase, to make pilot application users and developers responsible to extend and customize the general VCR to better fit their specific requirements and evolving needs. This is done both with our support and by means of generic tools that capture common basic requirements and are easily customizable.

The paper sketches were then refined into more detailed user interface mock-ups, which were then validated with users from the pilot applications.

4. THE VIRTUAL CONTROL ROOM

One of the key points for the successful integration of heterogeneous resources is to provide users with a uniform and easy access to them. The idea behind the VCR is to provide an environment where users:

- meet and collaborate by means of groupware tools;
- search, discover, and browse resources, such as people, instruments, and other cyberinfrastructure resources, like computing and storage ones;
- transparently operate with remote instruments using the GRIDCC middleware infrastructure.

A screenshot of a typical collaboration scenario for the VCR is depicted in Figure 3a, which shows a troubleshooting scenario in one of the project's pilot applications, i.e., the monitoring and control of an accelerator facility. The following sections describe the VCR architecture, user interface, and tools.

4.1 Architecture of the VCR

The architecture of the virtual control room is shown in Figure 2. The lower level is built upon the GridSphere framework [10], which provides a Web-based portal where modular components, called portlets [11] can be easily aggregated together. Portlets are reusable Web components that display relevant information into their own window panels, which are aggregated by a portal and displayed together in a Web page. GridSphere provides also a collection of portlets that offer basic functionalities (e.g., account management).

Upon the GridSphere layer, we implemented a set of services (i.e., low-level functionalities needed by the VCR, such as resources discovery) shown as the middle layer in Figure 2. The implemented services provide two sets of functionalities. The first one, which includes Awareness, Events and Notification, and Resources services, is used by the collaboration tools provided by the VCR. The second one includes the Remote Control service, which provides a unified interface to resources, such as instruments, to facilitate the development of pilot application specific functionalities of the VCR.

Upon the service layer, we developed a series of tools (each implemented by a portlet) shown as the top layer of Figure 2. These tools provide the user interface and leverage the service layer to retrieve and store information needed. For example, the People Browser tool is able to dynamically update and show presence of information using the Awareness service, while the Instrument Control tool can retrieve the status and parameters values of a remote instrument using the Remote Control service.

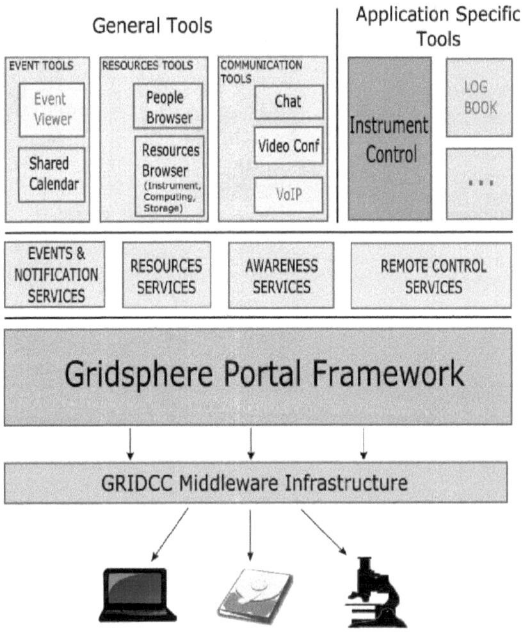

Figure 2. The VCR Architecture

Tools are divided into generic and application specific ones, following the distinction mentioned in the previous section. The portal framework makes it easy for developers to extend the VCR with tools (i.e., portlets) that are needed in a particular context, e.g., log books in scientific experiments. To help pilot applications in developing their own specific extensions, we try to identify general requirements and to provide generic portlets (like the Intrument Control Tool described later in this section) that act as a starting point.

4.2 The VCR User Interface

The VCR is a Web portal providing access to the typical functionalities of a collaboratory through a Web-browser. After the login phase, the VCR presents a page with a list of main tabs dedicated to different content:

- a Welcome tab, where the user can overview important and/or urgent information coming from the VOs (s)he participates in (e.g., events, scheduled meetings, urgent notifications), and change her/his profile and preferences (e.g., tabs layout);
- a set of tabs, each one devoted to a single VO the user participates in or to a specific application scenario, showing all information related to the current activities, as well as the collaboration and instruments-related tools (e.g., see Figure 3a).

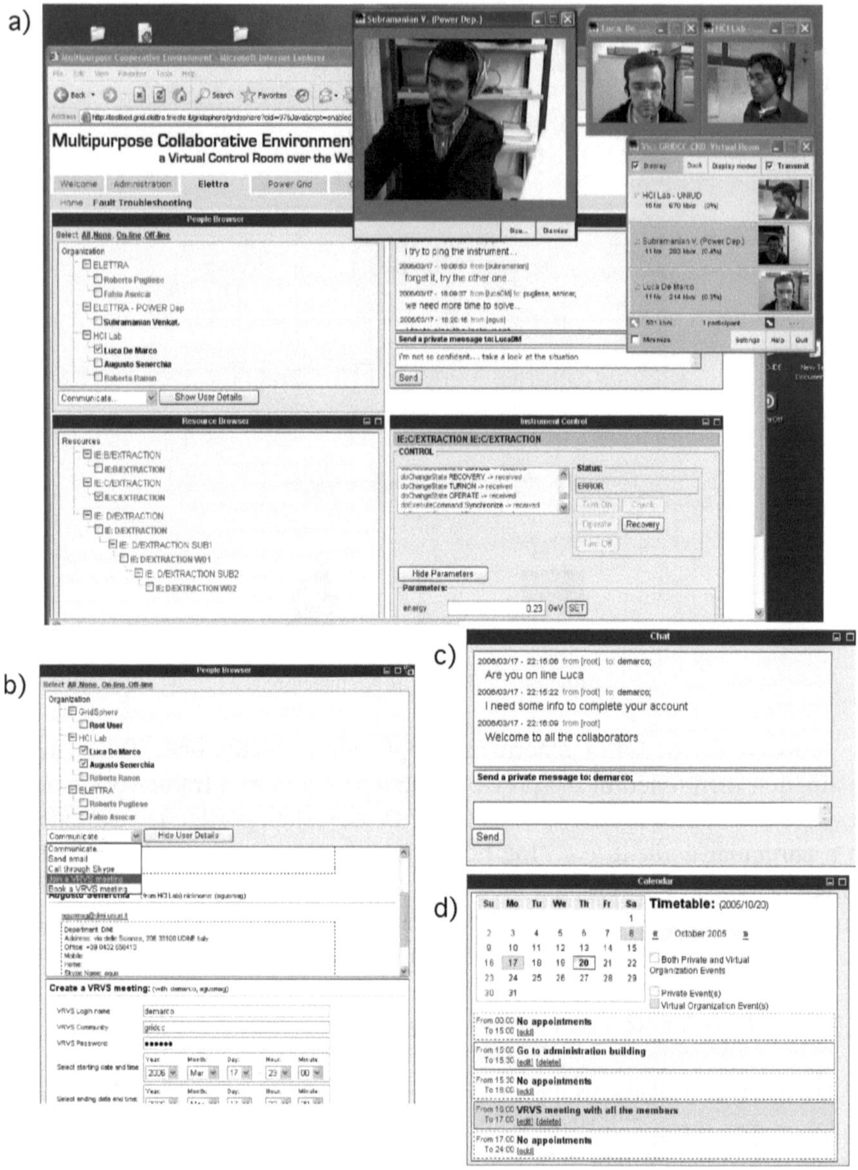

Figure 3. (a) A VCR VO workspace with communication and resources tools in a troubleshooting scenario; (b) the People Browser showing, from top to bottom, people, contact information and videoconference booking form; (c) the VCR chat; and (d) The VCR shared calendar

By selecting a particular tab, the user is able to access its associated workspace, which can be eventually arranged in different sub tabs. For example, the Welcome tab is arranged into a Home subtab showing calendar events and profile information, and a Layout subtab providing tools to customize the

VCR layout and appearance. VO tabs are, by default, arranged in a similar way, i.e., a Home subtab providing most important information and tools, and a set of task-specific or user-specific subtabs.

The creation of tabs and subtabs can be managed by either users or VO administrators, who can therefore create dynamic workspaces to support the execution of specific tasks. In each workspace, tools can be inserted and visually arranged as needed by the particular user or task.

VO administrators are provided with an additional administration tab, where they can manage users and groups, and organize workspaces by adding/ removing instruments and collaboration tools and arranging their layout.

4.3 Collaboration Tools

In the following, we describe the main VCR tools for remote collaboration, i.e., People Browser, Chat, Calendar, and Logbook.

People Browser. One of the key issues in distance collaboration is the availability of awareness information about remote collaborators. In particular, scientific collaboration often requires multiple persons and multiple specialties for carrying out experiments, or to get out the most from equipment. The People Browser provides a tool to browse VO participants, see their availability, and contact them.

For each VO workspace, a People Browser displays all VO members (grouped by the institution/department they belong to), in a tree-view (see Figure 3b). From the People Browser, the user has the opportunity to:

- look for a collaborator belonging to a particular department (e.g., maintenance personnel needed to repair a device) by clicking on the name of each institution to collapse/expand the members;
- be notified about changes of the collaborators' presence (users that are logged into the VCR are highlighted in bold). This is also useful to select the most appropriate communication tool to contact them (e.g., it makes more sense to contact an offline user via email or phone);
- for each collaborator, obtain detailed contact information (e.g., mobile or office phone);
- select one of the provided or integrated communication options to communicate with collaborators.

More specifically, the People Browser provides communication support among remote collaborators within a unique and consistent interface that integrates both VCR native communication tools (such as the chat) and external ones (such as email and Skype):

- text messaging to hold spontaneous and informal meetings with selected participants (or all VO people) by means of a chat (described below);
- send an email (selected contacts are automatically pushed to the default mail client as recipients);
- make an audio call using Voice Over IP through Skype [14] (if the participant has provided a valid Skype name in her/his profile, the VCR automatically loads Skype and starts the call);
- establish a videoconference through the virtual room videoconferencing system [3] (users can book a VRVS videoconference room within the People Browser, as in Figure 3b, and then enter the room).

The integration of existing and well-known communication tools also helps with users' acceptance, enabling remote collaborators to choose the communication tool they prefer or they feel comfortable with. For example, the VRVS conferencing system is very popular in the physics community (to which many of our pilot applications' users belong). Moreover, specific applications can add other technologies that will emerge in the future or are popular in their context.

Chat. The VCR provides also a chat tool for text messaging purposes (Figure 3c). The chat allows both to send VO-wide messages (when no participant is selected in the People Browser) and to support private conversations among two or more members (when participants are selected in the People Browser, messages are sent just to them). The private and public modes can be easily distinguished because the recipients' list is highlighted with different colors.

Public conversations are permanently stored in the VCR, and most recent messages are displayed when users log in. Such feature provides information on ongoing VO activities and conversations even for users that have been offline for a while.

Shared Calendar. The VCR provides a Shared Calendar tool (Figure 3d) which adds functionalities to facilitate meeting scheduling, project management, and coordination. In particular, the Shared Calendar allows one to:

- add, remove, or edit VO events, which are shared among all the participants of the current VO, or personal events, highlighted with different colors;
- overview the events defined for an entire month, and see the details of all personal and VO events of a given day; the Shared Calendar tool is, by default, available in the Welcome workspace.

Logbook. The VCR also provides an electronic notebook consisting of a forum-like utility for posting comments and holding discussions among the VO members and a wiki-like utility for creating a help system. Entries to the Logbook are added through a form with the help of a WYSIWYG text editor [6], while the postings from the instruments may be done programmatically through a Web-service interface.

4.4 Resources Tools

The resources tools provide generic support to browse, check the status, and access distributed resources (Resources Browser), as well as monitor and control distributed instruments (Instrument Control). These tools (and, in particular, the Instrument Control), while providing the functionalities requested by the GRIDCC project, are not meant to be used as they are, but rather to be customized and/or extended in the context of a specific application.

Resources Browser. The Resources Browser (depicted in the bottom-left part of Figure 3a) is the main entry point to access remote resources such as instruments, computing, and storage ones. It allows the discovery of resources beyond the boundaries of individual labs, ensuring that resources needed for a particular task can be reserved and accessed by users of the VO.

The Resources Browser uses a tree-view to display the list of available resources, following structural, and naming conventions that can be established by each application. Indeed, the logical organization of resources might change from one domain to another. For example, some applications prefer to view each node in the hierarchy as a physical instrument (leaf nodes) or subsystem (internal nodes), while others prefer to use the hierarchy to organize a set of instruments into operations modes.

Instrument Control. For the development of the remote monitoring and control functionalities of the VCR, we worked in close contact with the project partners to point out a common set of functionalities. Unfortunately, the user-interface requirements, as well as the preferred coordination mechanisms were too diverse to lead to a unique tool. Therefore, we developed a generic Instrument Control tool (depicted in Figure 4), which is general enough to satisfy basic needs of all the involved partners. In particular, the design process pointed out the need for four different functional sections:

- Control section, showing information to identify the remote instrument, allowing to monitor and change its status and configuration parameters, and providing the possibility to browse the history of the commands issued recently.

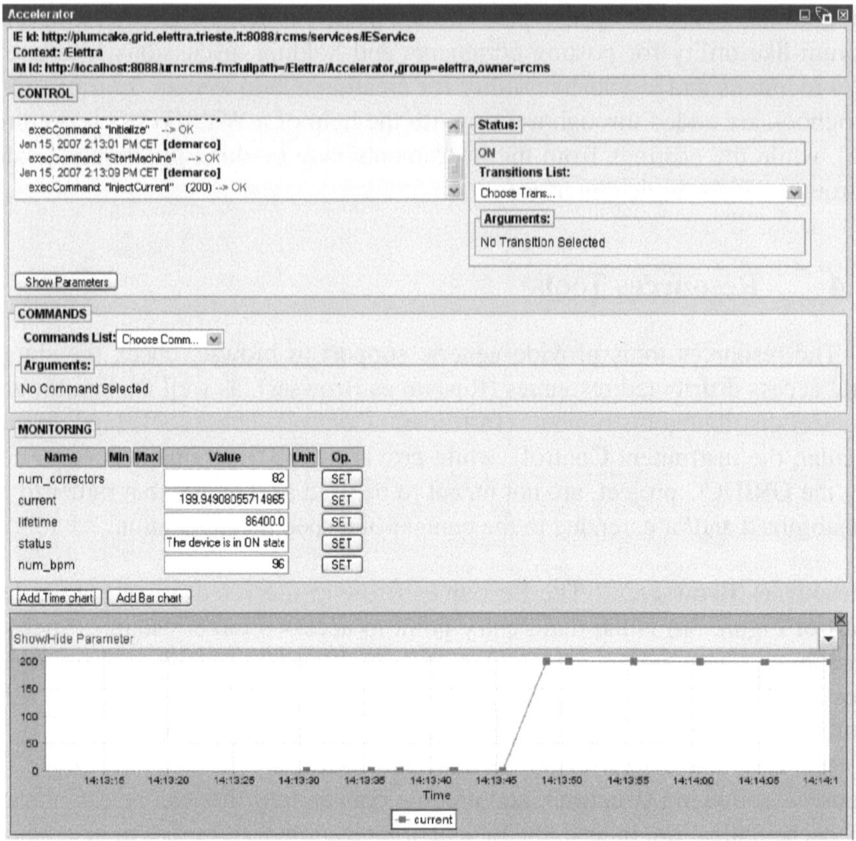

Figure 4. The Instrument Control tool

- Commands section, displaying the list of available commands that allow to change the status and perform operations on the remote instrument and the list of input arguments of the selected command.
- Monitoring section, allowing to monitor and change the values of the remote instruments' attributes.
- Graphical section, allowing for graphical monitoring of one or more instrument's parameters. Two types of charts are provided, a time chart, as the one shown in Figure 4 and a bar chart.

5. IMPLEMENTATION ISSUES

As mentioned before, the idea is that the different applications should extend and specialize the Instrument Control tool to meet their specific requirements.

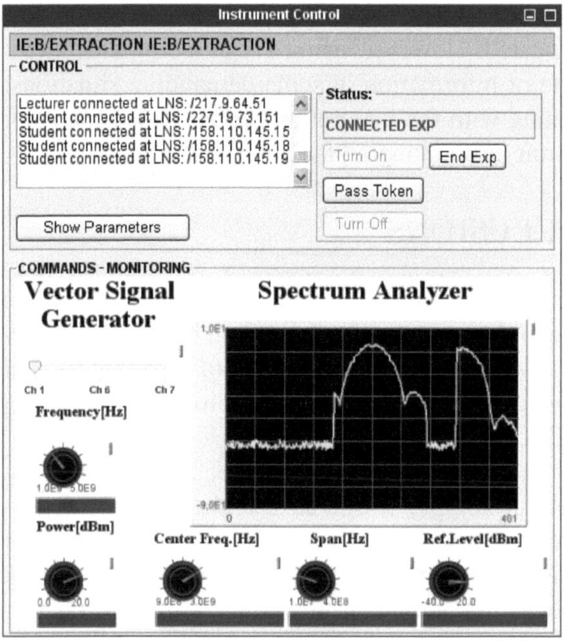

Figure 5. A possible Instrument Control Tool for a Signal Generator

Simple customizations involve the hiding of sections that are not needed for a particular operational context. We are now working with pilot applications to support them in more complex customizations. For example, Figure 5 shows a possible customized Instrument Control for an educational application in which a lecturer teaches students about Signal Theory and Practice by using a remote Signal Generator instrument. In this context, the Commands and Monitoring sections' interfaces need to be changed to reflect a real Signal Generator (using knobs to change values and a graph to display the signal waveform), and the Control section needs to be customized to allow the teacher to pass the instrument control token to any of the connected students.

The focus of the GRIDCC project on the interactive access and control of remote instrumentation poses some problems with respect to Web-based interfaces and technologies. For example, in the portlet paradigm, each time a users performs an operation, the entire page is entirely reloaded from the server and redisplayed by the browser, with the drawback that while the page is being reloaded and redisplayed, the user cannot interact with the system.

This problem is also common in most Web-based groupware systems. To overcome this limitation we have employed Javascript code to both implement some interface functionalities on the client side (e.g., hiding and showing parts of the interface), make asynchronous HTTP requests, whose XML answer is processed in the background, and change only parts of the interface that required to be updated. However, using HTTP communications still can limit real-time update of information in some demanding situations. We are therefore experimenting with the integration of distributed messaging frameworks to rapidly transmit data among all the connected clients.

6. CONCLUSIONS

This chapter has presented the issues, design process, and current prototype for a groupware tool to support distributed and heterogeneous teams of workers in the collaborative monitoring and control of instruments. This is embedded within a next-generation cyberinfrastructure for physics developed in the context of the GRIDCC European project.

With respect to future goals of this project, we intend to both extend and refine the current prototype, as well as to evaluate it in realistic scenarios, such as the ones provided by the pilot applications of the GRIDCC project.

More specifically, collaboration tools under development include an Event and Alarm Viewer for automatic notification of VO-related events (e.g., a safety critical fault of a remote instrument) and a remote desktop tool for accessing legacy instruments not provided with the Web-service interface.

Another functionality under development is to enable access from mobile devices, which would improve the support to collaborators in several scenarios (e.g., troubleshooting activities from an area where workstations are not available). However, due to limitations of mobile devices with respect to display size, it is likely that only parts of the VCR functionalities will be available. To address this issue, we need to study the operational contexts of our pilot applications to understand which tools and functionalities are most required in a mobile scenario.

REFERENCES

[1] G. Almes, J. Cummings, J. P. Birnholtz, I. Foster, T. Hey, B. Spencer, "CSCW and cyberinfrastructure: opportunities and challenges". *Proceedings of the 2004 ACM Conference on Computer Supported Cooperative Work (CSCW '04)*, ACM Press, New York, USA, 2004, pp. 270–273

[2] M. Bhandarkar, G. Budescu, W. Humphrey, J. Izaguirre, S. Izrailev, L. Kal, D. Kosztin, F. Molnar, J. Phillips, K. Schulten, "Biocore: a collaboratory for structural biology". *Proceedings of the SCS International Conference on Web-Based Modeling and Simulation*, SCS (The Society for Computer Simulation International) San Francisco, California, 1999, pp. 242–251.

[3] Caltech: Virtual room videoconferencing system, www.vrvs.org

[4] G. Chin, C. S. Lansing, "Capturing and supporting contexts for scientific data sharing via the biological sciences collaboratory". *Proceedings of the 2004 ACM Conference on*

Computer Supported Cooperative Work (CSCW '04), ACM Press, New York, USA, 2004, pp. 409–418

[5] EUROTeV: European design study towards a global tev linear collider, www.eurotev.org

[6] FCKeditor: The text editor for internet, www.fckeditor.net

[7] T. A. Finholt, G. M. Olson, "From laboratories to collaboratories: A new organizational form for scientific collaboration". *Psychological Science*, vol. 8, 2001, pp. 1–9

[8] I. Foster, C. Kesselman, S. Tuecke, "The anatomy of the Grid: Enabling scalable virtual organizations". *Lecture Notes in Computer Science 2150*, Springer-Verlag, Berlin, 2001

[9] GridCC: Grid enabled remote instrumentation with distributed control and computation, www.gridcc.org

[10] GridSphere: Gridsphere portal framework, www.gridsphere.org

[11] JSR-168: Java community process: Jsr 168 portlet specification, www.jcp.org/jsr/detail/168.jsp

[12] R. T. Kouzes, J. D. Myers, W. A. Wulf, "Collaboratories: Doing science on the internet". *IEEE Computers*, vol. 9, 1996, pp. 40–46

[13] G. M. Olson, D. E. Atkins, R. Clauer, T. A. Finholt, F. Jahanian, T. L. Killeen, A. Prakash, T. Weymouth, "The upper atmospheric research collaboratory (uarc)". *Interactions*, vol. 5, 1998, pp. 48–55

[14] Skype: Skype communication tool, www.skype.com

[15] W. A. Wulf, "The collaboratory opportunity". *Science*, vol. 261, 1993, pp. 854–855

THE WILAB TELEMEASUREMENT PLATFORM FOR DISTRIBUTED RESOURCES ON HETEROGENEOUS COMMUNICATION NETWORKS

R. Soloperto[1], A. Conti[1], D. Dardari[1], and O. Andrisano[1]

[1] *IEIIT-BO/CNR, CNIT, DEIS, University of Bologna, Viale Risorgimento 2, 40136 Bologna, ITALY. Andrea Conti is now with ENDIF, University of Ferrara, Italy*

raffaele.soloperto@cnit.it, a.conti@ieee.org, ddardari@deis.unibo.it, o.andrisano@ieee.org

Abstract This chapter presents a platform enabling the concept of cooperative telemeasurements involving distributed resources, such as laboratories with instrumentation, programmable devices based on FPGA (partially or fully programmable) or DSP architectures and network cameras. The platform allows an easy and fast remote configuration through virtual interfaces (VIs), implemented through both commercial software and Java; in the latter case VIs are designed in such a way that they do not depend on the particular resource, but on its functionalities. That means, for example, that the interface working for a specific spectrum analyzer also adapts itself to the same kind of instruments of different manufactures, whereas the traditional approach would require customized VIs for each specific instrument. New general-purpose data structures and communication protocols have been properly implemented in XML code, with high degree of scalability and transparency to user hardware and interface. The architecture has been implemented in the Wireless Communication Laboratories (WiLab) at the University of Bologna, Italy, where three different laboratories are interconnected through a heterogeneous communication network and properly cooperate to increase their functionalities.

1. INTRODUCTION

Nowadays, many research projects are related to the concept of telemeasurement with remote resources (such as instruments and programmable devices) to increase their intrinsic value by extending the applications range

This research is developed within SUMMIT project and UE RINGrid Project (contract number 031891).

through the interconnection of distributed laboratories. In [1] the concept of wide-send telemeasurement is introduced where both the instrumentation and the devices under test can be defined and controlled remotely. The concept of cooperative distributed laboratories is discussed in [18]. It consists in making use of distributed resources (instruments, sensors, devices, etc.) as a single aggregated measurement system. The cooperative distributed approach represents a geographical extrapolation of the basic structure of a typical measure instrument, where each unit can be spatially distributed but in cooperation with other units over the network. The realization of such scenarios presents several issues that have to be properly faced. In particular, the cooperation of heterogeneous devices, interfaces and networks requires the definition of a suitable middleware architecture with high level of scalability, transparency, manageability. Key aspects toward the remote configuration are given by the communication network characteristics and the development of virtual interfaces to export resource functionalities within the client–server [3, 7] or Grid architectures [3, 6, 7]. These topics are addressed in the SUMMIT project [19] with the aim of providing mobile users a platform for ubiquitous telemeasurement services. Traditionally, VIs were created through dedicated software libraries within proprietary development platforms (e.g., National Instruments LabVIEW [11] and HP VEE [8]). On one hand this solution allows a friendly graphical programming approach for the interconnection of instrumentation functional blocks (sub-VIs) made available for different communication protocols like general purpose interface bus (GPIB), LAN and RS232, on the other hand no further control can be performed on the driver implementation, unicast/multicast networking protocols, user equipment, licensing and the dimension of required plug-ins at the user terminal (of particular relevance when accessing the remote laboratory through a personal digital assistant (PDA)). As far as the required plug-ins at the user terminal are concerned, when the user decides to control a real instrument, he has to first install on his terminal a plug-in to emulate the proprietary environment, with dimension also reaching tens of MBs that can be relevant, depending on available interconnection bandwidth and terminal internal memory capacity. In the last years, several activities on these issues have been developed in the WiLab [23] at IEIIT/CNR (Istituto di Elettronica e Ingegneria Informatica e delle Telecomunicazioni del CNR, Consiglio Nazionale delle Ricerche) and CNIT (Consorzio Nazionale Interuniversitario per le Telecomunicazioni), University of Bologna, Italy. They regard telemeasurements through a heterogeneous network connecting three laboratories distributed around the Engineering Faculty, as depicted in Figure 1.

Within several national and European projects (e.g., [5, 6, 14, 19–21]) the telemeasurement concept has been extended in different directions related to

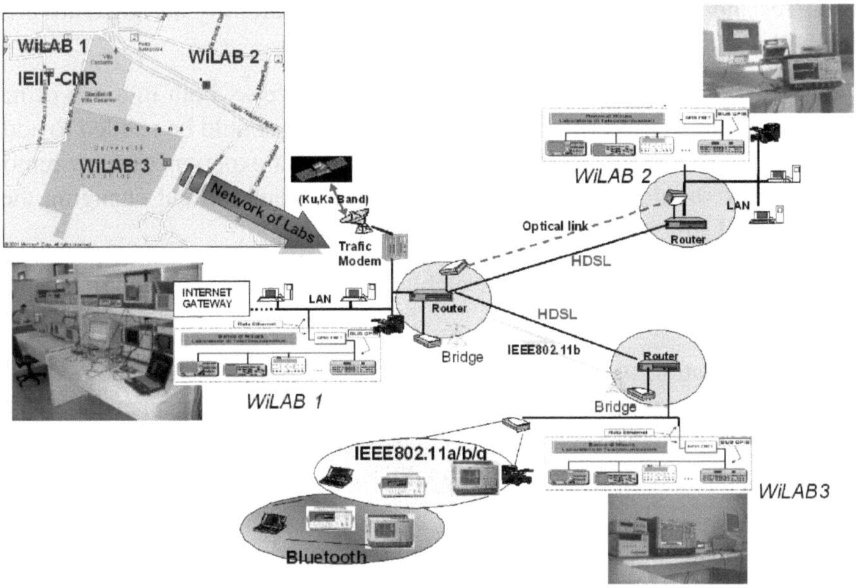

Figure 1. Distributed laboratories on heterogeneous communication network at WiLab, Bologna, Italy

the communication infrastructure, the remote VI in two and three dimensions, the access to experiences involving distributed laboratories, the cooperation among laboratories. Recently, within the SUMMIT project [19], WiLab was able to present a new architecture for telemeasurement on a heterogeneous network, based on Java (using JNI and XML tools) and compatible with NI LabVIEW. The chapter is organized as follows: in Section 2 the complete WiLab architecture for distributed and cooperative telemeasurements is presented, focusing the attention on each component of the implemented platform, underlining the advantages of Java and XML solutions. Section 3 and 4 present a possible realization of DSP remote control (in particular using TI C6711), using a web access to reserve this experience. Thus, the following section shows the web access and explains some menu and utilities implemented. Finally, an example of a telemeasurement experience, configuring instruments and DSP circuit is provided in Section 5 with some results, and our conclusion are given in Section 6.

2. WILAB ARCHITECTURE

The need of integration for distributed components involved in a telemeasurement experience and the goal of simple access to the distributed laboratories led to the choice of a client–server architecture. The heterogeneousness of resources and communication networks requires a carefully designed architecture with the following characteristics:

- inserting a new device in the network should be easy both in terms of network administration (no changes to the structure should be needed to add a device) and of its availability in telemeasurement experiences (new remote interfaces should be adaptable to the new element without changing their functional structure);
- the whole structure should be scalable even if an entire new laboratory network is added to the system;
- the organization of the laboratories and the management of the resources should be transparent for users to give them the idea of a super-laboratory, in which all the resources are aggregated as in a unique laboratory and this aggregation increases their functionalities.

To reach these targets the architecture we designed and implemented is based on a three-level hierarchical structure, as shown in Figure 2. This structure has been already implemented in the three laboratories of WiLab and is based on (bottom-up order)

- Level III: This level is composed of the single laboratories of the network in which the resources (instruments, circuits and programmable RF and baseband interconnection matrix) can be remotely controlled. Each resource

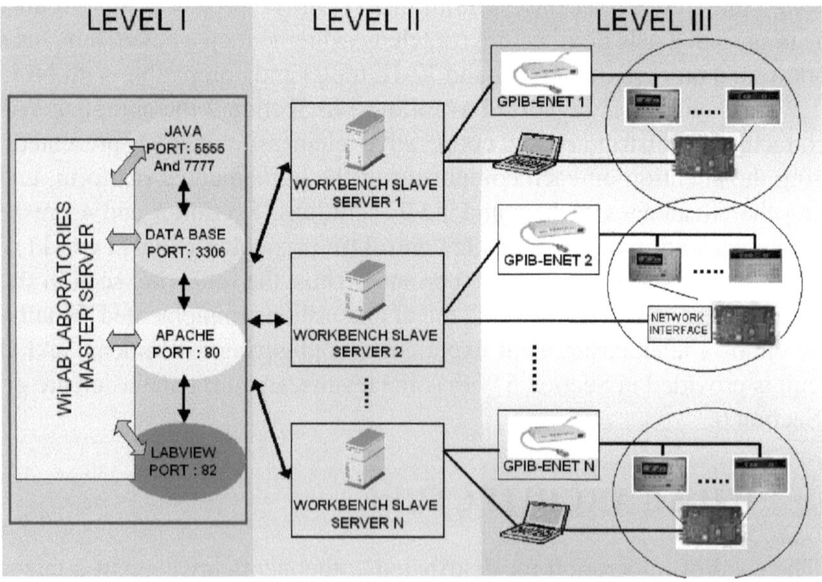

Figure 2. Distributed laboratories architecture at WiLab, Bologna, Italy

is connected to the network in a specific way: the measurement instruments are connected using the GPIB-ENET [9], a protocol converter from IEEE488 protocol to TCP-IP; programmable circuits based on DSP (i.e., digital signal processing) [22] are connected to the network by means of a PC host, and FPGA (field programmable gate array) based circuits, like Comblocks [4], are directly connected to the network using a LAN interface, which allows the chain of blocks to be addressed directly using a static IP address.

- Level II: This level is composed of a set of workbench slave servers each one controlling the laboratory they belong to. This level of servers (slave servers) has the role of managing the commands coming from the upper level to control the devices using the correct applications related to the particular resources involved in the experience. Slave servers also answer to upper level with information and data provided by controlled resources.
- Level I: At the upper level of the hierarchy there is the master server that is reached by the user through a common web access. It has the role of managing all the information about the working session of the user, interacting with the slave servers to perform the experience, retrieving information from the laboratories and presenting results to the user within the VI.

The master server is based on four different manager software tools to fulfill several needs of a complete telemeasurement experience:

- Apache [2]: Its role is to dialogue with the clients. The Apache server presents to the remote user both static pages (like HTML pages with documentation of instruments and technical notes) and dynamic pages (like PHP [16] pages containing information about the actual state of resources within the work session of the client). It is also able to report measurement results in the appropriate format according to clients' requests.
- MySQL [15]: Its role is to save, manage and upgrade information (about the instrumentation and the programmable devices present in the distributed laboratories) every time there is a request of control and the availability of the devices has to be checked. In a multiuser and distributed environment, the presence of the resource management system becomes strictly necessary to avoid conflicts among different contemporary users accessing the same resources. For this reason a scheduling system has been implemented to allow users to register themselves in the system and then to grant the reservation of the requested set of resources.
- Java: Its role is to allow the communication between the client interfaces and the controlled devices (when VIs are implemented in Java) using the

XML-based protocol to exchange commands and measurement results. It performs the translation of the device-independent commands into the specific communication protocol of the devices under control.

- LabVIEW: Its role is to provide the clients with executable VIs to control the chosen instruments (the LabVIEW server is used for downward integration with a previous platform).

These two last modules have been analyzed in order to investigate possible advantages/disadvantages in the control of resources present at WiLab (instruments and programmable circuits based on FPGA and DSP). The Java virtual interface (JVI) solution provides the following advantages:

- Portability: the Java code is interpreted by all machines and operating systems; it requires only the commonly installed Java virtual machine (JVM).
- Compatibility: Java is able to interact with other technologies like XML and LDAP [13], which are needed for a complete telemeasurement system to manage users and reservation for resources and experiences as it will be explained in the next sections.
- No licensing: Java applications are freeware and more than one user can have on their own PC the virtual remote interfaces at the same moment. It is obvious anyway that, being the only one real instrument, the control must be performed one user at a time.
- Multicast communication protocol: Java language completely supports multicast communication, which allows a more efficient traffic distribution through the MulticastSocket Java class.

In Table 1, the main characteristics of LabVIEW and JAVA platforms are reported for comparison; note that with additional plug-in we intend software usually not already installed in the client, such as LVRunTime Engine [12] for LabVIEW (about 30 MB).

Table 1. Main characteristics of LabVIEW-based and Java-based architectures

	LabVIEW	Java
Portability	Limited	Native
Purchase licenses	Yes	No
Multicast	No	Yes
Integration with other technologies	Limited	Native
Additional plug-in	Yes	No

One of the main targets of the activity related to networked laboratories in the project SUMMIT is the design and implementation of a more flexible telemeasurement platform exploiting a greater integration of different software, while reducing the impact of previously described disadvantages of proprietary platforms.

The telemeasurement platform in WiLab aims at remotely configuring more than 30 instruments and programmable devices of various brands for telecommunication system characterization, such as digital transmission analyzers, up to 40 GHz spectrum analyzers, vector signal generators and analyzers (with 50 MHz demodulation bandwidth), digital oscilloscopes, constellation analyzers, noise figure meters, noise and interference test set, RF channel simulators, switching matrixes, up to 50 GHz signal generators, network analyzers, RF antenna sets, WLAN access points, routers and DSP and FPGA evaluation modules.

From the above, it is possible to notice how resources are heterogeneous in their functionalities and, in addition, with respect to network connections

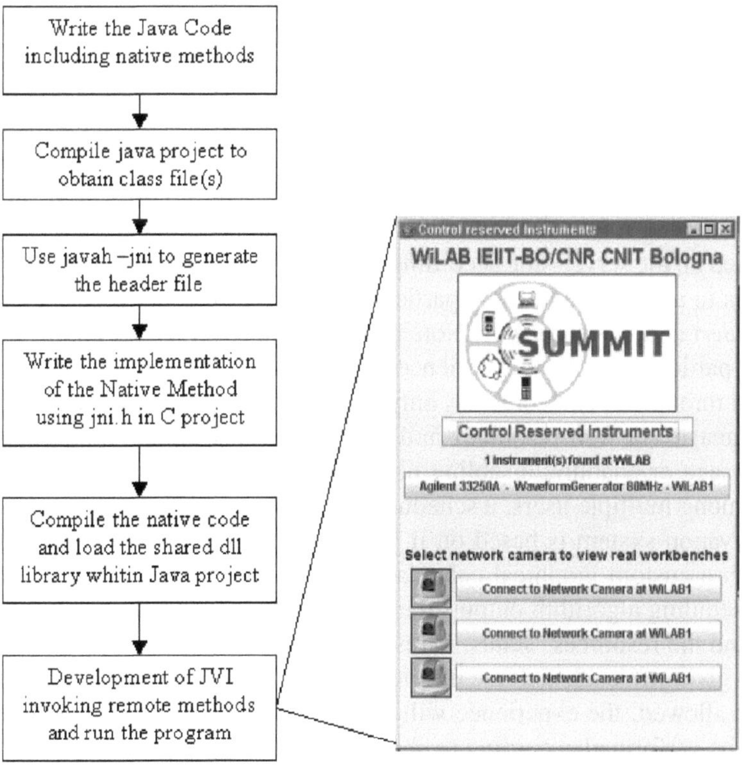

Figure 3. Operations needed for the creation of JVI starting from C drivers

(e.g., Ethernet access, RS232, IEEE488). For resources connected through the IEEE488 bus, a bridge between IEEE488 and TCP-IP protocols is needed; this is provided by the GPIB-ENET.

By using the remote communication syntax typical of each instrument, proper drivers (typically in C language) have been created for the management and the remote control of the devices. These drivers implement the commands characterizing each family of instruments, even of different brands. Hence, the conversion between C functions and Java methods declarations has been realized. This can be obtained using the Java Native Interface (JNI) [10] package, which is able to run the functions written in native language (C) using the declarations of the methods containing the functions and calling the DLL library obtained after compiling the C code.

The flow diagram in Figure 3 describes step by step how to realize a Java program by implementing the native C function by using JNI.

3. WEB ACCESS SYSTEM

A Web interface for web access to users has been implemented to grant the maximum portability of functionalities offered by the system: in particular, the website is based on Apache/PHP/MySQL. Since the access to resources must be controlled and regulated, a user is first required to login by inserting username and password (authentication is performed through a LDAP directory tree). If the login is successful, a new telemeasurement personal session is established on the server-side according to the user's group level; in fact each user belongs to a group featuring a particular expertise level, such as *guest, beginner, expert* and *administrator*. Note that, depending on the group, the user can access particular kinds of experiences and types of parameters that can be configured through JVIs. Moreover, only the members of the *administrator* group can manage via web resources and users, reserving some resources to only local usage or enabling/disabling users' accounts. To share the resources access among multiple users, a scheduling algorithm has been implemented and a reservation system is based on it. This is accessible from the user via web through a custom graphical calendar: once an experience has been selected, the scheduling algorithm outputs the calendar (Figure 4) which shows the time slots and the resources' status. If resources are available, a new reservation is placed giving an exclusive use of selected resources. Afterwards, whenever the user is allowed, the experience will be shown on a list of reserved ones, and it will be performed according to shared resource rules, not taking care of any sharing conflicts. A possible experience is to configure a transmitter on a DSP circuit as shown in the next section.

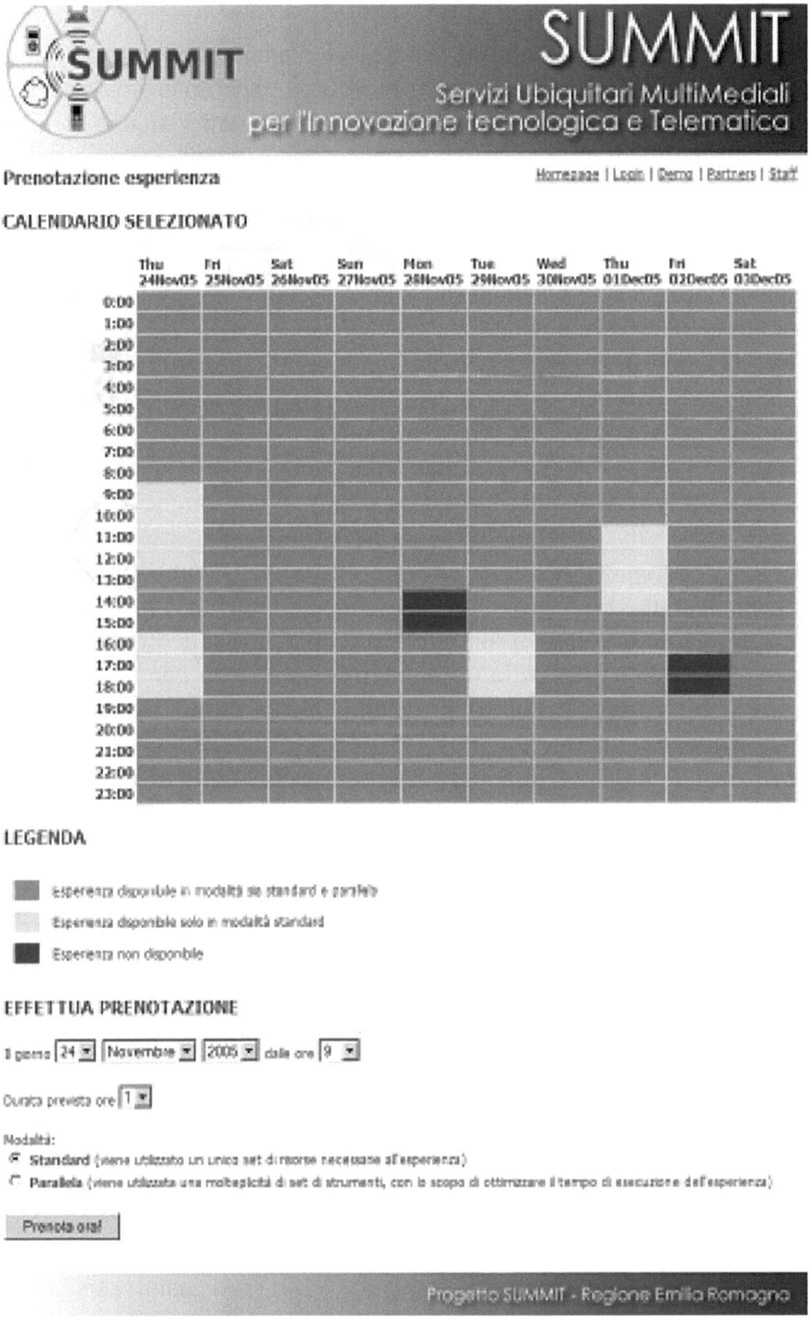

Figure 4. Scheduling for remote access

4. DSP REMOTE CONFIGURATION

In the last years, the meaning of telemeasurement has been strongly affected by a remarkable evolution: starting from the first experiments on remote control of measurement devices through GPIB cable, a new vision is coming, allowing the sharing of resources among several users, requiring the introduction of the concepts of resources availability and reservation.

The last steps of the research activities at WiLab concern the introduction of geographical distribution of the workbench laboratories as an enhancement of capabilities of a single laboratory. It implies a new concept of telemeasurements based on a geographical network of laboratories. Hence, the need to abstract the concept of resource, including not only measurement devices, but also DSP devices, FPGA partially and fully programmable circuits and wireless sensors networks, affects WiLab activities.

Thus, the concept of resource has been properly defined to implement a platform supplying different services by means of one unique access interface available and easily reachable through the Internet. Hereafter, with the term resource we will intend both measurement instruments and software-radio device such as DSP, FPGA programmable circuits, which can be configured for typical telecommunication system (toward software-defined radio paradigm). Thus, a software-defined radio (SDR) system is a radio communication system in which the user can control programmable hardware, by means of user-friendly software.

In particular, in this section a possible approach will be focused to perform a configuration of a L-QASK (quadrature amplitude shift keying) or L-PSK (phase shift keying) transmitter, based on a DSP boarded on TI C6711 circuit and sited in a laboratory remote from the user. Figure 6 shows the implemented Java interface in which users have the possibility to set some characteristic parameters related to modulation and then transfer them to remote circuit through

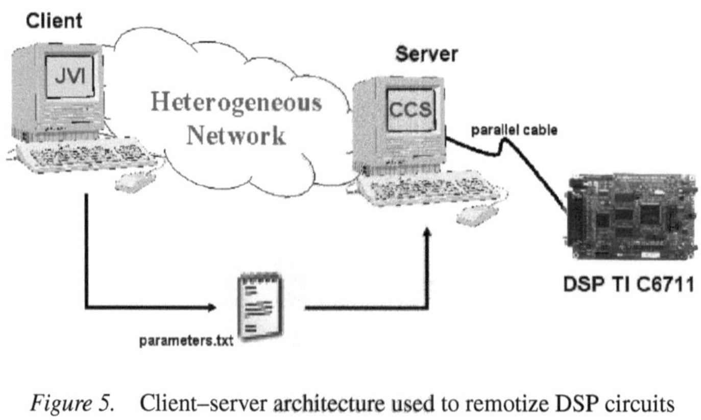

Figure 5. Client–server architecture used to remotize DSP circuits

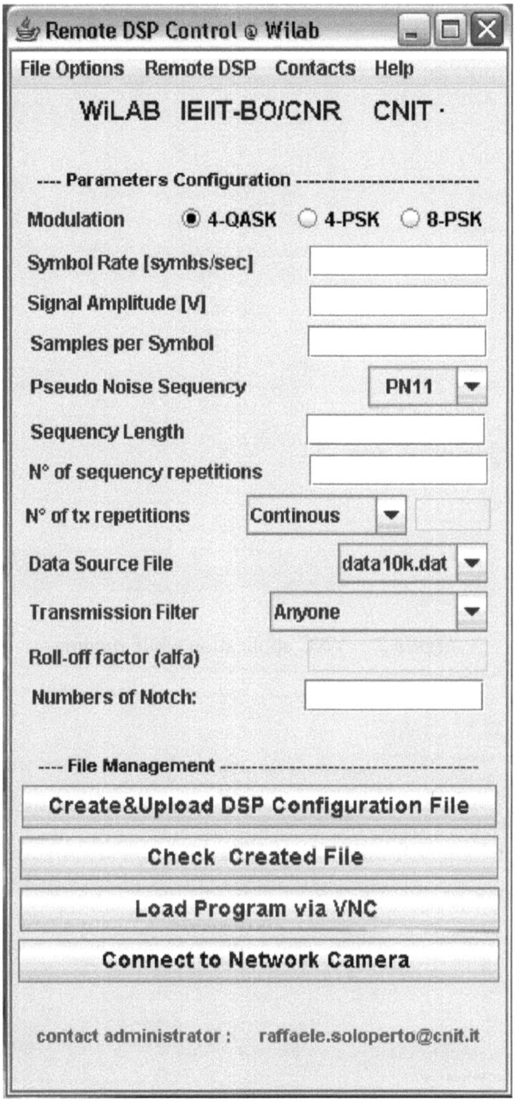

Figure 6. Java Virtual Interface to configure DSP C6711

the heterogeneous network. This is possible by a client–server architecture as depicted in Figure 5. The client, through the java interface shown in Figure 6, is able to write on the remote server a "parameters.txt" file, by pressing the "CreateAndUpload DSP Configuration File" button. In few seconds, all settings inserted by the user are captured and written to the server, according to the syntax defined by the administrator. After this operation, in order to assess the operation, the client could request the sent file to the server and control

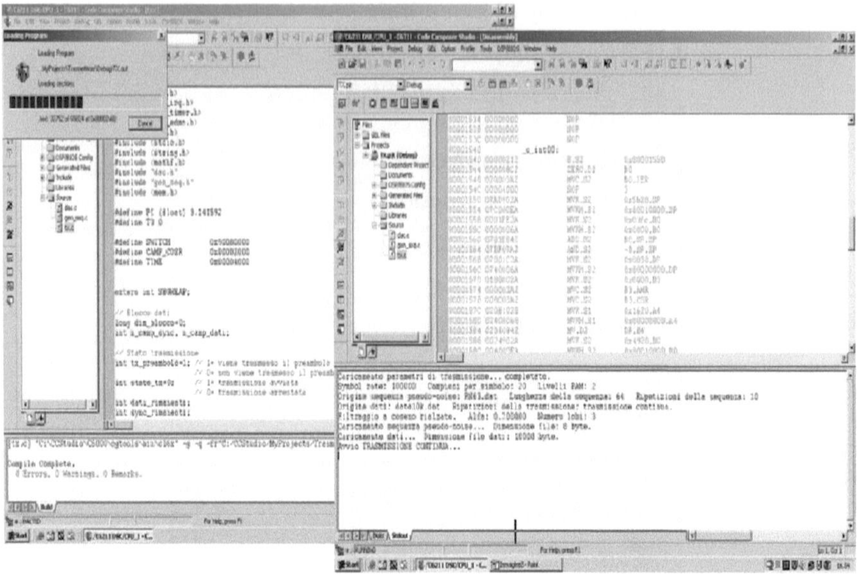

Figure 7. VNC application while running

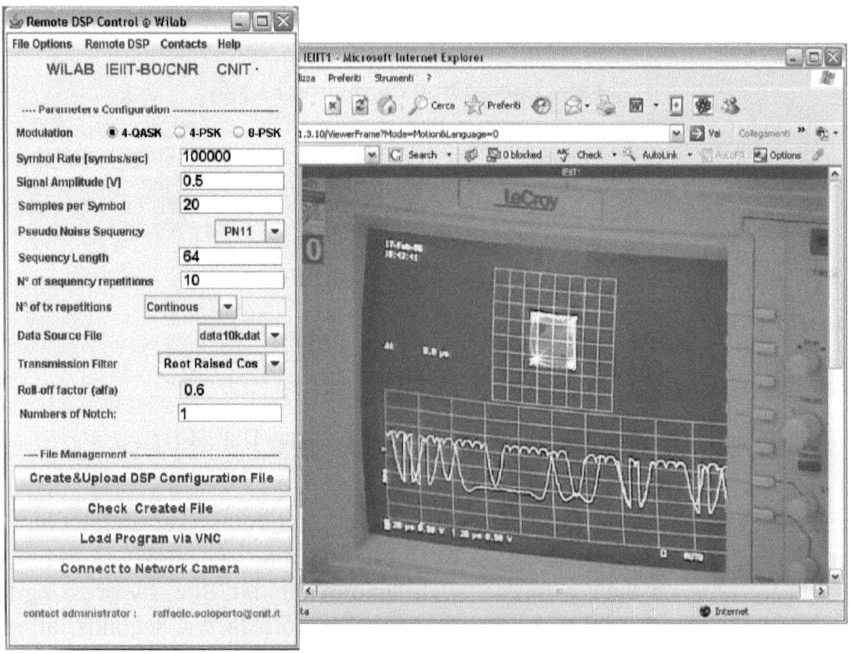

Figure 8. An example of realized transmitter

all the parameters (eventually the user can modify the requested file directly from the server, avoiding to run the java interface). When the loading operation terminates, another button will be enabled, in particular "Load Program via VNC", that allows to know the events on the server. The administrator that wants to release a new version of DSP software has to make four steps before reaching the full client remote control, in particular,

- Compile written c code;
- Build it;
- load file.out according to CodeComposerStudio (CCS) specifics;
- Run the implemented application to verify right funcionality.

Figure 7 shows an application while running, in which the main goal is to highlight what the server has effectively read from "parameters.txt", set up by the remote client, such as

- Modulation;
- Symbol rate;
- Samples per symbol;
- Transmission filter;
- Roll-off factor.

Moreover, when laboratory resources are running, it is possible to monitor them by means of networked cameras, also remotely controlled. In Figure 8 an example is reported of the realized transmitter with QPSK modulation (constellation diagram and I, Q signals).

5. EXAMPLE OF TELEMEASUREMENT EXPERIENCE

The architecture described in the chapter is currently used at WiLab for the characterization of telecommunication systems, but it can be extended to other fields. In Figure 9, a part of an experience is shown, in which programmable circuits are remotely configured through JVI and characterized in frequency with a spectrum analyzer, whose parameters are set with a JVI, and the measurement on real instrument is exported to the user's location by using a network camera, automatically zooming on the involved instrument. The programmable circuits are commercial FPGA-based modules, Comblock, which can allow the set-up of a complete telecommunication system on several radio

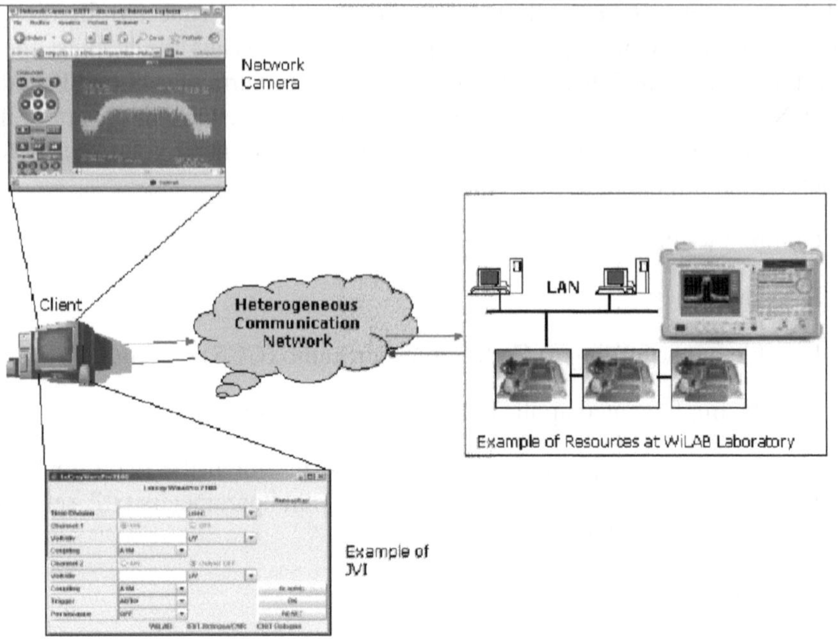

Figure 9. Example of telemeasurement experience

frequency bands (e.g., 902–928 MHz, 2025–2500 MHz, etc.). The spectrum analyzer is a Rohde&Schwarz [17] device that permits a frequency analysis from 20 Hz to 40 GHz. This is just part of an experience in which other circuits and instruments can be involved, interconnected and configured with our Java and XML-based platform.

Note that bit error rate (BER) measurements on a complete telecommunication systems have been performed by remote access and configuration of multiple devices using JVIs.

6. CONCLUSIONS

In this chapter we presented a new cooperative distributed telemeasurement platform based on Java and XML architecture, which has been realized at WiLab. Resources (instrumentations, programmable circuits based on DSP and FPGA, programmable RF and baseband interconnection matrix and network camera) are geographically distributed at WiLab laboratories and have been interconnected through a heterogeneous wireless/wired communication network. A system including a data structure definition and a communication protocol to access and control the resources has been defined in order to be the independent of the particular devices, operating systems and networks. By means of the implemented architecture, through a simple web access, users can download the executable software and run it on their terminals to manage and control the resources. Moreover, through a reservation system users are able to choose and reserve one of several experiences, independently of physical location of resources. Using networked cameras it is also possible to check the remote configuration made via JVI. The proposed architecture is used at WiLab for the characterization of telecommunication systems, an example is given in the chapter, but it can be extended to other kinds of experiences requiring remote configuration and measurements of both instruments and devices under test. The next step is to study the possibility to integrate this platform in a GRID architecture.

ACKNOWLEDGMENTS

The authors would like to thank Claudio Donzelli and Federico Marangoni for their contribution in the realization of the described telemeasurement platform.

REFERENCES

[1] O. Andrisano, A. Conti, D. Dardari, A. Roversi, "Telemeasurements and Circuits Remote Configuration Through Heterogeneous Networks: Characterization of Communications Systems", *IEEE Transactions on Instrumentation and Measurement*, vol. 55, no. 3, pp. 744–753, June 2006.

[2] Apache HTTP Server Project, http://www.apache.org.

[3] L. Benetazzo et al., "A Web-Based Distributed Virtual Educational Laboratory", *IEEE Transactions on Instrumentation and Measurement*, vol. 49, no. 2, pp. 349–356, April 2000.

[4] Comblock, http://www.comblock.com.

[5] CRIMSON, Cooperative Remote Interconnected Measurement Systems Over Networks.

[6] European Project RINGrid www.ringrid.eu.

[7] D. Grimaldi, L. Nigro, F. Pupo, "Java-Based Distributed Measurement Systems", *IEEE Transactions on Instrumentation and Measurement*, vol. 47. no 1, pp. 100–103, February 1998.

[8] Hewlett Packard, Easy Productive Instrumentation Control, January 1995 Edition.

[9] J. Humphrey, V. Malhotra, V. Trujillo, "Developing Distributed GPIB Test Systems Using GPIB-ENET/100 and Existing Ethernet Networks", *National Instruments*, Application Note 103, www.ni.com.

[10] Java Native Interface, http://Java.sun.com/docs/books/tutorial/native1.1/.

[11] LabVIEW 7 Express User Manual, National Instrument, April 2003 Edition.

[12] LabVIEW RunTime Engine, http://www.ni.com/support/LabVIEW/toolkits/appbuild/lv51ab.htm.

[13] LightWeight Directory Access Protocol, www.openldap.org.

[14] MultiMedialita – Interactive and Multimedia Network to extend Internet services.

[15] MySQL Database Server, www.mysql.com.

[16] PHP, www.php.net.

[17] Rohde&Schwarz, www.rohde-schwarz.com.

[18] A. Roversi, A. Conti, D. Dardari, O. Andrisano, "A WEB-Based Architecture Enabling Cooperative Telemeasurements", *2005 Tyrrhenian International Workshop on Digital Communications Distributed Cooperative Laboratories* – Issues in Networking, Instrumentation and Measurements – Sorrento, Italy, July 4–6, 2005.

[19] SUMMIT, Servizi Ubiquitari MultiMediali per l'Innovazione Telematica e Tecnologica, http://summit.aster.it.

[20] VICom, Virtual Immersive COMmunication, http://www.vicom-project.it.

[21] Teledoc2, www.teledoc2.cnit.it.

[22] Texas Instrument Digital Signal Processing, http://www.ti.com.

[23] WiLAB (Wireless Communication Laboratories), Bologna, Italy, www.wilab.org.

INSTRUMENTS IN GRID: THE NEW INSTRUMENT ELEMENT

C. Vuerli[1,2], G. Taffoni[1,2], I. Coretti[1], F. Pasian[1,2], P. Santin[1], and M. Pucillo[1]
[1]*INAF - Astronomical Observatory of Trieste, Via Tiepolo 11, 34143 Trieste, Italy,*
[2]*INAF - Information Systems Unit*

Abstract This work is focused on the interoperability aspects between the Grid and the scientific instrumentation. The IE (Instrument Element) makes possible the monitoring and the remote control of any kind of scientific instrumentation, although the test-bed of this first implementation is constituted of telescopes and related astronomical instrumentation. The first implementation of the IE deals with monitoring aspects; astronomers can remotely interface the telescope and related instrumentation and check the telemetric and scientific data when they are acquired. Future releases of the IE will include extensions, so that remote control capabilities will be also covered.

1. INTRODUCTION

Grid computing has emerged as the global resource for the next generation of e-science applications. The concept of grid computing originated as a solution to the problem of glueing geographically distributed computing and data resources belonging to different scientific institutions and bringing together communities with common objectives [5]. The first implementations of the grid concepts dealt with computational and data grids. This is for example the case of the High Energy Physics CERN LHC experiment [15] that contributes to the design and development of one of the biggest production grid infrastructure: the EGEE grid. However, the grid "idea" evolved with time and the grid environment can be now viewed as a seamless, integrated computational and collaborative tool that virtualizes resources of different kind such as clusters, storage, visualization devices, databases and scientific instruments and sensors.

Instruments and sensors are qualitatively distinct from computation and data, and thus impose some unique challenges and issues to current Grid technologies. For example, operational models may range from remotely-accessing a one-stop, full-service instrument site to dynamically composing distributed instrument, data and analysis services into a virtual instrument organization. Moreover, those new resources should be integrated with the computational and storage ones, so that the raw data produced by the Instrumentation can be processed and analyzed from a homogeneous environment.

This paper illustrates a novel approach in the integration of Instruments and sensors into grid production environments: the grid instrument element (IE). We present the general architecture of an IE and we focus our implementation examples on a simplified IE designed to interact with telescopes and astronomical instrumentation. Even if tested on astronomical telescopes, the IE can be exploited to "gridify" any kind of instruments. The goal is to make grid technology able to support tasks like accessing, configuring, getting the status and changing the status of these hardware resources.

The IE has been developed at the Italian Istituto Nazionale di Astrofisica [10].

It is worth mentioning that the first prototypal implementation of the IE has been built by extending a grid middleware based on Globus 2.4.1; it is fully embedded in the EGEE (Enabling Grids for E-sciencE) [13] environment.

2. RELATED WORK

The IE project dates back to 2002, when, in the framework of the Italian "GRID.IT" project funded by FIRB (Fondo Italiano per la Ricerca di Base), the porting in Grid of an application aiming at remotely monitoring astronomical telescopes and related instrumentation was attempted [11]. To carry out this task, a number of tools and services produced by similar projects were evaluated. In particular, the GRIDCC project (web services based) was carefully examined. Moreover, we took into account also the globus teleoperations control protocol and the common instrument middleware architecture.

In this chapter, we propose a different approach; we envisage a new "computational" element able to interact with Instruments on the basis of "Actions": the IE. The other projects developed a set of composable components that encompass access to instrumentation. They are actually new grid middlewares specialized for telecontrol and instrument integration. On the contrary, our approach is based on the idea that actual middleware implementations used to access computational resources should not be modified but integrated with new functionalities. In our vision, the interaction with an Instrument is the execution of a job, a special job: an "Action" command. Using the same commands, protocols and APIs are used to interact with the computational resources, we interface with instruments and sensors, providing a uniform access to different resources that can be directed and made interoperate together.

3. THE GRID-INSTRUMENT CONTROL SYSTEM ARCHITECTURE

From the point of view of the Grid architecture, the design of the G-ICS (grid-instrument control system) is closely related with what already done for the G-DSE (grid-data source engine) [1], the solution jointly proposed by INAF and INFN (Istituto Nazionale di Fisica Nucleare) to solve the problem of accessing databases via grid [3, 7–9].

The G-ICS does not represent a custom solution to meet specific astronomical requirements; it was thought and designed, instead, according to a general-purpose approach based on the concept of "Generalized Grid Element." The G-ICS can be used in all cases where a generic instrument has to be monitored and controlled via the grid. To produce the generalized grid element as a new grid element supported by the grid middleware, some elements of the grid architecture must be revised and integrated:

- The grid authentication and authorization protocols can be successfully used to map the grid user to the instrument user. The grid Security Infrastructure authentication protocol can be successfully used to authenticate the connections with instruments. However, authorization is often more sophisticated and fine-grained control can be provided over access on Instruments. Overall, a wide range of authorization constraints may be encountered in Instrument control systems.

- Resource availability is one of the key elements of the grid idea. Commonly a consumer makes a resource request on the basis of one or more resource properties. For example, a client may look for an Instrument that is able to detect a particular signal. The ability of a resource to describe what capability, capacity, quality and configuration it is able to offer and under what terms, is in charge of the information protocols. The *information protocols* are necessary also to clients in order to describe the desired resource capabilities and the way they will be used. To integrate an instrument in the resource description framework of the grid architecture [2], we need to extend the computational resource description with the peculiar attribute metrics of an Instrument. The new attributes metrics constitute the Information data model (IDM) for an IE. The IE IDM focuses on two different aspects: the Instrument intrinsic description and detection description. The virtual organization membership service (VOMS) [17] can be used to this purpose.

- Once we are able to find and successfully authenticate the desired resource, we have to be able to use it. The management protocols of the Resource layer are deputed to instantiate and monitor the interaction with the DSE resource. They provide a way to create a task (e.g., the execution of a job), monitor its status and manage the task data. The task type depends on the nature of the resource. In the case of instruments any task corresponds to an "Action," that is then executed by the Instrument. The IE is able to instantiate a task expressed in term of "Action" that the control system can understand, manage the life cycle of the "Action," and monitor the "Action" during execution as well as the status of the Instrument.

When a new instrument has to be interfaced to the grid a new specialization of the grid-embedded generic instrument will be produced in a way that the specific problem and kind of instrumentation to be monitored/controlled is taken into account.

It is worth pointing out here some key concepts that played an important role in approaching the G-ICS solution.

- In practice to produce the generalized grid element as a new grid element supported by the grid middleware, some components of the Grid middleware itself must be revised and integrated: namely, the BDII (Berkeley Database Information Index), the GIS (grid information service), the MDS (Monitoring and Discovery Service) and the RB (resource broker).

- The new generalized element (and each of its specializations) inherits from the classical CE (computing element) and extends it in terms of functionalities and intrinsic characteristics. An example of extended (special) job that exploits the new capabilities provided by G-ICS can be found in Section 3.6

- Such intrinsic characteristics are expressed by means of a set of new metadata, extending the set of metadata related to the traditional CE and SE;

- The new grid must be able to handle the new set of metadata, to ingest them in its Information System, and to publish them allowing users operating at client machines to discover these new resources.

3.1 Recycling G-DSE to Design G-ICS

The approach chosen for databases in grid and the related G-DSE solution has proved to be flexible enough to be adopted every time resources of different nature have to be ported in the grid. The problem of making the grid able to monitor astronomical telescopes (and more generally any kind of instruments) has triggered the design of the G-ICS.

To build the G-ICS, we used the same methodology and formal steps that drove the implementation of the G-DSE [1]. Both G-DSE and G-ICS are based

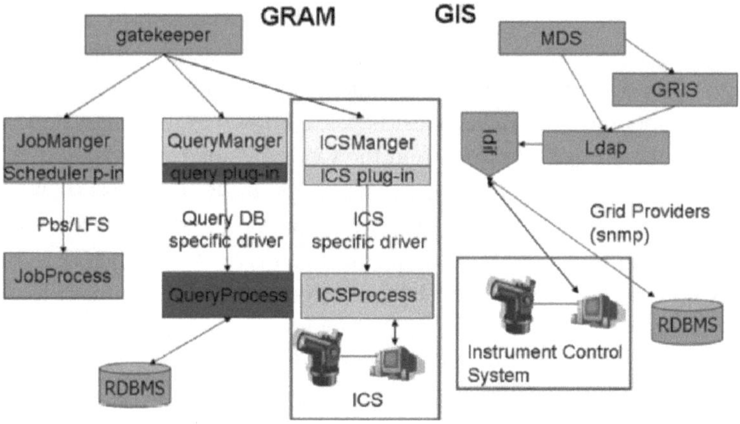

Figure 1. The modified GRAM and GIS components to enable G-DSE and G-ICS

on the concept of GACM (grid abstract computing machine). GACM is another example of virtual machine like JVM (Java virtual machine) and others. To implement a GACM, a formal specification language is necessary in order to define its syntax and the semantics like for any other virtual machine. To monitor and control instruments via grid, we started by defining the abstract machine ICS (instrument control system) thereafter we built the G-ICS (a new incarnation of the GACM) by integrating the ICS in the existing Grid RFL (resource framework layer). This integration has requested an extension of the semantics of the RSL (resource specification language) to take into account the semantics of the new GACM aimed at interfacing instrumentation and sensors. Another key point is the extension of the GIS to describe the new resources driven via G-ICS (scientific instrumentation), so that requests to the G-ICS (to monitor and control such instrumentation) can be understood and handled.

As shown in Figure 1, the GRAM (grid resource access manager) component of the middleware is modified in a way that two new managers are added to it besides the Job Manager, the first component of the GRAM that handles all jobs submitted to the grid for their execution. In a traditional computing-centric grid, the native Job Manager is the only component of the GRAM. The "Query Manager" and the "ICS Manager" make the grid able to access databases and generic instrumentation, respectively. What shown in Figure 1 is further explained in Sections 3.2 and 3.3.

3.2 The ICS Manager

The ICS Manager extends the capabilities of the original Job Manager and allows Grid users to write "Action" jobs to monitor the status of remote instrumentation; an “Action” may ask telemetry and scientific data to remote instruments at the same time of their acquisition. An example of such “Action” jobs is given in Section 3.6. When the gatekeeper of the grid receives a request to monitor a specific ICS, the request is passed down to the corresponding ICS Manager; as for the Job Manager and the Query Manager, a syntactical analysis of the job is performed and in case of success an ICS process (new instance of the parent process “ICS Manager”) is generated and the ICS is contacted through the specific driver. For this purpose, a driver for each ICS to be monitored must be provided. Because in the astronomical environment Instrument Control Systems are typically homemade and each of them has its own peculiarities, the driver is usually provided by people having the necessary knowledge of the specific ICS. Each ICS driver must be designed and implemented to be fully compliant with the common standards defined to interface any ICS and the Grid. The standards also specify the minimum set of functionalities to be implemented by each ICS driver.

These standards are currently under definition and strictly depend on the state of the art of actual instrument control systems currently in production, especially in terms of supported monitoring and control capabilities offered to final users.

The main task to be carried out in writing a specific ICS driver, then, is a mapping between the set of common functions and the functions provided by the specific ICS. The ICS Manager must be able to speak both the RSL to communicate with the gatekeeper and the ICSL (Instrument Control System Language) to communicate with the specific ICS driver. The concept of ICSL has been introduced in the context of the gridification of control systems. The full characterization of the ICSL requires the definition of the common set of functionalities that each ICS must provide for.

3.3 The Modified GIS

Modifications to the GIS in the G-ICS enabled grid middleware are of crucial importance. Without them it would be impossible to discover resources of the new type (databases and instrument control systems) in the grid infrastructure; users, therefore, would be forced to know in advance the physical location of such resources, wasting in this way one of the big advantages of using the grid. The key point in this case is to provide a new set of metadata to describe a resource of the new type (how to access it, some information about its architecture and implementation and above all what data are accessible through the new resource). Useful information for an observing structure could be the set of parameters characterizing the telescope, its location, the set of instruments interfaced to it and so forth.

The information service components of Globus Toolkit 2 are the monitoring and discovery services. In practice the MDS is built on OpenLDAP (Open Lightweight Directory Access Protocol) [6]. It uses a directory-based data model where data about each grid resource is published in the branches of the information tree. An information provider is a shell or Perl script that populates one LDAP branch with its specific data. Providers are executed on demand by the LDAP server. The existing Globus LDAP directory has been extended with the inclusion of a new LDAP schema defining object classes and attribute classes for the specific instrument. This, along with custom providers, allows publishing a complete new set of attributes and extending existing ones in order to export the equivalent of a standard computational resource object, but for Instruments purposes. Existing attributes have been preserved where possible.

3.3.1 The GLUE Extension for an Instrument Element

In the gLite Grid middleware the information service is based on the MDS and on the Berkeley Database Information Index (BDII) [4]. The BDII is a LDAP-based repository.

The Grid Laboratory Uniform Environment (GLUE) schema [2] is the standard adopted by gLite for the representation of resources, their state and functionalities. The GLUE is a collaboration effort among a number of European and American grid-related projects, which aims at producing a set of building blocks for the interoperability among different grid systems. GLUE is now an Open Grid Forum (OGF) [16] working group.

The GLUE conceptual design has been described by means of UML class diagrams supported by descriptive documents. The purpose of this approach is to provide a meta-schema from which mappings to different data models and technologies can be derived. In the case of gLite, the mapping is done to the LDAP directory-based data model. Actually, the GLUE schema is used to represent computational resources, storage resources, services and applications. To embed the IE in the gLite environment we integrate the GLUE with the representation of the instrument resource.

3.4 The G-ICS Architecture

We can describe in more detail the architecture of the new ICS-enabled grid middleware. At the "Collective services" layer, the RB is modified to make it able to understand and manage the set of metadata referring to the resources of the new type. At the "Basic services" layer, we redefined the information schema (the LDAP hierarchical tree) to insert the new set of metadata. At the "System software" layer, we find now the new systems that control observing structures (ICS). Telescopes and related astronomical instrumentation, finally, are new components that can be found at the "Hardware" layer. In summary, the modified components to make the grid able to support instruments are the Resource Broker (RB) at the "Collective Services" layer and the Information Schema at the "Basic Services" layer; those newly introduced are the ICS at the "System Software" layer and the Instruments at the "Hardware" layer. All other components were left untouched, although it is worth mentioning that RDBMS and data storage are new components that we provided in the framework of grid and databases projects, whose outcome was the production of the G-DSE and of the QE (Query Element).

In Figure 2 the details of G-ICS are shown. Within the GRAM protocol we find the Job Manager that, in turn, includes the ICSC (instrument control system connector) Manager. This is slightly different from what shown in Figure 1, where the Job Manager and the ICS Manager look like two separate entities. Figure 2 therefore well shows that the Manager (namely, the Job Manager) is extended to make it able to interact with an ICS. This solution has been chosen because we want to preserve the job-oriented approach; the proposed solution is totally based on the grid technology; other solutions combine the power of the grid with other technologies (web services). Jobs are generated and submitted to the grid quite apart from the kind of shared resources the user is accessing to.

Figure 2 focuses also on the GIS component, where the LDAP repository contains now a new set of meta-data for ICS resources. The core of the GIS subsystem is GANGLIA [14], a scalable distributed monitoring system for high-performance computing systems such as clusters and grids. It is based on a hierarchical design whose target is the creation of federations of clusters. It relies on a multicast-based listen/announce protocol to monitor the status within clusters and uses a tree of point-to-point connections among represen-

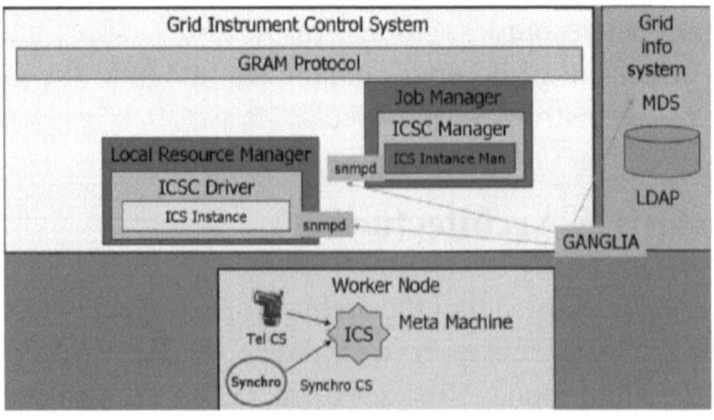

Figure 2. A different view of the ICS access-enabled Grid middleware

tative cluster nodes to federate them and aggregate their status. GANGLIA is used in this particular context to monitor the state of the various Grid resources through point-to-point SNMP connections; the so-collected state information is saved in the LDAP repository of the MDS.

3.5 From G-ICS to IE

Once the G-ICS-enabled grid architecture has been set up, the grid infrastructure is ready to include the new element "IE" of the grid. The IE is a CE able to handle monitor and control requests (in future releases) to interact with ICS resources via grid. The IE inherits the whole set of functions of a classical CE. In addition, the IE exploits all new functionalities offered by G-ICS to handle jobs whose target is the interaction with instrument control systems. Because the G-ICS is able to speak RSL when interacting with grid services, and ICSL when connecting to an ICS, these capabilities are translated from G-ICS to IE.

3.6 An Example of Special Job

In this section, an example of special job which contains directives to monitor and control remote instrumentation is given.

In practice, the standard Globus job submission commands can be used. For example, to get the telemetry data related to the status of the ccd by accessing the *telemetry* class of actions (*-queue* option), it is possible to use the command:

globus-job-run hostname:2119/iemanager-ascom -queue telemetry "get ccd status"

This example shows the modification done on the GRAM protocol: an *ie-manager* which uses the ASCOM [12] protocol to connect to the Telescope, a new queue concept now related to a specific Instrument property and a new job concept in the form of an "Action" statement. The advantage of this implementation is that all the GT2 functionalities (off line query, output redirection, interaction with a storage resource and co-allocation) and APIs can be used (Figures 3 and 4).

In the same way, it is possible to use the ldapsearch command to locate an IE. For example, if we want to find all the IEs accessible to the foo VO we execute (Figure 5):

ldapsearch -h bdii-fqhn -p 2170 -x -LLL -b "o=grid" '(&(objectclass= GlueIE) (GlueDSEAccessControlBaseRule=VO:foo))' GlueIEUniqueID

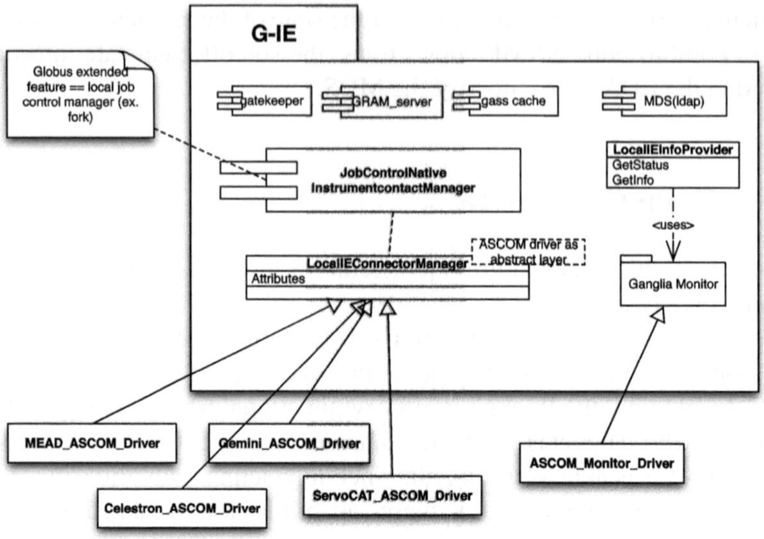

Figure 3. Resource level protocols overview for a GT2 IE. The picture presents the different components and layers involved in the "Action" processing. The key element that distinguishes the IE from the standard computational resources is the Instrument Manager

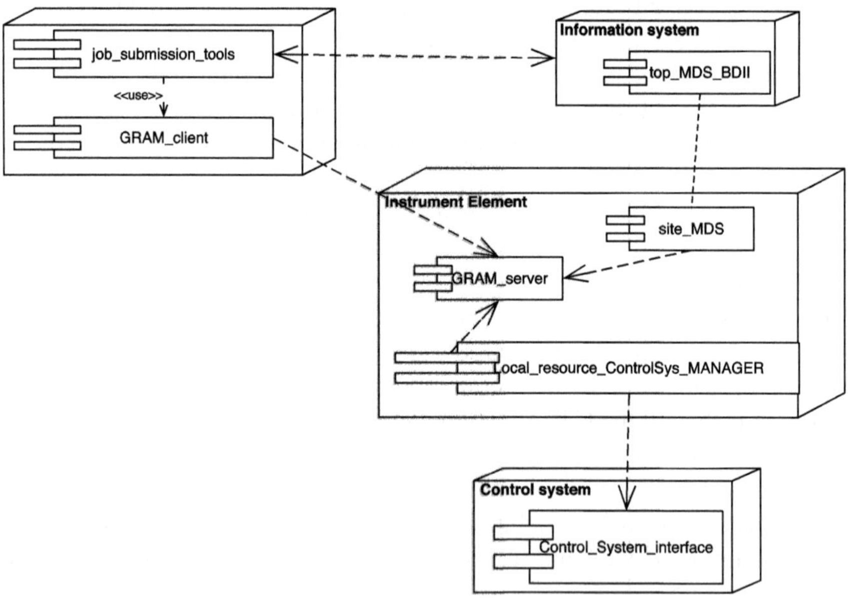

Figure 4. The resource allocation protocols. The picture presents some components of the GT2 implementation

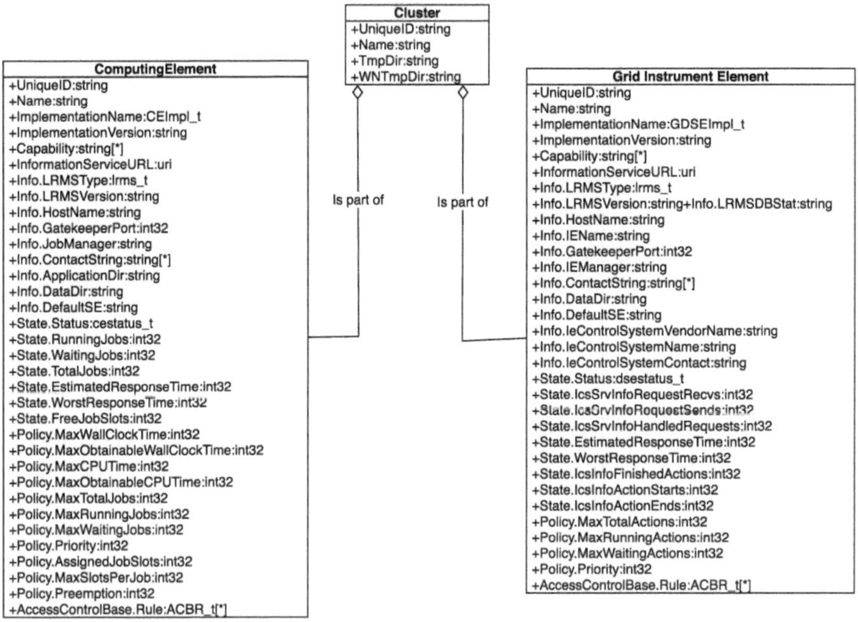

Figure 5. The GLUE computational resource UML description. We compare the CE and IE class view of the GLUE schema

4. CONCLUSIONS

Grid technology is not restricted to the field of distributed computing power within geographically spread VOs built over a grid infrastructure. Due to its nature, it goes beyond and can be used to share resources of different types on the geographical network. Starting from the early twin projects aiming at sharing astronomical databases in the grid as embedded resources of it, the rather ambitious goal of this project is to further extend the grid capabilities so that generic instrumentation can be configured as a new type of resource to share, monitor and control in the grid.

The G-ICS and the IE built on top of it are still in a prototypal phase. For this reason-limited tests have been carried out till now on the grid middleware where the prototypes were developed and using as test-bed a small telescope located at INAF-Trieste and operated for educational purposes.

The G-ICS approach provides a low-level integration of instrument resources in the grid. These new resources are fully embedded in the EGEE (Enabling Grids for E-sciencE) environment and can be totally managed and monitored using the same tools that we usually adopt to access standard computational resources. All this makes possible to use the standard grid workload system to submit and run mixed jobs that require both data from instruments and computations on them.

Although massive and comparative tests on G-ICS are not carried out so far we expect good performances by G-ICS as it naturally inherits all the scalability and flexibility features of the grid. As already highlighted in this paper, in fact, G-ICS is a tool entirely based on the grid technology avoiding in this way the overhead arising when the grid is interfaced to other technologies. We already verified this for the G-DSE. We expect the same good results for G-ICS, being it the counterpart of the G-DSE for what concerns Grid and instruments.

REFERENCES

[1] E. Ambrosi, A. Ghiselli, Taffoni G., "GDSE: A new data source oriented computing element for Grid", in: *Proceeding of Parallel and Distributed Computing and Networks - PDCN*, 517-523, 2006

[2] S. Andreozzi, M. Sgaravatto, M.C. Vistoli. "Sharing a conceptual model of grid resources and services". CoRR cs.DC/0306111, 2003.

[3] L. Benfante, A. Volpato, A. Baruffolo, L. Benacchio, "The OaPd System for web access to large astronomical catalogues", in: *ASP Conf. Ser.*, Vol. 238, 2001, F.R. Harnden Jr., F.A. Primini, H.E. Payne, eds, 2004.

[4] L. Field, "Berkeley Database Information Index", 2004 Retrieved September 18, 2007, from http://lfield.home.cern.ch/lfield/

[5] I. Foster, C. Kesselman, S. Tuecke, "The anatomy of the grid: Enabling scalable virtual organizations", in: *International Journal of Supercomputer Applications*, 15(3), 2001.

[6] T.A. Howes, M.C. Smith, G.S. Good, "Understanding and Deploying LDAP Directiory Services", in: *2003, Addison-Wesley (2nd Edition), ISBN 0-672-32316-8*

[7] S. Pastore, A. Volpato, A. Baruffolo, L. Benacchio, G. Taffoni, R. Smareglia, C. Vuerli, R. Crevatin, "Enabling access to astronomical databases through the grid: A case study", in: *Proceedings of Astronomical Data Analysis III Conference*, F. Murtagh ed., 2004.

[8] G. Taffoni, S. Pastore, R. Smareglia, A. Volpato, C. Vuerli, R. Crevatin, F. Pasian, "Astronomical data reduction on the grid", in: *Proceedings of Astronomical Data Analysis III Conference*, F. Murtagh ed., 2004.

[9] A. Volpato, G. Taffoni, S. Pastore, C. Vuerli, A. Baruffolo, R. Smareglia, G. Castelli, F. Pasian, L. Benacchio, E. Ambrosi, A. Ghiselli, "Astronomical database related applications in the GRID.IT project", in: *Astronomical Data Analysis Software and Systems XIV, ASP Conference Series*, Vol. XXX, P. L. Shopbell, M. C. Britton, and R. Ebert, eds, 2005.

[10] C. Vuerli, F. Pasian, A. Baruffolo, L. Benacchio, J. Alcalà, E. Cascone, U. Becciani, A. Fontana, G. Longo, R. Tagliaferri, "Grid related activity in progress at INAF", in: *Astronomical Data Analysis Software and Systems XIV, ASP Conference Series*, Vol. XXX, P. L. Shopbell, M. C. Britton, and R. Ebert, eds, 2005.

[11] C. Vuerli, G. Taffoni, A. Barisani, G. Capasso, E. Cascone, I. Coretti, M. Occhioni, P. Santin, R. Smareglia, F. Pasian, M. Pucillo, "Monitoring and remote control of scientific instrumentation through the Grid", in: *Astronomical Data Analysis Software and Systems XV, Publications of the Astronomical Society of the Pacific*, 351, 2006, p. 528.

[12] The ASCOM standards for Astronomy, http://ascom-standards.org/

[13] The EGEE Project, http://public.eu-egee.org/

[14] The GANGLIA Monitoring System, http://ganglia.sourceforge.net/

[15] The LHC Computing Grid, http://lcg.web.cern.ch/LCG/

[16] The OGF (Open Grid Forum), http://www.ogf.org/

[17] The VOMS (Virtual Organization Membership Service, http://voms.forge.cnaf.infn.it/

ON THE INTEGRATION OF TELECOMMUNICATION MEASUREMENT DEVICES WITHIN THE FRAMEWORK OF AN INSTRUMENTATION GRID

L. Berruti[1], L. Caviglione[2], F. Davoli[1], M. Polizzi[1], S. Vignola[1], and S. Zappatore[1]

[1] *CNIT – University of Genoa Research Unit, Via Opera Pia 13, 16145 Genova, Italy*

luca.berruti@cnit.it

[2] *Institute of Intelligent Systems for AutomationISSIA - CNR National Research Council of Italy, Genoa Branch, Via De Marini 6, 16149 Genova, Italy*

luca.caviglione@ge.issia.cnr.it

Abstract The possibility to perform distributed measurements, by means of complex tele-laboratory structures, is gaining momentum as a highly relevant aspect in numerous areas, including scientific research in general, environmental monitoring, distance education and training of staff operating over a vast territorial extension. Grid computing architectures or, more generally, service oriented architectures (SOA), viewed as platforms for the integration of distributed resources, play an ever-increasingly significant role, as witnessed by the large number of projects funded by the European Union (EU) on this and related topics. This role is no longer limited to the management of computational resources, and it extends nowadays to the integration of measurement instrumentation and large-scale distributed data acquisition platforms. In such a context, the functionality offered by the grid architecture allows unified control, maintenance, calibration and utilization of "traditional" instrumentation and laboratory equipment, as well as of remote data acquisition probes, by offering the user common standard interfaces and working environments. In this framework, the chapter introduces the design and performance evaluation of a test bench for telecommunication measurement instrumentation, interacting with the grid via the GRIDCC (Grid-enabled Remote Instrumentation with Distributed Control and Computation) middleware.

1. INTRODUCTION

Grid computing has emerged as a key infrastructure for "flexible, secure, coordinated resource sharing among dynamic collections of individuals, institutions, and resources" [7]. It is finalized to the aggregation of resources into virtual organizations [2,6]. Scientific communities embraced grid technologies to run projects involving distributed resources and using massive data sets, in the context of the so-called e-Science. Main goals are the improvement of the collaboration across distributed research groups, sharing experimental results that must be submitted to compute-intensive analysis and optimizing the use of computational and storage resources. So far, some data sources and sinks, such as measurement instrumentation and data acquisition devices, have not yet been considered as full Grid components. The interaction with the instruments is sometimes unmanaged and the instruments run off-line with respect to the Grid environment. Relevant research projects have been working or are currently working on these topics. The GRIDCC project [8, 10], one among the main European efforts in this direction, aims to extend grid technology to the real-time access and control of instrumentation. The CIMA (Common Instrument Middleware Architecture) project [4, 14], supported by the National Science Foundation in the United States, is oriented to the definition of a standard interface methodology across a broad range of instrument types and the abstraction of instrument capabilities. The final goal is to provide a uniform mechanism for registering, locating and interacting with instrumentation. At a national level, the CRIMSON and VLAB projects [9, 12] have investigated grid architectures, among other solutions, as concerns their capabilities of data gathering, storage and processing for tele-measurement applications. The ongoing RINGrid project [11] is exploring the impact of new emerging technologies to access remote instrumentation, in the light of user requirements and of the technological evolution in the mid-term.

Remote access to and management of measurement equipment has been considered extensively, though in scenarios not embedded within the Grid environment (see, among others [3] in the general measurement case, and [1,5] in the case of measurements for telecommunication systems).

In the present chapter, we highlight the features of a framework for tele-measurements on telecommunication systems (a "Device Farm" of telecommunication measurement systems), adapted to the general GRIDCC environment. The results of some performance evaluations of the ensuing test bed are also reported and used to discuss the opportunity, in some cases, of setting up data transfers "outside" the web services paradigm, e.g., by means of JMS (Java Message Service). It is worth noting that the integration of telecommunication measurement systems in a general setting of grid-enabled instrumentation may provide added value to more traditional approaches (such as those outlined in [1, 5]), in a number of respects: (i) the adoption of standardized interfaces; (ii) the complete abstraction of the instrumentation at the user level, giving the possibility of mixing real devices and software implementations of signal processing procedures; (iii) the capability – at least in some cases, where the underlying network transport capacity matches the sampling rate and the precision required of signal digitalization – of dynamically "composing" the measurement chain, by choosing devices within a pool of available ones, irrespective of their physical location; (iv) the advantage of using already available middleware functionalities for security, authorization, authentication, resource discovery and allocation.

The remainder of the chapter is structured as follows: in the next section we sketch the overall GRIDCC architecture. Section 3 is devoted to the integration of a telecommunication measurement systems' "Device Farm" (TMSDF) within the GRIDCC framework. A specific instance of integration is described in Section 4. Section 5 reports performance measurements, relative to synchronous data transfers. Section 6 contains the conclusions.

2. FRAMEWORK ARCHITECTURE

We provide a brief description of the general GRIDCC architecture and a slightly more detailed view of two architectural elements that are essential to our application.

2.1 The GRIDCC Architecture

Figure 1, taken from [15], depicts the principal components in the GRIDCC architecture. We summarize its main elements in the following.

The virtual control room (VCR) is the user interface of a GRIDCC application. Through the VCR users can either monitor and control instruments in real time or submit workflows to the execution services (ES). The VCR has been implemented with a grid portal technology in order to communicate with the underlying grid infrastructure.

The execution services consist of:

- workflow management system (WfMS): service that provides the capability of executing more tasks in sequential order. This is useful when the user has to work with more than one instrument, submitting a job to each one.
- workload management system (WMS), for logging, bookkeeping and service discovery.
- Agreement service (AS) for reservation management.

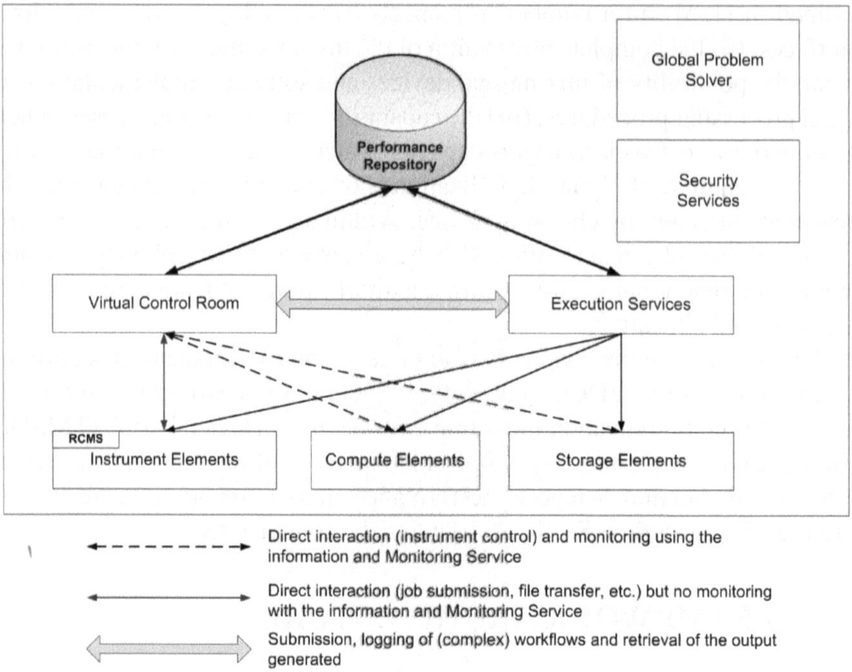

Figure 1. The GRIDCC architecture

Through the VCR, the user can define the workflows that must be executed by the ES. Moreover, the ES controls the quality of service, by making resource reservation agreements with the resources available (instrument elements, compute elements or storage elements).

GRIDCC provides two levels of automated problem solving: local problem solver (LPS) and global problem solver (GPS). The former, implemented locally to a given IE, works in order to solve malfunctions of the IE in which it is embedded; the latter solves problems related to the whole system, and it is embedded within knowledge-based services.

Security services are designed to guarantee security in terms of authentication, authorization and accounting.

GRIDCC uses an all-pervasive information and monitoring system (IMS): the run control and monitoring system (RCMS). The performance of this service is critical to the successful operation of the instruments within the GRIDCC architecture. RCMS works on a publish–subscribe model, by passing information to the different components, as they require it. The RCMS also passes information between the different services that make up the instrument element (IE).

2.2 The Instrument Element

The term instrument element describes a set of services, which provide the needed interface and implementation that enables the remote control and monitoring of physical instruments. In this context, the IE is a web service interface to monitor and control real instruments. Users view the IE as a set of web services. Web services provide a common language to the cross-domain collaboration and, at the same time, they hide the internal implementation details of accessing specific instruments.

That being so, the communications of external entities with the IEs are based on web services standards: WSDL (web services description language) files for describing the services, the information services for discovery and standards for describing workflows (e.g., BPEL4WS), agreements (WS-Agreement), etc. This of course leads to the requirement that the IEs themselves and all the supporting services, like ES and GPS, follow these specifications. On the other hand, the communication between the IEs and the corresponding instruments depends on the installation and can be handled by any network protocol or even by a physical connection, different from one instrument to another.

2.3 Instrument Managers

An instrument manager (IM) identifies the parts of the instrument element that perform the actual communication with the instruments. IMs act as protocol adapters that implement the instrument-specific protocols for accessing its functions and reading its status. Since the instruments are heterogeneous in nature, there is a need to support many instrument managers in the same container, one instance for each logical set of instruments.

Each IM is composed of three subcomponents: the command gateway, the monitor manager and the data mover:

- the *command gateway* accepts control requests from the virtual instrument grid service and forwards them to the instrument;
- the *monitor manager* is responsible for collecting monitoring data (status, errors, etc.) and providing it to the information and monitoring service in a common format;
- the *data collector* component gathers the flow of data (if any) from the instruments; it then passes it to the *Data Mover*, which provides the interface with any external storage or processing elements;
- the *finite state machine* reflects the collective state of instruments controlled by the instrument manager; it receives the status from the individual instruments asynchronously and updates its state accordingly.

If a need for transferring instruments from one instrument element to another arises, then all that is required is to remove the instrument manager instance from the instrument element and instantiate a new one on the other instrument element. The other services react to this action accordingly.

Generally, each instrument manager is collaborating locally with the interfacing services, to translate external requests for control or monitoring into requests that the instruments understand and can react to.

3. THE DEVICE FARM AND ITS INSTRUMENT ELEMENTS

The components of a TMSDF are illustrated in Figure 2. The minimal GRIDCC architectural elements required for its implementation are the IE and the VCR.

In relation to this specific application, it is important to highlight that the GRIDCC IE component can operate according to two main modes.

3.1.1 Synchronous Operational Mode

Preliminarily, the user has to design the experimental environment, by deciding how the various devices of the test-bench must interact among themselves. In the case of a telecommunication system test-bench used to study the characteristics of a certain DUT (device under test), during this phase, the user must plan which instruments have to be involved in the experiments, as well as their physical connections. Upon the completion of this preliminary step, the

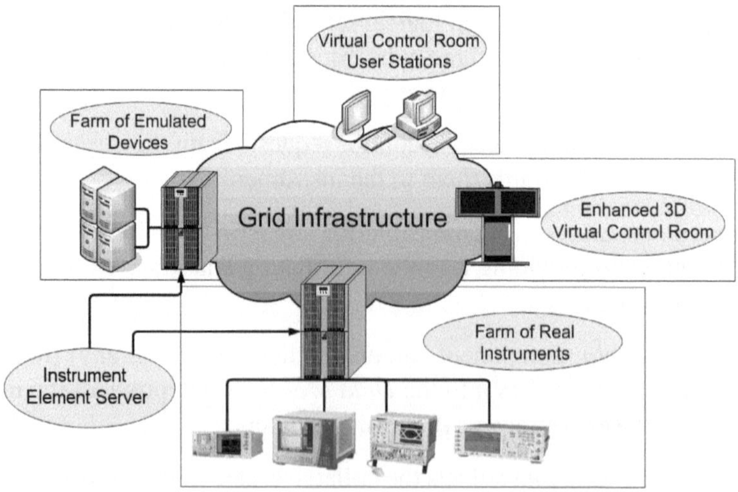

Figure 2. The "telecommunication measurement device farm" components

actual measurement phase can take place. Exploiting the facilities of the VCR, the user configures all the devices of the test-bench, by sending them a set of proper commands, with the appropriate parameters/working points. Then the IE waits till all the outputs are steady and notifies the user that the experiment results (i.e., the variables the user desires to observe) are available. Thus, the VCR downloads these results and displays them properly. If needed, while waiting for the outputs to become steady, the IE may record the behavior of a group of signals that the user intends to monitor. Successively, these records may be moved to the VCR off-line.

According to this operational mode, an experiment consists of a sequence of excitations and responses: by means of the VCR, the user synchronously forces a new set of excitations and, in turn, the IE returns the monitored (scalar and/or array) variables, which describe the resulting system's response.

3.1.2 Asynchronous Operational Mode

As in the previous case, the user must preliminarily set up the experimental environment. After starting the experiment, the user can quasi-continuously send commands to the equipment of the test-bench, thus changing the working conditions of the DUT. In turn, the IE gathers the (scalar and/or array) variables needed for monitoring the behavior of the DUT; after possible integration and adaptation, the IE returns the variables to the VCR, which properly displays them on the user's panel. In some sense, more than the previous one, this behavior resembles the usual operational mode with which a user manipulates real devices when he/she is actually present in a laboratory.

Which mode is more convenient depends on many factors and on the kind of experiment itself.

As a matter of fact, if the user is interested to know only the final state reached by a system after a certain excitation, the former operational mode is surely more appropriate; moreover, it calls for only a limited commitment of bandwidth resources.

On the contrary, if a high level of interactivity is required – for instance, in order to allow a user to have a feedback of what he or she is doing – the latter mode appears more appropriate. Obviously, the higher level of interactivity requires more bandwidth needs, as well as increased computational effort by all the elements involved in the experiment's management. In this case, the bandwidth requirement may be reduced, by using a principle borrowed from "delta-modulation": only the differences in the signal samples (e.g., representing a waveform or a power spectrum) that make up the content to be displayed, with respect to the previous ones, are actually transferred toward the user. An even more conservative option, which may be applied if the response time of the web service does not match the system's dynamics, may consist in sending

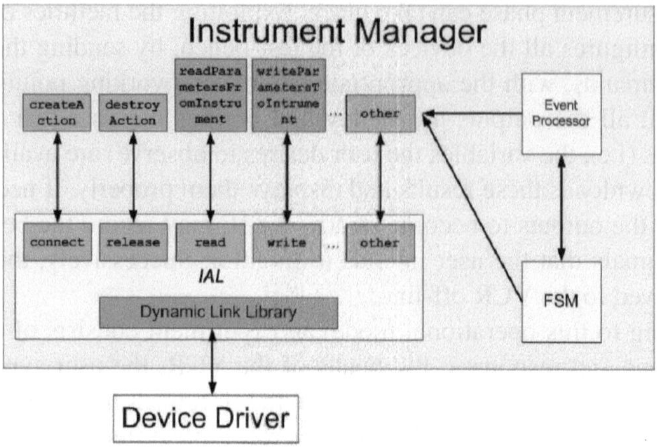

Figure 3. Instrument manager architecture

the new measurement samples only if they differ (in a given norm) from the previous batch by more than a certain threshold.

In order to embed real telecommunication measurement instruments in the GRIDCC framework, every instrument should be handled in a common way, by both the user and the developer, who has to write software to manage different devices. According to this view, we focused our attention on the implementation of a software module that could solve this problem and, at the same time, that could be simple to understand and to use. We have called this module "instrument abstraction layer" (IAL).

The basic idea for creating such a module is that each device should be identified by a proper set of variables, which can be written and read. Then, the IAL should act as an interface between applications that use the instrument and the driver of the instrument itself.

The IAL is written in Java language, and it implements a set of methods for reading/writing the variables, for connecting/releasing the device and so on. In order to use the device driver, the IAL can load one or more dynamic-link libraries (.dll) that represent the actual device, as depicted in Figure 3.

4. INTEGRATION OF A REAL MEASUREMENT INSTRUMENT WITHIN THE GRIDCC REFERENCE ARCHITECTURE

Basically, to integrate a real instrument within the framework, it is necessary to define

1. A finite state machine that controls the behavior of the framework, depending on the user's inputs.

2. An event handler that calls the appropriate functions, depending on the state reached.

3. The code that implements the instrument manager. This code specifies the right DLL to upload, the instrument address and the various operations typical of the physical instrument.

In this section we describe the exemplary integration in the framework of the digital oscilloscope Tektronix TDS 2024. Figure 4 depicts the finite state machine (FSM) used specifically for this oscilloscope.

Essentially, when a user accesses the IE, either through the web portal of the framework or through the VCR, the user can initialize the instrument within thc framework: this operation will cause the publication of the parameters that can be set on the real instrument.

In this case, parameters published for the oscilloscope are

- CH2_Waveform
- CH1_Waveform
- CH2_ChannelVerticalOffset

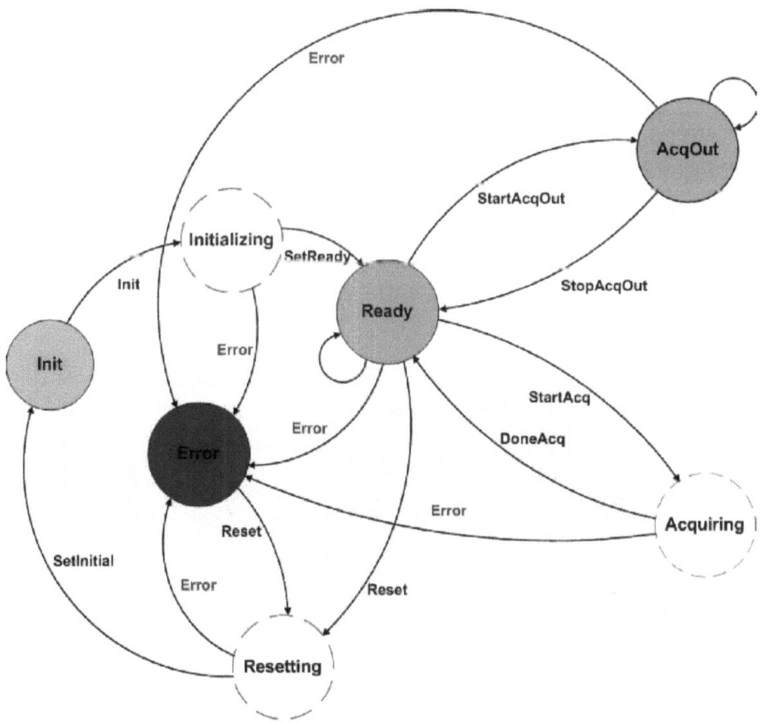

Figure 4. Finite state machine of the oscilloscope's IM

- CH1_ChannelVerticalOffset
- CH2_ChannelEnabled
- CH1_ChannelEnabled
- CH2_ChannelVerticalRange
- CH1_ChannelVerticalRange
- CH2_ChannelProbeAttenuation
- CH1_ChannelProbeAttenuation

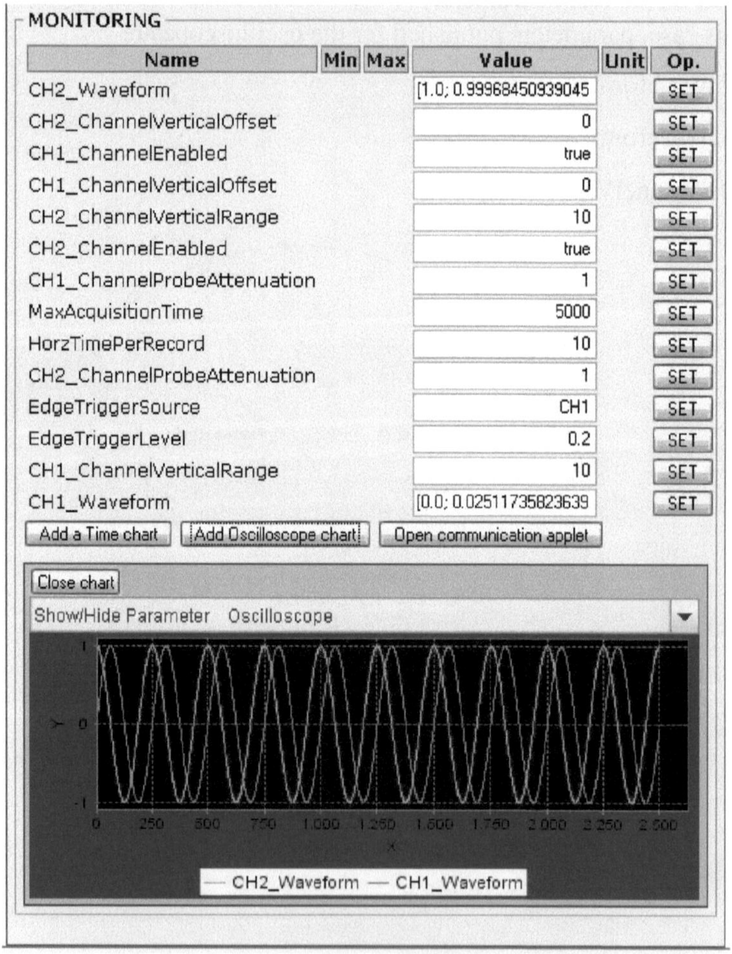

Figure 5. VCR screenshot of an oscilloscope's acquisition

- MaxAcquisitionTime
- HorzTimePerRecord
- EdgeTriggerSource
- EdgeTriggerLevel

Moreover, in the "ready" state, the user can modify and monitor all those parameters that have been published in the previous state.

Finally, through the "StartAcq" input, the user can make a real acquisition with the oscilloscope; this will produce a waveform that will be graphically visualized on the VCR.

Java code snippets illustrating the implementation of the finite state machine and of the way a specific event is handled are reported in the Appendix.

In order to graphically view the waveforms acquired, a java applet that plots the coordinates received from the framework over a Cartesian graph has been added to the VCR.

Figure 5 depicts a screenshot of the VCR that plots the waveforms acquired from both CH1 and CH2 channels of the oscilloscope.

5. PERFORMANCE EVALUATION

A number of tests were carried out in order to evaluate the performance of the IE, aimed at controlling a telecommunication system test-bench. In detail, our goal was to estimate (i) how many clients the IE can properly serve at the same time and (ii) the time the IE spends serving a specific request of the VCR. Obviously, the latter is strictly connected to the maximum throughput sustainable by the IE in terms of data exchange between a group of devices and a set of VCRs that have subscribed to view certain variables, describing the behavior of the experiment.

To this purpose, a suitable experimental set-up, sketched in Figure 6, was deployed.

The hardware adopted in the test bench (implementing the VCR and IE, as well as the client stations) is constituted by Fujitsu Scenico 300, based on Pentium 4 CPUs operating at 2.4 GHz. It is worth noting that, though the hardware type (and the software implementation) obviously affects the performance measures obtained, the general trend of the results that we present would not change drastically under a different setting. In other words, by increasing the CPU speed we could accommodate a larger number of clients, but the response time would increase similarly when approaching saturation.

Through all the tests, the set-up was configured to run according to the synchronous operational mode; furthermore, in order to highlight possible bottlenecks involved just in the software elements of the IE, no real instrument was

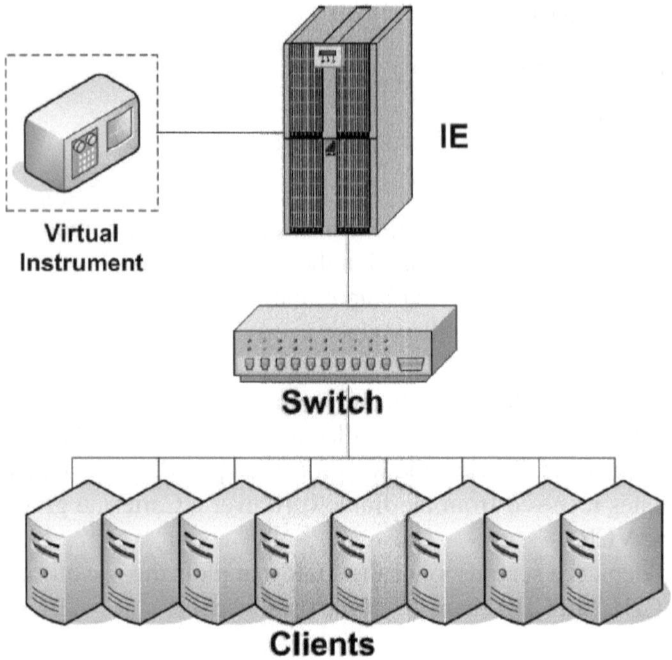

Figure 6. The experimental set-up used to evaluate the performance of the IE

employed: the virtual test-bench consists of a sole device that can suddenly export only an array of variables (e.g., related to the trace of an oscilloscope); in other terms, the IE can access instrument data with no time delay. Although the virtual instrument provides measurement data instantaneously, the IE's components spend a certain time to perform their tasks (finite state machine managing, data coding, packet assembling and transmission, etc.). It is evident that by doing so we can assess only a lower bound on the time the IE needs to serve a quite simple request (IE transaction) from the VCR. In a real test-bench, to the time estimated above, it is necessary to add the time involved in scheduling the actual device, the time required to send it the proper commands and the time needed to gather and check the data provided by the instrument.

At each user station, the VCR continuously polls the IE in order to obtain the array of data related to the virtual device. In this manner, it is possible to estimate the time spent by the IE to serve the request of each VCR, by changing the number of client stations involved in the experiment. Moreover, since the total net payload (consisting of measurement data) is known, the goodput and the related traffic load can be easily evaluated.

In detail, the experimental tests consist of some measurement campaigns: during each campaign the response time estimated in the presence of M client stations was achieved by averaging the times related to N consecutive IE transactions for each station. Then, the results obtained for each client were further averaged, in order to evaluate the response time of the entire campaign. For each campaign, let ΔT_{ij} be the time spent by the jth client to complete its ith IE transaction; then,

$$T = \frac{1}{M}\sum_{j=0}^{M}\frac{1}{N}\sum_{i=0}^{N}\Delta T_{ij}$$

where T is the estimated overall average response time.

The experimental results are graphically depicted in Figures 7 and 8. Figure 7 shows the behavior of the response time vs the number of client stations involved in the IE transactions. Specifically, the histogram reports the averaged response time evaluated through each campaign (the first three bars in each group of experiments) and the overall mean response time (the last bar in each group).

Figure 8 plots the behavior of the overall mean response time vs the number of active client stations (square marks) and the curve (dotted line) achieved by an exponential regression based on the measured data.

It should be highlighted that the overall response time quasi-linearly increases as the number of active client stations ranges from 1 to 15. At the other extreme, if the number of clients is greater than 15, the response time drastically increases and, consequently, the performance of the IE might become inadequate for many applications, even if they need only a medium–low level of interactivity. This behavior can be motivated by noting that beyond the limit of 15 clients the computational burden required to manage the stations involved in the experiment exceeds the maximum CPU capacity; in turn, this calls for a sort of severe queuing of the "threads" that handle the IE transactions.

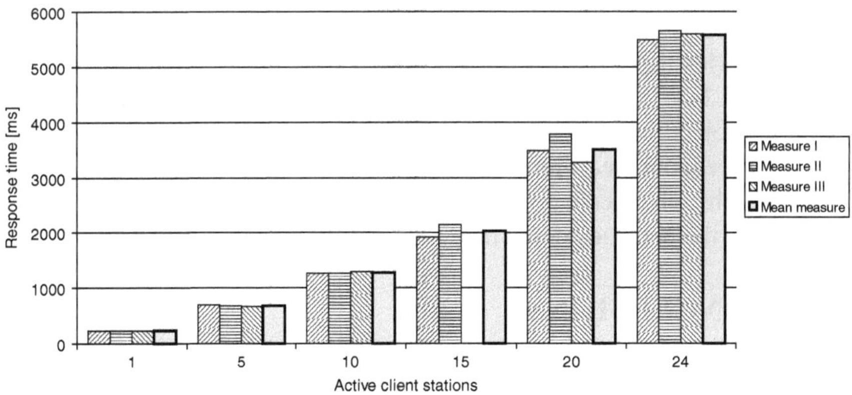

Figure 7. Behavior of the response time for different numbers of active client stations

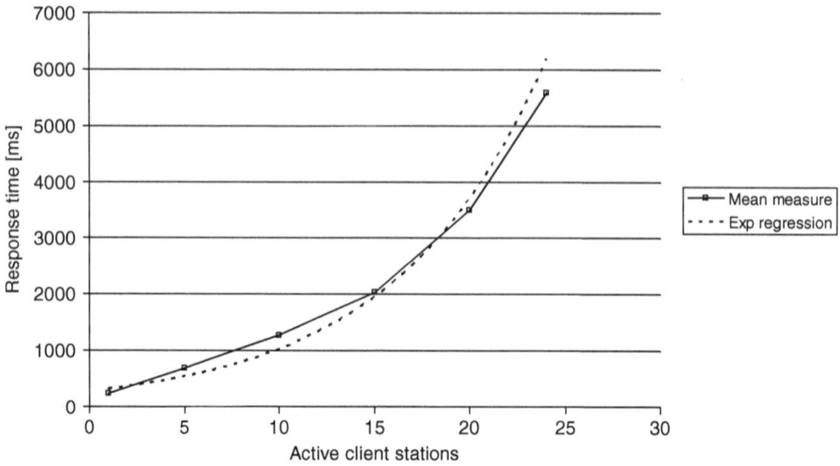

Figure 8. Behavior of the response time vs different numbers of active client stations, and curve achieved by an exponential regression of the estimated response time values

The results discussed above and the previous considerations affirm that whenever a medium–high level of interactivity is needed to fruitfully carry out an experiment and several client stations are involved, the IE operating in "synchronous mode" cannot represent an adequate solution to remotely control a test-bench of a laboratory. In this case, the IE should be assisted by a "dispatching facility", aimed at providing an efficient "out-of-band" mechanism to continuously update the array variables bearing data gathered from instrumentation to the clients. There are a number of tools that can play the role of message dispatcher: for instance, the data socket server by National Instruments can represent a possible solution, especially if the test-bench consists of only classical instrumentation (such as oscilloscopes, multimeters, and so on). If the communication network supports multicast and/or a heterogeneous pool of devices are part of the laboratory, more efficient facilities may consist of the Java Message Service (JMS) or the dispatching subsystem of the Lab-Net Server (LNS) [5, 13]. Indeed, the final implementation of the IE within the GRIDCC project includes a JMS module, which can be efficiently employed to dispatch frequent measurement results to a client population, on the basis of a publish–subscribe mechanism. The analysis of this mechanism and its comparison with the one considered here will be the subject of further research.

6. CONCLUSIONS

In this chapter we presented the design and performance evaluation of a prototypical test-bench for telecommunication measurement instrumentation, interacting with the Grid via the GRIDCC middleware. We have shown an example regarding the creation of the instrument element corresponding to a specific device, and the performance differences between synchronous and asynchronous operations. Future work will be aimed at investigating the performance of an enhanced version of the framework, which introduces a separate dispatcher, and at adding multiple devices with different characteristics, in order to prove the flexibility of the adopted solution.

ACKNOWLEDGMENTS

This work was partially supported by the European Union, within the framework of the GRIDCC (contract no. 511382) and RINGrid (contract no. (031891) projects.

APPENDIX

The code snippet that follows describes how the finite state machine of the oscilloscope is implemented in Java.

```
public class StateMachineDefinition
extends UserStateMachineDefinition {
  public StateMachineDefinition()
throws StateMachineDefinitionException {
      // Defines the steady States
      addState(States.INIT);
        addState(States.READY);
        addState(States.ERROR);
        addState(States.SIMULATION);
        // Defines the transition States
        addState(States.INITIALIZING);
        addState(States.RESETTING);
        // Defines the inputs
        addInput(Inputs.INIT);
        addInput(Inputs.SETREADY);
        addInput(Inputs.RESET);
        addInput(Inputs.SETINITIAL);
        addInput(Inputs.SETERROR);
        addInput(Inputs.STARTSIM);
        addInput(Inputs.STOPSIM);
        ...

        this.setInitialState(States.INIT);

        // Defines the State Transitions
```

```
        addTransition(Inputs.INIT, States.INIT,
                States.INITIALIZING);
        addTransition(Inputs.SETREADY, States.INITIALIZING,
            States.READY);
        addTransition(Inputs.STARTSIM, States.READY,
            States.SIMULATION);
        addTransition(Inputs.STOPSIM, States.SIMULATION,
            States.READY);
        addTransition(Inputs.RESET, States.READY,
            States.RESETTING);
        addTransition(Inputs.RESET, States.ERROR,
            States.RESETTING);
        addTransition(Inputs.SETINITIAL, States.RESETTING,
            States.INIT);
        addTransition(Inputs.SETERROR, State.ANYSTATE,
            States.ERROR);
  }
}
```

The following code shows how the event handler works in case the "Acquiring" state is reached, i.e., the user has launched a data acquisition.

```
public EventHandler() throws EventHandlerException {
// Entered Actions
  subscribeForEvents(StateEnteredEvent.class);
  subscribeForEvents(ParameterSet.class);
  // Adds callbacks action associated
// to a specific Function Manager State.
  addAction(States.ACQUIRING,"enteredAcquiring");
  ...
public void enteredAcquiring(Object obj)
  throws UserActionException {
    if (obj instanceof StateEnteredEvent) {
      try {
     oscilloScopeInstrumentManager.doAcq();
     logger.debug("Firing Event: " + Inputs.DONEACQ);
      oscilloScopeInstrumentManager.
fireEvent(Inputs.DONEACQ);
  logger.debug("Firing Event: " + Inputs.DONEACQ);
          } catch(Exception e){
        try {
              logger.error("enteredAcquiring: " + e);
              logger.error("Firing Event: " +
                Inputs.SETERROR);
              oscilloScopeInstrumentManager.
                fireEvent(Inputs.SETERROR);
```

```
                logger.error("FireEvent: " +
                  Inputs.SETERROR + '' done.");
          } catch (Exception e1) {
    logger.error("Execute Input SETERROR FAILED");
                throw new UserActionException(
                   "Execute Input SETERROR FAILED", e1);
          }
        }
    }
}
```

REFERENCES

[1] O. Andrisano, A. Conti, D. Dardari, A. Roversi, "Telemeasurements and circuits remote configuration through heterogeneous networks: characterization of communications systems", *IEEE Transactions on Instrumentation and Measurement*, vol. 55, no. 3, pp. 744–753, June 2006.

[2] M. Baker, R. Buyya, D. Laforenza, "Grids and Grid technologies for wide-area distributed computing", *Software: Practice and Experience*, vol. 32, no. 15, pp. 1437–1466, December 2002.

[3] M. Bertocco, S. Cappellazzo, A. Carullo, M. Parvis, A. Vallan, "Virtual environment for fast development of distributed measurement applications",*IEEE Transactions on Instrumentation and Measurement,* vol. 52, no. 3, pp. 681–685, June 2003.

[4] R. Bramley, K. Chiu, J. C. Huffman, K. Huffman, D. F. McMullen, "Instruments and sensors as network services: making instruments first class members of the Grid", *Indiana University CS Department Technical Report 588*, 2003.

[5] F. Davoli, G. Spanò, S. Vignola, S. Zappatore, "LABNET: towards remote laboratories with unified access", *IEEE Transactions on Instrumentation and Measurement*, vol. 55, no. 5, pp. 1551–1558, October 2006.

[6] I. Foster, "Service-oriented science", *Science Magazine*, vol. 308, no. 5723, pp. 814–817, May 2005.

[7] I. Foster, C. Kesselman, Eds., *The Grid: Blueprint for a New Computing Infrastructure*, 2nd Edition, Morgan Kaufmann, San Francisco, CA, 2004; ISBN: 1-55860-933-4.

[8] F. Lelli, G. Maron, "The GridCC project", in F. Davoli, S. Palazzo, S. Zappatore, Eds., *Distributed Cooperative Laboratories: Networking, Instrumentation, and Measurements*, Springer, New York, NY, pp. 269–277, 2006.

[9] Project CRIMSON (Cooperative Remote Interconnected Measurement Systems Over Networks) homepage: http://www.prin-crimson.org.

[10] Project GridCC (Grid Enabled Remote Instrumentation with Distributed Control and Computation) homepage: http://www.gridcc.org.

[11] Project RINGrid (Remote Instrumentation in Next-generation Grids) homepage: http://www.ringrid.eu.

[12] Project VLAB (Virtual Laboratory) homepage: http://vlab.psnc.pl.

[13] S. Vignola, S. Zappatore, "The LABNET-server software architecture for remote management of distributed laboratories: current status and perspectives", in F. Davoli, S. Palazzo, S. Zappatore, Eds., *Distributed Cooperative Laboratories, Issues in Networking, Instrumentation and Measurement*, Springer, New York, NY, pp. 435–450, 2006.

[14] The Grid Sphere: http://iri164.informatics.indiana.edu:8080/gridsphere/gridsphere.

[15] "The GRIDCC Architecture - *An Architecture of Services for a Grid Enabled Remote Instrument Infrastructure*" Version 1.2, GRIDCC Deliverable 1.2. http://www.gridcc.org/documents/D1.2-revised.pdf.

Part IV

Grid Resource Allocation, QoS, and Security

GRID INTEROPERABILITY BY MULTIPLE BROKER UTILIZATION AND META-BROKERING*

A. Kertész[1,2,4], Z. Farkas[2,5], P. Kacsuk[2,4,5], and T. Kiss[3,5]

[1] *Institute of Informatics, University of Szeged, H-6701 Szeged, P.O. Box 652, Hungary,*

[2] *MTA SZTAKI Computer and Automation Research Institute, H-1518 Budapest, P. O. Box 63, Hungary,*

[3] *Centre for Parallel Computing, University of Westminster, 115 New Cavendish Street, London, W1W 6UW,*

[4] *CoreGRID Institute on Resource Management and Scheduling,*

[5] *CoreGrid Institute on Grid Systems, Tools and Environments*

Abstract As grid technology matures the number of production grids dynamically increases. The management and the optimal utilization of these grid resources cannot be handled by the users themselves. To hide the lower level details of grid access, grid portals have been developed. Unfortunately, today's grid portals are typically tightly coupled to one specific grid environment and do not provide multi-grid support. The P-GRADE Portal is a workflow-oriented multi-grid portal with the main goal to support all stages of grid workflow development and execution processes. One very important aspect of utilizing multiple grids within one complex scientific workflow is to interact with multiple and potentially different grid brokers supported by the different grids. This paper describes how the portal was interfaced to grid brokers to reach resources of different grids in an automated way. This way a workflow can be brokered over several grids based on different underlying technologies but still providing optimal utilization of resources. To ease the addition and usage of different resource brokers, we introduce a meta-brokering concept. With this new grid middleware component portal users can utilize a growing number of grids for highly computation intensive applications in the future.

*This research work was carried out under the FP6 Network of Excellence CoreGRID funded by the European Commission (Contract IST-2002-004265).

1. INTRODUCTION

End-users typically access grid resources through portals that serve as both application developer and executor environments. As grid technology matures the number of production grids dynamically increases. Unfortunately, today's grid portals are typically tightly coupled to one specific grid environment and do not provide multi-grid support. Even if a portal is connected to multiple grids, applications that utilize services from these grids simultaneously are not supported. Though grids can be realized relatively easily by building a uniform middleware layer, such as Globus [4], on top of the hardware and software resources, the programming concept of such distributed systems is not obvious. Complex problems often require the integration of several existing sequential and parallel programs into a single application in which these codes are executed according to a graph, called workflow. The components of the workflows are jobs, which need to be transferred and executed on the Grid. This task requires special skills, such as how to find out the actual state of the grid, how to reach the resources, etc. To free the users from these efforts, various Resource Brokers [9] have been developed. Though most of these brokers can be used as stand-alone applications, connecting them with portals would provide an easier way to utilize their features.

There are many grid workflow management systems being developed, such as the Condor DAGMan [19], GridFlow [2] and GWES [12], which are used by different grid portals. Pegasus [17] is a web-based grid portal. Based on a special configuration file, filled up by the portal administrator with Globus GRAM and GridFTP [4] site addresses, Pegasus is able to map abstract workflows onto physical resources. At the same time – because of the centrally managed resource list and the single certificate the manager applies during workflow execution – Pegasus cannot be considered a multi-grid portal. The GridFlow portal [2] applies a more complex agent-based workflow executor subsystem. The workflow manager of GridFlow handles workflows at two levels: it manages workflows at a global grid level and schedules them at the level of different local grids. Though it has the ability to handle brokers of different grids, it is only coupled to the Globus middleware. In the K-Wf Grid Project [12] the GWES is used as a workflow enactor. They use GWorkflowDL, which is an extension of the GJobDL XML-based language that makes use of the formalism of Petri nets in order to describe the dynamic behavior of distributed Grid jobs. With this approach they are able to model arbitrary algorithms, including parallel branches or loops. It provides methods to initiate and analyze workflows and to coordinate and optimize the execution of these workflows. They use an own brokering service (Job Handler), which can be utilized on their underlying Globus middlewares.

2. THE P-GRADE PORTAL

The P-GRADE Portal [6] is a workflow-oriented multi-grid portal with the main goal to support all stages of grid workflow development and execution processes. It enables the graphical design of workflows created from various types of executable components (sequential, MPI, or PVM jobs), executing these workflows in multiple Globus-, and EGEE-based computational grids relying on user credentials, and finally, analyzing the monitored trace-data by its built-in visualization facilities.

Every workflow-oriented portal consists of a workflow GUI and a workflow manager part. A P-GRADE Portal workflow is a directed acyclic graph that connects sequential and parallel programs into an interoperating set of jobs. The nodes of such a graph are batch jobs, while the arc connections define data relations among these jobs. Arcs define the execution order of the jobs and the input/output dependencies that must be resolved by the workflow manager during execution.

The P-GRADE Portal contains a DAGMan-based [19] workflow manager subsystem which is responsible for the scheduling of workflow components in grids. DAGMan extended with the facilities of the P-GRADE workflow manager degrades workflows into elementary file transfer and job submission tasks and schedules the execution of these tasks. The semantics of the workflow execution means that a node (job) of the workflow can be executed if, and only if all of its input files are available, i.e., all the jobs that produce input files for this job have successfully terminated, and all the other input files are available on the Portal Server and at the pre-defined storage resources. Therefore, the workflow describes both the control-flow and the data-flow of the application. If all the necessary input files are available for a job, then the portal workflow manager transfers these files – together with the binary executable – to the computational resource where the job is allocated for execution.

3. RESOURCE MANAGEMENT THROUGH BROKERS

One very important aspect of utilizing multiple Grids within one complex scientific workflow is to interact with multiple and potentially different grid brokers supported by the different Grids. The P-GRADE Portal is interfacing several grid brokers to reach the resources of different grids in an automated way.

As the workflow managers do the actual job submissions, they should be set to utilize brokers. In P-GRADE Portal DAGMan is responsible for workflow execution. Although it itself cannot invoke grid services, it supports customized grid service invocations by its pre/scheduler/postscript concept [19]. One pre- and one postscript can be attached to every node (job) of a DAGMan workflow. DAGMan guarantees that it first executes the pre script, then the actual content script and finally the postscript when it reaches a workflow node. Consequently, the Portal Server automatically generates appropriate pre, content and postscripts for every workflow node when the workflow is saved on the server. These scripts – started by DAGMan according to the graph structure –, invoke the GridFTP and GRAM clients to access files and start up jobs in the connected grids. DAGMan invokes these scripts in the same way in both single- and multi-grid configurations, however, the contents of these portal scripts depend on the actual Grid. In general, when a broker is used for job submission, the pre script prepares the broker utilization, the scheduler script invokes the broker, and the postscript waits till the execution is finished. The broker provides information about the actual job status and the postscript notifies the portal about the status changes.

Currently the portal can utilize GTbroker [8] for Globus 2, 3 and LCG-2 grids, the WMS of the LCG-2 and gLite middlewares [11] and the broker of NorduGrid middleware [13]. The jobs of the workflow that require EGEE services can run on an EGEE type of grid; jobs that require only Globus services can be mapped to resources handled by GTbroker, and finally the NorduGrid Broker can be utilized to reach resources of the NorduGrid ARC [13]. Different Resource Brokers usually require different user job descriptions. In the Workflow Editor of the portal the users can choose a broker for each job of the defined workflow. According to this setting the Editor generates an RSL [4], JDL [11], or an xRSL [13] file from the job requirements, depending on the appropriate broker utilization. The scheduler script of DAGMan invokes the brokers with these descriptions. In case of Globus-based grids the file movements are also handled by GTbroker, so the scheduler script only needs to activate and run an instance of GTbroker. In case of EGEE WMS and NorduGrid broker, there are special commands for tracking job states and retrieving the output and log files, therefore, the scheduler script needs to call these services, too.

In case of remote files only the EGEE brokers use a so-called "close to file" policy: it means they try to place the job to a resource, which is the nearest to the location of its remote input files. Since none of these brokers handle remote file transfers, the portal workflow manager solves this problem again based on DAGMan services. DAGMan submits a wrapper script as the executable, carrying all the job files and descriptions. After this script is started on the selected computing resource, it handles the remote input file transfers and – after the real job execution – the remote output file transfers between the storage elements and the actual computing element. With this solution all kinds of file transfers can be carried out during broker utilization. This solution was described in detail for supporting MPI applications in [3].

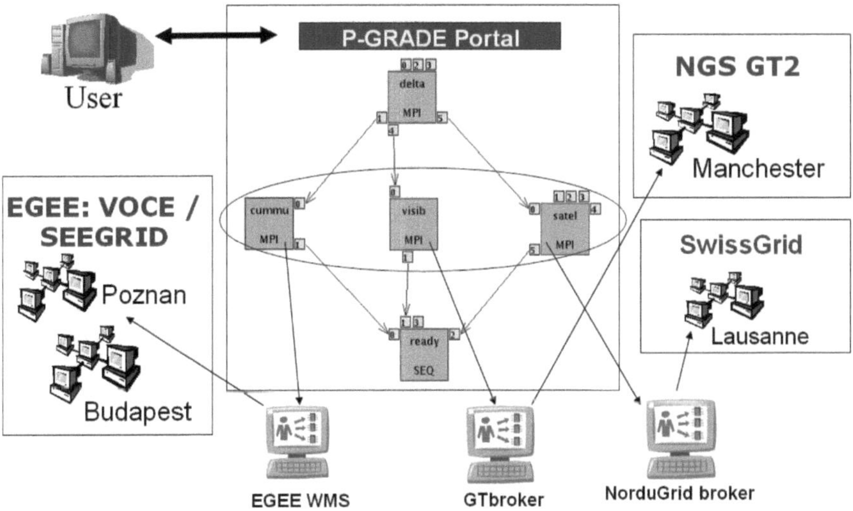

Figure 1. Multiple broker utilization

Figure 1 shows, how multiple broker utilization is carried out in the P-GRADE Portal. In this way a workflow can be brokered over several grids based on different underlying grid technologies and still providing optimal utilization of resources.

4. META-BROKERING APPROACH

Utilizing the existing, widely used and reliable resource brokers and managing interoperability among them could be new point of view in resource management. As we have seen in the previous section the P-GRADE Portal is able to connect to and exploit various brokers in order to reach resources of different middlewares. Users usually have certificates to access more VOs. A new problem arises in this situation: which VO, which broker to choose for my specific application? Just like users needed Resource Brokers to choose proper resources within a VO, now they need a Meta-Brokering service to decide which broker (and VO) is the best for them and also to hide the differences of utilizing them. The solution of this problem is particularly important when a workflow is executed as a parameter sweep application. Without a Meta-Broker the user (portal) is not able to dynamically and evenly distribute the various workflow runs among the connected VOs [7]. Figure 2 introduces the proposed architecture of a Meta-Broker that enables the users to access resources of different grids through their own brokers.

There are five major parts of this architecture. The Translator components are responsible for translating the resource specification language of the user

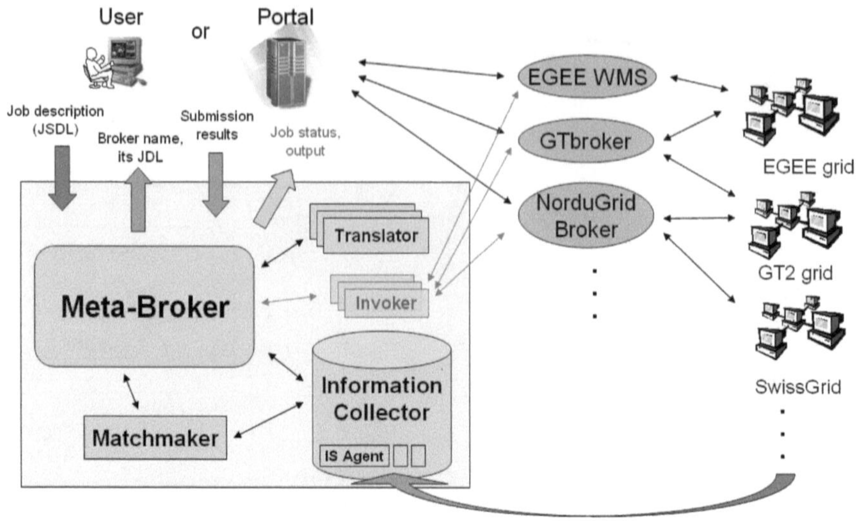

Figure 2. Meta-Broker architecture

to the resource specification language of the appropriate broker that the Meta-Broker selects to invoke. As we have seen in Section 3, different brokers use different resource specification languages like RSL, JDL, etc. The GGF has already developed a resource specification languages standard called JSDL [5]. As the JSDL is general enough to describe jobs of different grids and brokers, we have chosen this to be the language of the Meta-Broker. The Meta-Broker will accept the JSDL, and it is the task of the Translator components to translate JSDL to RSL, JDL, and so on. Once a broker will be capable of supporting the JSDL standard, the corresponding Translator will be removed from the Meta-Broker.

Besides describing user jobs, we also need to describe resource brokers in order to make difference among them. These brokers have various features for supporting different user needs. These needs should be able to be expressed in the users' JSDL and identified by the Meta-Broker for each corresponding broker. Therefore we propose a Broker Property Description Language (BPDL) [10] – similarly to the JSDL –, to store metadata about brokers. These two kinds of languages are used by the Meta-Broker to communicate with the inner and outer world.

The Information Collector stores the data of the reachable brokers and historical data of the previous submissions. This information shows whether the chosen broker is available, or how reliable its services are. The previously introduced languages are used for matching the user requests to the description of the interconnected brokers, which is the role of the Matchmaker component.

During broker utilization the successful submissions and failures are tracked, and regarding these events a rank is modified for each special attribute in the BPDL of the appropriate broker. The JSDL contains the user request – supposed to be an exact specification of the user's job, using the extended attributes. The load of the resources behind the brokers is also taken into account to help the Matchmaker to select the proper environment for the actual job. When too many similar jobs are needed to be handled by the Meta-Broker the so-called "best effort" matchmaking may flood a broker and its grid. That is the main reason, why load balancing is an important issue. In order to cope with this problem, there are IS Agents in the Information Collector, which regularly check the load of the underlying grids of each connected resource broker, and store this data. The IS Agents are connected to the public Information Systems of the grids in order to get an overview of the whole grid load. With this extra information the matchmaking process can adapt to the load of the utilized grids. In case of EGEE grids it is connected to the BDIIs [11] of the appropriate VOs. In the query an IS Agent gets the stored data of all the available resources, and computes a load percentage that is stored in the database of the Information Collector.

During matchmaking the following steps are taken to find the fittest broker:

- The Matchmaker compares the JSDL of the actual job to the BPDL of the registered resource brokers. First the basic attributes are matched against the basic properties: this selection determines a group of brokers that are able to submit the job.
- In the second phase those brokers are kept, which are able to fulfill the special requirement attributes of the job (these capabilities are looked up from the BPDL).
- Finally a priority list of the remaining brokers is created taking into account the ranks (stored for the requested features) and the load of the underlying grid of each broker. The first resource broker is chosen from the list.

Two different scenarios can be done with our proposed Meta-Broker. The first one allows the users or portals to invoke and track the connected brokers themselves. In this case only the JSDL document of a job needs to be submitted to the Meta-Broker, and it responds with the name of the matched broker and the translated job description (or with a message that none of the registered brokers is able to fulfill the specified job requirements). Finally the user/portal needs to provide the result of the actual job submission for the Meta-Broker (to modify the broker property ranks). This limited operation is useful for systems that already have reliable connections to resource brokers and would like to use this service as broker-filtering and inter-grid load balancing. Currently these issues are not taken into account in grid portals. Even multi-grid access is rarely supported, where the users need to choose from a list of resource brokers. Furthermore this scenario can be achieved with minimal adaptation efforts and requires less data transfers.

In the second scenario the utilization of the resource brokers is done by the Meta-Broker. The Invokers are broker-specific components, they communicate with the interconnected brokers, invoking them with job requests and collecting the results. Data handling is also an important task of this component. After the user uploaded the job, proxy and input files to the Meta-Broker, the Matchmaker component tries to find a proper broker for the request. If no good broker was found, the request is rejected, otherwise the JSDL is translated to the language of the selected broker. The responsible Invoker takes care of transferring the necessary files to the selected grid environment. After job submission it stages back the output files and upgrades the historical data stored in the Information Collector with the log of the utilized broker. The core component of the Meta-Broker is responsible for managing the communication (information and data exchange) among the other components. The communication to the outer world is also done by this part through its web-service interface.

The user's job description is independent from the execution environment, and the Meta-Broker does not need to know how to access resources of different grids. It is the task of the interconnected brokers to perform the actual job submissions and to find the best resource within their scopes (the VOs they have access to). The Meta-Broker only needs to communicate with the users (via its web-service interface) and the brokers (via the Invokers). In this sense meta-brokering stands for brokering over resource brokers instead of resources.

5. CONCLUSIONS

By exploiting the advanced workflow management features of the P-GRADE portal and the brokering functions of the interconnected resource brokers, users can develop and execute multi-grid workflows in a convenient environment [6]. Users who have access to several VOs can create multi-grid workflows that reach resources from different grids. Furthermore, the execution is carried out in an efficient, brokered way within each connected grid. The current version of P-GRADE portal (release 2.4.1) supports the following brokers: GTbroker for Globus 2, 3 and LCG-2 grids, the WMS of the LCG-2 and gLite middlewares and the broker of NorduGrid middleware. With the integration of these grid brokers the overwhelming majority of existing production grids is supported by the P-GRADE portal.

P-GRADE Portal [16] has been already connected to several European grids (LHC Grid [11], UK NGS [24], EU GridLab testbed [1], UK OGSA testbed [23], SwissGrid [21]) and several US grids (Open Science Grid [14] and TeraGrid [22]). P-GRADE Portal serves as the official portal for several production grids like SEE-GRID [18], VOCE [25] and HunGrid [20]. As P-GRADE the only portal that can support real multi-grid utilization so far, it has been chosen to be the resource-testing portal [15] of GIN (Grid Interoperability Now) VO.

Recently P-GRADE portal was extended with parameter sweep execution management of workflows [7]. This new and eagerly waited feature raised the need for dynamic, well-balanced distribution of workflows and workflow components among several grids. In order to fulfill this requirement we introduced the concept of meta-brokering service and showed in the paper how such a service can be realized based on the latest OGF standards. The Meta-Broker itself is a stand-alone web service that can serve both users and portals. The design and the abstract architecture of the Grid Meta-Broker Architecture enable a higher level brokering by utilizing resource brokers of different grid middlewares. The introduced meta-brokering approach opens a new way for grid interoperability support.

REFERENCES

[1] G. Allen et al., "Enabling Applications on the Grid: A GridLab Overview", *International Journal of High Performance Computing Applications*, Issue 17, 2003, pp. 449-466.

[2] J. Cao, S. A. Jarvis, S. Saini, and G. R. Nudd, "GridFlow: WorkFlow Management for Grid Computing", In *Proc. of the 3rd IEEE/ACM International Symposium on Cluster Computing and the Grid (CCGRID'03)*, 2003, pp. 198-205.

[3] Z. Farkas, Z. Balaton and P. Kacsuk, "Supporting MPI Applications in P-GRADE Portal", in *Proceedings of the 6th Austrian-Hungarian Workshop on Distributed and Parallel Systems (DAPSYS)*, Innsbruck (Austria), 2007, pp. 55-64.

[4] I. Foster and C. Kesselman, "The Globus Project: A Status Report", in *Proc. of the Seventh Heterogeneous Computing Workshop*, IEEE Computer Society Press, Orlando, Florida, USA, 1998, pp. 4-18.

[5] Job Submission Description Language (JSDL): http://www.ggf.org/documents/GFD.56.pdf

[6] P. Kacsuk and G. Sipos, "Multi-Grid, Multi-User Workflows in the P-GRADE Grid Portal", *Journal of Grid Computing*, Vol. 3, No 3-4, 2005, pp. 221-238.

[7] P. Kacsuk, G. Sipos, A. Tóth, Z. Farkas, G. Kecskeméti and and G. Hermann, "Defining and Running Parametric Study Workflow Applications by the P-GRADE Portal", *Krakow Grid Workshop*, 2006

[8] A. Kertész, "Brokering Solutions for Grid Middlewares", in *Pre-proc. of 1st Doctoral Workshop on Mathematical and Engineering Methods in Computer Science*, 2005.

[9] A. Kertész and P. Kacsuk, "A Taxonomy of Grid Resource Brokers", *6th Austrian-Hungarian Workshop on Distributed and Parallel Systems (DAPSYS)*, Innsbruck, Austria, 2006.

[10] A. Kertesz, I. Rodero, and F. Guim, "BPDL: A Data Model for Grid Resource Broker Capabilities", *CoreGrid Technical Report no. 74*, 2007

[11] LCG-2 User Guide, 4 August, 2005: https://edms.cern.ch/file/454439/2/LCG-2-UserGuide.html, gLite Middleware: http://glite.web.cern.ch/glite/

[12] F. Neubauer, A. Hoheisel, and J. Geiler, "Workflow-based Grid Applications", *Future Generation Computer Systems*, Volume 22, Issues 1-2, January 2006, pp. 6-15.

[13] NorduGrid Middleware: http://www.nordugrid.org/middleware/

[14] Open Science Grid: http://www.opensciencegrid.org/

[15] P-GRADE GIN VO Portal: https://gin-portal.cpc.wmin.ac.uk:8080/gridsphere

[16] P-GRADE Grid Portal: http://lpds.sztaki.hu/pgportal

[17] G. Singh et al, "The Pegasus Portal: Web Based Grid Computing" In *Proc. of 20th Annual ACM Symposium on Applied Computing*, Santa Fe, New Mexico, 2005.

[18] Southern Eastern European GRid-enabled eInfrastructure Development (SEE-GRID): http://www.see-grid.org/

[19] D. Thain, T. Tannenbaum, and M. Livny, "Distributed Computing in Practice: The Condor Experience", *Concurrency and Computation: Practice and Experience*, 2005, pp. 323-356.

[20] The HunGrid Virtual Organisation: http://www.lcg.kfki.hu/?hungrid

[21] The Swiss Grid Initiative: http://www.swiss-grid.org/index.html

[22] TeraGrid: http://www.teragrid.org/

[23] UK e-Science OGSA Testbed: http://dsg.port.ac.uk/projects/ogsa-testbed/

[24] UK National Grid Service: http://www.ngs.ac.uk/

[25] Virtual Organisation for Central Europe (VOCE): http://egee.cesnet.cz/en/voce/

ON QUALITY OF SERVICE SUPPORT FOR GRID COMPUTING

D. Colling[1], T. Ferrari[2], Y. Hassoun[1], C. Huang[4], C. Kotsokalis[3], A.S. McGough[1], E. Ronchieri[2], Y. Patel[1], and P. Tsanakas[3]

[1]*Imperial College London, London, UK*

d.colling@imperial.ac.uk, yhassoun@doc.ic.ac.uk, asm@doc.ic.ac.uk, yp03@doc.ic.ac.uk

[2]*National Institute of Nuclear Physics, CNAF, Bologna, Italy*

tiziana.ferrari@cnaf.infn.it, elisabetta.ronchieri@cnaf.infn.it

[3]*National Technical University of Athens, Athens, Greece*

ckotso@cslab.ece.ntua.gr, panag@cs.ntua.gr

[4]*Brunel University London, UK*

chenxi.huang@brunel.ac.uk

Abstract Computing Grids are hardware and software infrastructures that support secure sharing and concurrent access to distributed services by a large number of competing users from different virtual organizations. Concurrency can easily lead to overload and resource shortcomings in large-scale Grid infrastructures, as today's Grids do not offer differentiated services. We propose a framework for supporting quality of service guarantees via both reservation and discovery of best-effort services based on the matchmaking of application requirements and quality of service performance profiles of the candidate services. We illustrate the middleware components needed to support both strict and loose guarantees and the performance assessment techniques for the discovery of suitable services.

1. INTRODUCTION

Quality of service (QoS) is a broad term that is used, in this context, to denote the level of performance that a given client experiences when invoking an operation on a remote server. QoS support refers to the possibility of a given service instance to offer a performance level that satisfies the requirements of a given client.

QoS support is of paramount importance given the inherent shared nature of Grid services, and the limited capabilities of hardware and software resources available to satisfy client requests. In this environment, an aggressive best-effort usage of services can result in denial of service problems, as the non-policed use of services is against their effective provision. The more recent efforts for integration of scientific instrumentation, and the related requirements for high availability and responsiveness, urge for the support of differentiated services.

QoS requirements can be of many different kinds, taking under consideration the heterogeneous nature of Grid resources which the applications utilize. The support of QoS requirements needs the adoption of specific enforcement techniques in the service backend. Exclusive access typically requires resource-locking techniques, while other types of QoS parameters, such as responsiveness, are dependent on a number of different factors - both static and dynamic - such as the client and server physical location, the status of the communication infrastructure, the client/server implementations, the type of operation invoked, the number of concurrent clients served.

We propose a framework which supports two complementary types of guarantees: *strict*, involving digital contracts, and *loose*, implying best-effort usage based on statistical assumptions. First, we address the problem of establishing agreements (digital contracts) in Section 2 and the problem of delivering QoS guarantees for atomic tasks that need to be executed instantly in Section 3. Following, we analyze a list of performance metrics for loose QoS in Section 4 defined on the basis of the requirements of interactive applications, and propose their composition in Section 5. We continue to provide a framework for publishing and archiving relevant information in Section 6. Finally, we present a preliminary prototype in Section 7 and related work in Section 8. We conclude the chapter and elaborate on future work in Section 9.

2. STRICT GUARANTEES

The application of a specific QoS level to a Grid task requires the negotiation and establishment of a contract, named *Service Level Agreement* (SLA), between the customer or proxy and the service provider. A SLA quantitatively defines the performance level for the service requested and the obligations for the parties involved in the contract. The SLA is typically specified through a template containing both quantitative and qualitative information. In particular, a SLA specification supplies administrative information (e.g., the identity of the entities involved in the contract and the penalties applicable to the parties when the SLA guarantees are not honored), whereas the *Service Level Specification* (SLS) is a set of attributes and values describing the profile of the requested performance level [2]. Strict guarantees require the establishment of such a formal agreement via signalling and negotiation. This process involves both service consumers and service providers.

The Grid Resource Allocation and Agreement Protocol Working Group (GRAAP-WG) of the Open Grid Forum (OGF) proposes an architecture for grid SLA signaling comprising two functional levels: the *agreement layer* and the *service provider layer* [11]. The agreement layer implements the communication protocol used to exchange information about SLAs between the customer and the service provider, and it is responsible for ensuring that the SLA guarantees are enforced by a suitable service provider. In addition, the agreement layer exposes information about types of service and the related SLA templates offered, processes agreement creation requests to check the compliance of the so-called agreement offers (i.e., the SLA requests), and handles well-formed requests to the service provider layer. An agreement is successfully created if the corresponding guarantees can be enforced by one or more service providers, which are also responsible for supervising the status of their resource pools. Then the service provider layer concerns the terms, i.e., the content of the agreement, which binds the entities involved.

Reservation and allocation in Grids can be assisted by existing resource management capabilities. In this paper, we address the problem of the support of resource-independent agreement signaling by Grid workload and resource management services (WMS), and we describe a prototype implementation developed in the framework of the GRIDCC and EGEE EU projects. The WMS comprises a set of middleware components responsible for the distribution and management of tasks across Grid resources and services. In this context, the term *resource* refers to fabric-layer facilities, e.g., computing farms, data storage systems, and networking infrastructures. Conversely, we use the term *service* to refer to a general service instance regardless of the provided functionality level with respect to the Grid service architecture.

2.1 SLA Signaling

We propose an approach to SLA signaling based on the instrumentation of the WMS and on the availability of a new collective service, which is called the *Agreement Service* (AS). Resource broker instrumentation is needed in order to provide the capability of handling the allocation and reservation of Grid resources. The WMS supports the SLA-signaling phase, the discovery of service instances supporting the user's requirements based on data available within the Grid *Information System* (IS), and the monitoring of SLAs.

We borrow the notations of *Agreement Initiator*, *Agreement Offer* and *Service Provider* from the WS-Agreement proposed specification of the OGF.

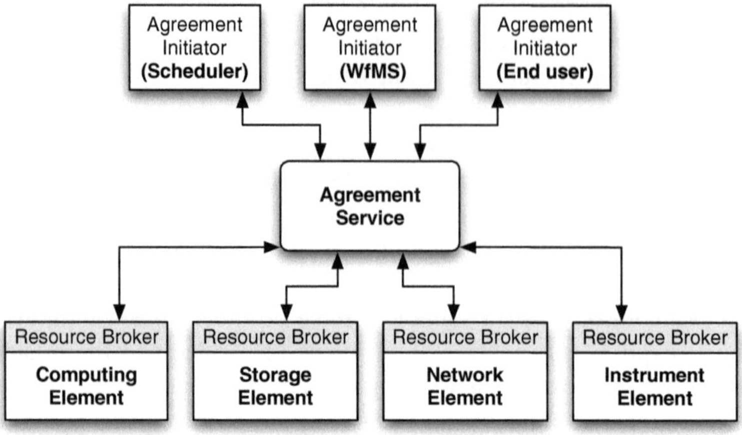

Figure 1. Agreement Service and the other components of the proposed architecture

In particular, as we focus on reservation and allocation, in this context the provider is called *Reservation and Allocation Service Provider* (RASP). The proposed architecture is shown in Figure 1, which details the interactions between the above components. Four types of resources are shown: the Computing Element (CE), the Storage Element (SE), the network (Network Service Access Point [3]) and the Instrument Element (IE) [5].

RASPs are fabric-layer resource-specific services that are responsible for the management of the availability of the local resource over time, for admission control based on the involved infrastructure's policies, for the control of the satisfiability of each authorized incoming requests and, finally, for the enforcement of the guarantees. We propose each reservation-capable resource to be associated to an individual RASP instance. User's resource requirements and preferences are part of the SLS and they consequently need to be specified as a list of *Service Description* and *Guarantee Terms*. We identify four term categories: *general* attributes (type of task requested and corresponding functionality), *time* attributes (information about the start time, end time and duration of the reservation), *functionality* attributes (reservation-specific information) and *guarantee* attributes (definition of the QoS profile).

The agreement initiator is the entity which triggers the SLA-signaling process. In our scenario, users, applications or grid middleware components can be initiators provided that the identity of the resources to be involved in the SLA signaling is known. If available, this identity is defined in the SLA, and the allocation and reservation task can be submitted directly to the AS. While direct negotiation with the AS is possible, nevertheless, very frequently users have no a priori knowledge of the grid infrastructure. When this applies, we propose the allocation and reservation task to be submitted to the WMS in order to take advantage of its discovery capabilities. After the submission the WMS determines both the RASPs, which can satisfy the task requirements, and the AS's, which are capable of supporting the corresponding RASP interfaces. This happens through the use of resource details fed by the RASPs into the IS. In this scenario, the WMS has the agreement initiator role and acts on behalf of the user or application.

When the AS receives an offer, it initially checks SLA compliance to the appropriate corresponding template, it extracts attributes from the SLA request and, finally, it dispatches them to a number of RASPs by invoking the appropriate interface. A given agreement can be successfully created only if one or more RASPs (depending on the nature of the agreement) accepted the responsibility of providing the guarantees associated with it.

2.2 Agreement Service

The AS is the novel service component that provides SLA management capabilities to the WMS. We propose the AS to be responsible for interacting with the agreement initiator (as a server) and with the RASPs (as a client) which can satisfy the initiator's request. In particular, the AS accepts agreement offers from initiators and is responsible for compliance checking as well as for the interaction with one or more RASPs in case of success. The AS exposes RASP capabilities through agreement templates [11]. The template is the SLA skeleton and includes a list of creation constraints applicable to the SLS. The templates advertised by one AS explicitly define the types of service the agreement service can handle, and consequently depend on the service providers the AS interacts with.

The AS can perform negotiation in order to assist initiators during the establishment of SLS whose characteristics depend on the current status of the RASP. For every well-formed offer, SLS attributes are translated to low-level RASP-specific terms according to the RASP interface. Finally, the AS supplies information about the status of agreements under negotiation and the attributes of the agreements successfully established.

3. LOOSE GUARANTEES

The provisioning of loose guarantees consists of the capability of clients to select service providers that, on a statistical basis, exhibit a best-effort QoS profile which meets the client's requirements. In this case, service discovery relies on the monitoring of critical performance parameters over time. Gathered results combined with appropriate statistical estimation methods can assist clients and schedulers during the selection phase.

Loose guarantees are delivered on a best-effort basis, and for this reason, they do not require the prior establishment of a SLA and the consequent negotiation of a contract. If consumers experience a service level that contradicts the past QoS profile exhibited by the service providers, compensation, adaptation and/or other self-healing operations are needed to cope with requirements that are not met in practice.

The usefulness of this approach lies in the requirements of lightweight interactive applications, which may be more easily addressed in a best-effort fashion via discovery and prediction of performance in the future. The performance experienced when invoking an operation on a remote service instance can largely benefit from the availability of loose guarantees, as it depends on a large number of factors. For example, asynchronous operations typically just exchange one-way messages from the client to the server, while synchronous operations imply the additional issuing of a reply message back to the client. Thus, the client/server interaction profile typically depends on the network path connecting the client and the server, and the processing overhead during the entire messaging process at both ends, this including request/reply message composition/decomposition and processing.

We propose a novel architecture and methodology for the delivery of loose guarantees, which relies on three functional components.

Monitoring: the availability of distributed sensors for the monitoring of the QoS performance profile of various Grid infrastructure components, both at the fabric and at the resource layer;

Publishing: data gathered from the sensors must be made available to distributed consumers;

Discovery: the availability of service discovery engines is required to allow clients to select the most appropriate service instance(s), by comparing the QoS profiles of the services of interest to the client requirements. Discovery and scheduling services used when selecting adequate service providers and agreement responders in case of strict guarantees can be easily extended to support this.

The provisioning of loose guarantees thus requires the complementing of the Grid middleware described in Section 2 with additional functional components. Consequently, the sections which follow will focus on the problem of information gathering and publishing.

We can identify a number of different stages in the process of completing a typical Web services-based client–server interaction, as shown in Figure 2.

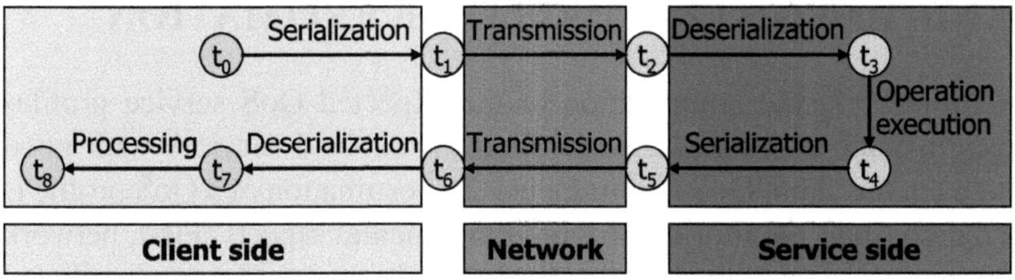

Figure 2. Decomposition of a synchronous operation

REQ refers to the request message exchanged during a session, RES identifies the corresponding response message, C represents the client side and S represents the server side of the interaction.

1. $\overline{t_0t_1}$ *(Message Generation Delay, mgd)*: composition of the request message REQ is taking place on the client side.

2. $\overline{t_1t_2}$ *(Message Transport Delay, mtd)*: REQ is being dispatched by the communication infrastructure connecting C to S.

3. $\overline{t_2t_3}$ *(Message Processing Delay, mpd)*: REQ is de-composed by S.

4. $\overline{t_3t_4}$ *(Execution Time, et)*: the operation called is being executed.

5. $\overline{t_4t_5}$ *(Response Message Generation Delay, rmgd)*: at this stage composition of the response message RES takes place.

6. $\overline{t_5t_6}$ *(Response Message Transport Delay, rmtd)*: it is the one-way delay experienced by the RES message after being issued by S in order to be transmitted by the network to C.

7. $\overline{t_6t_7}$ *(Response Message Processing Delay, rmpd)*: RES is de-composed by C.

8. $\overline{t_7t_8}$ *(Response Client-side Processing Time, rcpt)*: C processes the response message and acts accordingly. This component is not taken into account, as is also the case with $\overline{t_3t_4}$.

Composition and de-composition delays can be measured with hooks in modern service stacks, while networking delays can be measured with existing tools from the networking world. Different Grid infrastructure components contribute in the various stages identified above, for this reason we propose the estimation of end-to-end performance via the individual monitoring of each part (network, clients and servers), and the composition of the measurements gathered. The metrics of relevance and the composition methodology are described in what follows.

4. END-TO-END PERFORMANCE ESTIMATION

In our framework, the computation of the expected QoS service profiles is required in order to support the selection of the Grid services with good chances to meet the client QoS requirements. The estimation of a QoS profile is based on the constant monitoring of three fundamental aspects. First, network performance needs to be probed by specific network sensors that are distributed over the network infrastructure of interest. Second, we propose the monitoring of performance at the client side, as this component affects the time needed to construct a request message in the first place and, if applicable, to process reply messages from the invoked service. Finally, performance needs to be analyzed also at the server end, being the server a component that contributes to determine both the time needed to process a given request messages from remote clients and the time needed to construct reply messages. In this framework, the time needed to process a specific operation is not considered, as it is highly specific to the type of operation and the input parameters supplied to it.

4.1 Performance Metrics

The main components contributing to end-to-end performance are network paths, clients and servers. In this framework, the actual metrics that need to be subject to monitoring for loose guarantee support, necessarily depend on the application requirements considered. Starting from a set of requirements from applications for computing and instrument Grids, we have worked out a list of relevant metrics and defined different measurement approaches that are applicable to different metric categories [6].

Network performance metrics. Several network performance metrics can be relevant to the estimation of some Grid service performance parameters.

Message one-way delay (RFC2679) from the node-hosting client C to the node-hosting server S is of paramount importance, as it affects the completion and response time of a given operation. Measurement of one-way delay requires accurate clock synchronization in the client C and in the server S.

Round-trip Delay also known as round trip time (RTT) (RFC2681) complements one-way delay and corresponds to the time span from the instant when a given message is sent to the time when that message is received.

Throughput is particularly relevant in scenarios where applications require the exchange of large amounts of data. *Achievable throughput* is the average volume of application data that can be exchanged over time from C to S. The availability of achievable throughput [9] estimations, in combination with information about the file size to be exchanged, can help with the computation of the time needed to transfer the file from C to S.

Finally, *IP Packet Delay Variation* (IPDV, RFC3393) quantifies the variation of one-way delay for two subsequent IP packets. IPDV is critical for interactive and video/audio-streaming applications, whose performance is optimized in case of low IPDV.

Service metrics. The performance of a stand-alone service can be quantified via a number of metrics, all contributing in a complementary fashion to the estimation of the performance level which can be expected by a client when interacting with a given service. Several of these metrics strictly depend on a number factors, such as message size and its level of complexity, the type of operation invoked. In the list of metrics which follows, S denotes a generic server, while C the client.

Availability: we define it to be the parameter which expresses the probability that S can be effectively invoked by C at a given time. High availability is an indication of service robustness.

Accessibility: it denotes the probability that S can receive a request and return a response message. Essentially, this shows how well a service scales with an increasing number of consumers C. S can be available (serving some clients) but not accessible (rejecting or failing to serve additional requests).

Completion Time: given t – the time at which the composition (for example, its serialization if the Web Services technology is used) of a message starts in C, and T – the instant when the execution of the operation O triggered by the message, finishes at the server end, we define Completion Time $CT(C,S,O,t)$ and the time interval (tT). In other terms, considering the operation decomposition introduced in Section 3, $CT(C,S,O,t) = \overline{t_0t_4}$. Measurement of CT requires clock synchronization at both the client and the server.

Response Time: given t – the time at which the composition (for example, its serialization if the Web Services technology is used) of a message starts in C; and T – the time when the processing of the corresponding response message is terminated by C, Response Time $RT(C,S,O,t)$ is the time interval (tT), or equivalently $RT(C,S,O,t) = \overline{t_0t_8}$. RT is only applicable to synchronous operations, as RT takes into account the network latencies in both directions (C to S and S to C), while the Completion Time is only dependent on one-way delay and processing time at S).

4.2 Measurements Methodology

Network monitoring frameworks generally rely on a set of distributed sensors for the gathering of raw data, which is accomplished by end-to-end and/or edge-to-edge probing sessions. End-to-end measurements require the hosting of the sensors in customer networks, whereas edge-to-edge measurements refer to one or more network transit domains, and require the installation of sensors in network Points of Presence at the edge of provider's domains, giving the possibility to monitor the various intermediate network segments that compose an end-to-end path. The metrics that can be gathered in the two scenarios are complementary. The problem of client–server performance monitoring is addressed by adopting the data gathered by nearby network sensors as reference values.

The performance metrics described earlier in the text can be appropriately estimated via various methodologies at different levels of complexity.

Availability: When estimating availability, service performance has to take into account the quality of network connectivity experienced by the service. Monitoring of connectivity for every client–server pair is infeasible for scalability reasons, hence, we propose availability to be directly measured and logged by the service's clients. For every client–server communication successfully established on a socket-layer level, we propose the event to be logged and published according to the type of information service in use.

Accessibility: If a service is accessible, then operations can complete successfully. This event can be monitored differently for synchronous and asynchronous operations. In the former case, a response message is returned, and the successful completion can be logged and published, if the client receives a well-formed response message directly or indirectly indicating success, whereas completion of an asynchronous operation for a given client C needs to be recorded and notified at the server.

Completion Time: Measuring interval $\overline{t_0t_4}$, needed to estimate CT, is intrinsically difficult due to the requirement for clock synchronization between clients and services. Furthermore, operations experience different CT values, depending on the input and the type of operation. Because of this, we request a server to expose a "dummy" (empty) method for every time-critical operation, whose request processing overhead is null. For empty operations CT can be accurately approximated to the time interval $\overline{t_0t_3}$, and the response processing time $\overline{t_3t_4}$ can be neglected, as CT is reduced to the time for constructing, transferring and de-composing the request message:

$$CT = \overline{t_0t_1} + \overline{t_1t_2} + \overline{t_2t_3}$$

Response Time: Response Time RT, corresponding to the time interval $\overline{t_0t_8}$, is approximated by means of empty operations as described for parameter CT:

$$RT = \overline{t_0t_3} + \overline{t_4t_7}$$

5. METRIC COMPOSITION

While availability and accessibility can be measured relatively easily, accuracy of temporal metrics estimation is in general more difficult due to the reasons elaborated earlier, relevant to lack of clock synchronization. Unfortunately, a number of additional constraints come into play when estimating some of the metrics, such as CT and RT. First, the computation of one-way delay for every (C,S) pair via active measurements is infeasible for scalability reasons, and this requires the approximation of actual one-way delay to the value measured by nearby sensors. More important, both CT and RT vary according to the input sets considered, and forecasting is an additional source of inaccuracy. We propose CT and RT to be estimated via the composition of various estimated time intervals. In particular, time intervals mentioned in Section 3 are individually measured and then combined together. The accuracy achieved by this metric composition is currently under testing. We are considering a technique to model temporal overheads based on the Monte Carlo approach and application of Central Limit Theorem [4, 13]. The Central Limit Theorem states that the mean of any set of random variables with arbitrary distributions, having a finite mean and variance and defined on the same probability space, tends to the normal distribution. Moreover, if the *Probability Density Function* (PDF) is known, then finite integration could be performed to obtain tighter bounds.

6. INFORMATION PUBLISHING AND ARCHIVING

As explained in Section 4, different types of metrics are probed and published by various sensors in the infrastructure. The clients measure availability, accessibility for synchronous operations, message generation delay and response message processing delay. The services do the same for message-processing delay, response message generation delay, accessibility for asynchronous operations and completion time variation. The network probes measure message transport delay (simulated by message one-way delay), response message transport delay (also simulated by message one-way delay), IPDV and achievable bandwidth. Monitoring data are published by these producers, namely the clients with QoS requirements, servers and network sensors, as shown in Figure 3. Information is gathered in a repository, and the information consumers are the Grid middleware components that need to discover services matching the QoS requirements specified by client applications.

Monitoring data can be populated using a central or a distributed repository, and a push or a pull model, or a combination of the two. The most common

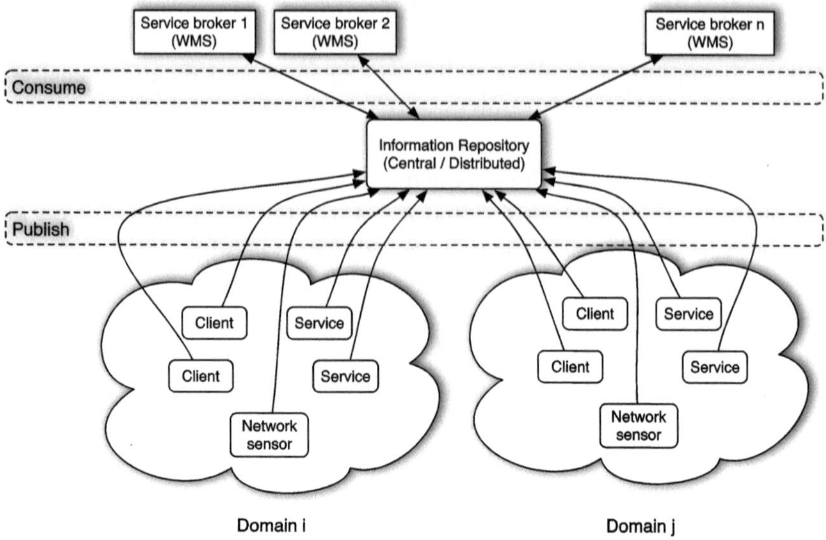

Figure 3. QoS performance information publishing and consuming

approaches are that of a central repository to which sensors push monitoring data, and that of a publish/subscribe mechanism where each client is subscribing for notifications from the services which are of interest, with each such service pushing data to the client whenever there is such a need.

7. PROTOTYPE

A prototype is being implemented to demonstrate the feasibility of the described framework. To this moment, it supports strict guarantees, with loose guarantees support being under development. It contains two components: the WMS and an Agreement Service with storage and instrument reservation templates exposed. This AS is the client of a *Storage Element* (SE) exposing a *Storage Resource Management* (SRM) 2.1 [7] interface, as well as an *Instrument Element* (IE) using a custom, novel interface for instrument reservations. The implementation of the web service part of the AS was based on the open source gSOAP v.2.7 toolkit [14]. The resulting AS (both the client and the server side) are entirely based on the WS-Agreement XML schema definitions. The operations exposed by the AS are similar to the Agreement Factory ones defined in [11]; however, the prototype WSDL exposed by the AS does not fully comply with the standard, as it is not currently based on the WS-Resource Framework [10].

8. RELATED WORK

Relevant work in the field of QoS for Web services and Grid Computing has taken place, by exploring different approaches.

In [1], the authors propose a framework that allows QoS criteria to be specified as part of a service interface, thus enabling service selection based on QoS profiles of candidate services. The framework defines three QoS layers with metrics similar to ours: the application layer (availability, accessibility, security, etc.), the middleware layer (resource requirements) and the network layer (networking QoS). All monitoring is performed by an Application QoS Manager (AQoSM) on every service execution invocation, where this includes the creation of sandboxes, transfer of data and execution startup. Thus, client-related delays are not included in the aggregate values that are archived and used to extract the QoS profile of a service. It is interesting to note that, again, execution time of the job is not measured (similarly to our framework not measuring operation execution time).

In [12], the author proposes a service discovery model which also takes QoS (non-functional) requirements as input, similarly to what has been mentioned up to this point. More specifically, the proposed framework includes roles for service providers, service consumers and QoS certifiers, as well as UDDI extentions to hold relevant information. The basic idea is that when a service is registered into the UDDI registry and claims a specific QoS level for itself, then the QoS certifier actually verifies these claims for correctness. The paper continues to define the new structures required and finishes with a list of QoS attributes, similar to the ones defined in our work.

In both cases, extensions to WSDL and UDDI are proposed. This is different from our work where such extensions are not suggested for reasons of compliance to current standards.

In [8], the authors propose a dynamic QoS registry and computation model for Web Services selection. The measurements are taken, similarly to our work, as a mixture of information sampled by either the service consumer or the service producer, depending on the metric which is being measured. In certain cases, it is the end-user herself that may provide the QoS feedback. The authors do not include availability or accessibility metrics, but rather cost, execution time, reputation (reliability), transaction support, compensation rate and penalty rate.

9. CONCLUSIONS

We presented a QoS framework for computing Grids, with emphasis on Web Services-based middleware implementations, which aims at supporting Grid sessions with various types of specific QoS requirements. The proposed framework relies on two complementary approaches. The former is designed to provide QoS guarantees on a deterministic basis, and relies on a novel Grid service, the Agreement Service, for SLA signaling. Conversely, the latter is a lightweight approach delivering probabilistic QoS guarantees for those use cases which cannot afford the additional delay of deterministic QoS enforcing techniques. Loose QoS guarantees are based on archived performance information and statistical models. Therefore, we propose an architecture for publishing of monitoring data, which is integrated with a QoS-aware Grid service for workload management, namely the workload management system. Several QoS performance metrics for network and Grid service profile characterization are defined starting from a number of application use cases for computing and instrument Grids, and a metric composition approach is presented for the estimation of some of the temporal QoS metrics of interest in the paper. Finally, preliminary implementation work is presented.

ACKNOWLEDGMENTS

This work has been partially funded by the GRIDCC (Grid-enabled Remote Instrumentation with Distributed Control and Computation) EU project under contract number 511382 and is presented on behalf of the GRIDCC consortium. It was also supported by the EU project EGEE sponsored by the European Union under contract number INFSO 508833, by the INFN project INFN-GRID, and by the Greek Secretariat for Research and Technology.

The authors would like to thank Peter Hobson and Paul Kyberd (Brunel University, UK), Panagiotis Louridas (Greek Research and Technology Network, Greece), Gaetano Maron and Francesco Lelli (National Institute of Nuclear Physics, Italy), Luke Dickens, Marko Krznaric and Janusz Martyniak (Imperial College London, UK) for their work and feedback on the chapter.

REFERENCES

[1] R. Al-Ali et al., "G-QoSm: Grid Service Discovery Using QoS Properties", *Computing and Informatics Journal*, 21(4), 2002

[2] EGEE, Institution of SLAs and Appropriate Policies, Deliverable DSA2.2, 2005

[3] EGEE, Specification of Interfaces for Bandwidth Reservation Deliverable DJRA4.1, 2005

[4] W. Feller, "Introduction to Probablity Theory and its Applications, Vol II". *John Wiley and Sons*, New York 1966

[5] GRIDCC, GRIDCC Architecture, Deliverable 1.2, 2005

[6] GRIDCC, Definition of the QoS Parameters for a Real-Time and Interactive Environment. Deliverable 2.2, 2005

[7] J. Gu et al., "The Storage Resource Manager Interface Specification", 2005

[8] Y. Liu et al., "QoS computation and policing in dynamic web service selection In *WWW Alt. '04: Proceedings of the 13th international World Wide Web conference on Alternate track papers & posters*, pages 66–73, ACM Press, New York, NY, USA. 2004

[9] B. Lowekamp et al., "A Hierarchy of Network Performance Characteristics for Grid Applications and Services", OGF Publication GFD23 2004

[10] OASIS, Web Services Resource Framework (WSRF), 2004

[11] OGF, Web Services Agreement Specification (WS-Agreement), 2007

[12] S. Ran, "A model for web services discovery with QoS", *SIGecom Exch.*, 4(1):1–10, 2003

[13] H., Tijms, "Understanding Probability: Chance Rules in Everyday Life", *Cambridge University Press*, 2004

[14] R. van Engelen, "Pushing the SOAP Envelope with web services for scientific computing", In *Proceedings of the international Conference on Web Services* (ICWS), Pages 346–352, Las Vegas, 2003

VICTOR-SPACES: VIRTUAL AND REMOTE EXPERIMENTS IN COOPERATIVE KNOWLEDGE SPACES

S. Cikic[1], S. Jeschke[2], N. Ludwig[1], U. Sinha[1], and C. Thomsen[1]

[1]*MuLF-Center, Technische Universität Berlin, Straße des 17. Juni 136, 10623 Berlin, Germany*
[2]*RUS - Center of Information Technologies, University of Stuttgart, Allmandring 30a, 70550 Stuttgart*

Abstract Cooperative knowledge spaces create new potentials for the experimental fields in natural sciences and engineering because they enhance the accessibility of experimental setups through virtual laboratories and remote technology, opening them for collaborative and distributed usage. A concept for extending existing virtual knowledge spaces for the means of the technological disciplines ("ViCToR-Spaces" – Virtual Cooperation in Teaching and Research for Mathematics, Natural Sciences and Engineering) is presented. The integration of networked virtual laboratories and remote experiments will be described.

1. INTRODUCTION

Focusing on the social aspects of communication, coordination and cooperation, cooperative knowledge spaces possess a high potential to support the learning, teaching and research processes at universities by means of new media and new technologies. "ViCToR-Spaces", currently under development at Technische Universität Berlin, focus on the enhancement of virtual cooperation in teaching and research, in the fields of mathematics, natural sciences and engineering, presenting novel collaborative working environments for knowledge acquisition and research as well as supporting natural forms of scientific and technological cooperations.

Cooperative knowledge spaces use a generalized "room metaphor" [4] as a guideline. They provide a virtual meeting point where interaction, communication and collaboration take place. The environment as a whole is defined by the combination of its dynamically linked objects, i.e., members, documents, tools and services [7]. Hence, the design of those components and the way they are linked is the key issue to successful cooperation and knowledge acquisition.

eLearning and eResearch environments can be categorized into two main types. Within *content-oriented system architectures*, the content defines the center point of the system design; communicative and cooperative scenarios are (missing or) developed around the content objects. In *community-oriented system architectures*, communication and cooperation processes, as well as workflows between the users, are the main focus of the system design, while content objects are embedded into the cooperation infrastructure. As of now, content-oriented systems present the more "common" approach, forming the basis of most eLearning platforms currently available. Community-oriented systems can be regarded as part of the research field of computer-supported cooperative work/learning (CSCW/CSCL) [1, 11]. Within this approach, social processes, represented by communication, coordination and cooperation, form the basis for successful knowledge acquisition in education, as well as in research. Since this view is becoming increasingly accepted for computer-supported models of education and scientific work, community-oriented systems are currently under intense development.

Experiments play a vital role in natural and engineering sciences, thus their presence in the form of virtual and remote laboratories in a cooperative knowledge environment is strongly desirable. We will have a closer look at cooperatively performed experiments in Section 4.2, explaining why they are so essential to ViCToR-Spaces.

2. RELATED WORK

A wide variety of CSCW and CSCL environments has been developed during the last 15 years and many noteworthy results have been achieved. Generic cooperative knowledge space platforms following the room metaphor include *opens*Team [7] and CURE [6]. *opens*Team and CURE likewise are based on the concept of virtual rooms which serve as shared workspaces for user groups and as meeting places for students and teachers. Both platforms support shared viewing, exchanging and re-arranging of items, links between them and discussions between users. In both platforms, new items may be created either by upload or using a built-in wiki engine. Moreover, both *opens*Team and CURE facilitate spontaneous forming of groups by providing awareness information and by managing access permissions on rooms and documents. Both platforms are being developed actively and are available under an open source license.

Since presently CURE adheres more closely toward established standards and already provides a basic support for mathematical notations throughout its components, it is used as a technological basis to implement the first prototype of the ViCToR-Spaces concept.

The Virtual Math Teams [14] project (VMT) addresses virtual cooperation in the field of mathematics. Its most outstanding component is the *ConcertChat* system, which combines a (virtual) shared whiteboard and a chat. These synchronous forms of communication and cooperation are in the main focus of the VMT project. Asynchronous cooperation, however, lies outside the scope of VMT; rather, it is one objective of the MathForum project which VMT is embedded into. Then again, MathForum only supports ASCII-based circumscriptions of mathematical formulae, i.e., x^2 for x^2, etc.

Thus, the specific requirements for cooperative knowledge spaces for mathematics and natural sciences [2] are not – or only rudimentally – supported by the systems presented in this section. This is probably the main reason for why there are still no satisfactory developments promoting community building and virtual cooperation in those disciplines.

3. THE CONCEPT OF VIRTUAL ROOMS

Most community-based systems follow a room-based approach, as it supports synchronous as well as asynchronous and mixed scenarios of cooperative learning [13]. These systems are composed of multiple rooms offering different functions. The function of a room derives from the number and character of the room's users, the attached tools and the assigned rights in the room. Different combinations of these components form different types of rooms (see Figure 1). Following this principle, it is possible to design and create rooms for work groups, rooms with teaching material or laboratories with special virtual equipment, according to the respective requirements. The users of such a system can decide on their own if they want to work in cooperation with other users or if they just want to solve tasks and study in their own work rooms. In addition, there is the possibility to build rooms as a mandatory element of a lecture series. Students will have to pass some of the rooms (e.g., a laboratory) on their way through a seminar before the successful completion of the lecture series is accepted.

Special function rooms such as a room solely for teaching material or for a laboratory provide a clearly arranged structure of offered learning units in contrast to rooms containing all material of a complex lecture series. For example, a course consisting of a lecture, a seminar and a hands-on experiment can be distributed over three different rooms, which are presented in sequence or in parallel. By assembling rooms with different functions in sequence or in parallel, a whole virtual university with lecture, seminar, workshop or laboratory rooms with virtual and remote labs can be built up.

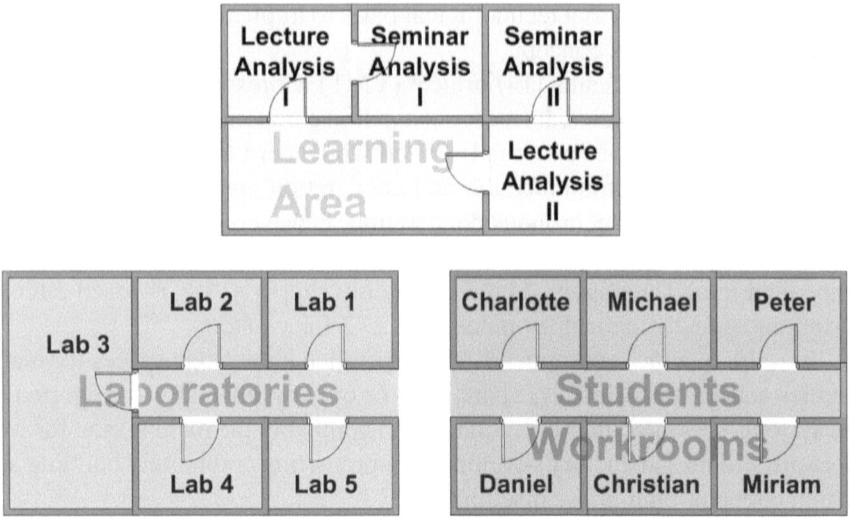

Figure 1. Schematic presentation of different functional rooms

By this means, a user can be virtually supported in all phases of a course; scientists from different institutions and locations can discuss their research achievements and approaches or perform experiments in virtual and remote laboratories – all in one virtual building.

The "ViCToR-Spaces" room concept does not provide the rooms in a hierarchical structure as it is characteristic for currently deployed cooperative knowledge spaces [6,7]. We are not designing a static root-to-leaf structure but rather a dynamic, heterogeneous network, known in philosophy as a rhizome [3], such that each user will obtain a different view on the room structure of the system – according to the rooms which the user is interested in and related to (Figure 2).

The users can be assigned to certain rooms supporting their individual profile, e.g., to the rooms of a course the student is attending in this semester. Additionally, users can delete and add rooms, constructing their individual network of rooms. The different perspectives of the users provide many different views on the room structure. The rooms define a new way of knowledge management.

The room will be built in a successive way. Thus, the function of a room will be constructed from three components (Figure 3):

- Room-related rights
- Specification of the room users
- Used tools and communication interfaces

These room components will be elaborated in the next three sections.

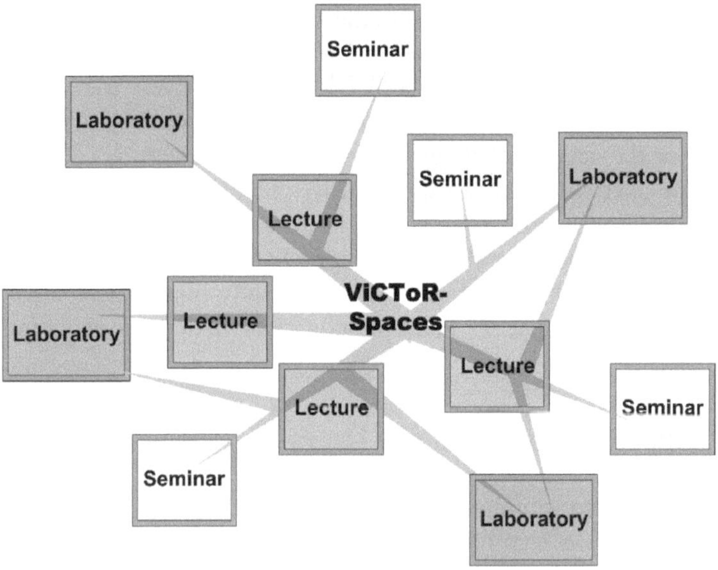

Figure 2. The rooms in ViCToR-Spaces as a network

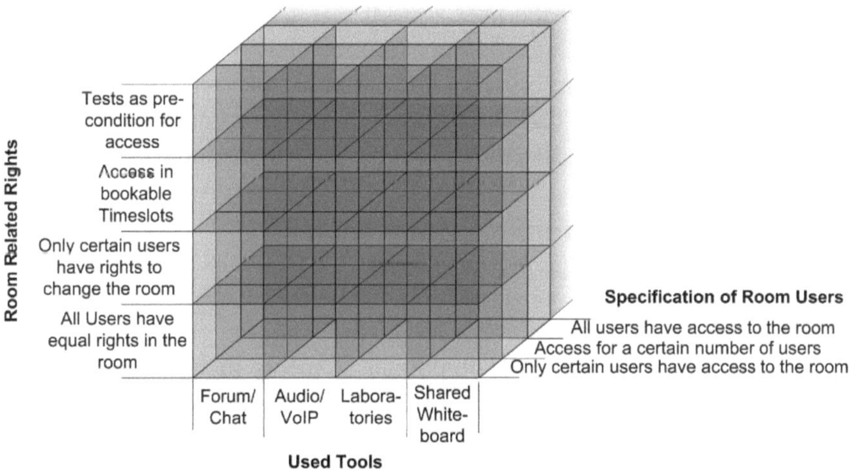

Figure 3. Components of a functional room

3.1 Room-Related Rights

Unlike a physically existing room a virtual room can be built and completely deleted in short time. It is obvious that not every user can be allowed to change or delete every room in the system. Room-related rights are needed and have two different aspects. On the one hand, they refer to creating, deleting and changing a room. On the other hand, they define pre-conditions for access to the room. For rooms that can or should only be used by a limited number of users, access can be controlled by bookable timeslots. Every user is assigned to limited timeslots to accomplish an experiment in a laboratory. As another pre-condition for the access to rooms a special test room can be added upstream such that the users first have to successfully complete the test before entering the next room. Following this principle, i.e., the serial operation of completing a test, booking a timeslot and finally accomplishing the hands-on experiment, an artificial bottleneck is created. This is especially important in cases of special laboratories where users need previous knowledge for accomplishing a certain experiment.

3.2 Specification of Room Users

In most “real-life seminars”, three or four students form a work group to solve the tasks of the seminar together. In this case the room should “know” the identity of the group members to restrict the access to the room to this user group alone. All participants have equal rights in the room concerning the modification of the room. The users’ own work room would represent such a user-restrained room. Only the user himself can access his own work room. There are also rooms that should only be accessed by a certain number of users, e.g., in remote laboratories. In this case only the number of users who have access to the room is restrained without specifying certain users. A third possible type of room user specification is the access to a room by all users.

3.3 Tools and Communication Interfaces

As a third function component the room can be previously filled with tools and communication interfaces, according to the room function.

Communication interfaces are important elements in virtual, cooperative work. In our room concept we provide chats, discussion forums, email and a VoIP function. Not all interfaces need to be attached in all types of rooms. Small groups can manage a clear information exchange in a chat or VoIP session. In larger groups such communication sessions can get unmanageable. In turn, smaller groups might not require a discussion forum.

Shared whiteboards provide a consolidated graphical work area for all room users. All participants are able to work with the same set of objects on the board. Similarly to a chat client the written and drawn objects are shown identically on all participants’ boards. Shared whiteboards can be used both in synchronous and asynchronous manner. Geographical and chronological distances are bridged with version-controlled sessions and documents. Scientists from Berlin and Shanghai can thus work cooperatively on scientific problems.

Virtual and remote labs are mainly used in education for engineering and natural sciences. These installations are an important issue for providing access to laboratory equipment independently of time or financial means. As in both laboratories parameters can be manipulated, controlled and read off, different measurement instruments can be attached and combinations of virtual and remote labs for the same physical effect can exchange parameters and other information. In this way, output from a virtual experiment can be verified by the physically existing experiment. Experiments – be they virtual or remote – integrated in a virtual knowledge space can exchange any kind of information. The integration of such experimental settings in ViCToR-Spaces will be described in the next section.

4. EXPERIMENTS IN VIRTUAL SPACES

Experiments form an important part of learning, teaching and research within the technological disciplines. Integrated into a cooperative knowledge space, they provide better access to experimental setups for all students,

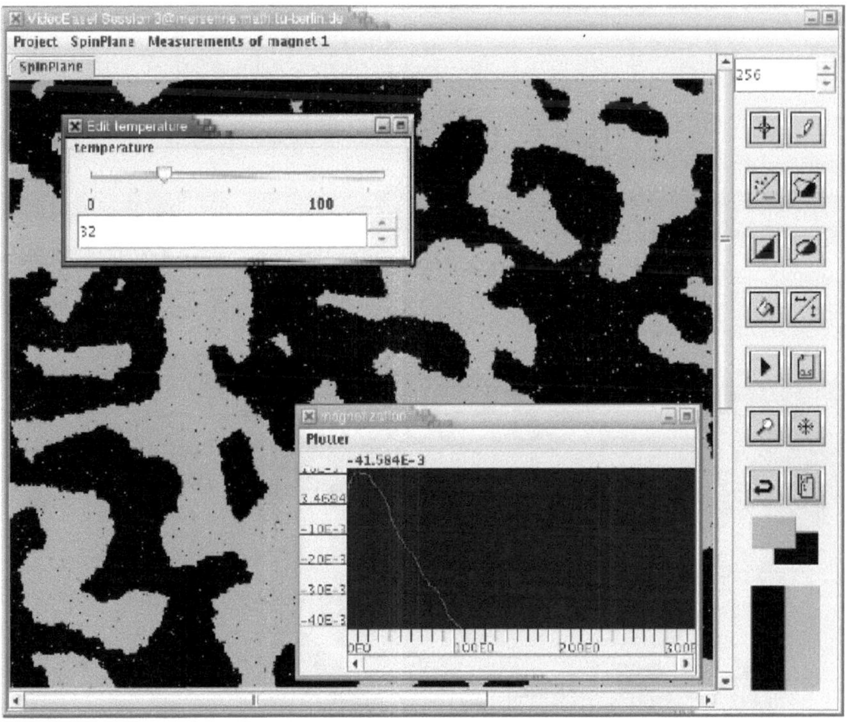

Figure 4. Screenshot of VideoEasel

independent of limitations in time, budget or access to classical laboratories – thus forming one of the most important parts of the ViCToR-Spaces concept.

To enhance access to experimental setups, there are two principal alternatives [9]: Virtual laboratories and remote experiments. Virtual laboratories use the metaphor of a "real" scientific laboratory as a guiding line [8, 10]. The

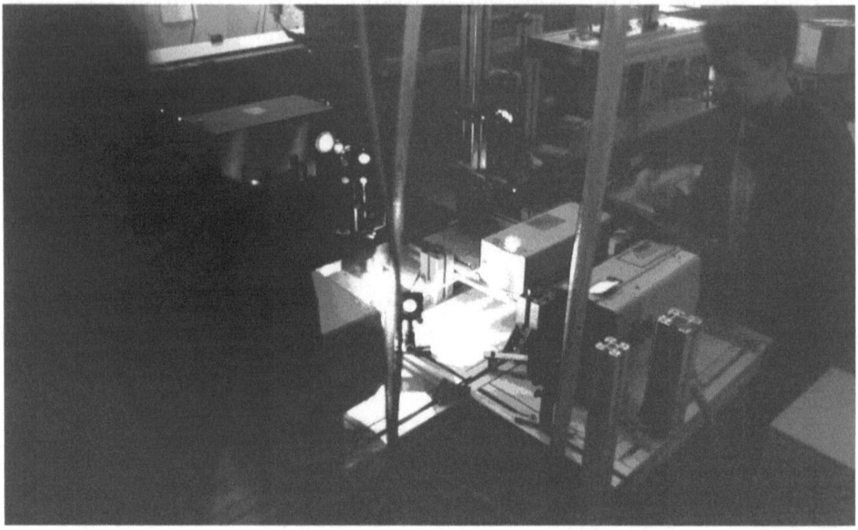

Figure 5. The remote laser in the lab...

Figure 6. ...and the experimenters outside the lab using Tablet-PCs to perform experiments

software design focuses on emulating scientific hands-on experience in virtual spaces (see Figure 4 for an example). In theoretical fields such as mathematics and theoretical physics, virtual laboratories have revolutionized education and research, as they allow an intensive experimental access to abstract objects and concepts. They are capable of building bridges between the theoretical fields and experimental sciences.

Complementary to virtual laboratories, remote experiments are real experiments, remotely controlled by the experimenter from outside the laboratory [12]. They are based on a technology which allows true experimenting from a remote location at almost any given time; this is illustrated in Figures 5 and 6. The different approaches have a number of similarities and enrich each other through their differences: remote experiments allow the investigation of real objects including hands-on measurement experience, which obviously does not hold for virtual laboratories. On the other hand, virtual laboratories are capable of constructing an experiment, whereas this kind of flexibility is hard to imagine or implement in remote experiments.

4.1 System Architecture

As said in [2], ViCToR-Spaces, in its first prototype, will be built on top of the CURE system developed at FernUniversität Hagen. CURE ("Collaborative Universal Remote Education") is a web-based platform for collaborative learning, which is based on combining the room metaphor with WIKI concepts and communication tools [5, 6]. CURE is implemented as a set of Java Servlets. The implementation strictly follows the "everything is an object" philosophy; i.e., rooms, documents, users are realized as plain Java objects to which a simple, but efficient persistence mechanism backed by a DBMS has been added. Basic operations in CURE are made available to external applications through a SOAP-based web services interface.

The core concept of CURE – and hence of ViCToR-Spaces – is the room metaphor [4]. A virtual knowledge space as defined by CURE is subdivided into virtual rooms. In CURE, rooms are arranged in a hierarchical, recursive, tree-like structure. The root of this room structure is a so-called entrance hall, which a user sees first after logging in; the other rooms are arranged "below" this root. Each room offers communication channels such as chat, threaded discussions, email or instant messaging among group members, as well as documents provided as a Wiki page or created by external applications and integrated into the room. The communication interfaces and user awareness features are always tied to their room as well as documents are always associated with the room there are created in. It is possible, however, to simply move or copy documents from one room to another. Managing different versions of documents is essential for cooperative work. CURE's version control allows branching and is able to detect versioning conflicts in documents edited concurrently.

Being geared toward mathematicians, natural scientists and engineers, ViCToR-Spaces unconditionally have to support formulae not only in documents, but in all components, most notably communication tools. CURE allows rendering formulae as PNG images.

Cooperative work requires an appropriately flexible rights management for group workspaces and documents. CURE supports this through its concept of virtual keys. The user who created a room may pass on one or more of these keys to one or more of the other users. Keys may be associated with three groups of rights: room-related rights, communication- and document-related rights and key-related rights. Consequently, groups are defined implicitly by ownership of a virtual key to a room.

4.2 Integration of Remote and Virtual Labs

At the core of our concept to adapt CURE for the special requirements of mathematics, natural sciences and engineering there is the integration of experimental settings. Experiments play a vital role in these disciplines, thus their presence in ViCToR-Spaces is strongly desirable. As a rationale, virtual [10] and remote [12] laboratories must be treated and integrated uniformly (i.e., as a "black box") wherever possible. Experiments – be they virtual or remote – are to be held in a so-called "warehouse". There is only exactly one instance of each experiment present in the virtual knowledge space. Using links, experiments can be integrated into each room. Quality of experiments uploaded to ViCToR-Spaces is assured in a two-stage process. First, they are uploaded to what might be called a "submission hall". There, they are checked automatically for technical correctness and propriety of their semantic classification. This is to prevent, e.g., an experiment on electronics to be classified as an optics experiment, rendering it undiscoverable for an electronics teacher. For providing this automatically, developers of experiments must provide them with appropriate metadata (s.b.). If the automated checks are passed, a human user assesses the experiment for scientific soundness. It is up to him/her to finally transfer the experiment from the "submission hall" to the "warehouse", where it is made available to other users.

4.3 Java Web Services and Modularization of Components

Service Broker and Java Web Services. In order to provide "lab services" we will implement a "service broker" in the current knowledge space system for java Web Services. The service broker will manage different tasks between the integrated components and the system. As one task we will provide the communication between the different laboratories – virtual and remote labs – as Web Services. The labs will be able to exchange information with each other. In case there is a virtual and a remote lab investigating the same physical effect the devices are considered to exchange measurement results to, e.g., prove the effect from the virtual lab in the real experiment of the remote lab. Also, we will supply Web Services between the knowledge space and the laboratories – for example to assign a laboratory to a certain group of users while in use. As another implementation there will be a time booking system for remote laboratories. A time slot reservation will be communicated to the knowledge space system through a Web Service routine. The Web Services will also be used for saving the measurement results in the knowledge space. All offered services will be registered in the service broker integrated in the knowledge space system.

Use of Metadata Standards. Over the last years many different types of virtual or remote laboratories have been developed in almost every university or research institute. One of the main aims of our project is to integrate these software laboratories in our knowledge space system in a way of a "black box". This will provide reusability and interoperability as well as a modularity for integrating both our components in other systems and components from other institutions in our system. We will use approved eLearning standards to assure interoperability and reusability. In the last years only a few standards have been established. For our project we shortlisted the Dublin Core metadata element set and learning object metadata (LOM). At the moment we are working with both standards and we will decide which standard works best for our requirements and the components to be integrated and the virtual knowledge space system. The corresponding standard will then be combined with a semantic description to create ontologies between the integrated components and other appropriate documents.

Semantic Description of Objects and Creating Ontologies. In the current knowledge space system different types of documents can be integrated and made available for other users. Our further developments will provide these documents, our integrated remote and virtual labs and other possible components with a semantic description using RDF, XML and/or SPARQL. As, e.g., SCORM 2004 already uses XML binding for the metadata description (defined in the Content Aggregation Model (CAM)) and for building sequencing rules (defined in the Sequencing and Navigation Book (SN)), we will combine such types of standards to derive an eLearning object with completely machine readable semantic descriptions. Furthermore, we will build ontologies between the corresponding components and documents. Users will be able to find respective documents or remote experiments in the system after, e.g., completing a virtual experiment in a laboratory.

5. SAMPLE SCENARIO – eRESEARCH

As a sample scenario for our vision of a knowledge space system for collaborative eResearch we will describe the work of scientists in the knowledge space system during a research session with a remote laboratory (Figure 7). We categorize this scenario as eResearch since the scientists are researching a specific physical effect. In difference to experiments performed by students, the results of the experiment might either be known (thus serving as certification) or be completely new.

The Room. The room will be created by a system administrator or one of the scientists as a laboratory room. The remote experiment required by the scientists can be found in the "warehouse" of our system and has only to be selected for integrating in the appropriate room. The experiment is already physically set up and can be run immediately. In this scenario the scientists work on a problem from nanotechnology. Hence, the laboratory consists of a clean room with an AFM (atomic force microscope) and a mass spectrometer. There is also a shared whiteboard and the VoIP function activated. Only four specific scientists have access to the room. They do not have to fulfill any preconditions such as a test or booking timeslots. All scientists have equal rights in the room. In case the experiment does not lead to the expected results, the equipment can be changed, certain measuring devices can be attached or disconnected.

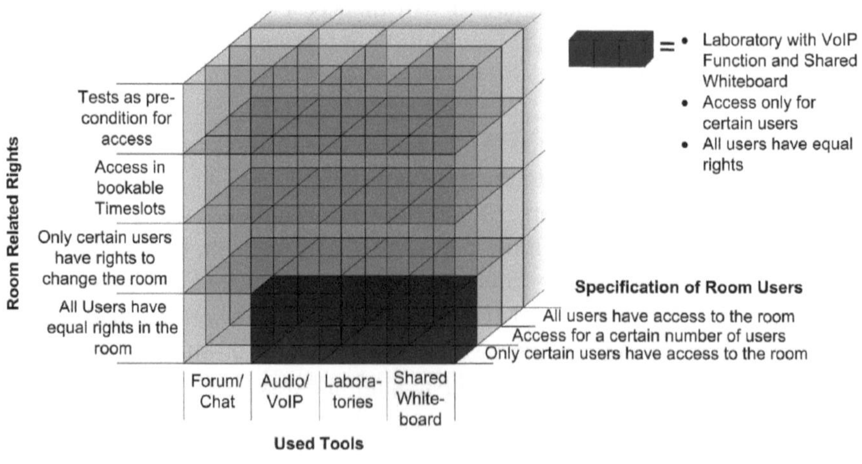

Figure 7. The components of the sample room from scenario eResearch

Interaction Between the Users. All participants have the same view on the laboratory. Of course, only one participant can manipulate a certain controller at a time, as in a real-world lab. Suppose one participant can control parameter A then another participant controls parameter B, but one parameter cannot be controlled by both participants at the same time. Only one user can manipulate the laboratory. Also, the shared whiteboard is used for the joint development of new theories and algorithms the scientists conceive while performing the experiment. The actual experimentation and the corresponding discussion on the whiteboard can be held asynchronously and are stored under version-control in the room.

Completion and Data Storage. As already mentioned the data will be stored under a version-control system. The results can also be stored in the scientists' work rooms so they can be used in future measurements and experiments. Imported in a shared document the results can be used for a jointly written publication. Under version-control all participants of the experiment can bring in their experiences and knowledge in the paper.

6. CONCLUSIONS

In the past years, the main focus in developing eLTR-technologies (learning, teaching and research) has been on stand-alone applications and solutions for specific tasks. Today, modern approaches in the design of the architectures required show that the integration and interconnection of independent, single components play a central role in providing diverse, comprehensive functionality and addressing a broad, heterogeneous user spectrum. As a result, we face two serious challenges. First, the next period in information technology will be dominated by demands for application integration. Research and applications are increasingly oriented toward semantic content encoding as a prerequisite for interconnectedness on a content level and toward integrative technologies for software components as "Web Services/Semantic Web". Second, integration on a social, community-oriented level, that is, support of communication and cooperation structures and shared workflows, is becoming more and more important. The human desire for communication and cooperation – as a basis for successful knowledge acquisition – has been largely deferred in favor of self-determined learning and work, independent of time and location. Not realizing that communication and cooperation in a virtual world will even facilitate one's need for mobile and freelanced access to knowledge we currently face a knowledge landscape of widely spread developments and unequally distributed knowledge. Crosslinking existing knowledge repositories and developments will open knowledge and technologies to students, teachers and researchers beyond geographical limitations by advancing the building of virtual communities. Instancing ViCToR-Spaces the project reveals the high potential of networked learning, working and researching. Democratic sharing in an open-source environment (where open-source is not only related to software developments but as an ideal for community building) will save precious time and money for the people, organizations and institutions involved and increases the value of all the numerous and great developments already made.

REFERENCES

[1] U. M. Borghoff and J. H. Schlichter. *Computer-Supported Cooperative Work: Introduction to Distributed Applications*. Springer, Berlin/Heidelberg, 2000.

[2] S. Cikic, S. Jeschke, and U. Sinha. "ViCToR-Spaces: cooperative knowledge spaces for mathematics and natural sciences". In T. C. Reeves and S. F. Yamashita, editors, *Proceedings of the E-Learn 2006*, pages 2358–2365. Association for the Advancement of Computing in Education, Chesapeake, 2006.

[3] G. Deleuze and F. Guattari. *A Thousand Plateaus*. University of Minnesota Press, Minneapolis, MN 1987.

[4] S. Greenberg and M. Roseman. "Using a room metaphor to ease transitions in groupware". In M. Ackerman, V. Pipek, and V. Wulf, editors, *Sharing Expertise: Beyond Knowledge Management*, pages 203–256. MIT Press, Cambridge, MA, Jan. 2003.

[5] J. M. Haake, T. Schümmer, A. Haake, M. Bourimi, and B. Landgraf. "Two-level tailoring support for CSCL". In J. Favela and D. Decouchant, editors, *Groupware: Design, Implementation, and Use*, volume 2806 of *LNCS*, pages 74–81, Springer, Berlin, 2003.

[6] J. M. Haake, T. Schümmer, A. Haake, M. Bourimi, and B. Landgraf. "Supporting flexible collaborative distance learning in the CURE platform". In *Proceedings of the 37th Annual Hawaii International Conference on System Sciences*, Los Alamitos, CA, USA. IEEE Computer Society, Alamitos, 2004.

[7] T. Hampel and R. Keil-Slawik. "sTeam: structuring information in a team – distributed knowledge management in cooperative learning environments". *J. Educ. Resour. Comput.*, 1(2):3, 2001.

[8] S. Jeschke and T. Richter. "Mathematics in virtual knowledge spaces – user adaptation by intelligent assistants". In R. Kaschek and K.-P. Jantke, editors, *Intelligent Assistant Systems*. Idea Group, Hershey, 2006.

[9] S. Jeschke, T. Richter, H. Scheel, R. Seiler, and C. Thomsen. "The experiment in eLearning: magnetism in virtual and remote experiments". In *Conference Proceedings ICL 2005, Interactive Computer Aided Learning*, Villach/Austria, Kassel/Germany, September 2005. Kassel University Press, Germany, 2005.

[10] S. Jeschke, T. Richter, and R. Seiler. "VideoEasel: architecture of virtual laboratories for mathematics and natural sciences". In A. Méndez-Vilas, B. González-Pereira, J. Mesa González, and J. A. Mesa González, editors, *Recent Research Developments in Learning Technologies*, page 874 ff. FORMATEX, Badajoz, 2005.

[11] G. Stahl. *Group Cognition: Computer Support for Building Collaborative Knowledge*. MIT Press, Cambridge, MA, 2006.

[12] C. Thomsen, H. Scheel, and S. Morgner. "Remote experiments in experimental physics". In *Proceedings of the ISPRS E-Learning 2005*, June 1–3, Potsdam, Germany, June 2005.

[13] M. Wessner. "Kontextuelle Kooperation in virtuellen Lernumgebungen", volume 8 of *Schriften zu Kooperations- und Mediensystemen*. Josef Eul Verlag, Lohmar–Köln, 2005.

[14] M. Wessner, W. Shumar, G. Stahl, J. Sarmiento, M. Mühlpfordt, and S. Weimar. "Designing an online service for a math community". In *ICLS '06: Proceedings of the 7th International Conference on Learning Sciences*, pages 818–824. International Society of the Learning Sciences, 2006.

ENABLING SCIENTISTS THROUGH WORKFLOW AND QUALITY OF SERVICE

A.S. McGough[1], A. Akram[1], D. Colling[1], L. Guo[1], C. Kotsokalis[2], M. Krznaric[1], P. Kyberd[3], and J. Martyniak[1]

[1] *Imperial College London, London, UK*

asm@doc.ic.ac.uk, aakram1@doc.ic.ac.uk, d.colling@imperial.ac.uk, liguo@doc.ic.ac.uk, marko@doc.ic.ac.uk, janusz.martyniak@imperial.ac.uk

[2] *National Technical University of Athens, Athens, Greece*

ckotso@cslab.ece.ntua.gr

[3] *Brunel University, London, UK*

Paul.Kyberd@brunel.ac.uk

Abstract There is a strong desire within scientific communities to Grid-enable their experiments. This is fueled by the advantages of having remote (collaborative) access to instruments, computational resources and storage. In order to make the scientists experience as rewarding as possible two technologies need to be adopted into the Grid paradigm: those of workflow, to allow the whole scientific process to be automated, and Quality of Service (QoS) to ensure that this automation meets the scientists' needs and expectations. In this chapter we present an end-to-end workflow pipeline which takes a user's design and automates the processes of workflow design, resource selections and reservation through to enacting the workflow on the Grid, thus removing much of the complexity inherent within this process.

1. INTRODUCTION

In order to perform constructive science, a scientist will in general need to perform multiple tasks in order to achieve their goal. These may include such things as configuration of an instrument, collecting and storing relevant data from the instrument, processing of this information and potentially further iterations of these tasks. The use of services such as computational resources and computer-enabled instruments has been seen as an advantage to the achievement of this goal. Many scientists now seek to automate the process of bringing their tasks together through the use of workflows. Here we define a workflow as a set of tasks, along with a definition of how these tasks should interact.

The Grid [13] is an enabling paradigm bringing different services into a collaborative environment. The use of Web Services [21] is emerging as the de facto communication mechanism, with most Grid projects now supporting them to some degree. Workflow languages such as BPEL [4] and WS-Choreography [7] are powerful for developing workflows based on Web Services. However, the development and execution of workflows within the Grid is a complex process due to the mechanisms used to describe them and the Web Services they are based upon. The selection of resources to use within the Grid is complicated due to its dynamic nature-resources appearing and disappearing without notice and the load on them changing dramatically over time. These are issues that the scientist will, in general, not wish to be concerned about. The use of advanced reservations can help to reduce the amount of uncertainty within the Grid. However, in general, the selection of the "best" service to reserve is a complex process.

In this work we construct an end-to-end workflow pipeline, based on the work of McGough et al. [16] as part of the ICENI [15] project. This is exemplified through the Grid architecture used within the Grid-enabled remote instrumentation with control and computation (GRIDCC) [10] project. GRIDCC is a European Union-funded project aimed at Grid-enabling instruments along with extending Grid middleware such that instruments may be used as part of a set of orchestrated tasks, thus allowing the whole of the scientist's workflow to be performed in an automated manner with interaction provided through a virtual control room [20]. The GRIDCC project is building instrumentation, and the end-to-end workflow pipeline, as compatible services to those of the EGEE [11] gLite [12] project. However, as our interaction with gLite is through Web Service interfaces it is conceivable that this work could be developed on top of other Grid infrastructures.

The conceptual workflow pipeline is illustrated in Figure 1. The scientist will develop the workflow within the Workflow Editor. Although a workflow is developed at this stage, the view presented to the scientist is abstracted away from the workflow language. Rather than presenting a generic Workflow Editor the scientist is presented with a more tailored view, instruments are presented as instrument entities rather than generic Web Services. The Workflow Editor (software) will then generate appropriate workflow language documents based on the scientists design. This helps to alleviate the scientist from needing to become a workflow specialist. The user may also have a number of requirements

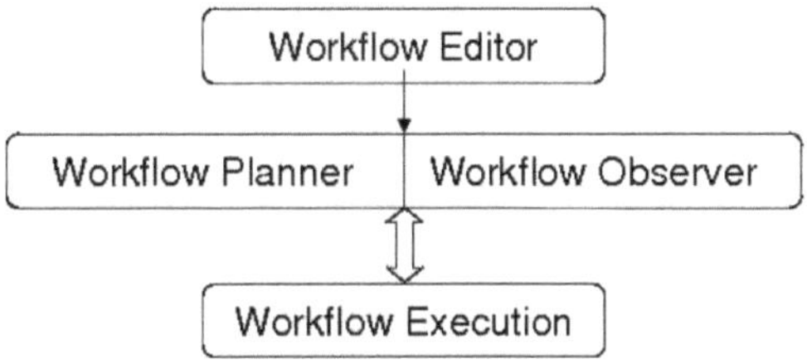

Figure 1. An end-to-end workflow pipeline

on how the workflow should execute – a required level of *Quality of Service* (QoS) provided by the system – this may be specified alongside the workflow description. The Workflow Editor is discussed further in Section 2.

Selection of the most appropriate resources based on workflow and QoS constraints is now performed. Within the QoS document three types of constraints can be specified on resources:

- *Specified Resource*: The scientist has specified the resource that should be used. This may be due to the fact that the scientist will have a sample placed onto a specific instrument at a given time. This informs the planner of fixed requirements on the workflow.
- *Specified Requirement*: The scientist may not know the best resource to use for a given task though they may know that a reservation will be required. This could be for ensuring enough space is made available to store data generated from an experiment. It is the role of the planner to convert this into a specified resource and make reservations as appropriate. Requirements will describe specific features of the resource which need to be matched, such as the amount of disk space required.
- *Unspecified*: The resource is not specified nor is there a specific requirement on this resource. The planner may choose to convert this into a specified resource in order to achieve some overall QoS requirement specified by the user.

Constraints may also be defined over the whole or parts of the workflow, for such characteristics as execution time, latency on communication, time before execution starts or amount of data which can be transmitted in a given time interval. These QoS requirement may be strict (hard) thus considered mandatory – a hard constraint of “this must take less than five minutes” must be adhered to or the whole workflow rejected. Alternatively the requirement may be soft (loose) in which case a value is given indicating the adherence required – a soft constraint of “this should complete in less than five minutes in 80% of cases” can be accepted as long as the planner believes that it can achieve this. A full breakdown of the Workflow Planner can be found in Section 3.

The observer monitors the progress of the workflow, defined by the planner, and if the tasks deviate from the plan it will cause the workflow to change in order to increase the chance of the workflow completing successfully. This allows for dynamically changing the workflow during execution both to bring the workflow back to the plan and to make use of new and potentially better resources unknown at the time of planning. This is an area of active research.

1.1 The Grid Architecture

For our architecture we adopt the design from the GRIDCC [10] project as illustrated in Figure 2. The user interface to the Grid is the *virtual control room* (VCR), which contains the Workflow Editor. The *Workflow Management service* (WfMS) contains the Workflow Planner and the Workflow Observer along with the Workflow Engine, in the case of GRIDCC this is the ActiveBPEL [3] engine. The Workflow Engine may communicate with various Grid services such as Compute Elements (CE), Storage Elements (SE) and, as defined within the GRIDCC project, the Instrument Elements (IE) – a Web Service abstraction of an instrument – along with the Agreement Service (AS) for brokering reservations. The Performance Repository (PR) contains information about previous executions of tasks on resources and models derived from testbed experiments.

Figure 3 illustrates the internal structure of the WfMS. Tuples containing a BPEL4WS and QoS document are validated and converted into an object representation. The workflow and QoS documents can then pass through various stages (outlined further in Section 3) before the BPEL4WS document is submitted to a BPEL4WS engine and the revised QoS submitted to the Observer. The WfMS must decide, based on information from the PR and an understanding of the workflow, if the submitted request can be achieved. The WfMS may choose to manipulate the workflow/QoS to achieve this, for example making reservations on tasks and/or changing the structure of the workflow [16]. In order to allow the BPEL4WS engine to communicate with the existing Grid resources inline facades are used to manipulate the Web Services calls.

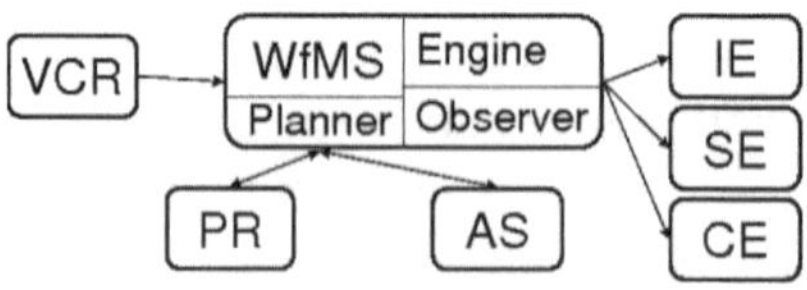

Figure 2. The GRIDCC architecture

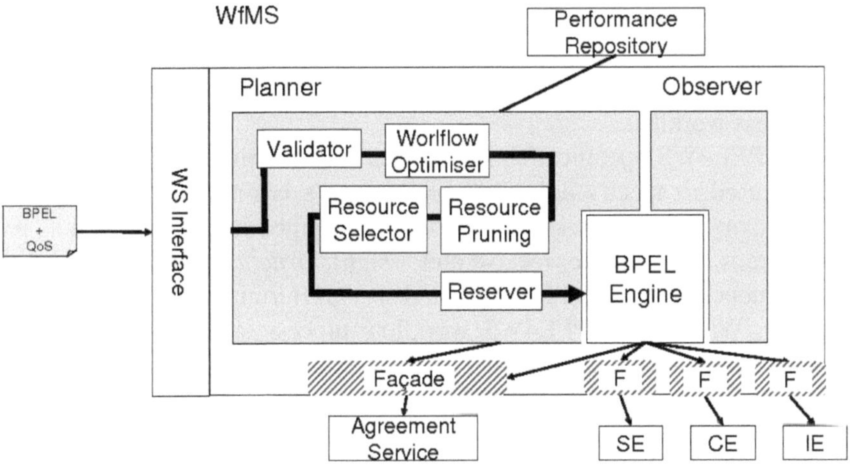

Figure 3. The architecture of the WfMS

2. WORKFLOW EDITING

Service-oriented architecture (SOA) [18] encourages software components development for on-demand consumption with Web Services establishing itself as the popular "connection technology". SOA principles also influence the business logic of services by encouraging good design, i.e., promoting modular and dispersed components that can be separately developed and maintained. The interface required for Web Services is described in a WSDL file (Web Service Description Language [8]). Services exposed as Web Services can be integrated into complex services that may span multiple domains and organizations.

The GRIDCC project is providing a Web-based Workflow Editor based on Business Process Execution Language for Web Services (BPEL4WS) [4]. BPEL4WS is generally regarded as the de facto standard for composing and orchestrating workflows. The goal of the BPEL4WS specification is to provide a notation for specifying business process behavior based on Web Services. BPEL4WS builds on top of the Web Service stack and is defined entirely in XML, is compatible with other emerging technologies including WS-Reliable Messaging [5] and WS-Addressing [6] (BPEL4WS, however, is not limited to these technologies).

2.1 Business Process Execution Language

BPEL4WS provides a rich vocabulary for the description of business processes supporting two different types of process declaration. An "abstract process" specifies the messages exchanged between participating services without exposing the internal details of the process. An "executable process" extends an abstract process to allow specification of the exact (internal) details of a process workflow.

The BPEL4WS specification defines various building blocks, called "activities", used to specify an executable process's business logic to any level of complexity. The BPEL4WS activities are grouped into *basic*, such as procedural steps (e.g., invoke, receive and assign); *structured*, which control the flow (sequence, switch and foreach); and *special* (terminate, validate and fault handlers). Within the BPEL4WS workflow process, activities exchange information using global data variables. Though BPEL4WS was developed for business processes there are properties which make it useful for scientific processes, such as modular design, exception and compensation handling. Scientific services are often subject to change, especially for data types and service-endpoint locations. Data flexibility is supported using "generic" or "loose" data types while workflow adaptivity, by the use of "empty" activity element, provides a mechanism for future extensions.

Reusability of primitive and structured activities in BPEL4WS is limited – it is not possible to re-execute an activity that is defined earlier in the script, it must be re-defined each time. Nor is it possible to inject, say, Java code directly into a workflow script, except by wrapping it as a Web Service with the associated performance impact. BPEL4WS does not provide any mechanism to support management and monitoring of a workflow script, nor Web Service security issues, such as passing of credentials through a BPEL4WS engine. Here WfMS uses "deanonymization" technique in which the users identity is passed in the message header. The outgoing message is intercepted to attach user credential and to perform a secure call to the external service.

Generating a workflow is a process of bringing these elements together – usually is a complex and tedious process. A number of BPEL4WS editors exist, such as the Active BPEL Editor [2] and the Oracle BPEL4WS Process Manager [19]. However, these are developed for computer scientists and an extensive knowledge of BPEL, WSDL, etc., is required. In general scientists will not wish to develop such skills.

2.2 Aim of GRIDCC Editor

The flexibility and richness of the BPEL4WS specification bring unavoidable complexities, which hinder its use by scientist and researchers. We can identify two categories: (1) users executing existing workflows and (2) expert users engineering new or existing processes as workflows. For the first type of users Web Portlets are appropriate as this allows selecting existing workflows, setting parameters and constraints. The second type of user needs rich client software with advanced functionality such as a polished user interface, drag-and-drop, provision to save partial workflows and integrate them as building blocks in other workflows. The purpose of GRIDCC editor is to provide a hybrid solution. The GRIDCC editor differs from existing editors in the following aspects:

Portal Based Editor: All workflow editors require installation and configuration, requiring access to the local file system and potentially admin rights – which is not always available. This JSR 168 [1] compliant workflow editor allows access on demand. Users can edit and save the workflow on the server, accessible wherever and whenever desired. Use of JSR 168 complaint portal allows a mixing of the editors presentation layer with the back-end business logic. Browser-based clients do not need to be upgraded and provide a pervasive access mechanism.

Drag and Drop: The editor provides drag-and-drop facilities to drag various BPEL4WS activities, Web Services or operations from the Web Service registry, quality of service (QoS) constraints from QoS panel and variables from XML schema registry into workflow designer pane. Dragging of different components on to the designer pane updates either the QoS or the BPEL4WS script. The Workflow Editor is based on the ActionScript 3.0 [14] and MXML [9], the core components of Macromedia Flash. ActionScript is a scripting language supporting various features normally only available to desktop applications and MXML is the XML-based markup language for the presentation layer.

Hiding Complexities: A generic BPEL4WS workflow designer demands advanced skill from workflow designers, i.e., thorough understating of Web Services architecture and different types and styles of Web Services, expertise in managing XML files from various namespaces, experience of any XML query specification such as XPath or XQuery, and familiarity with the BPEL4WS specification. Web Services and BPEL4WS specification have dependencies on other related specification such as WS-Addressing, which further complicates the designing process. The GRIDCC editor hides these complexities from the scientist and researchers by automating different tasks in following ways:

1. A workflow process requires global variables for its proper functioning. Whenever a new process is created a global corresponding `<variables>` element is created at specific location as required by BPEL4WS specification. The `<variables>` element is a registry of different XML-specific data types and variables.
2. A workflow orchestrates various web services which are wrapped within `<partnerLinkType>`, `<partnerLinks>` and

<partnerLink> elements. Each <partnerLinkType> element wraps a <portType> of the Web Service, which itself wraps various operations. The GRIDCC editor creates the <partnerLinkType>, <partnerLinks> and <partnerLink> when any Web Service is added in the Web Service registry.

3. Operations of the partner Web Services are called through an <invoke> activity. The <invoke> activity specifies the <partnerLink>, <portType>, <operation>, <input> and <output> elements defined in WSDL or BPEL4WS script. An operation of the Web Service can be dropped directly on a workflow process from a Web Service registry and the editor itself creates the corresponding <invoke> activity by relating various required elements.

4. An <invoke> element specify <input> and <output> variables to store the values of outgoing and incoming messages. If these variables do not exist the editor creates the <variable> required for successful invocation of the operation in the <variables> element (see the first item above).

5. The GRIDCC editor also adds "exception handling" activities with <empty> activities as a template for flexibility and extensibility. The designer of the workflow can replace these <empty> activities with actual business logic.

Ease of use: The GRIDCC editor is easy to use due to its built-in error handling. The Workflow editor validates the actions of the designer and makes sure different workflow activities are arranged according to pre-defined specifications. Different activities are arranged in the logical groups and different activities are enabled only when they can be used. The Designer creates structured activities as a container and other activities can be dropped into the container to visually assist the modular designing of the workflow.

Web Service Registry: The Editor provides a very basic Web Services registry to arrange web services for later use. When a user adds a new Web Service to the registry, the designer creates corresponding <partnerLinkTypes>, <partnerLinks> and <partnerLink> elements for later use in the <invoke>, <reply> or <receive> activities. The Web Services registry hides inner details and complexities of BPEL4WS script and WSDL extensions (required by BPEL4WS specification) from the user. This allows the designer to concentrate more on the business logic rather than wrapping the partner services in various elements. The Web Service registry can also be used as pool of semantically equivalent but geographically dispersed Web Services.

XML Schema Registry: When any Web Service is added in the Web Services registry, the editor parses the WSDL and extracts data types declared in the `<wsdl:type>` and `<wsdl:message>` elements and builds the registry entry. XML data in the registry is namespace aware and user does not need to address the namespace issues and conflicts. At design time the user only drags the required data type and editor creates the respective variable with correct structure in the BPEL4WS script.

QoS Constraint: The GRIDCC editor provides a pallet grouping various QoS constraints. The BPEL4WS specification does not define the quality issues related to overall workflow or individual Web Services. QoS constraints can be coupled with different BPEL4WS activities particularly the `<invoke>` activity. QoS constraints for the workflow are specified in the separate file rather than embedding them in the BPEL4WS script.

Workflow Monitoring: Different workflow engines describe the state of a workflow in XML though not all. If a workflow state is available in XML format then it can be exposed and can be queried by the client application. It is also possible to retrieve a workflow snapshot at run time in XML, which can be transformed using XSLT into user acceptable formats. However, it must be stated that extracting an XML snapshot of the current state of the executing workflow requires some understanding of the underlying workflow engine. Consequently, it is easier to create a workflow monitoring Web Service as part of the workflow script that can be used by workflow clients to monitor or track the execution of a workflow in a portable manner. Development of a custom monitoring service can be especially important when status of a workflow and different activities are crucial to decide the execution path of workflow. If workflow monitoring can be integrated with existing management capabilities, it will help to give a more complete picture of workflow and control on the execution progress at any given moment.

3. WORKFLOW WITH QUALITY OF SERVICE REQUIREMENTS

Quality of service (QoS) is a broad term that is used in this context to denote the level of performance and service that a given client will expect to experience at a specific point in time, when invoking a given specific operation on a Web Service instance.

QoS support is paramount given the inherent shared nature of the Grid. This is compounded through the limited capability of certain highly demanded services. Furthermore, QoS is essential to deal with the requirements for real-time interactions, such as guaranteeing the streaming and storage of data from an instrument. In this environment an aggressive best-effort usage of services will eventually fail to satisfy clients on the first come first served basis. As the service will not be able to prioritize the requests that it receives thus completing low priority jobs in favor of higher priority jobs. This effect can be alleviated through client specifying loose (soft) and strict (hard) QoS requirements.

The provisioning of loose guarantees consists of the capability of clients to select service instances that can potentially provide a best-effort QoS profile meeting the client's requirements. This is based on previous measurements of that service. Loose guarantees are delivered on a best-effort basis, and for this reason, they generally do not require the prior establishment of a service level agreement (SLA).

Strict QoS requires the certainty of the delivery of the prescribed level of service, which needs to be agreed upon through signaling and negotiation processes involving both the client and the service provider, for example, the reservation of a given portion of a resource such as RAM, disk space, network bandwidth or an instrument. The reservation service provider is responsible for keeping information about resource availability over time, for ensuring that no resource is over-committed and for supporting resource-specific locking mechanisms.

QoS provisioning of our work relies on both strict and loose QoS guarantees. These QoS requirements can be made either by a user through the client interface as discussed in Section 2 or as part of the workflow manipulation process. While hard QoS requires the making of reservations on the resources to be used, soft QoS requires the user (or planner acting on the users behalf) to model the execution of the services they require. These models may vary from the simple when only a single service is required to the extremely complex when several services are required as part of a workflow. In such cases it may be necessary that a reservation is required even to satisfy loose QoS constraints.

A user submits a BPEL4WS document, which is likely to contain a number of service invocations – referred to as an activity. A separate document is used to describe the QoS requirements placed onto this BPEL4WS document. Although a number of languages exist for describing QoS requirements, none exist which is suitable for describing QoS requirements with a workflow description.

As such a simple QoS language has been defined which uses XPath [22] references into the BPEL4WS document to tag activities, both basic and structured as shown in Figure 4. These XPath tags are then annotated with the QoS requirement such as time to execute, disc space required, memory required.

QoS requirements fall into three categories according to the range of activities in BPEL4WS models that it specifies: global, single invoke and multiple invoke.

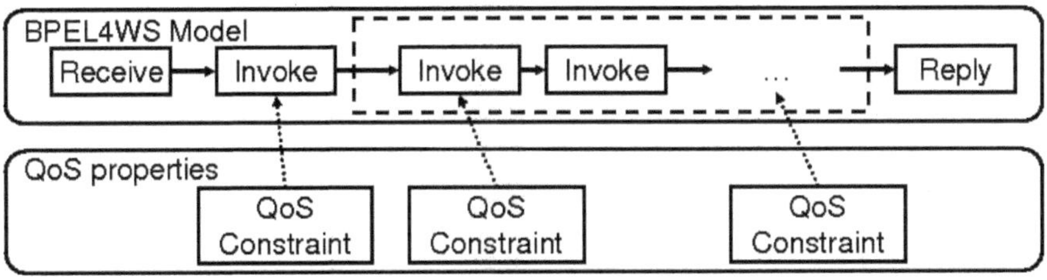

Figure 4. Connecting QoS document and BPEL4WS model

A **global requirement** is a QoS element that specifies single global QoS requirements. A simple example is given in Figure 5 (for simplicity, all the unnecessary technical details are omitted).

```
<?xml,version="1.0",encoding="UTF-8"?>
<QoSRequirements>
    <QoSConstraint>
        <XPathReference>process</XPathReference>
        <Resources>
            <CPUSpeed>2000000</CPUSpeed>
        </Resources>
        <MaxDurationTime>100</MaxDurationTime>
        <Reliability>100</Reliability>
    </QoSConstraint>
</QoSRequirements>
```

Figure 5. A global requirement

In the above example, there is a single *XPathReference* pointing to the entire process of the BPEL4WS document. Thus everything within this process must match these QoS requirements. In this case the overall time should be less than 100 seconds, all CPUs should be 2 GHz and all resources should be fully reliable.

Single invoke activity requirement is a QoS element that specifies requirement on a particular BPEL4WS invoke activity. Only that invoke activity needs to satisfy the QoS requirement specified, as shown in Figure 6.

```
<?xml\,version="1.0"\,encoding="UTF-8"?>
<QoSRequirements>
    <QoSConstraint>
        <XPathReference>
          /process/sequence[1]/invoke[1]
        </XPathReference>
        ...
    </QoSConstraint>
</QoSRequirements>
```

Figure 6. A single invoke activity requirement

```
<?xml\,version="1.0"\,encoding="UTF-8"?>
<QoSRequirements>
    <QoSConstraint>
        <XPathReference>
          /process/sequence[1]/invoke[1]
        </XPathReference>
        <XPathReference>
          /process/sequence[2]/invoke[2]
        </XPathReference>
        <MaxDurationTime>100</MaxDurationTime>
    </QoSConstraint>
</QoSRequirements>
```

Figure 7. Multiple invoke activity requirements

In this case there is a single *XPathReference* pointing to a single invoke element of the BPEL4WS document. Thus everything within this invoke must match these QoS requirements.

Multiple invoke activities requirement is a QoS element that specifies requirement over a set of invoke activities – i.e., all must satisfy (jointly) the QoS requirement. In a single QoS constraint several *XPathReferences* pointing to different invoke activities in a BPEL4WS document are defined. See Figure 7 where the time for both invoke activities must be less than 100 seconds.

It should be noted that multiple QoS elements may exist within the same QoS document and overlaps may exist between QoS elements.

3.1 QoS Components

The Planner is responsible for ensuring QoS adherence. This is achieved through the use of a number of stages (see Figure 8), namely *constraint resolver, basic resolver, performance repository and QoS reserver*. These components are chained by the QoS and BPEL4WS documents that are passed through. We explain each of these in detail in the following subsections.

Performance Repository (PR) is central to being able to perform any quality of service decisions. The Performance Repository will serve data of two types: modeling and dynamic. Model information provides information about how the service performs under different conditions. This may include such information as the reliability of an instrument or the amount of time it takes to bring the instrument on-line. Dynamic information, provided by the information service in many Grids, indicates the current status of the resources, such as load, network bandwidth or if the service is currently in use. Eventually, the PR will also be able to make predictions about future usage. Initially this will be

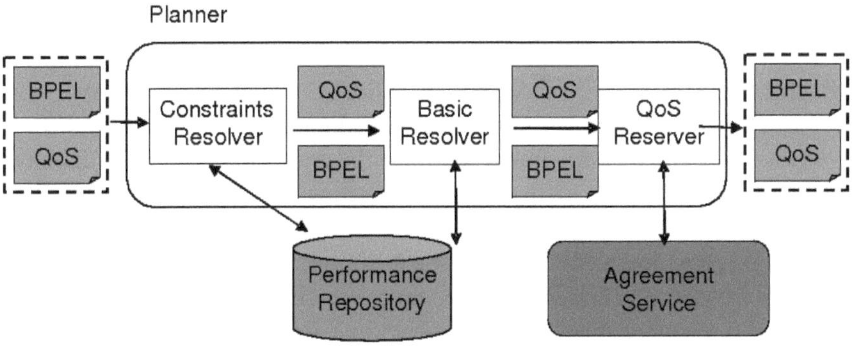

Figure 8. QoS components

through using knowledge of advanced reservations and ultimately using tools such as Network Weather Service [17]. There are two customers for the PR's information, the individual user and the WfMS planner. When making a reservation the PR is interrogated in order to acquire a list of the resources available that can satisfy the QoS demands.

Constraint Resolver is the first component to provide scheduling functionality into the planner. This implementation is based on a constraints equation method. The workflow along with the QoS requirements are converted into a set of constraint equations. Information from the Performance Repository is used to solve these constraint equations.

Basic Resolver (BR) takes a QoS document which states that a resource needs reserving; though the selection of the resource has not been determined, it will query the Performance Repository to select a resource to make a reservation on. No inspection of the BPEL4WS document is done at this stage. The QoS element requesting a reservation without a named resource is then changed into an element requesting a reservation with a named resource. This can then be passed onto the QoS Reserver for making the actual reservations.

QoS Reserver inspects incoming QoS documents looking for requests for making reservations with known resources. The Agreement Service is then contacted in order to make these reservations. The QoS document is then updated to indicate that the reservation has been made and records the unique token used to access the reservation. All requests for reservations are processed here. In the current implementation if reservations cannot be satisfied then the whole document will be thrown back to the user to select new reservation times. Once we have components capable of selecting timings for reservations internally, the workflow and QoS will be returned to this component.

4. CONCLUSIONS

In this paper we have presented an end-to-end workflow pipeline which takes a user's design and implements it within the Grid. The workflow is augmented with a QoS document defining the user's expectations for the execution. This is developed through a Workflow Editor which abstracts the user away from the technical complexities of dealing with WSDL and BPEL. The WfMS provides a mechanism for building QoS on top of an existing commodity-based BPEL4WS engine, thus allowing us to provide a level of QoS through resource selection from a priori information along with the use of advanced reservations.

The end-to-end workflow system is currently being deployed onto the GRIDCC testbed and we are developing workflows with scientists from within the project. This will allow us to evaluate the benefits of this approach against existing approaches which, for example, only allow for the use of workflows without reservations or planning.

We are working to extend the workflow editor and WfMS. We are working with the application scientists from the GRIDCC project to make the workflow editor more tailored to the community. We are also developing the WfMS to allow the Observer to modify the running workflow within the BPEL4WS engine. This will allow us to change the flow of the workflow and the resources used.

REFERENCES

[1] A. Abdelnur and S. Hepper. "Portlet specification (jsr-168)". http://jcp.org/aboutJava/communityprocess/review/jsr168/

[2] "Active endpoints". *ActiveBPEL Designer.* http://www.active-endpoints.com/products/activebpeldes/index.html

[3] "Active endpoints". *ActiveBPEL Engine (2.0).* http://www.activebpel.org

[4] T. Andrews, F. Curbera, H. Dholakia, Y. Goland, J. Klein, F. Leymann, K. Liu, D. Roller, D. Smith, S. Thatte, I. Trickovic, and S. Weerawarana. "Business process execution language for web services v. 1.1 (BPEL4WS)". http://www.ibm.com/developerworks/library/ws-bpel/, May 2003.

[5] R. Bilorusets, D. Box, L. F. Cabrera, D. Davis, D. Ferguson, C. Ferris, T. Freund, M. A. Hondo, J. Ibbotson, L. Jin, C. Kaler, D. Langworthy, A. Lewis, R. Limprecht, S. Lucco, D. Mullen, A. Nadalin, M. Nottingham, D. Orchard, J. Roots, S. Samdarshi, J. Shewchuk, and T. Storey. "Web services reliable messaging protocol (WS-ReliableMessaging)". http://www.ibm.com/developerworks/library/specification /ws-rm/, Feb. 2005.

[6] D. Box, E. Christensen, F. Curbera, D. Ferguson, J. Frey, M. Hadley, C. Kaler, D. Langworthy, F. Leymann, B. Lovering, S. Lucco, S. Millet, N. Mukhi, M. Nottingham, D. Orchard, J. Shewchuk, E. Sindambiwe, T. Storey, S. Weerawarana, and S. Winkler. "Web services Addressing (WS-Addressing)". *Technical report*, Aug. 2004.

[7] D. Burdett and N. Kavantzas. "WS-Choreography model". http://www.w3.org/TR/ws-chor- model/

[8] E. Christensen, F. Curbera, G. Meredith, and S. Weerawarana. "Web services description language (WSDL) 1.1". http://www.w3.org/TR/wsdl, Mar. 2001.

[9] C. Coenraets. "An overview of MXML: The Flex markup language". http://www.adobe.com/devnet/flex/articles/paradigm.html, Mar. 2004.

[10] D. Colling, L. Dickens, T. Ferrari, Y. Hassoun, C. Kotsokalis, M. Krznaric, J. Martyniak, A. McGough, and E. Ronchieri. "Adding instruments and workflow support to existing grid architectures". Volume 3993 of *Lecture Notes in Computer Science*, pp. 956–963, Reading, UK, Apr. 2006.

[11] EGEE project. "Enabling grids for e-sciencE". http://public.eu-egee.org/

[12] EGEE project. "Lightweight middleware for grid computing". http://glite.web.cern.ch/glite/

[13] I. Foster and C. Kesselman, editors. *The Grid: Blueprint for a New Computing Infrastructure*. Morgan Kaufmann, July 1998.

[14] G. Grossman and E. Huang. "ActionScript 3.0 overview". http://www.adobe.com/devnet/action script/articles/actionscript3_overview.html, June 2006.

[15] A. McGough, W. Lee, J. Cohen, E. Katsiri, and J. Darlington. "Workflows for eScience: Scientific Workflows for Grids", June 2006.

[16] A. McGough, J. Cohen, J. Darlington, E. Katsiri, W. Lee, S. Panagiotidi, and Y. Patel. "An end-to-end workflow pipeline for large-scale grid computing". *Journal of Grid Computing*, vol. 3, no. 3–4, pp. 1–23, Feb. 2006.

[17] Network weather service. http://nws.cs.ucsb.edu/ewiki/, Apr. 2006.

[18] D. Nickull and F. McCabe. "SOA reference model". http://www.oasis-open.org/committees/tc_home.php?wg_abbrev=soa-rm

[19] Oracle. "Oracle BPEL process manager". http://www.oracle.com/technology/products/ias/bpel/index.html

[20] F. Asnicar, L. Chittaro, L. De Marco, L. Del Cano, R. Pugliese, R. Ranon, and A. Senerchia. *Collaborative Environment for the Grid – Multipurpose Collaborative Environment*. Distributed Cooperative Laboratories, Springer 2006

[21] W3C. "Web Service". http://www.w3.org/TR/ws-arch/

[22] Xml path language (xpath) version 1.0. http://www.w3.org/ TR/xpath, Nov. 1999.

IMPROVING THE SECURITY PERFORMANCE IN COMPUTER GRIDS

Architecture and Results

A. Moralis[1], V. Pouli[1], M. Grammatikou[1], S. Papavassiliou[1], and V. Maglaris[1]
[1] *National Technical University of Athens, Athens, Greece*

Abstract Security in computational Grids is mainly based on Grid Security Infrastructure (GSI) for authentication and Virtual Organization Membership Service for authorization. Although these mechanisms provide the required level of security, they lack in performance due to their dependence on public key cryptography. In our proposed security architecture we use a Kerberos-based approach (symmetric cryptography) to establish common secrets between grid services (exposed as web services) and clients. The architecture does not nullify GSI and VOMS, but allows a full mapping of GSI-VOMS to Kerberos credentials. The security architecture was designed to meet the specific quality of service (QoS) for nearly real-time control of distributed instruments that belong to different organizations by minimizing the impact of security processing. It is based on GSI and VOMS certificates for the initial login, translates them into Kerberos credentials for authentication and provides message level security implementing the OASIS Kerberos Token Profile. The security performance of our implementation, as shown in our measurements, outperforms the one when X509 Token Profile is used.

1. INTRODUCTION

Distributed Kerberized Access Architecture (DKAA) presented in [14] aims at extending Grid middleware to provide improved performance and also secure Authentication and Authorization Infrastructure (AAI) for distributed systems. A remote, distributed over the net, interactive and multiuser system

[1] This work was partially supported by the EC GRIDCC Project (IST 511382). The authors performed this work within IASA (Institute of Acceleration Systems & Applications), a joint Institute of the University of Athens & NTUA.

is subject to threats encountered in resource-sharing systems. These include unauthenticated and unauthorized access to resources, session hijacking, replay attacks, denial of service (DoS) attacks, etc.

DKAAs' main goal is to provide a balance between a mechanism that offers access control, confidentiality and non-repudiation by authentication, authorization and encryption – thereby eliminating most of the threats mentioned above – for low latency usage of large distributed Grid resources. Therefore, the security architecture should specify a light-weight scalable Authentication and Authorization Infrastructure (AAI), along with efficient procedures for secure message exchanges.

DKAA was designed as the AAI and secure data exchange of the GRIDCC project [7]. GRIDCC aims at developing a Grid framework that enables the remote control of instruments with real-time or near real-time constraints. Their successful operation often requires rapid interaction with computing – storage resources and a large number of instruments. The real time and interactive nature of instrument monitoring and control requires the provisioning of acceptable QoS among various Grid components. Furthermore, since the shared instruments may belong to different organizations, the implementation of a high-performance security scheme that allows dynamic management of virtual organizations is of paramount importance.

In this chapter we extend the work presented in [14] by introducing the implementation details of the DKAA and some measurements that justify the performance improvement of our security implementation based on the OASIS Kerberos Token Profile [21] against the X.509 Token Profile [22].

The remaining of this chapter is organized as follows. Section 2 provides a brief overview of existing popular AAI architectures that can be applied in a collaborative Grid environment, Section 3 presents the proposed distributed kerberized access architecture, while in Section 4 the implementation details of the DKAA applied to a simple GRIDCC use case scenario are provided. Finally in Section 5 some performance measurements are presented and in Section 6 we conclude the paper.

2. AAI INFRASTRUCTURES FOR GRIDS

Grid Security Infrastructure [6] (GSI) is the de facto authentication protocol used in Grid environments [17]. It is based on public key cryptography [11] (PKI), using X.509 certificates. GSI also utilizes an important extension of the PKI, namely the proxy certificates. Through proxy certificates, except for single sign-on, delegation is also achieved. The proxy certificate profile is documented in [12]. In GSI the users present their X509 certificates, which need to be signed by a trusted third party, the Certification Authority (CA). Thus GSI is easily deployed and managed, as there are numerous CAs [19] that issue and manage user and host certificates. Secure message exchange can be achieved through encryption.

Concerning authorization, as it is not part of GSI, it can be performed independently at the level of Computing Elements (cluster of Grid resources, e.g. instrument controller in GRIDCC) using CAS [18]. Alternatively, authorization in Grids can be achieved by the employment of Virtual Organization Membership Service (VOMS) [1]. VOMS provides to users an attribute proxy certificate that exposes the roles and the capabilities of each user within his/her Virtual Organization (VO) [2]. Then the user uses this certificate to access the resources. The service acknowledges the user's authorization by checking the attributes in the VOMS proxy certificate.

During the design phase of the DKAA, we thought of maximizing the performance of the grid security infrastructure (GSI), which uses public key cryptography (PKI). In PKI, an asymmetric scheme establishes a secure session after a computationally expensive handshake procedure (SSL - Secure Socket Layer) imposing a penalty factor that varies from 3.5 to 9, as shown in [5]. Although resuming a session via caching of the results of a previous SSL handshake helps reducing the problem, it does not alleviate it completely, as the initial setup introduces performance penalties, especially under heavy load. In some GRIDCC use cases, thousands of instruments might be involved, each requiring secure message exchanges. The problem is more intense if sessions are terminated in a proxy (e.g. Computing Element or Instrument Element) instead of the terminal resources (e.g. "worker nodes" or instruments) whereby a large number of sessions are concentrated in a single point.

In this respect, GSI as all PKI-based solutions, if used exclusively, due to the heavy overhead of public key cryptography that it employs, presents an important drawback, especially in applications that require low latency. If an entity that has specific QoS requirements accepts numerous requests, the substantial higher CPU load of asymmetric (public) key cryptography can reduce the overall performance to a great extent. Nevertheless, GSI, as a PKI-based system, is suitable for a Grid environment, with a multitude of users and resources.

3. DISTRIBUTED KERBERIZED ACCESS ARCHITECTURE (DKAA) OVERVIEW

DKAA is designed to protect web service-based services and offers the same functionality that GSI presents [20]. The handling of authorization and authentication is performed in the SOAP message layer and not in the application layer or transport layer (http). This allows DKAA to be relatively easily deployed in front of any Web Service. An important feature of the DKAA is the interoperability with the GSI credentials, thus it supports interoperability in the security level. In Figure 1, the general Grid architecture deployed in the

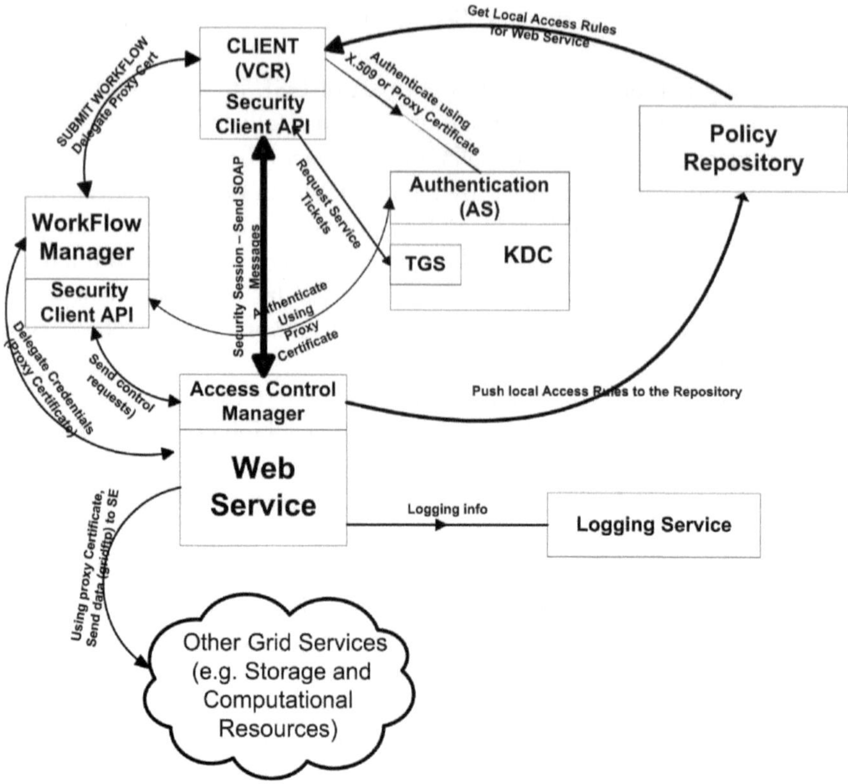

Figure 1. Architecture overview

GRIDCC project, with respect to the DKAA architecture is presented. All the basic elements and components are described below.

3.1 Non-security Related Architectural Components

The architecture components, excluding the security-related ones, are:

- **The Client (Virtual Control Room):** is a portal that exposes the client functionality to the users.
- **Workflow Manager (WfM):** it executes complex workflows that compose multiple web services. These workflows are submitted by the user through the Client.
- **Web Service (e.g. Instrument Element):** The web service that provides the functionality of a grid resource. It could also communicate with other web services or other services in general within the Grid.

- **Logging Service:** It holds the audit logs. Every transaction with the service providers is recorded, along with its status. The status could be "success" or "failure". A transaction may include, but not limited to, the User/group, the command, the status and the reason.

3.2 Security Related Architectural Components

The security architectural choices stem from the need to cover distributed environments that give emphasis on providing secure real-time (or near real-time) access to resources (e.g. instruments) for monitoring and controlling purposes, with acceptable Quality of Service (QoS) guarantees. The design of our security-related architectural components was guided by the following criteria and objectives, which comprise the guidelines of a baseline security architecture:

1. An end-user is authenticated by using an X.509 or proxy certificate issued by a trusted Certification Authority to the KDS. An authenticated user is authenticated once per session and accesses many resources. This provides single-sign-on functionality.

2. The core system consisting of clients and web services employs a ticket-based authentication mechanism and symmetric cryptography. Tickets are acquired via centralized key distribution centers.

3. Authorization is performed at the service level employing local access rules.

Specifically, DKAA provides the required security functionality, including message encryption or message signatures, focusing on providing performance on the server side (Web Service). It utilizes the Kerberos Protocol to distribute the required by the security functionality symmetric keys. Using Kerberos, as a centralized authentication system, renders the authentication procedures relatively easy via the Key Distribution Center (KDC) acting as a Third Trusted Party. Management of users and roles is significantly simpler than a distributed scheme that relies on several PKIs with revocation lists difficult to maintain.

In order to cope with the fact that KDC may be a single point of failure, failover KDC systems should be used, so that the availability of the distributed system is guaranteed. Similarly, security of the KDC itself is critical, thus it must be well protected and monitored. Based on Kerberos' single sign-on provisioning capability, a user is authenticated once (gives his/her credentials) and uses multiple resources and services over a period of time (e.g. 8 hours). This function is implemented by the limited lifetime of the Kerberos tickets, issued by a Ticket Granting Server (TGS). After expiration of a ticket, re-authentication is needed. Depending on the application, lifetimes can vary from minutes to even a day. This duration should be determined, based on time statistics of a control session.

Following this approach, in the remaining of this section, we present and describe the various security components of the DKAA architecture:

- **Kerberos Key Distribution Center (KDC):** A Kerberos [9] system that authenticates users and provides them with service tickets is required to call the web services (e.g. Instrument Elements). The authentication is performed by the authentication server (AS) in conjunction with the Ticket Granting Server (TGS). Both of these servers run as services within the Key Distribution Server (KDS) and will be described in detail below. Kerberos is based on the Needham–Schroeder Protocol [15] and provides single sign-on features using symmetric key cryptography. In addition, there are mechanisms that map X.509 certificates to the Kerberos authentication system, in order to maintain interoperability with certificate-based systems.

- **Authentication Server (AS):** AS handles the authentication procedure. The client sends a request to AS, along with the user's credentials. These could be a trusted certificate or username–password. Other types of credentials (like one-time passwords) could be used as well. The authentication procedure is encrypted, using SSL. The AS checks the user's credentials and if they are found correct, a special ticket named **Ticket Granting Ticket (TGT)** is returned. This TGT among others contains user-specific information, referred to as principal. The principal contains the user name and the subgroup that he/she belongs to (called instance in Kerberos terminology). The user credentials are also stored in the AS.

- **Ticket Granting Server (TGS):** TGS grants tickets for a requesting service so when a user wants to access a Web Service, a DKAA-aware client requests a ticket from the TGS transparently to the user by presenting the TGT received from the AS. After checking, the TGS grants a ticket for the required service. The ticket contains the user principal and, when presented to the Web Service, performs mutual authentication. Ticket-based authentication is based on symmetric key cryptography and thus the authentication setup imposes minimum overhead to the control clusters in contrast to public key cryptography. Moreover, a ticket defines a security session that has a limited duration restricted by the duration of the ticket.

- **Security Client API:** DKAA provides a client API that hides the complexity of the Kerberos protocols from the Client. It manages user credentials, authenticates to the Kerberos system, requests service tickets and calls the web services offering secure message exchange, with different levels of security (requested by the application). In the secure message exchange, the client and the service provider can use the session key (incorporated within the ticket) to encrypt the message or part of the message. By default DKAA appends a digital signature to the message, thus authenticating and preventing alterations by other parties. The level of encryption can play a significant role to achieve the performance goal that each system based on DKAA poses.

- **Access Control Manager (ACM):** Authenticates and authorizes user messages at the web service side. It is a software layer similar to an application layer firewall. It filters all incoming connections to the service in two stages:
  - Authentication of the received message. The first time a user accesses the end-entity, within a control session, he/she sends the ticket he/she owns to access the end-entity. The ACM decrypts it with the end-entity key and stores locally the session key along with the user principal (username/subgroup) and the expiration time of the session key. This is done only once for the duration of the session key. After that, authorization of the messages is performed with the decryption of the content with the session key.
  - Authorization to the requested service. The end-entity checks the local access rules set by the site owner, on the requested service; in conjunction with the user's subgroup it accepts or denies access to the service. This is done per request.
- **Policy Repository:** It collects local policies for all the web services, easing the authorization information management. The policy of each service is set by the owner of the site where the service resides. It is used to locate the existence of specific subgroups within the various services. This is especially important for a single sign-on procedure.

3.3 Users and Authorization

DKAA is designed to allow users to authenticate to services that permit access and control on distributed resources (e.g. instruments). Therefore, the overall infrastructure has to support communication and interactions between two types of entities: Users and Resources (e.g. instruments). The resources are exposed as web services. The users control resources by calling the corresponding operations of the web services.

User authorization is tightly coupled with the policy of each site. It is handled by the service provider via ACM. The ACM checks the users' credentials against a local access list created by the web service administrator. In order to minimize the access rules, users are divided into groups and the access rules are referred to these groups instead of users. All the access rules of all the service providers are uploaded to the Policy Repository. This global view helps to trace problems regarding the authorization in the distributed environment.

The identity of a user or a service within the Kerberos Domain is uniquely identified in the simplest form with the use of a principal/instance@REALM (following Kerberos V5 terminology). The use of a REALM permits the existence of the same principals in different domains. A REALM can be viewed as analogous to a domain or a virtual organization (VO) in Grid terms. The principal value refers to the username and the instance value refers to the group that the user belongs to.

These identities are valid within the DKAA. Outside, the user is identified by his/her X.509 certificate. PKINIT [13] allows authenticating the user with his/her X.509 Certificate or proxy certificate to the DKAA, and as a result the user gains the Kerberos identity as described above. To permit the transparent and automatic mapping between X.509 certificates or proxy certificates, we utilize the certificate or proxy certificate Distinguished Name (DN) to extract the user's DN. Then, this DN is hashed and this hash is used as a Kerberos principal.

Within the DKAA a user can belong to various groups (instances). These groups are defined when the user identity is created at the Kerberos KDS. Groups are used in DKAA for authorizing the user to the resources. If we map these groups to VOMS groups, mapping of the user's roles can be achieved.

The creation of a group and the specification of its users and services do not automatically provide authorization to the resources. Access to the resources and services must be explicitly requested from the resource owners (administrators of the web service).

The authorization to invoke a service offered by a specific web service must be granted by its resource owner. This authorization is given at the group level, and is not provided explicitly to each individual user. This approach is adopted in order to limit the number of access rules that exist within each end-entity and therefore improve the efficiency of the authorization process. However, if granularity down to the user level is required by a use case, a separate group can be created for each user. In this way, the access control can be extended to support access to the user level. Nevertheless, following this approach renders the mapping of the VOMS roles to groups non-applicable.

The rules format that we have defined is given in Table 1.

The previous rule allows/denies a user belonging to a specific Instance (group) and REALM (VO) to call a web service. The default access policy

Table 1. Rules format

Instance (group)	REALM (VO)	Operation	Service Principal

is deny. This means that if specific access rules are not explicitly defined, the operation on the web service is denied to any group requesting it. Therefore, effectiveness and simplicity are maintained in enforcing these access rules.

4. DKAA IMPLEMENTATION AND RESULTS

4.1 DKAA Implementation Details

DKAA is being developed as the Authentication and Authorization Infrastructure (AAI) of the GRIDCC project [7]. To this end, a new resource has been defined, namely the Instrument Element (IE), similar to the Computing and Storage Elements (CEs, SEs) of the batch Grid. The IE acts as the middleware that enables the remote control of instruments over the network as services. With the addition of supporting services, like the Workflow Service and the Agreement Service, a user of the system will be able to automatically control instruments and define the quality of service characteristics of his/her interactions. To offer simplicity to the users, a Virtual Control Room (VCR) has also been defined that acts as the User Interface of all the services in the GRIDCC collaborative environment.

The IE is exposed as a web service and the software is written in Java using Apache AXIS [3]. The VCR is also developed in Java. Thus, security should be able to integrate easily with these components. For all these we chose Java as our implementation language.

Consequently, we have chosen to use the OASIS Web Services Security [16]. We have developed an implementation of the WS Security Kerberos Token Profile [21], by extending the Apache WSS4J Java library [4]. WSS4J supports the web services Security and specifically implements the X.509 Certificate and Username Token Profile, but it does not implement the Kerberos Token Profile yet.

In Figure 1 the general architecture was presented. To meet these designed goals we have implemented the following components:

- KrbClient: it is a Java object that provides the client security API needed to login to Kerberos server, acquire service tickets and apply signatures or encrypt the messages. It sends the messages as SOAP messages. Specifically, it inserts the WS Security Header, implementing the Kerberos Token Profile, into the SOAP header, it attaches XML signatures and encrypts the body using XML Encryption. The goal of the KrbClient is to facilitate the development of DKAA clients by hiding the complexity of the security processing from the application developer. This involves hiding all the Kerberos and WS Security details. In addition, it holds all the security state, such as user tickets and session keys with the various services. It uses the

Java GSS API [10], PKINIT [13] that allows the use of certificates for authentication to the Kerberos system [8] and WSS4J for supporting all the security processing.

- ACMHandler: it performs the access control and all the security service logic of the IE and it is deployed as an AXIS handler. It preprocesses the SOAP messages destined to the service and it specifically checks the security header, authenticates the messages, checks the possible signatures and decrypts the body, if it is encrypted. Handlers are deployed at the request flow of a web service and are transparent to the service. Finally the ACMHandler authorizes the message by checking against the locally stored access rules. If the security processing is successful it forward the processed message to the actual service.
- Policy Repository: it is a web service with a Relational Database backend that stores all the local policy rules, allowing the monitoring of the global authorization scheme distributed among the IEs.

4.2 A Simple Use Case Example

In the following we demonstrate the operation of the basic elements of the DKAA architecture in the GRIDCC security Use Case, by identifying and describing the various interactions and exchanges involved during the authentication and authorization phase. In a typical GRIDCC scenario there are at least two different organizations that have IEs. Let us consider the scenario illustrated in Figure 2. In this scenario we assume the existence of two organizations: Organization1 and Organization2. All other supporting GRIDCC entities have been omitted for simplicity in the presentation. These two organizations are independent in every aspect, including policy and users. Organization1 has two Instrument Elements, IE1 and IE2, while Organization2 has only one, IE3. The two organizations have agreed to share part of their resources to users belonging to Virtual Organization 1 (VO1). Specifically, Organization1 provides IE1 and Organization2 provides IE3. VO1 also sets up its own Key Distribution Server (KDC) that has all the credentials (GSI certificates) of the users belonging to VO1.

Passwords are created for IE1 and IE3. These passwords are only known to their corresponding IEs and the KDC. Supposing that user@VO1 tries to send control commands to IE1 and IE3, the steps involved in this exchange are

1. The user logs in the Control Room. The user submits his/her credentials (GSI certificate) and the subgroup that he/she wants to use. Transparent to the user, the KrbClient class of the client API logs in to the KDS1 using the user's certificate. In case that the credentials are valid, the log in is successful and a TGT is returned.

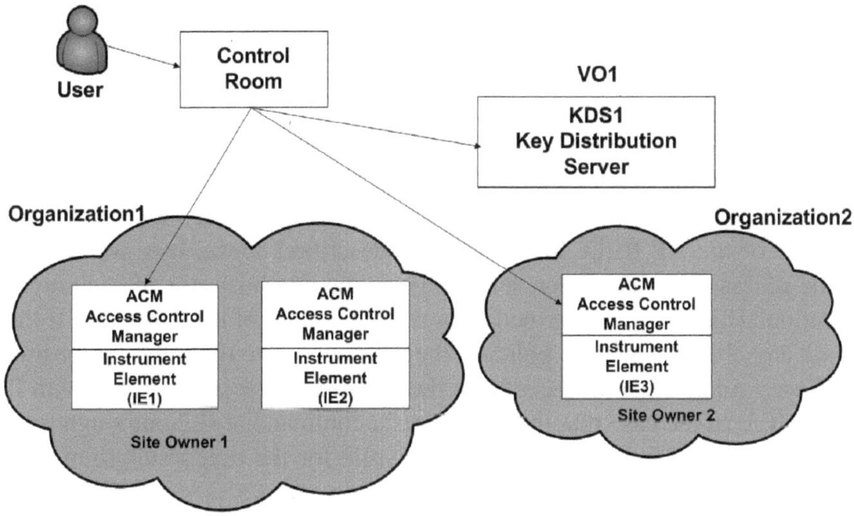

Figure 2. A Dual Domain operation Example

2. Let us assume that the user sends commandA and commandB to IE1 and IE3, respectively, through the Control Room interface, choosing the security scheme (whether the body of the SOAP message is encrypted or a signature is attached). Then, transparently to the user the following actions take place:

   - The KrbClient requests tickets for IE1 and IE3 from the TGS by providing the TGT that was acquired in step 1.
   - The KrbClient sends the service request (CommandA and CommandB) as a SOAP message, attaching a security header and the tickets as Binary Security Tokens, to the IE1 and IE3. The tickets are contained in structures called AP_REQs, which hold the ticket along with an authenticator.
   - The ACMHandlers of IE1 and IE3 process the header of the incoming SOAP message and extract the ticket. If it is accepted the user is authenticated and the Session Key (SK1 for IE1 and SK3 for IE3) is stored locally at the ACM of IE1 and IE3, along with the user and the user's subgroup. The Session Key (SK) is contained inside the ticket and is valid until its lifetime expires. After storing the SK, the ACMHandler, using the SK, decrypts the SOAP body if encrypted or validates the signature. When using symmetric cryptography, digital signatures are Hashed Message Authentication Codes (HMAC). Finally, before forwarding the message to the actual service, the ACMHandler authorizes the user requesting the service by checking against the locally stored access rules.
   - The service processes the request and sends back the reply.

3. The Control Room displays the result of the user's commands (success or failure).

4. In the subsequent messages of the user to IE1 or IE3 the above procedure is performed using the Session Key for encryption and signature. It is not needed that the ticket be attached (in the form of an AP_REQ), but for performance reasons a reference to the ticket is used. This reference is the hash value of the AP_REQ. In the scenario described above, it is assumed that the access rules have been set locally by the administrator of each organization. This step is performed prior to the creation of the subgroup. If there are any changes in local policies, they are pushed to the Policy Repository. Access rules that allow access for the subgroup of users are set only in IE1 and IE3. In the case that the user sends a command at IE2, although he/she is authenticated, the lack of an access rule for the user's subgroup at IE2 results in denying access to it.

4.3 Measurements

In the framework of this work we measured the performance of the ACMHandler, by flooding it with requests, and compared it with the default handler that WSS4J provides. The default handler supports the X.509 Certificate Token Profile, whereas ACMHandler implements the WS Security Kerberos Token Profile.

In our test bed, we used five clients and one server, where the clients send messages constantly with a specific size. Specifically, every client sends 10000 messages (50000 total). All clients start sending messages at the same time to both handlers by encrypting only the body, using AES128 symmetric algorithm. The measurements are repeated with changing the size of the message payload (string) to 60, 200, 400 and 800 bytes each time. It should be mentioned that the message payload size is before the encryption. The encryption of the body of the SOAP request is done afterward. The X.509 handler uses RSA (public key cryptography) to encrypt an attached symmetric key that is used for decryption. The CPU utilization of the service process was over 95%. For testing purposes, we have chosen an echo web service, which accepts a string and replies sending that string. This service is of low complexity, as the main objective is to picture the performance of the WS security by minimizing the load from other factors, like the ones from the service. The payload of the request, that was encrypted, was 200 bytes. In Table 2 we present the results.

The performance gain of our Kerberos implementation is 27.9%. This performance difference is mainly attributed to the way that each method acquires the key that is used for decrypting the message. Our implementation uses the

Table 2. Performance of WS Kerberos Token Profile compared with X.509 Certificate Token Profile

WS Security Profile	Average packets/s
Kerberos Token	104.86
X.509 Token	82
Gain of Kerberos against X.509	27.9%

session key that is inside the Kerberos ticket. This is extracted using symmetric cryptography. In contrast, the implementation of the X.509 Token derives the symmetric key by decrypting the attached – encrypted with RSA (public key cryptography) – key in the SOAP message. After acquiring the symmetric key, both implementations decrypt the body that is encrypted with AES128.

Additionally, by changing the packet size, as shown in Figure 3, we can observe the greater packet throughput achieved by our implementation in contrast with the X.509 implementation for different body sizes. Both performances of the ACMHandler and the X.509 handler degrade slightly with the performance difference sustained at the same level. We have not tested body sizes greater than 800 bytes, as we wanted to avoid the IP fragmentation of the SOAP message. The size of messages destined to control devices should always be kept to a minimum, so that latency is avoided.

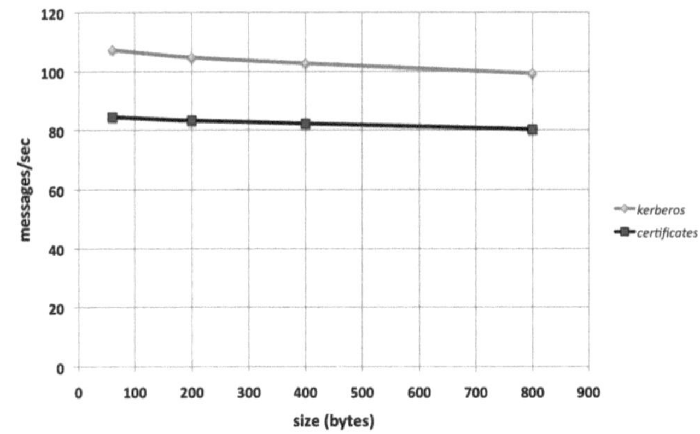

Figure 3. Average packet rate for different body sizes

5. CONCLUSIONS

In this chapter we presented a distributed security infrastructure, namely DKAA that aims to minimize the overhead of cryptography and improve the QoS of the overall system. It achieves that by using exclusively symmetric cryptography for the message exchanges, when interacting with a large number of Grid resources (e.g. distributed monitoring and control of time-critical instruments). The architecture implementation has been adapted with the web services paradigm. DKAA provides authentication, integrity, confidentiality and authorization at the SOAP message layer. Furthermore it is interoperable with GSI security due to the PKINIT and the mapping of certificate DN to Kerberos principals. Mapping of roles could also be supported provided that the DKAA groups are mapped to VOMS roles.

The performance improvement regarding encryption of SOAP body messages is based on the fact that the symmetric key cryptography approach has lower computational complexity than public key cryptography. Therefore, it allows the overall service to offer lower response time and better performance compared to alternative Public Key Cryptography approaches. This performance improvement reflects to our measurements and shows clear advantage in overloaded servers when DKAA is used.

REFERENCES

[1] R. Alfieri et al., "VOMS, an authorization system for virtual organizations", *Presented at the 1st European Across Grids Conf.*, Santiago de Compostela, Spain, Feb. 14, 2003.

[2] R. Alfieri, R Cecchini, V. Ciaschini, F. Spataro, L. Dell'Agnello, A. Frohner, K. Lorentey, "From gridmap-file to VOMS: managing authorization in a Grid environment", *Future Generation Computer Systems*, Vol. 21, no. 4, pp. 549–558. Apr. 2005.

[3] Apache AXIS – http://ws.apache.org/axis/

[4] Apache WSS4J – http://ws.apache.org/wss4j

[5] C. Coarfa, P. Druschel and D.S. Wallach, "Performance analysis of TLS web servers", *9th Network and Systems Security Symposium*, pp. 553–558, 2002.

[6] I. Foster, C. Kesselman, S. Tuecke: "The anatomy of the grid: enabling scalable virtual organizations", *International Journal of Supercomputer Applications*, Vol. 15, no. 3, pp. 200–222, 2001.

[7] GRIDCC Project web site – www.gridcc.org

[8] Heimdal Kerberos Server – http://www.pdc.kth.se/heimdal/

[9] IETF RFC 1510 – The Kerberos Network Authentication Service (V5).

[10] IETF RFC 1508 – Generic Security Service Application Program Interface.

[11] IETF RFC 2459 – Internet X.509 Public Key Infrastructure Certificate and CRL Profile.

[12] IETF RFC 3820 – Internet X.509 Public Key Infrastructure (PKI) Proxy Certificate Profile.

[13] IETF RFC 4556 – Public Key Cryptography for Initial Authentication in Kerberos (PKINIT).

[14] A. Moralis, A. Lenis, M. Grammatikou, S. Papavassiliou, V. Maglaris, "A distributed Kerberized access architecture for real time grids", *4th International Workshop on Security in Information Systems WOSIS*, 2006.

[15] R. Needham, M. Schroeder, "Using encryption for authentication in large networks of computers", *Communications of the ACM*, Vol. 21, no. 12, pp. 993–999, Dec. 1978.

[16] Oasis WS Security Standards – http://www.oasis-open.org/specs/index.php#wssv1.1

[17] Open Grid Forum – http://www.gridforum.org

[18] L. Pearlman, V. Welch, I. Foster, K. Kesselman, S. Tuecke, "A community authorization service for group collaboration", *IEEE Workshop on Policies for Distributed Systems and Networks*, 2002.

[19] The European Policy Management Authority for Grid Authentication in e-Science – http://www.eugridpma.org/

[20] W3C Web Services Activity – http://www.w3.org/2002/ws/

[21] WS Security Kerberos Token Profile – http://www.oasis-open.org/committees/download.php/16788/wss-v1.1-spec-os-KerberosTokenProfile.pdf

[22] WS-Security X509 Token Profile – http://www.oasis-open.org/committees/download.php/16785/wss-v1.1-spec-os-x509TokenProfile.pdf

SUPPORTING INTERACTIVE APPLICATION REQUIREMENTS IN A GRID ENVIRONMENT

A. Di Stefano[1], G. Pappalardo[2], C. Santoro[2], and E. Tramontana[2]

[1]*Dipartimento di Ingegneria Informatica e Telecomunicazioni,* [2]*Dipartimento di Matematica e Informatica, Università di Catania, Viale Andrea Doria, 6, I-95125 Catania, Italy.*

adistefa@diit.unict.it, pappalardo@dmi.unict.it, csanto@diit.unict.it, tramontana@dmi.unict.it

Abstract When using Grids as infrastructures for interactive application scenarios, such as the implementation of virtual laboratories, new concerns arise, which are absent in traditional Grid usage, basically oriented to the execution of batch processes. In particular, the main requirement arising in such new application scenarios is the ability to meet soft real-time and other QoS constraints. However, current Grid middlewares are not designed, and cannot cater, for this purpose. In this respect, the main inadequacies are related with application characterization and resource discovery and reservation.

As a solution, this chapter puts forward a suitable software architecture and a set of techniques. To begin with, a component is proposed, capable of analysing a target application in order to automatically extract from it information on the amount of resources that will be needed during execution. This is because, for an application to satisfy soft real-time and QoS constraints when executed in a Grid system, its resource requirements should be known in advance of execution. Then, we introduce a resource discovery algorithm, based on peer-to-peer concepts, to find a Grid node that meets application requirements "at best". Once such a node is found, the resources needed are reserved in advance, in order to make resource availability predictable during application execution. Finally, since knowledge gathered about the application could be incomplete, and moreover some execution flow (caused, e.g. by unknown input data) could turn out to be unpredictable, the runtime behaviour of the application is monitored by a suitable software component, so that proper activities can be started to handle further needs and cope with anomalies that might manifest themselves during runtime.

1. INTRODUCTION

Traditionally, Grid systems have been employed as infrastructures to support the execution of data- and computation-intensive applications, mainly run as batch processes for simulation purpose, data crunching, etc. However, a recent trend is towards extending Grids in order to better exploit their computational power, by making them capable of supporting not only batch processing, but also other kinds of applications, especially *interactive* ones, i.e. those characterized by a significant level and amount of interaction with their environment. In particular, the use of Grids as infrastructures to build virtual laboratories is a scenario that is emerging as a new and interesting research and development application area.

In such cases, the batch paradigm is surely not appropriate, and different software architectures and solutions from the conventional ones should be employed, capable of taking into account various features of the new application scenarios, including interactivity, quality-of-service requirements and (soft) real-time facilities. As for the last two features, issues to be addressed involve various parts of a Grid system: to begin with, initially, it is necessary to know the requirements posed by the target application (to be executed on the Grid); this knowledge can be obtained thanks to an explicit declaration provided by the programmer or can be extracted on-the-fly, at job submission time, by means of code instrumenting tools. Then, before execution, a discovery phase has to be started, in order to conveniently find the node (the so-called computing element) capable of offering (and possibly guaranteeing) the needed requirements. Finally, in order to enforce requirements during execution, application behaviour should be observed and monitored, in order to trigger suitable corrective actions when one or more requirements are not being met.

In this context, this chapter describes some algorithms and techniques, developed by our research group at Catania University, which can be employed to support the scenario described above. These solutions have been integrated into a Grid software infrastructure mainly consisting of two parts.

The first part has the twofold task of identifying application requirements and exploiting resource discovery algorithms to (i) find the Grid node able to offer adequate resources for executing the intended application and (ii) reserve such resources, before the application starts execution, so as to make resource availability predictable.

The second part of the proposed software infrastructure observes the runtime behaviour of the application and takes appropriate actions to handle the needs revealing themselves at this stage. It is worthwhile to note that having a platform and service available to monitor the runtime behaviour of the application is essentially unavoidable, for any static analysis aimed at predicting it is bound to be imprecise to some degree, due to the varying input data and the relative unpredictability of the execution flow. Thus, we think it fit to propose a supporting layer which, properly deployed through aspect-oriented technology, can transparently observe application execution and handle issues related with resource reservation and allocation, timing requirements, recovery activities, etc. on behalf of the application.

This chapter is structured as follows. Section 2 describes how application code is statically analysed to identify resource requirements posed by the application. Section 3 deals with discovering resources throughout the Grid and reserving them. Section 4 describes techniques and architectural components dealing with the runtime monitoring of application behaviour and the enforcement of the related resource requirements. Section 5 evaluates the proposed resource discovery strategy. Finally, Section 6 draws the authors' conclusions.

2. EXTRACTING AND IDENTIFYING REQUIREMENTS

In order to make the execution time of an application predictable and guarantee an agreed response time, a reservation management system is the necessary support that can avoid overload and improve resource availability. Reservation support can be used effectively, i.e. without either wasting resources or exceeding the boundaries established for response times, when it is possible to rely on a proper estimation of application needs.

For the sake of efficiency and to avoid passing users the burden of estimating applications, we have developed a distributed service, called *Java Job PROfiler (JJPRO)* [5, 8], to which applications whose job executable is Java bytecode can be submitted. JJPRO will perform a static characterization of the submitted application through an automatic analysis and will build and store an application static profile (*ASP*), returning a unique identifier whereby the ASP can be retrieved afterwards. The ASP encodes the characterization that our reservation support (cf. Section 3) needs to efficiently use Grid resources.

Specifically, JJPRO estimates, for each thread t of the application, the degrees of use by t of processor, disk and network (du_t, nu_t, pu_t), as well as the worst case execution time (WCET) [10] ($wcet_t$) for t.

The degrees of processor, disk and network use are estimated, for a method, by an analysis of its bytecode. Three totals called op_{tot}, op_{disk} and op_{net} are calculated by adding up all weighted occurrences of, respectively, all opcodes, method invocations to packages related to disk access and method invocations to network-related packages. The weight attributed to each opcode occurrence depends on the associated JVM instruction complexity, the nature of the application (e.g. a file copier application is given different weight than an image recognition application) and its context (including e.g. the loop nesting level).

Then we set

$$\mathrm{du} = \frac{\mathrm{op_{disk}}}{\mathrm{op_{tot}}} \qquad \mathrm{nu} = \frac{\mathrm{op_{net}}}{\mathrm{op_{tot}}} \qquad \mathrm{pu} = 1 - \mathrm{du} - \mathrm{nu}$$

which are intended to describe the use rate of disk, network and processor, respectively. For example, du represents how much disk operations weigh related to the total amount of computations carried out. The higher the value of du, the faster the disk needed for such a class to perform well.

3. RESOURCE DISCOVERY AND RESERVATION

Once the resource parameters and requirements, i.e. (du, nu, pu, wcet), have been derived for the application to be executed on the Grid, the next step is to find the Grid node able to offer such a required amount and to guarantee its availability for the duration of application execution. This amounts to performing a *resource discovery* phase, followed by a *resource reservation* on the target node that has been found.

As for the first phase, resource discovery, the traditional approach used by existing middlewares for the Grid systems are based on distributed repositories that store and maintain information on available resources in the overall Grid. A widely used solution is Globus Monitoring and Discovery Service MDS [1, 11, 12, 16], which is based on a hierarchical scheme whereby some nodes are designated to host an *Index Service* and each assigned a set of nodes to monitor; each Index Service node maintains a database describing current resource availability on its assigned nodes. Resource availability is thus (periodically) recorded into such a repository, while fulfilling a resource request on behalf of a job allocation implies a query to the repository, for the node capable of fulfilling the job's requirements.

An MDS-like approach has three main disadvantages. First, it is not very efficient—there is an obvious latency between the time instant in which a resource is monitored and the instant in which the relevant value is read from the repository. Second, it is unsuitable for (dynamic) large-scale environments, since increasing the Grid dimension implies a growth of the repository information size and the complexity of its management [13]. Finally, a repository-based approach is not fault tolerant, since a fault occurring in a node hosting an Index Service prevents the nodes managed by that node from being considered for matchmaking (i.e. job allocation).

In any case, for resource allocation to be as effective as possible, the discovery of suitable resources should be complemented by reserving them. However, reservation support is unavailable in current Grid middlewares. This limitation is adverse for interactive applications, especially real-time ones. For these, once a node has been found, which holds enough resources to execute the application fulfilling its requirements, it should be guaranteed that such resources will indeed be available when the application is launched on that node. Only reservation can ensure that resources will not be preempted by another job.

The solution we propose to deal with resource discovery and allocation is along the lines we put forward in [6, 7]. Its key aspect lies in the Grid architecture we consider: the Grid is modelled by arranging its nodes in an overlay network, a structure typical of P2P systems, whereby each node is virtually connected to one or more neighbours and able to interact only with them. In such a system, a job request is submitted to one of the Grid nodes and then forwarded from neighbour to neighbour using an appropriate routing strategy; request forwarding ends when the node able to fulfil the request has been found. In this way, resource information is gathered as neighbouring nodes are visited, and this task can be accomplished without needing a hierarchy of repositories. Of course, the specific routing strategy adopted is particularly sensitive, since it remarkably affects not only the effectiveness of the algorithm (i.e. its ability to find the "right" node) but also its cost.

To this aim, we employ a resource-finding algorithm that exploits some concepts proper of "spatial computing" [15]. This is a new paradigm, mainly applied to multi-agent self-organizing systems that takes inspiration from the laws governing *gravitational fields* and applies gradient-descending techniques in order to, for example, find an object, meet a peer, obtain an auto-organizing system. We extended such concepts for our purpose and proposed an approach, for resource finding, based on navigating a 3D surface formed by considering Grid nodes, together with the overlay network, as distributed along a virtual XY plane. By way of abstraction, we suppose that each Grid node i possesses a *mass* $\mathcal{M}(i)$ whose value is proportional to the amount of resources *in use*. As a result, Grid nodes, with their mass values, come to form a *gravitational field*, where potential lines are generated on the basis of the links of the overlay network. In such a surface of potential, valleys will represent nodes with a large quantity of available resource, while hills will correspond to low resource availability (see Figure 1).

In the framework set out, finding the most suitable node to fulfil a new job request amounts to navigating the surface of potential trying to reach its global minimum, which in turn represents the node with the highest quantity of available resources. The navigation algorithm proposed mimics the laws of kinematics. A resource request is seen as a *capsule*, with a *mass* and a certain amount of initial *kinetic energy*. The mass is a numeric value proportional to the quantity of the resources that are being requested to correctly execute the application; the energy represents the kinetic capability of the capsule to cover a certain distance along the grid surface and, as intuition suggests, is somewhat related to the probability that it will actually reach the global minimum. This capsule moves along the surface (from node to node), gaining energy while descending and losing energy while ascending. When the energy value is zero

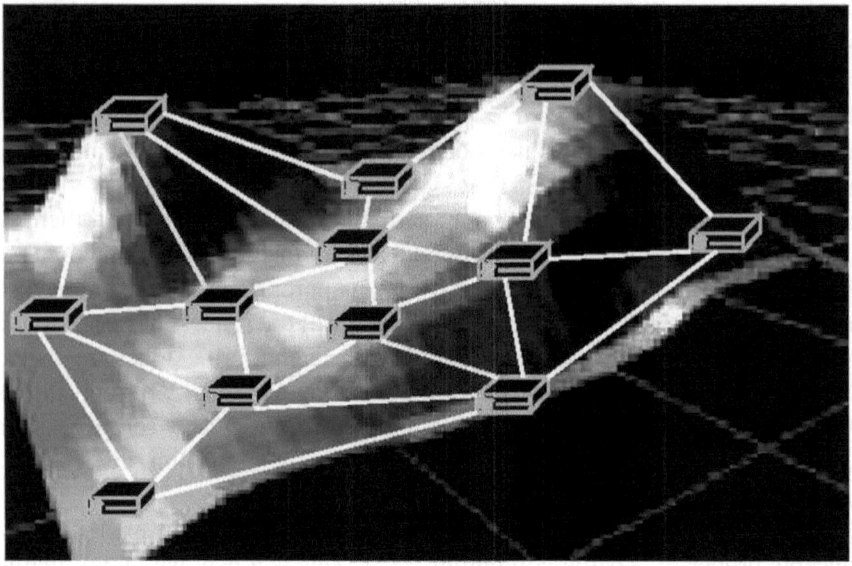

Figure 1. A potential field generated by the amount of resources

or below, the capsule is in a minimum of the surface, hence the current node is selected for the job allocation.

The navigation algorithm operates by evaluating the following two functions:

- A function $\mathcal{PD}(i_n, i_m)$ evaluates the *difference of potential* between neighbour nodes i_n and i_m; it reflects the difference between mass values, i.e. $\mathcal{M}(i_n) - \mathcal{M}(i_m)$. $\mathcal{PD}(i_n, i_m)$ models the attraction force that guides the capsule towards one or the other node: if its value is greater than zero, attraction is from i_n to i_m, and symmetrically. According to the physical concept of difference of potential, function $\mathcal{PD}(\cdot,\cdot)$ must obey the "circuitation law", i.e. the sum of all differences of potential along a closed path must be equal to zero.

- A *friction* is considered, which implies that, for each migration from i_n to i_m, the capsule always loses a quantity of energy $\delta(i_n, i_m)$ (always greater than zero) which depends on the surface gradient, and thus once again on mass difference between the two nodes. Such a friction ensures that the capsule will sooner or later stop, allowing the search algorithm to terminate.

Given these premises, in accordance with mechanics, the basic behaviour of the capsule is to always go towards valleys, provided that it possesses sufficient energy to overcome hills found along its path, The capsule starts its navigation from a given node[1] i^* and, based on the value of $\mathcal{PD}(i^*, i')$ computed for each neighbour i', chooses the next (neighbour) node to visit. Such a choice is based on two heuristics, called H_{down} and H_{up}: the former is used during capsule descending, while the latter performs neighbour choose when the capsule is climbing a hill. To assess the effectiveness of the algorithm with respect to such heuristics, we proposed some strategies for neighbour choose that will be detailed in Section 5, where a performance evaluation of the algorithm will be reported.

During capsule navigation, the capsule's energy is updated on the basis of its direction: when the capsule goes down a descending slope, it gains energy; when the capsule goes up a hill, it loses energy; when, while going up, its energy reaches zero, the capsule changes its direction and starts a descending path (see Figure 2). The energy gained (or lost) is evaluated by using function $\mathcal{PD}(i_n, i_m)$. In addition, for each migration from i_n to i_m, the capsule always loses a quantity of energy equal to the friction $\delta(i_n, i_m)$. Searching stops when two conditions hold:

1. A minimum is reached in the surface.

2. The capsule has no more energy.

This situation represents the achievement of a global minimum or a local minimum that should, however, be close enough to the global one (provided the capsule possessed a reasonable initial kinetic energy).

When the algorithm stops at a node, which thus represents the computing element where the job will be run, the amount of resource(s) required has to be reserved and allocated, in order to avoid that further requests lead to overloading this node. This is performed simply by *adding* the capsule's mass (which

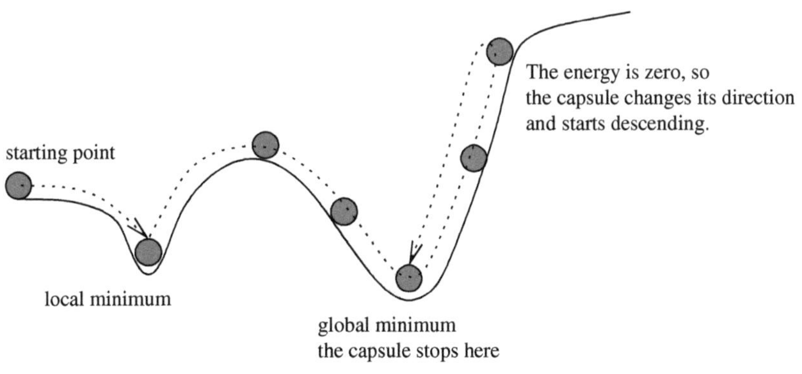

Figure 2. A capsule navigating over a surface of potential

[1]All nodes are peers, so the algorithm does not depend on the starting node, which can be freely chosen.

represent job's requirements) to the node's mass: the amount of resource in use increases and the node becomes "heavier", thus forming a higher hill or a less deep valley. This condition holds for the whole duration of the job execution, therefore no other reservation/allocation request can be fulfilled if the overall resulting utilization (current usage plus requested amount) would exceed the total quantity of resources present in the node.

4. RUNTIME MONITORING AND REQUIREMENT ENFORCING

Even though the above algorithm performs an advance reservation of resources, when the application executes, a variation of the amount of such granted resources could be possible, due to e.g. unpredictable loading conditions of the Grid node running the application. For this reason, provision of a suitable runtime monitoring mechanism, intended to enforce resource requirements posed by applications, is mandatory. The aim of such a mechanism is twofold.

On one hand, it is instrumental in comparing the amount of resources (e.g. disk, network bandwidth, processor) reserved in advance and then allocated for an application, with the amount actually used by the application at runtime. Differences between two such amounts are essentially caused by the application behaviour when unpredictable input data are processed. A relevant difference between allocated and used resources could overload the host and also diminish the amount of resources available to other applications co-allocated on the same host.

On the other hand, for a (soft) real-time application, our monitoring system can contribute to the goal of meeting the deadlines. Such an application is monitored to detect in advance whether soft deadlines are going to be met or missed. If there is danger of a deadline miss by any part of the application, appropriate actions can be started, aimed at either making the execution smoother—by e.g. preempting extra resources for the benefit of the application—or starting recovery activities to avoid faulty conditions to be met.

Pursuing the approach of our works [2, 5, 8], we have designed a software architecture to support monitoring and requirement enforcing. It comprises two main components: one, called `Remon`, deals with application and resource monitoring; the other, called `DeadlineSensor`, with deadline control. Figure 3 shows the outline of the proposed architecture.

4.1 Monitoring Resource Use

Architecture component `Remon` is intended to monitor resource use at runtime. When an application class spawns a new thread at runtime, by invoking

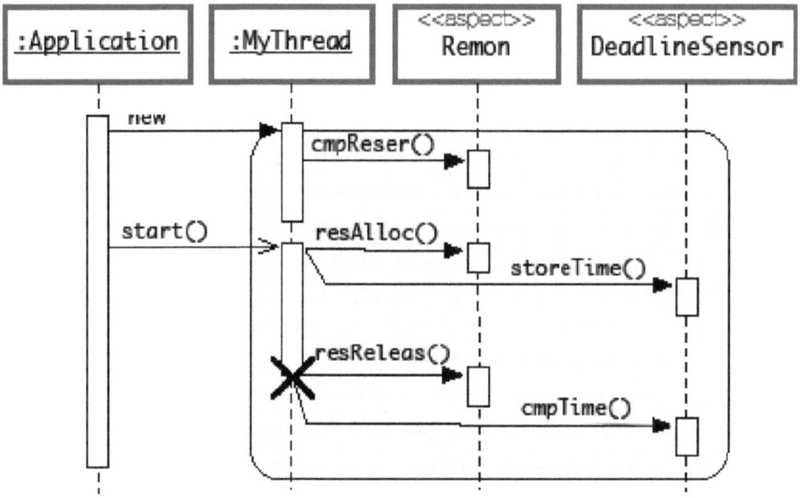

Figure 3. Interactions between an application and the monitoring components

`new` on classes extending `Thread` or implementing `Runnable`, a need arises for more resources, for the CPU will be asked to operate in parallel for both the new and the invoking threads. Thus, it seems appropriate that, upon detecting such a `new`, `Remon` should check whether resources that have been previously reserved suffice for the needs of the new thread (see `cmpReser()` in Figure 3).

`Remon` has access, through the service described in Section 2, to the ASP profile calculated for the application and specifically to the tuple (du_t, nu_t, pu_t, $wcet_t$) describing the needs of the new thread t in terms of disk use du_t, network use nu_t, processor use pu_t and worst case execution time $wcet_t$.

Moreover, `Remon` has taken notice of the previously reserved amount of resources, as triple (du_r, nu_r, pu_r) (this will of course vary in time, depending on the *wcet* of previous reservations and as jobs terminate). Finally, through monitoring, `Remon` holds knowledge of the currently used resources, as the triple (du_c, nu_c, pu_c).

`Remon` can thus check whether, within the timeframe $wcet_t$, the reserved amount of each resource is less than the needed amount (e.g. $pu_r < pu_c + pu_t$). Should this be the case, the reservation of additional resources is requested, which will be granted when some degree of availability is left for any desired resource, e.g. $pu_r + pu_c + pu_t$ should stay below 1[2].

Although `Remon` requests resource reservation upon intercepting a `new` for a thread, it postpones resource allocation until `start()` is actually invoked

[2]A value equal to 1 for any of resource usage indices indicates a 100% usage.

on the newly created thread (provided of course the requested reservation has been granted in the meantime). The values of the currently used resources (du_c, nu_c, pu_c) are then increased to reflect the appearance of the thread that has been just started (see `resAlloc()` in Figure 3).

A thread finishing its execution should free the resources that had been allocated for it. Accordingly, `Remon` takes this into account by decreasing the values of the currently used resources (du_c, nu_c, pu_c) as soon as a thread terminates (see `resReleas()` in Figure 3). Note that, for a thread t, the worst case execution time $wcet_t$ is usually longer than (or at most equal to) the actual execution time. As a result, a terminating thread t will typically release some resources that had been conservatively reserved for its whole $wcet_t$. `Remon` will then take actions to have such resources reused (without going through reservation again) for other threads that will be started later on.

For its very nature, `Remon`'s concern crosscuts across several application classes. We have therefore undertaken its modular design and implementation by means of aspect-oriented programming. In our experiments, `Remon` is an *aspect* that uses *pointcut*s to capture (i) all the invocations to `new` on classes extending `Thread` or implementing `Runnable` and (ii) all the invocations to method `start()` by class `Thread` and its subclasses. Finally, the termination of threads is captured by having a dedicated thread, created by the `Remon` aspect for this sake, synchronize with Java `join()` method calls. As aspect-oriented technology dictates, the said pointcuts will hand control to two *advices*, which handle resource reservation and allocation, respectively.

4.2 Monitoring Deadlines

A component called `DeadlineSensor` is used to monitor both the thread spawn and termination times, to determine the actual thread duration at runtime [4].

`DeadlineSensor` can then calculate the difference between the actual execution time and the statically estimated worst case execution time $wcet_t$ for the thread. When such a difference exceeds a threshold, `DeadlineSensor` performs checks to determine whether adjustment activities are necessary for the application. Moreover, for a thread that actually lasts significantly longer than the time between its estimated end and its deadline, it seems reasonable to monitor execution time at method level, so as to have the chance to take actions with a finer granularity, along its execution as well. In such a case, `DeadlineSensor` will compare the actual execution time and the statically estimated time *also for the longer methods.*

Possible adjustments that can be carried out when a deadline has not been met depend both on the characteristics of the application itself (criticality of the deadline, whether the application is distributed, fault tolerant, etc.) and on those of the underlying system (availability of a fully dedicated host, lower level support for scheduling, checkpointing, etc.).

When threads of different applications are executing, we increase CPU availability for the application of interest by increasing the priority of those of its threads whose deadline is nearer.

Another useful measure to help meeting deadlines could be to "inject" into the running application alternative, faster implementations of some methods (typically the slower and/or more invoked ones). These alternatives typically produce similar results with a lower precision. This concept of adaptation is essentially based on N-version programming. For details on how to characterize and dynamically introduce such alternatives refer to [3, 9, 14].

Similarly to `Remon`, `DeadlineSensor` has been designed and implemented as an *aspect* having a *pointcut* that captures thread spawn (see `storeTime()` in Figure 3) and termination (see `cmpTime()` in Figure 3), while an *advice* performs the above-mentioned calculations and checks.

5. PERFORMANCE EVALUATION OF RESOURCE SELECTION

In order to validate the resource-finding algorithm described in Section 3, we designed a P2P Grid simulator, written in Java, and used it to perform a series of tests. We introduced the following indicators aiming at evaluating the main aspects of the algorithm:

- **Cost.** It measures the number of steps required (nodes visited) before finding the node in which to allocate the job.
- **Optimality.** It aims at evaluating the ability of the algorithm to find the optimal solution (i.e. the global minimum); for an allocation request, this indicator is computed as the difference between the mass of the global minimum and the mass of the node found by the algorithm. So the lowest the value the highest the optimality.
- **Fairness.** It evaluates the ability of the algorithm to make the system self-organizing by measuring its capability to fairly distribute the computational load among the various Grid nodes; for an allocation request, this indicator is computed as the variance (σ^2) of the masses of all the nodes of the Grid. Therefore, also in this case, the lowest the value the highest the fairness degree.

Since the algorithm depends on the heuristics used to perform the selection of a node from the set of neighbours N, we devised three main different strategies, which are summarized in Figure 4. They use the same heuristic for node selection during capsule descending (H_{down}), that is, they always select

Deepest slope (DS)	$H_{\text{up}}(N) = \text{random}(\{i \in N \vert \forall n \in N, n \neq i, \mathcal{M}(i) > \mathcal{M}(n)\})$ $H_{\text{down}}(N) = \text{random}(\{i \in N \vert \forall n \in N, n \neq i, \mathcal{M}(i) < \mathcal{M}(n)\})$
Minimum energy (ME)	$H_{\text{up}}(N) = \text{random}(\{i \in N \vert \forall n \in N, n \neq i, \mathcal{M}(i) < \mathcal{M}(n)\})$ $H_{\text{down}}(N) = \text{random}(\{i \in N \vert \forall n \in N, n \neq i, \mathcal{M}(i) < \mathcal{M}(n)\})$
Random (R)	$H_{\text{up}}(N) = \text{random}(N)$ $H_{\text{down}}(N) = \text{random}(\{i \in N \vert \forall n \in N, n \neq i, \mathcal{M}(i) < \mathcal{M}(n)\})$

Figure 4. Strategies and heuristics

the neighbour node with the lowest mass; instead they differ for the heuristic used during hill climbing; in particular, **Deepest Slope (DS)** selects the highest hill, i.e. the node neighbour with the highest mass; **Minimum Energy (ME)** selects the neighbour with the mass closest to that of the current node; while **Random (R)** performs a (uniform) random selection.

To evaluate the advantages of using one of the noted selection strategies, we compared them against a **First Fit (FF)** approach, which stops the capsule at the first occurrence of a local minimum, so it does not use the H_{up} heuristic.

By means of the simulator developed, we tested a real-case scenario of a P2P Grid composed of 400 nodes, aggregated in an overlay network, with an average number of neighbours of 6. The value for the mass of both nodes and capsule has been selected in the interval $[0,1]$, thus representing the percentage of utilization of a given resource (0 = no utilization, 1 = maximum utilization, no availability). As for the $\mathcal{PD}(\cdot,\cdot)$ and $\delta(\cdot,\cdot)$ functions, we used the following ones:

$$\begin{aligned}
\mathcal{PD}(i_1,i_2) &= \mathcal{M}(i_1) - \mathcal{M}(i_2) \\
\delta(i_1,i_2) &= |\mathcal{M}(i_1) - \mathcal{M}(i_2)| + 0.01
\end{aligned}$$

We evaluated the behaviour of the algorithm with respect to the amount of initial energy; this parameter has been set in the interval $[0,50]$, with an increment of 1. For each value of the initial energy, the network is started by assigning a random initial value to the mass of each node; this initial mass distribution has been chosen to evaluate the ability of the algorithm of "flattening" the surface, that is to transform the network into a fairly balanced one (from the resource utilization point of view). In this network, 10000 allocation requests have been generated and allocated by using the algorithm; then the average values of cost, optimality and fairness have been computed and plotted.

A first remark is that, even if the **FF** approach has the lowest cost, it is the worst from the point of view of fairness and optimality. This proves that an approach that does not stop at the first occurrence, but tries to always find a better solution, provides some advantages. As for the other three strategies, **DS**, **ME** and **R**, basically they behave in quite the same way. Sensible differences are reported in the cost, for which the **DS** strategy is the winner, but the same cannot be said for the other indicators. Indeed, for both optimality and fairness, the **ME** is the best strategy, given that the initial energy is less than a threshold (about 25); if this value is overcome, the best strategy seems **R**, meaning that, in these cases, the way in which hills are treated is no more sensible.

The final remark is that, among **DS**, **ME** and **R**, there is not a winner in absolute, but the adoption of one or another strategy should be based on suitably weighting the importance of the three factors—cost, fairness and optimality—in the Grid that is being designed.

6. CONCLUSIONS

This chapter tries to overcome the difficulties existing when an interactive application, in particular a soft real-time one, is to be run on a Grid system, i.e. the unpredictability of appropriate resources availability and, as a result, the actual duration of application execution. This is the case, in particular, of virtual laboratories and systems aimed at monitoring physical experiments and phenomena.

We have proposed a few services that find application, respectively, to deal with (i) application characterization in terms of needed resources, (ii) resource discovery and advance reservation within a large distributed system, and (iii) resource monitoring and dynamic reservation and allocation during the execution of an application.

We characterize an application, before its execution, by statically analysing its code so as to determine the number of threads it is composed of, the degree of use of disk, network and processor for each thread and the worst case execution time for each thread.

In order to find and reserve the needed resources, we employ a discovery process that consists in propagating a request from node to node within an overlay network. This can be depicted as a capsule travelling on a surface, where a valley represents a node with a large amount of resources available and a hill represents a node with low resource availability. This propagation ends when a surface minimum is reached or when the capsule has no more kinetic energy to pursue its exploration.

After the initial reservation and allocation of resources has been performed, the application can start executing. It will now be entrusted to a supporting monitoring layer, properly deployed through aspect-oriented technology, which will transparently observe execution. Such a layer is able to react to the performance figures (specifically timing) observed, by initiating reservation and allocation of additional resources, when the application needs more than those estimated in advance.

REFERENCES

[1] K. Czajkowski, S. Fitzgerald, I. Foster, and C. Kesselman. "Grid Information Services for Distributed Resource Sharing". In *10th IEEE International Symposium on High-Performance Distributed Computing (HPDC 2001)*, San Francisco, August 6–9, 2001.

[2] A. Di Stefano, M. Fargetta, G. Pappalardo, and E. Tramontana. "STREGA: A Support for Transparently Handling Resources for Grid Applications". In *Proceedings of Workshops on Enabling Technologies: Infrastructures for Collaborative Enterprises (WETICE'04)*, Modena, Italy, June 2004. IEEE, pp. 249–254.

[3] A. Di Stefano, G. Pappalardo, and E. Tramontana. "An Infrastructure for Runtime Evolution of Software Systems". In *Proceedings of the Symposium on Computers and Communications (ISCC'04)*, volume 2, Alexandria, Egypt, June 2004. IEEE, pp. 1129–1135.

[4] A. Di Stefano, M. Fargetta, G. Pappalardo, and E. Tramontana. "Separating Soft Real-Time from Functionality Concerns in Complex Object-Oriented Applications". In *Proceedings of OTM Federated Conferences*, volume 3292 of Lecture Notes in Computer Science, Ajia Napa, Cyprus, October 2004. Springer Verlag, pp. 406–417.

[5] A. Di Stefano, M. Fargetta, G. Pappalardo, and E. Tramontana. "Supporting Resource Reservation for Unaware Applications in Grid Systems". *Journal of Concurrency & Computation: Practice & Experience (CCPE)*, 18:851–863, Published Online (in advance of print): 17 Nov 2005, DOI: 10.1002/cpe.980 2006.

[6] A. Di Stefano and C. Santoro. "A Decentralized Strategy for Resource Allocation". In *Proceedings of Workshops on Enabling Technologies: Infrastructures for Collaborative Enterprises (WETICE'05)*, Linkoping, Sweden, June 2005. IEEE.

[7] A. Di Stefano and C. Santoro. "A Peer-to-Peer Decentralized Strategy for Resource Management in Computational Grids". *Journal of Concurrency & Computation: Practice & Experience (CCPE)*, Published Online (in advance of print): DOI: 10.1002/cpe.1110 2006.

[8] A. Di Stefano, E. Tramontana, and D. Zito. "Strategies for Reserving Resources on a Grid Environment". In *Proceedings of Workshops on Enabling Technologies: Infrastructures for Collaborative Enterprises (WETICE'06)*, Manchester, UK, June 2006. IEEE, pp. 207–213.

[9] M. Milazzo, G. Pappalardo, E. Tramontana, and G. Ursino. "Handling Run-Time Updates in Distributed Applications". In *Proceedings of Symposium on Applied Computing (SAC'05)*, Santa Fe, New Mexico, USA, March 2005. ACM, pp. 1375–1380.

[10] P. Puschner and C. Koza. "Calculating the Maximum Execution Time of Real-Time Programs". *Journal of Real-Time Systems*, 1(2):159–176, September 1989.

[11] J. M. Schopf, M. D'Arcy, N. Miller, L. Pearlman, I. Foster, and C. Kesselman. "Monitoring and Discovery in a Web Services Framework: Functionality and Performance of the Globus Toolkit's MDS4". *Technical Report ANL/MCS-P1248-0405*, Argonne National Laboratory, 2005.

[12] X. Z. J. Schopf. "Performance Analysis of the Globus Toolkit Monitoring and Discovery Service, MDS2". In *International Workshop on Middleware Performance (MP 2004) at IPCCC 2004*, April 2004.

[13] D. Talia and P. Trunfio. "Toward a Synergy Between P2P and Grids". *IEEE Internet Computing*, 7(4):94–96, 2003.

[14] E. Tramontana. "Reflective Architecture for Changing Objects". In *Proceeding of the ECOOP Workshop Reflection and Metalevel Architectures (RMA'00)*, Sophia Antipolis, France, June 2000.

[15] F. Zambonelli and M. Mamei. "Spatial Computing: An Emerging Paradigm for Autonomic Computing and Communication". In *1st International Workshop on Autonomic Communication*, Berlin (Germany), October 2004.

[16] X. Zhang, J. Freschl, and J. Schopf. "A Performance Study of Monitoring and Information Services for Distributed Systems". In *12th IEEE International Symposium on High-Performance Distributed Computing (HPDC 2003)*, Seattle, Washington, June 22–24, 2003.

THE COMMON INSTRUMENT MIDDLEWARE ARCHITECTURE

Experiences and Future Directions

D.F. McMullen[1], R. Bramley[2], K. Chiu[3], H. Davis[1], T. Devadithya[2], J.C. Huffman[4], K. Huffman[5], and T. Reichherzer[2]

[1]*Pervasive Technology Labs at Indiana University,*
[2]*Indiana University Computer Science Department,*
[3]*SUNY Binghamton Computer Science Department,*
[4]*Indiana University Department of Chemistry,*
[5]*Indiana University School of Informatics*

Abstract It is clear from the number of e-science projects involving remote access to instruments that remote access and facility sharing is an emerging area of concern. In this chapter we discuss the Common Instrument Middleware Architecture (CIMA) project funded by the US National Science Foundation (NSF) through the NSF Middleware Initiative. CIMA aims at providing a middleware set to facilitate remoting instruments and sensors as community resources and building collaborations around shared facilities. We provide an overview of CIMA, discuss its relationship to some other instrument remote access projects, examine our experiences in developing and applying the architecture and look forward to future development directions for CIMA and its applications.

1. INTRODUCTION

Instruments and sensors are critical sources of experimental data for exploration theory building. However, instruments are still largely "off-line" as far as downstream software analytical components are concerned and instruments are not yet fully first-class members of grid computing environments with respect to location, allocation, scheduling and access control. This is a serious problem as three issues continue to grow in importance in research: (1) investments in geographically extended national and international collaborations organized around large shared instrument resources; (2) increasingly "real-time" use of instruments by remote researchers both for "first look" activities and pipelined data acquisition, reduction, event detection and response; and (3) sensor networks with hundreds to thousands of nodes being deployed using equipment from multiple vendors and with equipment replacement and upgrade (technology refresh) cycles measured in decades.

Several development efforts have taken on these challenges to greater or lesser degree. These efforts are characterized as device access mechanisms, device capability descriptions and system integration efforts. Examples of device access mechanisms include IEEE 1451.1 [18, 27] access to smart sensors by IP network, Universal Plug and Play [24, 25], WS-Discovery and the NEESGrid Telecontrol Protocol [13]. Device capability description projects include the OpenGIS Consortium's sensorML [1, 31], the W3C Composite Capabilities/Preferences Profile, the IEEE 1451.0 Transducer Electronic Data Sheet (TEDS) and the Monterey Bay Aquarium Research Institute "plug and work" instrument puck. Integration efforts include projects like the ORNL SensorNet, SDSC/CALIT2 RoadNet based on Antelope [6], GRIDCC instrument grid architecture, MIT's iLab project [16] in education applications of remote access to instruments and the DARPA/DoD Sensor Web Enablement community's suite of activities. These projects each have an approach to remote access to instruments and sensors and collectively provide a composite picture for a middleware for supporting remote access. The purpose of the Common Instrument Middleware Architecture (CIMA) middleware project is to build on these ideas to provide instrument and sensors with a service-oriented abstraction layer that facilitates the location and use of instruments, sensors and other real-time data sources and supports their integration into scientific workflows including computing and storage elements.

2. PROJECT GOALS

Scientific researchers by and large consider experimental or observed data to be the raw material from which knowledge is extracted through a process that might involve reduction, fusion and analysis. This approach on data per se focuses on the latter part of the chain in the reduction and analysis phases that are performed "off-line" with respect to the instrument that collected the data and ignores possible interactions between the analyst and the instrument itself. Large-scale projects organized around instruments such as detectors at the Large Hadron Collider or aggregations of weather data for severe weather prediction have yielded useful application test beds, but still the focus is on data as separate from the instrument that produced them and at some time after their acquisition. The possibility of designing and executing an experiment in real time in response to current conditions is not directly supported by this mode of interaction. For many major instrument facilities that provide advanced, complex or even experimental observational hardware, researchers with the expertise to judge the quality of appropriateness of the data may not have access to the proper tools to perform this analysis until well after the data have been collected and not at critical initial stages of the experiment when adjustments or corrections to the hardware can make a difference in the outcome. This focus on data as separate from their source and the operations necessary to control instruments remotely in real time in order to generate, acquire and filter data, combined with improved prospects for making instruments and sensors network-accessible has left a gap in capabilities for e-Research.

CIMA [10] is an approach to making instruments and sensors network-accessible with the following objectives: (1) use service-oriented architecture principles to integrate instruments and sensors as real-time data sources into distributed computing and storage environments; (2) improve accessibility and throughput in instrumentation investments; (3) promote sharing across institutions and disciplines; (4) develop a methodology for describing instrument capabilities and functions and embedding these in the hardware to improve the flexibility and lifetime of data acquisition and analysis applications and to improve the findability of instrument resources; and (5) automate and move the production of metadata as close to instruments as possible in order to facilitate data management, provenance and reuse. A further goal is to leverage expertise in the user community for a given instrument for collaborative problem solving around instrument-related problems

A common middleware layer makes it possible for instrument users to build, test and deploy new software to implement their experiments as needed and to continue to use existing software without modification when instruments are redesigned or upgraded. Such middleware also allows hardware developers to open up their products to take advantage of a broader range of control applications and facilitates the fusion of data from multiple sources in real time.

Some requirements for an instrument middleware are explored by Devadithya [10] and McMullen [20]. These include self-describing hardware and control software [23]; functional transparency; resource-oriented stateful services [21]; interoperable with other data acquisition and transport systems; efficient data transfer, lightweight with respect to CPU, memory and network resources; support for intermediaries for signal processing and data aggregation; support for authorization; multiple modes of interaction (e.g., streaming, polled and callback). In the next section we will discuss the architecture for the CIMA middleware developed to meet these and further requirements.

3. COMMON INSTRUMENT MIDDLEWARE ARCHITECTURE

CIMA consists of three main components, the service architecture, the channel protocol and parcel XML schema and the instrument description as shown in Figure 1. The service architecture is the instrument-independent service code that provides life-cycle functions and the core communications package,

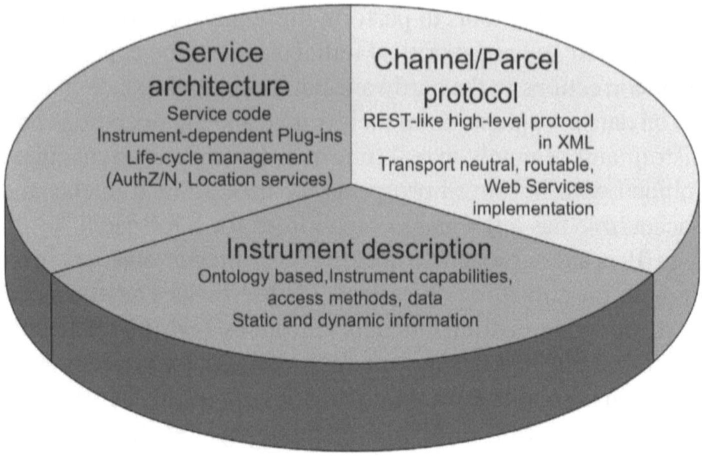

Figure 1. Components of the CIMA system

a Web Services endpoint for clients to communicate with the instrument. In addition to the instrument-independent parts there are instrument- and function-specific plug-ins that accompany the service code and handle requests for data from an instrument or sensor or requests for a change in the status of an actuator on the instrument. Service life-cycle management functions include loading of plug-in code, registration of the service instance with directory services such as a UDDI service or the GRIMOIRE [12] registry and communications transport start-up and shutdown.

The second component of the CIMA (Figure 1) is the channel protocol and the parcel XML schema. Although SOAP [34] is used as the primary transport the encoding style of the SOAP messages are document/literal rather than RPC encoded [15] to minimize the dependence on Web Services and to make asynchronous communications possible between clients and the service. The general approach taken is to view the instrument service as a RESTful Web Service [29] that could be implemented using a variety of schemes. The forward-looking potential of this decision will be discussed in more detail in Section 5.

The CIMA channel protocol is contained in the parcel XML schema. This schema defines a "type" tag, the values of which enumerate operations that clients can perform on the instrument and data types that can be returned by an instrument. Additional sections provide routing information (source and destination), data and metadata. Communication between a client and an instrument service is a stream of parcel XML documents. Typically a client will register to receive data from the service, either at regular intervals or when data are available. The service responds with SOAP messages to a client endpoint provided in the registration parcel with data. Protocol variants include RPC-like *get* and *set* operations on a sensor or actuator name handle which return an immediate parcel-valued responses. Alternatively endpoints can be specified for responses to get and set calls so that asynchronous operation is possible. The *describe* operation returns a description of the instrument. Plug-ins map logical name handles for sensors and actuators, possibly provided by the channel's describe function, to routines that read and write the actual devices. Output from a plug-in may include the sensor value plus additional metadata such as a time stamp or error code.

The default communications protocol and transport are document/literal-encoded SOAP over Secure Hypertext Transport Protocol (HTTP/S) over Transmission Control Protocol (TCP), etc. However, all levels are decoupled and essentially any transport can be used. Alternative transports we have implemented include Java Messaging Service (JMS) over TCP and Antelope [6,35], a system for carrying data from seismic ground motion sensors.

The third component is the instrument description, provided in RDF as Ontology Web-Language – Description Logic (OWL-DL) [19] instances. The description is based on an ontology (controlled vocabulary and relationships between terms) that provides information about instrument capabilities, access methods and input and output data formats. Some real-time information about the state of the instrument can also be provided if the description is associated directly with the CIMA service, rather than a registry.

Figure 2 illustrates the main components of CIMA and their relationship to the underlying instrument. In Figure 2 a user application interacts with the CIMA channel service through the following sequence of steps: (1) user application registers to receive streaming data from a given sensor; (2) the

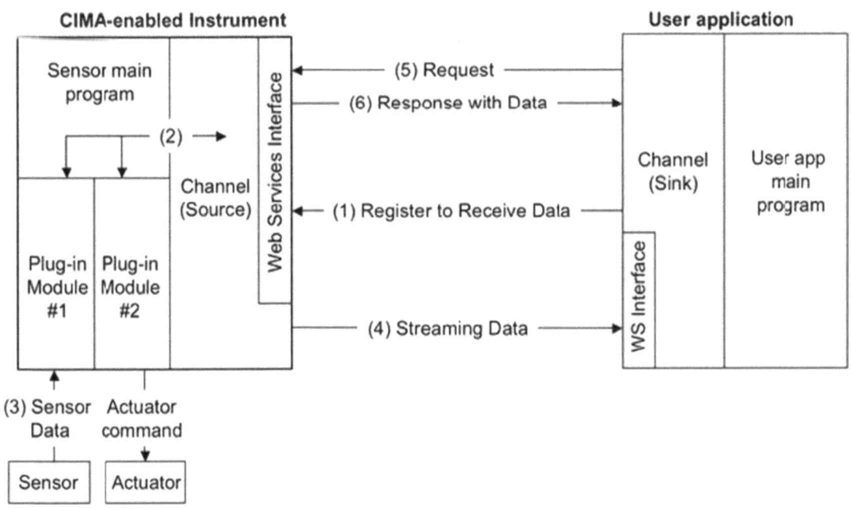

Figure 2. CIMA operational overview

channel service calls the appropriate plug-in to set up the data stream; (3) the plug-in repeatedly returns the sensor's data according to the specification in the request; (4) SOAP calls are made by the channel to an endpoint given by the user application. All messages sent and received by the channel are instances of the parcel schema. In addition to streaming data, the application can make requests to return or get one sensor reading or to issue an actuator set command (steps 4 and 5 in Figure 2). Other "immediate" commands include retrieving a description of the instrument's sensors and actuators as OWL-DL instances (describe) and authorization of the user application (session). Additional functionality such as message-level security and authorization can be added by extending the service or using additional WS-* layered specifications and message handling code at the Web Services interface.

An important design element is to make the instruments and sensors self-describing so that software and end-users can obtain information about them as and when needed, such as what data the instruments and sensors provide and how they are accessed and configured. This capability also allows analytical software components elsewhere in the data acquisition and reduction process to understand and manage data from instruments and sensors more effectively by applying unit conversions and calibrating data in real time. The capabilities of a device can be discovered by an application through the instrument description available through the channel protocol's *describe* operation, or through an external registry, if the CIMA service supports this. The instrument description can also be used by applications to help parse data messages from and build control messages to CIMA-enabled instruments and sensors. Appropriately constructed descriptions allow applications to interact with instruments at a high level, to determine utility for a given application, then providing the details of how to acquire, frame and convert data for a particular purpose.

Although an instrument description could be placed in a registry, it is primarily intended to be embedded in the device. Under these conditions the device itself is able to modify and update components of the description so as to reflect state in real time for that particular instance of hardware. This allows device class, device instance and temporal state information about an individual device to be made available that is not possible with static, external, device-class type descriptions held in a registry. Also, if this information is embedded in the device itself the process of discovery can be made independent of directory services if the service instances are members of a peer-to-peer network.

Another use for self-description is to provide the basis for automatic annotation of instrument data, supporting activities related to interpretation, organization and curation of the data. The development of a self-description capability is predicated on a "CIMA ontology" for instruments and sensors. The current implementation of the CIMA ontology is encoded in OWL-DL, a subset of OWL [19]. OWL-DL was chosen for several reasons: it makes the description amenable to machine reasoning tasks, it facilitates distributed development and extension of the CIMA ontology and, through inferencing, makes it possible to check the consistency of the ontology across multiple developers and sites. Reasoners such as Pellet [32] accessed through query languages such as SPARQL-DL [33] are available to use as a means of manipulating the descriptions in applications. We have developed several applications for manipulating CIMA OWL-DL descriptions using available off-the-shelf Semantic Web [4] tools.

As the quality, stability and complexity of these reasoning tools improve, it is increasingly possible to reason with complex data types as predicates. This allows us to leverage within our OWL-DL instrument descriptions existing XML schema for describing complex data such as SensorML and ISO schema for location (ISO-19107) and time (ISO-19108). Using existing system- or domain-specific complex data types helps to bridge the gap between using "plain old XML" and OWL-DL/RDF, which we believe will be increasingly relevant as Semantic Web applications drive the development and improvement of tools for manipulating OWL.

A functional model of a CIMA service and its relationship to the instrument and sensors/actuators it represents are shown in Figure 3.

At the outer level is the observatory, the physical or logical aggregation of one or more instruments with related functionality. The prototypical example

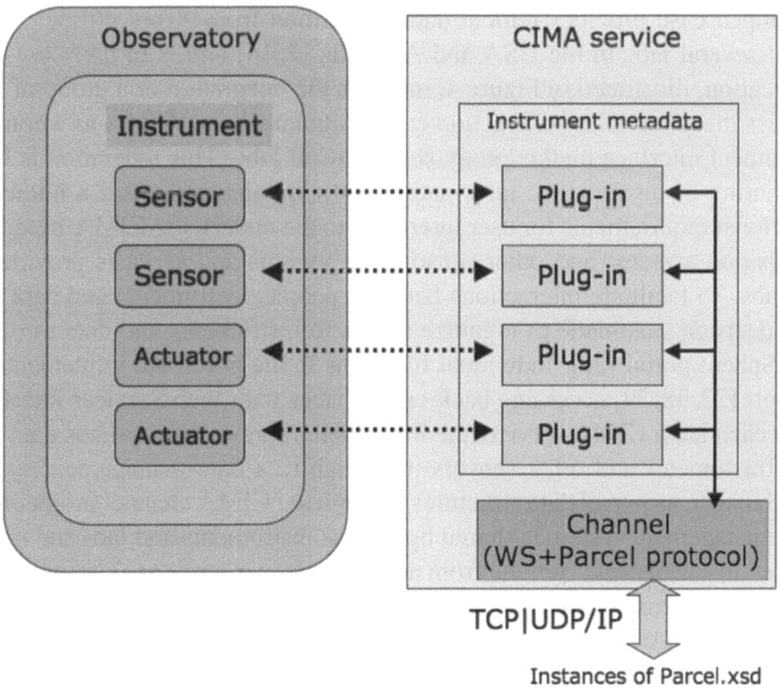

Figure 3. CIMA instrument descriptive model

would consist of a telescope, optics, CCD detector and controls for moving telescope and dome. An alternative example of an observatory is a crystallography bay containing a goniostat, a CCD array and several temperature probes used to collect data for an X-ray diffraction crystallography experiment [5]. The CIMA ontology in OWL-DL consists of classes to model the physical components such as observatories, instruments, sensors and actuators as well as the observable products of sensors and the service characteristics of the instruments. To separate types of descriptive elements there are separate ontologies for base classes (e.g., observatory-instrument-sensor model, units, observables and communications protocols), details of types of instruments (e.g., what is a telescope and diffractometer) and a third for specific models, though not instances, of hardware. The description for a given instrument consists of an OWL-DL "instance" containing descriptors from the ontology along with their values or other quantifiers.

4. EXPERIENCES WITH CIMA

The CIMA components described earlier, the channel protocol, parcels and plug-ins, seek to address the requirements for an extensible approach to remote access to instruments and sensors. To evaluate this approach we have developed several instrument implementations based on CIMA. One well-developed case provides remote data acquisition from X-ray diffractometers [5] at several labs in the USA and Australia [2, 14] and is in daily use. This application, illustrated in Figure 4, involves the integration of a group of laboratories that does X-ray diffraction crystallography using CIMA as a common instrument interface methodology shared by all labs. This federation is based on sharing of instruments, information and expertise and poses a number of functional requirements for user interfaces to the underlying CIMA-based data acquisition systems and other sensor data streams and services provided by the labs. To facilitate interactions between people, instruments and data a design decision was made to organize access to instruments and data through a GridSphere portal [28]. Individual functions in the portal are implemented as portlets [22, 38, 39] accessing back-end services with Web Services interfaces.

In each lab a CIMA service, the instrument representative, sends data from a diffractometer and other sensors in the lab to a data management system, MyManager, as parcel data structures through the CIMA channel protocol. The data management system is shared by several instruments and labs and may be located in a data center remote from any of the labs it supports. Two of the sites are at synchrotron beamlines at the Advanced Photon Source (APS), Argonne National Lab. As a part of these beamline collaborations we have developed a methodology for making EPICS process variables produced by the synchrotron and beamline control systems available as CIMA data streams.

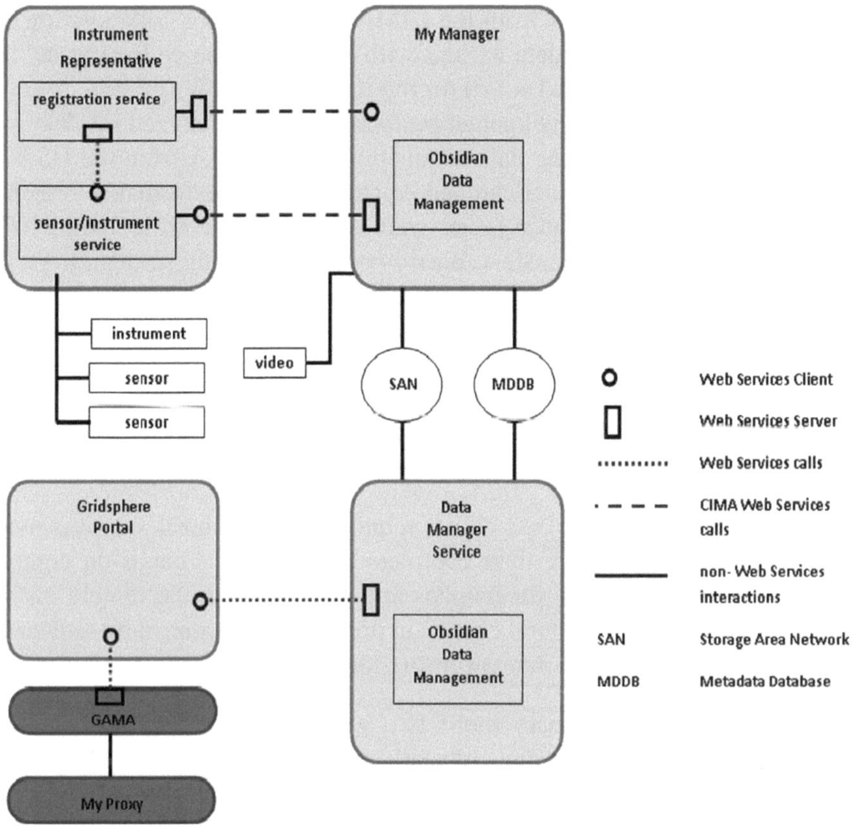

Figure 4. Components of a CIMA-based application for X-ray crystallography

A related capability inherent in CIMA is the ability to compose its Web Services interface with other services in a workflow environment. In an application under development by a research group at the University of Sydney, Australia, CIMA service endpoints are supported by Web Services containers (Axis) in JSP containers (Tomcat) and interact with portlets in a portlet container [2]. Using this approach CIMA data flows can be cached in a local portal container for near real-time retrieval and presentation in portlets using browser pull technologies such as Ajax or JavaScript and JavaServer Pages.

Acquisition and storage of data from CIMA-enabled instruments are also drivers for the IU Data Capacitor (DC) project (CNS-0521433) in which the DC is used as a high-speed disk cache for a workflow that includes multiple instruments and sites. Data are acquired and sent to the DC by CIMA and grid jobs process and manage the data after it is collected. In another large-scale application CIMA was adopted as core technology for two Australian Research Council projects (DART and Archer), for which a demonstrator project is being implemented [3]. Outcomes from the CIMA-DART/Archer collaboration include integration of CIMA data streams with Storage Resource Broker and the development of CIMA-based actors for the Kepler data flow system.

Performance of the channel/parcel protocol has been analyzed by Zhang et al. [40] in a paper describing data acquisition using CIMA from the USA to Australia through North American and Australian high-performance research networks. Several performance issues were explored in detail including TCP tuning for long haul data transfers, binary payload encoding strategies, use of SOAP attachments vs. including large data blocks in a parcel data structure and use of a "pull model" to transfer data using a protocol other than HTTP. Chiu has provided additional insight into performance improvement through the use of binary XML and schema-specific parsing [8, 9].

Some of the lessons learned and observations made during the development of CIMA and applications include the following:

- Many remote access projects do not require remote control, i.e., it is more about remote monitoring than complete replication of hands-on control. Considerations of safety for humans and equipment at the remote instrument site often preclude total control in preference for control by sequencer and validated program or human-in-the-loop control.

- As a corollary, the primary motif is a workflow involving data acquisition, processing and storage, typically with a human in the loop: Set-up→control→measure → compute/store/visualize→shut down.

- Data acquisition requirements can be met at many levels in the design and with appropriate technologies. The top-level architecture does not have to explicitly guarantee a level of service if these can be met at lower levels by selecting the right technology for that function. For instance in high data volume, data rate or low latency applications, the CIMA protocol allows for information to be sent from server to client by means other than Parcel XML protocol and Web Services.

- Remote access systems must have low impact on existing hardware and software and lab practices.

- A design with well-described high-level functionality where the implementation of lower level components can change as better technologies are introduced is preferable to one that is deeply dependent on particular low-level technologies.

- Performance of the XML-based CIMA protocol and simple strategies such as base64 encoding of binary data are sufficient to support a wide range of applications in remote access to instruments and sensors. The use of XML as the basis of the protocol greatly simplifies the development of code to interface with CIMA-enabled instruments over protocols that are proprietary, binary or both.

As we continue to apply CIMA to remote access problems we extend our collection of lessons learned and will include these in future versions of CIMA.

5. FUTURE DIRECTIONS

As a natural outgrowth of previous activities we continue to extend and build new CIMA-based application solutions and to extend the capabilities of CIMA components. Work continues on several other fronts:

5.1 Semantic Web for Instruments and Sensors

The current approach to making a CIMA instrument findable is to place location and description information in a registry that is shared and managed by the community of interested users. If instances of CIMA services representing instruments could instead form a peer-to-peer (P2P) network then addition or removal of instruments would be "automatic" by the presence or absence of a service instance in the network and finding available instruments would amount to issuing a query to the P2P network. This approach allows a bottom-up growth of instrument and sensor networks without a centrally managed registry.

5.2 Beyond Portals – Instruments and Sensors in a Web 2.0 Context

CIMA as implemented requires Web Services-aware clients, with an instance of a Web Service built in to or associated with the client to provide a sink for the service to call back with new instrument data. Our strategy [14] for user interfaces to CIMA instruments has been based on heavy weight JSR 168 [17] portlets in a GridSphere portal environment [28], for which a considerable programming effort is required to develop. Mirroring the rapid development of Web 2.0 technologies and APIs, work is in progress to develop RESTful [29] URI-based interfaces for CIMA in addition to the SOAP-based services now used. A Web 2.0 approach to services, however, is attended by loss of the use of WS-* layered products such as message security. Since the CIMA protocol is entirely contained in the parcel XML schema a CIMA parcel message could be delivered by GET or POST http transaction without the additional overhead of transporting, parsing and handling the SOAP envelope. A CIMA instrument or sensor service that operates in this manner can then be composed of other Web 2.0 offerings such as Google desktop [11], NetVibes [26] or Yahoo widgets [36] to rapidly assemble e-Science portals using a mix of community developed and commercial offerings. For example, recent experiments using RSS [30] as a protocol for sending CIMA parcel "events" to users and to Yahoo Pipes [37] shared, distributed applications have been highly successful at leveraging commonly used communications channels for distributing the output of instruments and sensor networks to a community of interested users.

5.3 Support for Sequencer-Based Experiments

Many of the security concerns for remote access to instruments can be mitigated if control over the device is local to the lab hosting the instrument and if user control was provided indirectly through a programmed sequence of instructions that had been validated perhaps by simulation prior to execution. We are developing a model for including sequencers such as Tango [7] within the CIMA service model so that programs can be sent to and executed by CIMA services as a data request.

5.4 Education and Outreach Through Remote Access to Instruments

CIMA interfaces to instruments are also proving to be invaluable in establishing educational outreach programs for students at colleges and universities that do not have major research instrumentation or expert users on-site. One such program allows undergraduate students to monitor or conduct X-ray diffraction experiments remotely with expert help. Faculty at the participating colleges receive additional training in how to include remote access to research quality instruments in their curriculum and how to coordinate with hosting labs to best connect students to expert scientists and technicians.

6. CONCLUSIONS

The Common Instrument Middleware Architecture is a specification for remote access to instruments and sensors that can be implemented in a flexible and extensible manner. The implementation is highly modular and many of the functional blocks can be re-implemented as needed with newer technologies as they become available. We described the core CIMA components of the service, channel protocol and its parcel data structures and the ontology used to develop instrument descriptions. We have also listed some of the applications for which CIMA has been used to provide remote access to instruments and concluded with lessons learned and some new areas of development for the project.

ACKNOWLEDGMENT

Support for this work by the National Science Foundation through cooperative agreement SCI-0330568 is gratefully acknowledged.

REFERENCES

[1] Aloisio, G., D. Conte, C. Elefante, I. Epicoco, G.P. Marra, G. Mastrantonio, G. Quarta (2006) "SensorML for Grid Sensor Networks." *Proceedings of the 2006 International Conference on Grid Computing and Applications*, GCA 2006, Las Vegas, Nevada, USA, June 26–29, pp. 147–152. CSREA Press.

[2] Atkinson, I.M., D. du Boulay, C. Chee, K. Chiu, T. King, D.F. McMullen, R. Quilici, N.G.D. Sim, P. Turner, M. Wyatt (2006) "CIMA Based Remote Instrument and Data Access: An Extension into the Australian e-Science Environment." *Proceedings of IEEE International Conference on e-Science and Grid Computing (e-Science 2006)* Amsterdam, The Netherlands, December.

[3] Atkinson, I., D. du Boulay, C. Chee, K. Chiu, P. Coddington, A. Gerson, T. King, D. McMullen, R. Quilici, P. Turner, A. Wendelborn, M. Wyatt, D. Zhang (2007) "Developing CIMA Based Remote Access for Collaborative e-Research." *5th Australasian Symposium on Grid Computing and e-Research (AusGrid 2007)*, Ballarat, January.

[4] Berners-Lee, T., J. Hendler, O. Lassila (2001) "The Semantic Web". *Sci. Am.*, 284, 34–44.

[5] Bramley, R., K. Chiu, T. Devadithya, N. Gupta, C. Hart, J.C. Huffman, K.L. Huffman, Y. Ma, D.F. McMullen (2006) "Instrument Monitoring, Data Sharing and Archiving Using Common Instrument Middleware Architecture (CIMA)." *J. Chem. Inform. Model.*, 46(3), 1017–1725.

[6] BRTT (2007) Boulder Real-time Technologies, Inc., http://www.brtt.com (accessed 1/15/07).

[7] Chaize, J.-M., A. Götz, W.-D. Klotz, J. Meyer, M. Perez, E. Taurel (2000) "TANGO – An Object Oriented Control System Based on CORBA." *Proceedings of the International Conference on Accelerator and Large Experimental Physics Control Systems (ICALEPS'99)*, Trieste.

[8] Chiu, K. (2004) "XBS: A Streaming Binary Serializer for High Performance Computing." *Proceedings of the High Performance Computing Symposium 2004*, Society for Computer Simulation International.

[9] Chiu, K., W. Liu (2004) "A Compiler-Based Approach to Schema-Specific XML Parsing." *First International Workshop on High Performance XML Processing*, WWW 2004, New York, NY, May 17–22.

[10] Devadithya, T., K. Chiu, K. Huffman, D.F. McMullen (2005) "The Common Instrument Middleware Architecture: Overview of Goals and Implementation." *Proceedings of the First IEEE International Conference on e-Science and Grid Computing (e-Science 2005)*, Melbourne, Australia, December 5–8.

[11] Google (2007) "Google Desktop – Getting Started," http://desktop.google.com/mac/gettingstarted.html.

[12] GRIMOIRE Grid RegIstry with Metadata Oriented Interface: Robustness, Efficiency, Security (2004) http://twiki.grimoires.org/bin/view/Grimoires/ (accessed 1/15/07).

[13] GTCP (2007) https://globus.org/toolkit/docs/development/4.1.0/execution/gtcp/gtcp-release-notes.html (accessed 1/15/07).

[14] Huffman, J., D.F. McMullen, K. Huffman (2006) "Remote Monitoring and Access to Instruments and Data. The CIMA Crystallography Portal." *Proceedings of the 2006 Annual Meeting of the American Crystallographic Association*, Honolulu, Hawaii, July 22–27.

[15] IBM (2005) "Which Style of WSDL Should I Use?," http://www.ibm.com/developerworks/webservices/library/ws-whichwsdl/.

[16] iLab (2001) http://www.swiss.ai.mit.edu/projects/icampus/projects/ilab.html (accessed 1/15/07).

[17] Java Community Process, Sun Microsystems (2004) "JSR 000168 Portlet Specification – Final Release," http://jcp.org/aboutJava/communityprocess/final/jsr168/index.html (accessed 1/15/07).

[18] Lee, K.B., R.D. Schneeman (2000) "Distributed Measurement and Control Based on the IEEE 1451 Smart Transducer Interface Standards." *IEEE Trans. Instrum. Meas.*, 49(3), 621–627.

[19] McGuinness, D., F. Harmelen (2004) Owl Web Ontology Language Overview, W3C Recommendation. World-Wide Web Consortium, http://www.w3.org/TR/owl-features/.

[20] McMullen, D.F., I.M. Atkinson, K. Chiu, P. Turner, K. Huffman, R. Quilici, M. Wyatt (2006) "Toward Standards for Integration of Instruments into Grid Computing Environments." *Proceedings of IEEE International Conference on e-Science and Grid Computing (e-Science 2006)*, Amsterdam, The Netherlands, December.

[21] McMullen, D.F., T. Devadithya, K. Chiu (2005) "Integrating Instruments and Sensors into the Grid with CIMA Web Services." *Proceedings of the Third APAC Conference on Advanced Computing*, Grid Applications and e-Research (APAC05), Gold Coast, Australia, September 25–30.

[22] McMullen, D.F., K. Huffman (2005) "Connecting Users to Instruments and Sensors: Portals as Multi-user GUIs for Instrument and Sensor Facilities." *Proceedings of GCE 2005: Workshop on Grid Computing Portals* held with SC05, Seattle, WA, November 18.

[23] McMullen, D., T. Reichherzer (2006) "Identity and Functionality in the Common Instrument Middleware Architecture." *Appl. Ontol.*, 3.

[24] Microsoft (2000) "Universal Plug and Play Device Architecture," Microsoft Corporation, http://www.upnp.org/download/UPnPDA10_20000613.htm.

[25] Microsoft (2000) "Understanding Universal Plug and Play: A White Paper," Microsoft Corporation, http://www.upnp.org/download/UPNP_UnderstandingUPNP.doc.

[26] NetVibes (2007) http://www.netvibes.com/.

[27] NIST IEEE-P1451 Draft Standards Home Page (2005) http://ieee1451.nist.gov (accessed 1/15/07).

[28] Novotny, J., M. Russell, O. Wehrens (2004) "GridSphere: A Portal Framework for Building Collaborations", *Concurrency Pract. Exp.*, 16(5), 503–513.

[29] Richardson, L., S. Ruby (2007) *RESTful Web Services*, O'Reilly Media, Inc., May 8. ISBN-10:0596529260.

[30] RSS (2007) "RSS Specifications and RSS Feeds," http://www.rss-specifications.com/.

[31] SensorML (2005) http://vast.nsstc.uah.edu/SensorML/ (accessed 1/15/07).

[32] Sirin, E., B. Parsia, B. Cuenca Grau, A. Kalyanpur, Y. Katz (2007) "Pellet: A Practical OWL-DL Reasoner." *J. Web Semantics*, 5(2).

[33] Sirin, E., B. Parsia (2007) "SPARQL-DL: SPARQL Query for OWL-DL," *3rd OWL Experiences and Directions Workshop (OWLED-2007).*

[34] SOAP Simple Object Access Protocol (2004) http://www.w3.org/TR/soap (accessed 1/15/07).

[35] Vernon, F. (2002) "Overview of the Usarray Real-Time Systems." *EOS Trans. AGU*, 81, S318.

[36] Yahoo (2007) "Yahoo Widgets," http://widgets.yahoo.com/.

[37] Yahoo (2007) "Pipes: Rewire the Web," http://pipes.yahoo.com/pipes/.

[38] Yin, H., S. Brenes-Barahona, D.F. McMullen, M. Pierce, K. Huffman, G. Fox (2006) "A PERMIS-Based Authorization Solution Between Portlets and Back-End Web Services." *Proceedings of GCE 2006: Workshop on Grid Computing Portals held in conjunction with SC06*, Tampa, FL, November 12.

[39] Yin, H., D.F. McMullen, M.A. Nacar, M. Pierce, K. Huffman, G. Fox, Y. Ma (2006) "Providing Portlet-Based Client Access to CIMA Enabled Crystallographic Instruments, Sensors, and Data." *Proceedings of the 7th IEEE/ACM International Conference on Grid Computing (Grid2006)*, Barcelona, September 28–29.

[40] Zhang, D., P. Coddington, A. Wendelborn (2006) "Data Transfer Performance Issues for a Web Services Interface to Synchrotron Experiments." *Proceedings of the 2007 Workshop on Service-Oriented Computing Performance: Aspects, Issues, and Approaches*, Monterey, California, USA, pp. 59–66.

WEB 2.0 FOR GRIDS AND E-SCIENCE
(Invited Contribution)

G.C. Fox[1,2,3], R. Guha[3], D.F. McMullen[4], A.F. Mustacoglu[1,2], M.E. Pierce[1], A.E. Topcu[1,2], and D.J. Wild[3]

[1] *Community Grids Laboratory,*
[2] *Department of Computer Science,*
[3] *School of Informatics,*
[4] *Knowledge Acquisition and Projection Laboratory, Indiana University*
{gcf, rguha, mcmullen, amustaco, marpierc, atopcu, djwild}@indiana.edu

Abstract Web 2.0-style services and capabilities collectively define a comprehensive distributed computing environment that may be considered an alternative or supplement to existing Grid computing approaches for e-Science. Web 2.0 is briefly summarized as building upon network-enabled, stateless services with simple message formats and message exchange patterns to build rich client interfaces, mash-ups (custom, composite, Web applications), and online communities. In this chapter, we review several of our activities in these areas: service architectures for chemical informatics; Web 2.0 approaches for managing real-time data from online experiments; management and federation of digital entities and their metadata obtained from multiple services; and the use of tagging and social bookmarking to foster scientific networking at minority serving institutions. We conclude with a discussion of further research opportunities in the application of Web 2.0 to e-Science.

1. INTRODUCTION: BROADENING THE DEFINITION OF GRID COMPUTING

Distributed computing research has been greatly influenced by developments in Internet computing and e-Commerce. The key concept of these systems is the service-oriented architecture [13], in which network accessible services expose well-defined programming interfaces and communicate with well-defined, application-layer network protocols and message formats. The global scale of electronic commerce and online communities provides an unprecedented opportunity to investigate research problems in scalability, reliability, system robustness, system federation, and security.

The general Web Service Architecture [1] concepts that build upon service-oriented architecture abstractions are typically specified using XML-based standards and recommendations. The Web Service Description Language (WSDL) is used to describe a service's capabilities. It is also a guide for requestor agents to construct valid messages for interacting with the service. SOAP is the most common network message format and encapsulates the specific communications between a requestor agent and a service (and also possibly routing intermediaries). SOAP is also an extensible format: additional directives for security, reliability, addressing, and similar higher-level qualities of service are added as supplemental information to the SOAP header.

Grid computing [7], as it is normally defined, is aligned closely with Web Service Architecture principles. The Open Grid Forum's Open Grid Computing Architecture (OGSA) [6] provides, through a framework of specifications that undergo a community review process, a precise definition of Grid computing. Key capabilities include the management of application execution, data, and information. Security and resource state modeling are examples of cross-cutting capabilities in OGSA. Many Grid middleware stacks (Globus, gLite, Unicore, OMII, Willow, Nareji, GOS, and Crown) are available. Web and Grid Services are typically atomic and general purpose. Workflow tools (including languages and execution engines) [8] are used to compose multiple general services into specialized tasks. Collections of users, services, and resources form virtual organizations that are managed by administrative services. The numerous Web Service specifications that constitute Grids and Web Service systems are commonly called "WS-*".

There are alternatives to this approach. As we have shown in previous discussion [21], Web 2.0 collectively represents a comprehensive distributed computing paradigm that, while compatible with Web Service Architecture principles, is a challenge to the usual WS-* implementations. The initial applications of Web 2.0 to e-Science are surveyed at the workshop, "Web 2.0 and Grids", at Open Grid Forum 19 (see http://www.semanticgrid.org/OGF/ogf19/). See also keynote talks by Hey and Goble and also the panel discussion moderated by Fox at Grid 2007 (http://www.grid2007.org/). As we survey in this chapter, Web 2.0 approaches can meet the requirements of cyberinfrastructure activities (such as chemical informatics and online laboratories). It furthermore suggests a broader view of cyberinfrastructure to include support for user and community-driven metadata, social networking and online communities of scientists.

1.1 Key Web 2.0 Concepts

Web 2.0 is a diverse and uncoordinated activity, but we may characterize it generally as building on the following key concepts.

1.1.1 Representational State Transfer (REST) Services

REST services [5] stringently avoid issues of exposed state and provide a single programming interface (essentially, HTTP GET, PUT, DELETE, and POST) for all services. REST services operate on URLs, which may respond with XML messages as well as other formats. XML payloads may be in any format. News feeds using RSS and Atom formatting are typical examples. These are best known as news feed formats, but they can be used as general-purpose envelopes for conveying sequential data (similar to SOAP). JavaScript Object Notation (JSON) is an alternative to XML for message exchange. The combination of clients to multiple services in a single application is called a mash-up (see www.programmableweb.com). REST services are purposefully simpler and have less sophisticated interaction patterns than WS-* services, relying on human developers rather than tooling to construct applications.

1.1.2 Rich User Interfaces Based on JavaScript, AJAX, and JSON

Rich user interfaces make use of client-side Web browser scripting coupled with REST-style server calls to enable browsers to provide enhanced interactivity that overcomes the limitations of the HTTP request/response invocation pattern. Updated content is delivered from the server to the user based on interactive events such as keystroke and mouse events rather than through HTML FORM actions. Numerous JavaScript libraries and tools exist for creating these applications. These range from high-level JavaScript libraries (Yahoo's YUI, Prototype, Scriptaculous) to JavaScript and HTML generators (Google's GWT and Ruby on Rails, for example). Related commercial environments (Adobe Flash/Flex and Microsoft Silverlight, for example) are also available.

1.1.3 Virtual Communities and Social Networks

Web 2.0 services are often designed to enable users to establish networks and share content (such as uploaded images, movies, and interesting Web sites). Online communities can develop in a number of ways, but a common example is through shared tags of online resources. Tags are single word descriptions of URLs. Collections of tags that describe a particular online resource or type of resource are known as a "folksonomy", a coined term that indicates the emergent and user-driven nature of the description.

1.1.4 Widgets, Gadgets, and Badges

The content and metadata of Web 2.0 services such as Flickr and del.icio.us may be accessed through their Web sites, via RSS feeds, and via REST API calls. They may also be accessed via exportable badges (typically XHTML and JavaScript) that may be embedded by a user into other Web pages. Start pages such as iGoogle and Netvibes aggregate these small, self-contained user interfaces to Web 2.0 services. Such pieces are known as widgets or gadgets and are conceptually similar to Java portlets that are often used to build science gateways [26]. Architecturally, badges and gadgets are very different from portlets, as the former are aggregated on the client side and are (essentially) unrestricted by the Web container, while the latter are aggregated on the server side and are under the control of the portal service provider. Gadgets can be shared and tagged to create user environments.

In the remainder of this chapter, we examine a range of applications and case studies using Web 2.0 for e-Science. Section 2 describes the application of service architectures to chemical informatics, in which we are supplementing WS-* services with Web 2.0 approaches. Section 3 describes the application of Web 2.0 approaches to instruments in online laboratories. Section 4 discusses our work in federating multiple online services for searching scholarly journals. This research addresses important gaps (federation and synchronization) in unrelated services. Section 5 describes our efforts to adapt the concepts of tagging and online bookmarking to the problem of helping researchers at minority serving institutions identify collaborators. Finally, Section 6 summarizes the work and describes further research opportunities.

2. CHEMICAL INFORMATICS AND WEB 2.0

The field of chemical informatics, or cheminformatics, has its origins in the 1960s with the development of the first representations of 2D and 3D chemical structures and structure–activity models [10] to relate chemical features to biological activity. Since then the field has increased significantly in terms of scope and techniques. The field is highly multidisciplinary – many methods used in cheminformatics have first appeared in the statistical, mathematical, text searching, machine learning, and artificial intelligence literature. In addition, in recent years, there has also been an increased intersection of biology and chemistry (chemical biology [20, 24], chemogenomics [11], etc.).

Commercial tools and private, proprietary data controlled by industry have historically dominated cheminformatics. However, the growing field of academic cheminformatics has brought more open source tools and libraries (such as Openeye, Daylight, CDK [25], and JOELib) to the community, and the National Institutes of Health's open research (PubMed) and open community data (PubChem) online services are spearheading a greater amount of open information. Thus, as we review in this section, the time is ripe for the application of service-oriented computing architectures and distributed infrastructure to the field.

To this end, we have developed the *Chemical Informatics Cyberinfrastructure Collaboratory (CICC)* to develop state-of-the-art cheminformatics tools and techniques and provide access to these via cyberinfrastructure [4, 9]. In this context cyberinfrastructure includes facilities such as computational grids, databases, and Web Services. The goal is to make cheminformatics functionality available in such a manner that non-expert users can use the infrastructure as prepackaged tools but will also be able to combine different tools and methods to create applications that are personalized for their specific problem. In the following sections we discuss aspects of the infrastructure that allow us to achieve these goals.

One of our primary activities has been to build a service-oriented computing framework, implementing Web Services using the CDK and other libraries. We highlight below one of these services, which wraps the R statistical package, and discuss the natural extension of this service into a Web 2.0 approach.

2.1 R Statistical Services and Community Models

A common task in cheminfomatics is the development of predictive models. These may be simple linear regression models or more complex models such as support vector machines or random forests. The traditional approach has been to develop the model and report its statistics in a publication. Such a model of dissemination precludes any form of machine (or even human) interaction with the model. An alternative approach is to provide a web page front end to the software that can evaluate the model. Such an approach is certainly an improvement, but still faces the restriction that one must manually navigate to the web page to make use of the model for predictive purposes. Another aspect of such models is that they may require calculation of features that are not publicly available. Thus even if one did have access to the underlying model, one could not calculate the features required as input to the model.

Our approach to this problem is the development of an R-based web service infrastructure, in which one can deposit a previously built model (currently a R binary file representation of the model) that is then made available as a Web Service. The infrastructure is sufficiently general that any model support by R can be deployed. Once deployed, one can access the model via SOAP-based Web Services to obtain predictions and various statistics. However, the current limitations include the fact that it is difficult to interrogate the model without actually loading the model in R, models are restricted to R, and the feature calculation problem is not addressed.

Future work aims to address these issues by way of developing an XML-based model specification document. Such a document will include information regarding the type of the model (linear regression, SVM, etc.) as well as various model statistics. In addition the model must include provenance data such as author, data of development, and so on. As we discuss in more detail in Section 5, these descriptions are suitable for keyword tagging. An important component of such a description will be a specification of the input features. This is especially challenging due to the wide variety of features that one may calculate using different software.

A key feature of any solution to this problem will involve community consensus on how one specifies features. This may be done using a top-down ontology approach, but it is also a good test case for tagging and folksonomy-style approaches. We envisage folksonomy-based approach using keyword tagging, where features will be characterized by what they represent, the software (and version of the software) required to calculate them, and so on. Such an approach has the advantage of being open-ended and usage-driven. Given such a model description, one need not load models into an R session (thus saving time) but, more importantly, models can now be searched. This represents a significant advantage over the current state of the art. In addition to model specification, we also plan to investigate the issue of model exchange. As noted above, our infrastructure is based on R. But there are many other popular statistical platforms, and models built in R cannot be run on them. We plan to use Predictive Modeling Markup Language (PMML) as a means of converting our R models to a platform-agnostic XML format. A number of platforms such as SPSS and DB2 Miner support this format.

With a combination of model specification and model exchange, we believe that the R-based computational infrastructure will provide a flexible and open approach to the deployment and searchability of predictive models in cheminformatics.

2.2 Databases and Data Handling

Though various computational services are very useful a key component of cheminformatics development is easy access to data as well as methods to manipulate data sets. To this end we have developed a number of databases that are derivatives of PubChem. The PubChem database is a collection of nearly 10 million molecular structures along with bioassay results for 549 assays. Along with developing our infrastructure we have built databases that provide add-on functionality to the data in PubChem. Examples include performing docking with the PubChem structures and generating 3D structures for 99% of PubChem. All our databases are relational databases based on the PostgreSQL platform. Though we have provided direct SQL access to these databases, we strove to provide a mode of access that is similar to our other services. Thus each database can be accessed via SOAP-based Web Services. In addition to providing access to databases we also provide services that allow one to manipulate data sets. One example is the VOTables services. VOTables are an XML specification for handling tabular data. Our services allow one to convert various forms of tabular data (plain text and Excel formats are supported) to a VOTables document and vice versa. Many of the statistical Web Services can accept VOTables data directly.

The above discussion has focused on accessing the various services via SOAP calls using standard Web Service techniques, which constituted the majority of our initial work. However, we have subsequently examined the use of Web 2.0 techniques for building services. Given that the services are effectively remote function calls, we can consider other ways to expose their functionality. One example is the use of RSS feeds and REST (specifically HTTP GET) invocations. Traditionally, news sites have used RSS feeds to syndicate news items. More generally, RSS feeds can be used to syndicate any type of "new" item or state change in an existing item. In our case we have made available a number of database searches in the form of RSS feeds. For example the user can specify a query for the PubDock database, asking for docking results for molecules that have a score above a certain specified value. The query is executed and the return value is an RSS feed (in RSS 2.0) format that can be viewed with any RSS reader. Furthermore our RSS feeds are able to embed chemistry within them by the use of Chemical Markup Language. Thus, if a given RSS item has an associated 2D or 3D structure, the structure can be embedded as a CML [17] fragment within the item. Such combinations are termed CML-RSS [18] feeds and can be viewed in a variety of environments such as Jmol or Bioclipse. Example of such CML-RSS feeds can be found at http://www.chembiogrid.org/cheminfo/pcrss/dockrss/form and http://www.chembiogrid.org/cheminfo/rssint.html.

2.3 Workflows and Mashups

Workflow tools and mash-ups offer alternate approaches to aggregating multiple sources of information and computation into new and ad hoc applications. Workflow tools are generally restricted in the kinds of links that can be made between services (for example, one can usually follow a linear path through a set of services, the output of one serving as the input to another, but more complex constructs like looping and non-sequential interaction fit less well into the paradigm), although they usually are built with user-friendly interfaces that allow development by non-programmers. The Yahoo! Pipes service is an attempt to allow such workflow development on the Web, specifically for processing RSS and Atom feeds. Completed workflows can usually be shared with others. Mash-ups permit a much higher degree of flexibility in the intercommunication of services, but generally require scripting or programming using APIs to achieve this (although there is no intrinsic reason why mash-ups could not be created in a graphical user interface, and indeed the latest version of blogging programs like BlogSpot permit the use of computational components in a page). In the pharmaceutical industry, one particular workflow tool called Pipeline Pilot (www.scitegic.com) has been widely used in cheminformatics. This tool uses a graphical interface for workflow development and includes extensive pre-packaged computation and database functionality for cheminformatics and bioinformatics. In the non-commercial sector, Taverna (taverna.sourceforge.net) and Knime (www.knime.org) have been used. While mash-ups have been developed for cheminformatics, they have not revolutionized or commoditized the field in the same way that has happened with online mapping. The reason for this could be lack of availability of simple, easy-to-use APIs or the underlying complexity and proliferation of the data structures in the field. However, given the growing use of distributed computation in both cheminformatics and bioinformatics, mash-ups, as well as workflow tools, seem like an excellent way of integrating these two fields.

2.4 User Interfaces

RSS feeds are one way to access and view information provided and generated by our cyberinfrastructure. Another mode of access to our functionality is the use of Userscripts, which are examples of rich client interface techniques. Userscripts are small scripts that can run in browsers when particular URLs or groups of URLs are visited in the browser. What makes them unique is that they can modify the content of a Web page on the fly, by altering the page components in the script. They can therefore be used to make user-customized variants of Web pages. We have applied userscripts, particularly Greasemonkey scripts for Firefox (https://addons.mozilla.org/en-US/firefox/addon/748) that use our Web service infrastructure to enhance existing chemical and cheminformatics resources on the Web [27]. These scripts include, inter alia, the visualization of 3D representations of molecules when visiting PubChem (using our Pub3D database) and the automatic mark-up and hyperlinking of chemical structures in journal articles.

3. SCIENTIFIC INSTRUMENTS AND WEB 2.0

3.1 Online Instruments and Services

Instruments and sensors provide observational data needed to drive the process of discovery in science. Along with simulation, experimental observation forms the basis for forming and testing theories. Availability and accessibility of appropriate instrumentation for a given research program can be a rate-limiting factor in discovery. Furthermore sensors and sensor networks are playing an ever-increasing role in longitudinal studies in ecological, earth, and biological sciences. Remote access to instruments and sensors shared by a research community is a highly desirable or critical capability in many fields of research and is becoming a requirement for many new large-scale instrument installations such as synchrotron light source beam lines [2].

There are numerous solutions to the problem of remote access to instruments, ranging from extending an instrument's console to users via screen sharing (e.g., VNC [23] or NX, http://www.nomachine.com/) to specifications for embedded "smart sensors" such as the IEEE 1451 suite [12]. There are also domain-specific solutions such as the Antelope messaging system (http://www.brtt.com) used in seismology to network ground motion sensors and the Monterey Bay Aquarium Research Institute "plug and work" instrument, Puck (http://www.mbari.org/pw/puck.htm). Each of these solutions provides part of the solution for remote access but individually is only suitable for a narrow range of applications.

We are developing the Common Instrument Middleware Architecture (CIMA) [3], a broad, middleware-based approach that can be used to solve remote access and control problems in many areas and on many scales. Use of a common middleware specification enables instrument users to build, test, and deploy software to implement their experiments as needed and to continue to use existing software with minimal modification when instruments are redesigned or upgraded [14]. Such middleware also allows hardware developers to open up their products to take advantage of community-driven application development efforts. CIMA provides application-level access to remote instruments and sensors through interface and protocol standards based on Web Services and a service-oriented architecture (SOA) approach.

We are now investigating approaches for the next generation of CIMA. As discussed in Section 1, Web 2.0 is a counterpoint to Web Services that shows great promise for increasing the diversity and rate of production of new services as well as the accessibility and reuse of existing services. The immediate challenges for CIMA middleware are to re-implement a Web Services product as a URL-based RESTful service and to develop gateways to existing CIMA instrument services. CIMA in its current form can potentially leverage WS-* layered products such as WS-Security, WS-Addressing, and WS-Attachment, so in migrating to a service based on a Web 2.0 approach, the functionality provided by these WS-* specifications need to be considered.

The advantages for adopting the Web 2.0 approach are clear with hundreds of new services being created and made available as Web 2.0 APIs (see, for example, www.progammableweb.com). As a matter of development cost and capability, the pool of skilled developers and integrators of these services is very large and growing rapidly, representing a potentially huge marketplace of developers worldwide, contrasting sharply with the availability of developers for more traditional Web Services. In e-Science applications, where specifications can be fluid, development time horizons are close, and funding is often aimed at science, not system development, using Web 2.0 approaches may provide a path to high-quality, low-cost software systems.

In sections below we will discuss the impact of Web 2.0 technologies and applications on the evolution of the CIMA architecture and the design of RESTful services for instruments, sensors, and other real-time data sources.

3.2 CIMA Service Architecture

An instance of a CIMA instrument service has three main components: the service code consisting of instrument-independent communications and management functionality and plug-ins for specific instruments, sensors, and actuators; the Channel protocol and Parcel XML Schema that define operations on instruments available to clients; and the instrument description which allows clients to understand the functionality and data provided by the instrument.

The service component provides a Web Services (WS) endpoint for clients to communicate with the instrument using the Channel protocol. Service lifecycle management functions include loading and execution of hardware device drivers (plug-ins), registration of the service instance with directory services, and communications transport start-up and shutdown.

The second component, the Channel protocol, is used by clients to interact with remote CIMA-enabled instruments by sending and receiving chunks of XML based on the Parcel XML Schema. SOAP over HTTP is used as the primary transport for parcels, the encoding and style of the SOAP messages is document/literal rather than RPC/encoded to reduce dependence on Web Services and to make asynchronous communications possible between clients and the service. A complete client–server Channel consists of the server's WS endpoint for commands and data requests in Parcel XML chunks to the instrument, and a second WS endpoint at the client to receive replies asynchronously from the instrument. The parcel schema defines a tag that enumerates operations that clients can perform on the instrument and data types that can be returned by an instrument. Additional elements provide source and destination information for routing, and data plus metadata.

The third CIMA component is the instrument description, provided in RDF as OWL-DL instances. The description is based on a controlled vocabulary that provides information about instrument capabilities, access methods, and input and output data formats [16].

CIMA is intended to be a component in a larger service-oriented architecture in which it provides real-time instrument or sensor data to downstream components. SOAs tend to be "pull" oriented where clients make RPC-like requests to the service. Grid systems also tend to be stateful in which clients use operations to query or change the state of an instance of the target service or associated resource dedicated to their use. CIMA is stateful in that clients can have a relationship with an instrument service that can last many transactions, but it also supports a "push" model in which information can be provided asynchronously to clients as it becomes available. This push capability requires Web Services-aware clients, with an instance of a Web Service built in to or associated with the client to provide a sink for the service to call back with new instrument data. A publish/subscribe approach is an alternative to this WS-based callback mechanism and we have implemented callbacks based on Java Messaging Service and Antelope.

As an alternative to WS callbacks and publish/subscribe approaches, a Web 2.0 approach could use Ajax on the user's Web browser to poll for new data continuously. If supported by the server and the client data could also be streamed directly to code running on the browser via server push. Since the CIMA protocol is entirely contained in the Parcel XML Schema a parcel message could be delivered by GET or POST HTTP transaction without the additional overhead of transporting, parsing, and handling the SOAP envelope. Message delivery latency would now also depend on the frequency of polling by the client in addition to factors present in the server push scenario.

3.3 REST Services for Instruments

Our objective is to re-implement the SOA version of CIMA services so that it can coexist and work well with Web 2.0 applications and services. We expect that a "resource-oriented architecture" (ROA) approach using RESTful interfaces will provide a suitable mapping from our current SOA to a form that can leverage existing REST-based services and Web 2.0 technologies [22]. The current callback scheme can be preserved if the client provides a Web server to which callbacks can be made but in general the client's role must change from message sink to polling the data source. The server must also add buffering for each client to compensate for polling latency, and additional semantics for requesting buffered data must be added to the polling sequence but these are relatively minor changes and suitable to an SOA version of the polled service as well.

There are other considerations in moving from SOA to ROA for real-time data services beyond the basic shift from event-based callbacks to polling. In particular, there is a stark contrast between the processes used to create WS-∗ specifications and Web 2.0 APIs: WS-∗ specifications are usually aimed at solving a more general enterprise-level process problem and are carefully vetted through a standards body such as OASIS (http://www.oasis-open.org/) and receive much consideration before the first evaluation implementations are available. Web 2.0 APIs on the other hand tend to be focused on providing specific functionality rapidly and in a somewhat more ad hoc manner. The result then is that classic SOAs are stable, evolve very slowly, tend to require considerable developer knowledge, and are often difficult to develop without supporting development tools.

In theory Web Services specifications can also be layered and functionality added incrementally to an application by adding the appropriate handler and client code for a given Web Services function to the client's code and server's handler stack. Development and testing of Web Services generally requires both client and server test harnesses and a complex development and test environment. The hallmark of Web 2.0 is rapid development and implementation of APIs that conform to a simple (i.e., REST) interaction method. This approach allows developers to prototype and refine a service quickly and the results to be tested and evaluated using only a Web browser or program implementing HTTP GET calls.

The economic tradeoffs are stark. It seems clear that developing REST services that can be integrated with other Web 2.0 offerings is a winning strategy where timeliness or development costs are at issue, and Enterprise Web Services are appropriate where use of WS-* is required or the client or service under development must interact with other Web Services.

3.4 Portals for CIMA Instruments

Although clients that acquire data to storage media do most of the work when it comes to performing an experiment with a CIMA-enabled instrument, on-line instruments and sensors are more interesting and useful when provided with user interfaces. We also provide access to instruments and data through a GridSphere portal which hosts JSR 168 portlets and provide portlets with a user log-in context for identity management and access control [15, 28]. Individual user interface functions in the portal are implemented as portlets accessing back-end CIMA services with Web Services interfaces. Although GridSphere is very functional and suits the current requirements for the CIMA project, the community of GridSphere and JSR 168 users and developers is limited and further development of both environments is in question. A further issue is that development for GridSphere is nontrivial and combining this with the need for portlets to bind data from Web Services and to provide Web Service sinks, the complexity of the result is high and maintainability is consequently a problem.

If we look to Web 2.0 technologies though, the situation seems brighter. A CIMA service that operates in a RESTful manner can be composed of other Web 2.0 offerings such as Google desktop, NetVibes, or Yahoo widgets to rapidly assemble e-Science portals using a mix of community developed and commercial offerings. As an experiment we were able to reproduce some of the functionality of our GridSphere portal in iGoogle using off-the-shelf widgets in a matter of a few minutes using RESTful versions of CIMA services.

3.5 Research Opportunities for Real-Time Instruments

There are further opportunities for accessing real-time data from instruments and sensors using Web 2.0 technologies. In a recent experiment we used the news distribution protocol Atom (http://tools.ietf.org/html/rfc4287) as a method for sending CIMA parcel "events" to interested users through common applications for reading Atom feeds. These systems provide a well-supported, almost zero development approach for real-time condition monitoring and alerting. This approach further enables us to integrate Atom-based services into mash-up tools. As described in Section 2, highly functional signal processing applications that consume and process the feed data streams from CIMA-enabled sensors can be developed using Yahoo Pipes (http://pipes.yahoo.com/pipes/) and shared broadly both as "code" and as functioning service instances. Yahoo Pipes offers a compelling model for future community-based distributed programming efforts and particularly for creating value-added real-time data products from instruments and sensors embedded in our environment.

Emerging Web 2.0 technologies and APIs offer a flexible, rapidly evolving environment for developing real-time applications for sensors and instruments. Moving from a Enterprise Web Service approach to REST for instrument services presents some challenges with respect to client design but doing so makes a broad range of Web 2.0 components and environments available for rapid and cost-effective development of software systems for instrument-driven e-Science.

4. FEDERATING ONLINE DIGITAL ENTITIES

Many Web 2.0 services such as social bookmarking sites and scholarly journal search services provide overlapping capabilities. Unfortunately there has been little work (and little incentive from the service providers) on integrating or federating these services. We describe in this section a federating architecture for such services with a specific application to managing online digital entities. In Web 2.0 terminology, this may be viewed as a server-side mash-up.

4.1 Managing State in Distributed Digital Entities

We have been working on providing consistent service-based architecture based on the event-based model for reconciling the dynamic and possibly inconsistent multiple sources of metadata information obtainable from Web

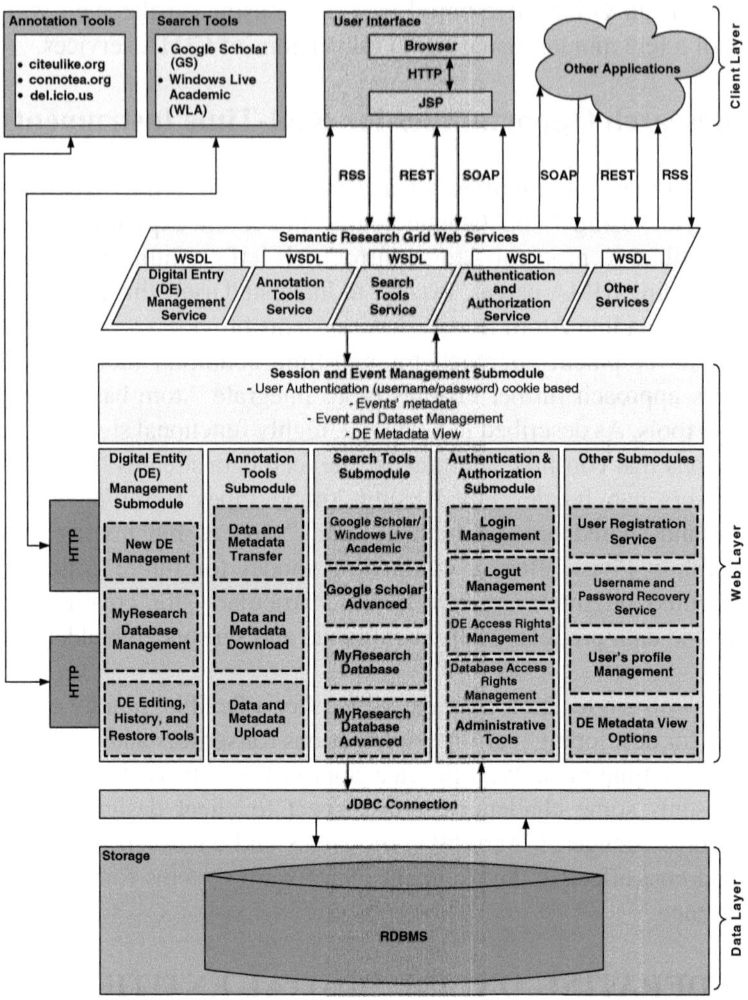

Figure 1. Semantic research grid architecture

2.0 services for resource annotation and tagging. As part of our research, we have designed, developed, and implemented the following modules to our prototype project, the Semantic Research Grid (SRG). The system architecture is shown in Figure 1.

4.1.1 Session and Event Management

This module provides a mechanism for storing data about the current user. These user-specific data include user authentication credentials, any modifications to a digital entity (called minor events), and the selected "view options", which control the level of detail with respect to the metadata fields displayed for each digital entity. Once the user has logged in the SRG system, the user's authentication credentials, all minor events for each digital entity, and view options for metadata fields of a digital entity are all maintained in the user session. When a user logs out from the SRG system, all unused minor events (modifications to a digital entity) for a data set creation are removed from the current user's session. To solve the consistency problem in the SRG system, we have designed a novel event-based consistency model on the concepts of *event* and *dataset* [19]. In our model, we adopt the view of an event as a time-stamped action on a digital entity, which only maintains the modifications to an object. We distinguish between *minor* and *major* events: insertion of a new digital entity into the system or deletion of an existing digital entity from the system is considered a major event; modifications to existing digital entities are considered minor events. Each digital entity has a reproducible state that is the sequence of the events that have been applied to it.

4.1.2 Digital Entity Management and Update Model

This module integrates PubsOnline software (an open source tool for management and presentation of databases of citations via the Web) into the SRG system and provides an interface for searching the local/remote databases of SRG. It also provides a user with an interface (1) to manually insert a digital entity into one of the local/remote SRG databases; (2) to access to the history of a digital entity, from its entry into SRG system to present, and ability to rollback to a previous state; (3) to view detailed information about a digital entity; (4) to update any metadata fields of a digital entity, which is saved into the session as a minor event for this digital entity.

In our system, we need to have a well-defined update model, which is built on top of an event-based model, to provide the user with flexible choices to manage their digital entities. Our update model uses events for updating digital entries and it is based on the following concepts: (a) keep the existing version; (b) replace the existing version with the new one; and (c) merge the existing and the new versions.

In our update model, we can provide the user with an ability to select an option from the above update model to be applied for all matching digital entries or each individual digital entry. In this manner, updates can be applied to each individual or all digital entries as a default based on the selected choice. We have also designed and developed "Periodic Search for Updates" module that allows a user to search and retrieve updates for digital entities in the system.

4.1.3 Annotation Tools

This module implements an interface that allows a user to manage the social bookmarking tools such as Delicious, CiteULike, and Connotea. Through this module a user can (1) upload digital entity data and metadata to one of these social bookmarking Websites; (2) download digital entity data and metadata from one of the social bookmarking Websites into one of the SRG research databases; (3) transfer digital entity data and metadata between these social bookmarking Websites.

4.1.4 Search

This module provides an integration layer that allows searching over multiple databases. This is an extensible module that includes implementations for searching Google Scholar, Microsoft Windows Live Search, and multiple local databases.

4.1.5 Consistency Model

We have also been working on the consistency mechanism of the system. We have two mechanisms to provide consistency: push and pull. We push updates to annotation Websites from our system once they occur by using the remote sites' Web API. We also periodically check annotation Websites for updates and pull them into our system if there is any. Our consistency model is a data-centric consistency model. It implements strict consistency: once updates occur in the system, they are propagated right away. Our model is based on the primary copy consistency protocol. Since updates are treated as events, a user can always roll back to a previous state at any time by undoing events.

4.1.6 User Management

The SRG system has the following modules for managing user interactions: (1) user registration module; (2) username and password recovery module; (3) user profile management module, where a user can edit personal information, modify system password, and request subscription to available SRG system groups; (4) digital entity metadata "view options" mechanism, which allows a user to define the metadata fields of a digital entity to be displayed or hidden.

Figure 1 shows the overall architecture of the SRG system. This system consists of three main layers: (a) the *client* layer; (b) the *Web* layer; and (c) the *data* layer.

The client layer is implemented using Java Server Pages (JSP) but can be built with any Web Service client. The client layer communicates with the Web layer over the HTTP protocol through SOAP messages encapsulating WSDL-formatted objects. The Web layer consists of several Web Services that handle communication with the existing online tools. As with other projects described in this chapter, we are investigating alternative protocols and message formats (such as REST APIs and RSS/Atom feeds) for managing interactions with the Web layer. Our goal is provide all major Web 2.0 access modes (see Section 1) in the Web layer. The Web layer communicates with the data layer through JDBC connections. Finally, the data layer is composed of several local and remote databases.

4.2 Authentication and Authorization Issues

In collaboration systems such as SRG, users may access group-editable metadata. We must provide mechanisms for establishing privileges to view and possibly edit portions of digital entities within the SRG system.

4.3 Access Rights

We have adopted an access-control matrix model that supports multiple groups and multiple users for each object. The model should support any changes for a group of people. There are set of objects (digital entity or database, access rights) pairs for users. Digital entities may be created by the users of the SRG system in several ways: (a) using annotation tools (Delicious, CiteULike, and Connotea); (b) using search tools; (c) manually, through the "Insert New Citation" Web interface.

The access policy has two basic modes. *Public:* Digital entity creation must be associated with at least one group, including possibly universal membership groups. *Private:* the digital entity can only be assessed by its owner.

The owner can modify the access mode. For both public and private operations, read and write (i.e., modify or edit) permissions are given to user or group for the inserted digital entity.

The SRG system distinguishes between three kinds of users: *Owner* is initiator of the digital entity and database creation; *Group* is any available groups in SRG system; and finally *Other* users are all remaining users. The owner of digital entity and database can specify the digital entity and database permissions for all three kinds of users mentioned above.

We have used a permission model similar to UNIX file system. In SRG system, digital entity corresponds to the file element, and its home database corresponds to the directory element. For each digital entity and database, there are three types of access rights defined in the systems. Access rights, summarized in Table 1, are used in the implementation of the authorization module. Authenticated users can create databases dynamically. A user needs to have write permission for the database in order to put digital entities into the database.

The access matrix describes data protection and permissions that each subject holds for each object. As an example in Table 2, users and groups are shown in the rows and digital entities in the columns. In this example, Bob is the owner of DE_1, DE_2, and DE_3. He has read (R), write (W), and delete (D) accesses for these digital entities. $Group_1$ has write access for DE_1 and read

Table 1. Access rights summary

Access right	Digital entity (DE)	Database
Read	Read DE	Access to database for viewing digital entities
Write	Modify DE	Insert digital entities into database
Delete	Delete DE	Remove entity from system

Table 2. Access-control matrix for digital entities

Name	DE_1	DE_2	DE_3	User
Bob	R,W,D	R,W,D	R,W,D	Owner
$Group_1$	W		R	Group
$Group_2$	R,W	R,W	R	Group
Alice		R		Other
Henry	R,W			Other

access for DE_3. Alice is another user and has only read access to DE_2. The system allows having only one owner of a digital entity and database. However, there might be more than one group for Digital Entity and database. Bob can modify DE_1, DE_2, and DE_3user access rights or give them permissions to groups and other users.

The owner of the digital entities may give access rights to other users. We have defined two methods for this operation. First, the owner may give permissions for all digital entities stored in the database. A user can give permissions to any user or groups that can be accessed by all users of that group for selected database. However, users and groups need to have required permissions for owner's selected database. Second, owners can modify their digital entity permissions for users and groups.

4.3.1 Group Administration

Having permissions for digital entities and databases for users and groups does not provide a complete authorization scheme. We need to have levels of control of the users and groups to complete the authorization module. We have defined the following levels of administration control: Super Administrator and Group Administrator. The system allows having more than one Super Administrator. An existing Super Administrator can add other Super Administrators to the system. A Super Administrator can assign any user to become a Group Administrator and remove a Group Administrator from group. Each group has at least one Group Administrator. If a user creates a new group, that user automatically becomes Group Administrator. Users are allowed to belong to more than one group. User can make a request to members of any group in the system.

5. RESEARCHER TAGGING AND MATCHMAKING

As we discussed in the introduction and expanded in Section 4, online bookmarking services provide a very simple way for users to share interesting URLs with each other. This sharing may be amongst friends and collaborators in networks, or it may be anonymous and democratic: by selecting a particular site, one "votes" for it. Bookmarked content is commonly described with keyword tags as well as free-form comments provided by the user. Bookmarking services often provide tag recommendations based on previous tagging to encourage the user to avoid synonyms and alternative spellings.

Although simple, tagging systems provide a very fundamental capability: they enable community-driven metadata descriptions of URL-identified Web resources. One may envision the extension of this to URIs generally. Seen from this point of view, we may readily recognize that tags and their associated URLs can be represented as graphs and are a variation on the graph-based Semantic Web relationships more commonly represented with RDF and OWL. We may treat either all tags as nodes and URLs as arcs or vice versa. This allows us (with sufficient amounts of data) to apply interesting computational graph techniques such as cluster identification and ranking.

To seed these research issues as well as provide a useful service, we are developing the MSI-CIEC portal (Figure 2), a bookmarking and social networking service to help faculty at minority serving institutions find interesting research opportunities and collaborators in their fields.

The MSI-CIEC portal's basic function is to provide a user interface to bookmarks. We use Connotea as a backend service for storing bookmarks: users have accounts created simultaneously in the MSI-CIEC portal and in Connotea when they register. While Web browsing (particularly journal browsing as described in Section 2), users may choose to bookmark and tag a particular journal article or NSF calls for proposals (available via RSS feed). The portal provides a browser toolbar button (installable by dragging the installation link from the portal interface to the browser's toolbar area) to allow sites to be tagged in a process that is not disruptive to browsing.

Figure 2 illustrates various "tag cloud" views. Other portal views include displays of selected RSS news feeds, such as NSF funding opportunities. To further simplify the tagging process for these particular feeds, we are developing a "click tagging" process: a user can indicate interest, disinterest, or ambivalence toward a particular proposal call by clicking an icon. This allows users to quickly create a set of tags interesting to them. Other users, examining the same link, can see other users interested in the same proposals.

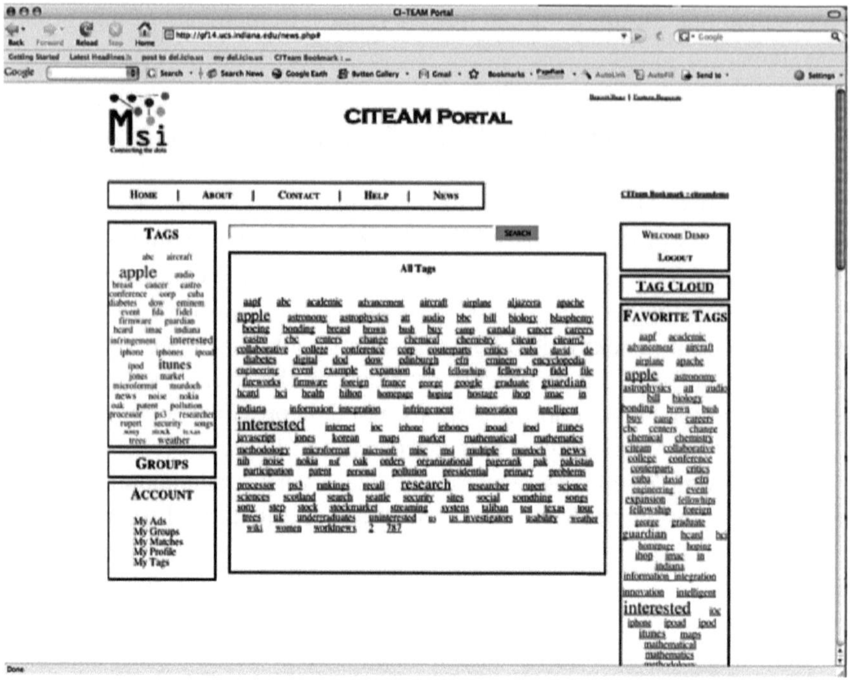

Figure 2. The MSI-CIEC portal is designed to help users manage online bookmarks and identify researchers with similar tagging profiles

To further populate the system, we are incorporating public data obtained from NSF and TeraGrid user databases. We represent data such as project award number, principal investigator name, NSF division, and directorate to tags. In addition, principal investigators' public information is used to generate profile pages within the system. This allows the user to treat other users as tags for navigating through the system (to see, for example, someone's list of collaborators or funded projects in various NSF directorates). Alternatively, the user may want to view the profile entries of others in the system.

The MSI-CIEC portal may be used for bookmarking, managing proposal call links from the NSF RSS feed, or navigating through databases. However, its full purpose (when sufficiently populated with data) is to investigate social networking strategies. These strategies are intended to help researchers find other researchers with similar or complementary interests. We are basing this functionality on tag profiles that users build up through normal use of the portal as well as seeded information from NSF databases. As we have observed, tag profiles and their associated resources may be treated as graphs. This allows us to examine techniques of varying sophistication for determining similarity between users. These may range from simple link counting to clustering, machine learning, and Page Rank-style algorithms.

6. CONCLUSIONS

We have reviewed the application of Web 2.0 technologies to e-Science. Sections 2 and 3 reviewed the evolution of two Web Service-based cyberinfrastructure projects (chemical informatics and instruments) into Web 2.0-style distributed systems. Common themes include the use of REST-style services communicating with message formats adapted from news feed standards. We also discussed the use of Web 2.0 for less traditional cyberinfrastructure: the federation of multiple bookmarking and scholarly search services (Section 4) and the use of tagging to build up scientific communities (Section 5). By adopting Web 2.0 approaches, we hope that these systems may leverage the considerable investment by major software and service providers. Mash-up building tools (which work primarily with RSS and JSON) are an important example.

Challenges facing Web 2.0 for e-Science include the lack of standards to enable portability of user identity and metadata between services. Typical Web 2.0 services such as Flickr, Facebook, and del.icio.us provide portable capabilities (such as JavaScript badges and RSS feeds) but they do not provide mechanisms for sharing or federating identity, preferences, and other metadata. We have made initial efforts to synchronize digital entity representations in multiple, unrelated services, but this needs further investigation.

Web 2.0 parallels Grid and Web Service systems in most capabilities (see for example [21]). Major exceptions include security and notification. OpenID is one important Web 2.0 security standard for creating distributed identity for enabling single sign-on systems, but the problem of distributed access controls (for example) are open issues. Notification and messaging systems are another fundamental problem, since they do not match well with the request/response messaging style used by REST services. One can simulate event pushing with client polling, at the cost of either increased network traffic or increased message latency. These issues are particularly important to instrument and real-time grids (Section 3).

ACKNOWLEDGMENTS

The Chemical Informatics and Cyberinfrastructure Collaboratory is funded by the National Institutes of Health. CIMA is supported by the National Science Foundation's Middleware Initiative program. The Semantic Research Grid is supported by Microsoft Research. The MSI-CIEC portal is supported by the National Science Foundation's CITEAM project.

REFERENCES

[1] Booth, D., Haas, H., McCabe, F., Newcomer, E., Champion, M., Ferris, C., Orchard, D., "Web Service Architecture", *W3C Working Group Note*, 2004.

[2] Bramley, R., Chiu, K., Devadithya, T., Gupta, N., Hart, C., Huffman, J.C., Huffman, K.L., Ma, Y., McMullen, D.F., "Instrument Monitoring, Data Sharing and Archiving Using Common Instrument Middleware Architecture (CIMA)", *Journal of Chemical Information and Modeling*, 46(3), 1017–1025, 2006.

[3] Devadithya, T., Chiu, K., Huffman, K., McMullen, D.F., "The Common Instrument Middleware Architecture: Overview of Goals and Implementation", In *Proceedings of the First IEEE International Conference on e-Science and Grid Computing* (e-Science 2005), Melbourne, Australia, Dec. 5–8, 2005.

[4] Dong, X., Gilbert, K.E., Guha, R., Heiland, R., Kim, J., Pierce, M.E., Fox, G.C., Wild, D.J., "Web Service Infrastructure for Chemoinformatics", *Journal of Chemical Information and Modeling*, 47(4), 1303–1307, 2007.

[5] Fielding, R.T., Taylor, R.N., "Principled Design of the Modern Web Architecture", *ACM Transactions of Internet Technology*, 2(2), 115–150, 2002.

[6] Foster, I., et al., "The Open Grid Services Architecture, Version 1.5", Open Grid Forum GFD-1.080.

[7] Foster, I., Kesselman, C., Tuecke, S., "The Anatomy of the Grid: Enabling Scalable Virtual Organizations", *International Journal of Supercomputer Applications*, 15(3), 200–222, 2001.

[8] Fox, G.C., Gannon, D., "Special Issue: Workflow in Grid Systems". *Concurrency and Computation: Practice and Experience*, 18(10), 1009–1019, 2006.

[9] Guha, R., Wiggins, G.D., Wild, D.J., Baik, M.H., Pierce, M.E., Fox, G.C., "Improving Usability and Accessibility of Cheminformatics Tools for Chemists Through Cyberinfrastructure and Education", *Cheminformatics*, Accepted September 2007.

[10] Hansch, C., Maloney, P.P., Fujita, T., Muir, R.M., "Correlation of Biological Activity of Phenoxyacetic Acids with Hammett Substituent Constants and Partition Coefficients", *Nature*, 194, 178–180, 1962.

[11] Harris, C.J., Stevens, A.P., "Chemogenomics: Structuring Drug Discovery to Gene Families", *Drug Discovery Today*, 11, 880–888, 2006.

[12] Lee, K.B., Schneeman, R.D., "Distributed Measurement and Control Based on the IEEE 1451 Smart Transducer Interface Standards", *IEEE Transactions on Instrumentation and Measurement*, 49(3), 621–627, June 2000.

[13] MacKenzie, C.M., Laskey, K., McCabe, F., Brown, P.F., Metz, R. (eds) "A Reference Model for Service Oriented Architecture", *Reference Model for Service Oriented Architecture 1.0*, 2006.

[14] McMullen, D.F., Atkinson, I.M., Chiu, K., Turner, P., Huffman, K., Quilici, R., Wyatt, M., "Toward Standards for Integration of Instruments into Grid Computing Environments", In *Proceedings of IEEE International Conference on e-Science and Grid Computing* (e-Science 2006), Amsterdam, The Netherlands, December 2006.

[15] McMullen, D.F., Huffman, K., "Connecting Users to Instruments and Sensors: Portals as Multi-user GUIs for Instrument and Sensor Facilities", *Concurrency and Computation: Practice and Experience*, 18, 1–11, 2006.

[16] McMullen, D., Reichherzer, T., "Identity and Functionality in the Common Instrument Middleware Architecture", *Applied Ontology*, 3, 2006.

[17] Murray-Rust, P., Rzepa, H., "Chemical Markup, XML, and the Worldwide Web. 1. Basic Principles", *Journal of Chemical Information and Computer Science*, 39, 928–942, 1999.

[18] Murray-Rust, P., Rzepa, H., Williamson, M.J., Willighagen, E.L., "Chemical Markup, XML, and the World Wide Web. 5. Applications of Chemical Metadata in RSS Aggregators", *Journal of Chemical Information and Computer Science*, 44, 462–469, 2004.

[19] Mustacoglu, A.F., Topcu, A.E., Cami, A., Fox, G., "A Novel Event-Based Consistency Model for Supporting Collaborative Cyberinfrastructure Based Scientific Research", In *Collaborative Technologies and Systems CTS 2007*, Orlando, 2007.

[20] Nicolaou, K.C., "Joys of Molecules, 2, Endeavours in Chemical Biology and Medicinal Chemistry", *Journal of Medicinal Chemistry*, 48, 5613–5638, 2005.

[21] Pierce, M.E., Fox, G., Yuan, H., Deng, Y., *Cyberinfrastructure and Web 2.0 Proceedings of HPC2006*, Available from http://grids.ucs.indiana.edu/ptliupages/publications/CIWeb20Chapter.doc

[22] Richardson, L., Ruby, S., *RESTful Web Services*. O'Reilly Media, Inc. May 8, 2007.

[23] Richardson, T., Stafford-Fraser, Q., Wood, K.R., Hopper, A., "Virtual Network Computing", *IEEE Internet Computing*, 2(1), 33–38 January/February 1998.

[24] Schreiber, S.L., Kapoor, T.M., Gunther, W. (eds.) *Chemical Biology: From Small Molecules to Systems Biology and Drug Design*, Wiley-VCH, Wienheim, Germany, 2007.

[25] Steinbeck, C., Hoppe, C., Kuhn, S., Floris, M., Guha, R., Willighagen, E.L., "Recent Developments of the Chemistry Development Kit (CDK) – An Open-Source Java Library for Chemo- and Bioinformatics", *Current Pharmaceutical Design*, 12, 2110–2120, 2006.

[26] Wilkins-Diehr, N., "Special Issue: Science Gateways – Common Community Interfaces to Grid Resources", *Concurrency and Computation: Practice and Experience*, 19(6), 743–749, 2007.

[27] Willighagen, E., O'Boyle, N.M., Gopalakrishnan, H., Jiao, D., Guha, R., Steinbeck, C., Wild, D.J., "Improving Dissemination of Scientific Literature and Data with Userscripts", *BMC Bioinformatics*, submitted September 2007.

[28] Yin, H., McMullen, D.F., Nacar, M.A., Pierce, M., Huffman, K., Fox, G., Ma, Y., "Providing Portlet-Based Client Access to CIMA Enabled Crystallographic Instruments, Sensors, and Data", In *Proceedings of the 7th IEEE/ACM International Conference on Grid Computing* (Grid2006), Barcelona, September 28–29, 2006.

Part V

Applications

INT.EU.GRID PROJECT APPROACH ON SUPPORTING INTERACTIVE APPLICATIONS IN THE GRID ENVIRONMENT

B. Coghlan[3], E. Fernández[2], E. Heymann[2], P. Heinzlreiter[5], S. Kenny[3], M. Owsiak[1], I. Campos Plasencia[6], M. Płóciennik[1], H. Rosmanith[5], M.A. Senar[2], S. Stork[4], and R. Valles[7]

[1] *Poznan Supercomputing and Networking Center, IChB PAS,*
[2] *Universidad Autonoma de Barcelona,*
[3] *Trinity College Dublin,*
[4] *University of Stuttgart,*
[5] *Johannes Kepler Universität Linz,*
[6] *Instituto de Física de Cantabria,*
[7] *CSIC, Santander,*
[8] *University of Zaragoza*

Abstract In this chapter we describe the Interactive European Grid project approach to handling interactive grid applications. Interactivity is an important feature that gives users the possibility to interact with applications in a natural way, giving the possibility to change parameters while the application is running. There are only few initiatives that address those issues in the grid environment. Execution of interactive applications on a grid environment is a challenging problem that requires the cooperation of several middleware tools and services. Additional problems arise when this interactive support is intended for parallel applications, which may run remotely across several sites. Our project aims to provide transparent and reliable support for such applications. We present the general architecture and interactive services, as well as technical details. We are proving its great added value and usefulness on a very demanding example. We also emphasize the open nature of the proposed solution and ready-to-use infrastructure, encouraging new applications and user communities to make use and profit of it. The int.eu.grid project supports interactive applications not only at the technical level. There is a special activity "Identification and Support for Interactive Grid-Enhanced Applications" that takes care of defining an adequate support in the e-Infrastructure for grid-enhanced interactive applications, so that the corresponding research communities can clearly benefit from the resources and mechanisms offered.

1. INTRODUCTION

The objective of the EU-funded Interactive European Grid project is the deployment of an advanced grid-empowered infrastructure in the European Research Area specifically oriented to support the execution of interactive demanding applications. While guaranteeing interoperability with existing large e-Infrastructures like EGEE by providing basic common middleware services, the initiative exploits the expertise generated by the EU CrossGrid project to provide researchers with interactive and simultaneous access to large distributed facilities through a friendly interface with powerful visualization.

Grid environments constitute one of the most promising computing infrastructures in computational science. One of their main goals is to tackle current and upcoming complex, large-scale problems, by enabling seamless sharing of computational resources. Most of today's grids are used for executing sequential applications that follow a batch processing approach. However, some users require interactive access when running jobs on grid sites. Execution of these applications on a grid environment is a challenging problem that requires the cooperation of several middleware tools and services. Additional problems arise when this interactive support is intended for parallel applications, which may run remotely across several sites. The Interactive European Grid project aims to provide transparent and reliable support for such applications. Our solution extends existing grid middleware oriented to batch applications with new services that cope with the requirements of interactive and parallel applications.

2. ARCHITECTURE OVERVIEW

The architecture of the Interactive European Grid project evolves from previous experience from other grid projects such as EU CrossGrid, EU DataGrid and LCG (the LHC Computing Grid). Additionally, it is influenced by the architecture of the EGEE project because basic services to build the grid infrastructure use gLite (the EGEE Grid Middleware). gLite Grid services follow a service-oriented architecture (SOA), which facilitates interoperability among grid services and allow for easier compliance with upcoming standards, such as OGSA, that are also based on these principles. The architecture constituted by this set of services is not bound to specific implementations of the services and although the services are expected to work together in a concerted way in order to achieve the goals of the end-user they can be deployed and used independently, allowing their exploitation in different contexts. Services and tools developed in our project follow the same design principles of SOA. The architecture of the entire project is presented in appropriate documents [12].

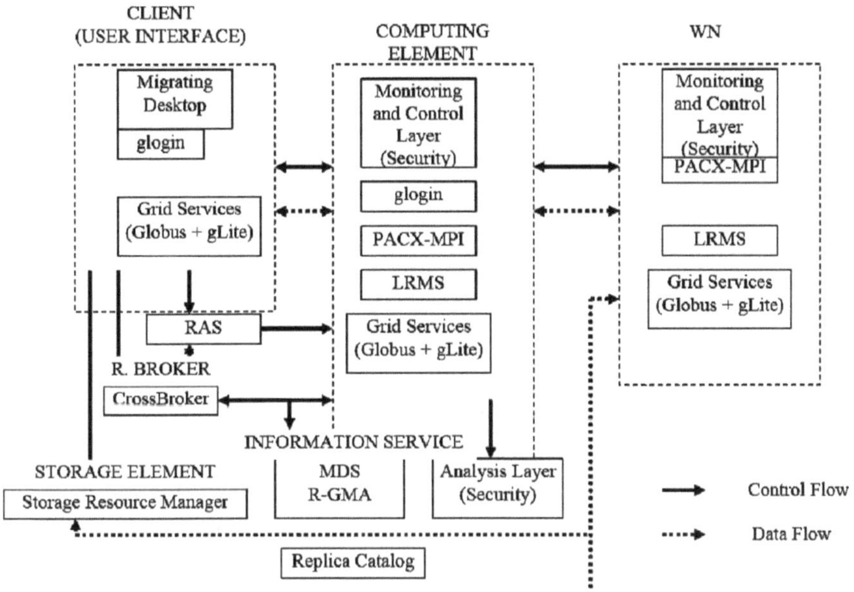

Figure 1. Overall architecture

For the sake of this chapter we present only the services components that are used to support interactivity (Figure 1).

The components presented in Figure 1 are services with very clear defined roles:

- The User Interface is the access point to int.eu.grid. The User Interface includes the following services: Migrating Desktop Platform, Complex Visualization Services, glogin and basic Globus and gLite middleware.
- The Computing Element (CE) is an entry point to the local cluster and runs the Gatekeeper module, which is the gateway through which jobs are submitted to local farm nodes. The Gatekeeper is an interface through which grid-enabled systems can use the local resources. The CE includes the following services: glogin, PACX-MPI/Open MPI, the LRMS and Globus and gLite middleware.
- The Worker Node (WN) is a local farm computing node where jobs are actually executed. Jobs received by the Gatekeeper are sent to the WNs through the local scheduling system. The services included in the WN are PACX-MPI, the LRMS and basic Globus and gLite middleware.
- Storage Element provides uniform access to storage resources. Storage resources are managed by the Storage Resource Manager.

- Resource Broker includes Advanced Scheduling Services. The goal of this component is to be able to execute and manage the different user tasks (interactive, parallel) across the available grid resources. After a matchmaking process, the Resource Broker passes the job to the Computing Element chosen for its execution. The user will be able to know at any time the status of his/her submitted job.

- Logging and Bookkeeping (L&B) is responsible for storing and managing logging and bookkeeping information generated by various components of the Workload Management System (WMS): bookkeeping information refers to currently active jobs, i.e., jobs within the WMS; logging information concerns the WMS itself.

- Information Service provides information about grid resources and their status. gLite uses two systems: MDS and RGMA. The Information and Monitoring Subsystem provided by gLite – R-GMA (relational grid model architecture) offers a global view of the information as if each virtual organization had one large relational database. It is suitable both for information about the grid (primarily to find out what services are available at any one time and then to find details of those services) and for application monitoring. Entities that want to share information publish it via a "Producer" service and those seeking information obtain it via a "Consumer" service.

In Section 5 we present how all these components interact and work together. Some parts of the services are the results of previous projects work. One of the aims of this project is to extend the already existing interactive approach by integrating the most important interactive services like Migrating Desktop Platform, Complex Visualization Services, Advanced Scheduling Services, Active Security Services and MPI Grid Proxy and provide production level e-Infrastructure. We put special emphasis on achieving fast, reliable mechanisms with low latency. This combination of services with new unique possibilities will give a clear added value to grid technology.

3. INTERACTIVE GRID SERVICES

- The Migrating Desktop Platform (MD) constitutes the whole platform that supports users during their work with the grid environment [10]. The MD environment is divided into several parts; it is based on a client–server architecture. The client part (graphical user interface) is a Java application that is independent of the running platform and hardware. This framework allows the user to access grid resources; run, monitor and visualize batch and interactive applications; and manage data files. MD provides a front-end framework for embedding some of the application mechanisms and interfaces, and allows the user to have virtual access to grid resources from other computational nodes. This product as one of the core service will enable simple usage of Complex Visualization Services, Advanced Scheduling Services and MPI Grid Proxy. Generic API for applications will be provided for the job submission phase, remote interaction with application and visualization of the results.

- Complex Visualization Services consist of a set of tools for supporting interactive visualization on the grid (glogin, GVid and Visualization Services based on VTK) [9, 11]. For terminal-based applications the ssh-like tool glogin is a solution that allows for interactive usage in a grid environment. The glogin tool provides a light-weight solution for bidirectional, secure and interactive connections between arbitrary grid nodes and clients on and off the grid. Applications with a graphical user interface (GUI) profit from advanced grid visualization tools such as GVid or VTK, which are capable of securely transporting visual data originating from arbitrary OpenGL and X11 applications. GVid executes all the visualization tasks except for the final display on the grid. Only an efficiently and thus highly compressed video stream is transferred through the network and displayed on one or more thin video clients, while the feedback is from a single master user who interacts with the application. These services will bring new quality and possibilities for the researcher desktops in the process of interaction with applications that profit from complex visualizations. Moreover, the added value doubles if we take into account combined usage with other interactive oriented services in the int.eu.grid project.

- Advanced Scheduling Services are provided by CrossBroker, the middleware service responsible for scheduling user jobs over the testbed infrastructure [4, 5]. When users submit their applications, the CrossBroker is responsible for optimizing scheduling and node allocation decisions on a user basis. The CrossBroker includes a streaming facility based on interposition agents and applies them transparently to both sequential and parallel jobs. It also has a multiprogramming mechanism that enables both interactive and batch jobs to share a single machine so that the interactive application starts its execution as soon as it is submitted.

- MPI Grid Proxy will be based on the PACX (PArallel Computer eXtension) and the Open MPI middlewares [7, 8]. PACX is an optimized MPI implementation for running MPI-compliant applications transparently in heterogeneous grid environments. PACX-MPI supports the complete MPI-1 standard and many features of the MPI-2 standard. The Open MPI Project is a collaborative effort by several international members to create a new MPI implementation. In the int.eu.grid project synergy of MPI and interactivity will considerably widen application abilities. This is a very important fact if we take into account a very computing-intensive interactive application.

- Active Security Services. The Active Security considers security measures for an interactive framework, and subsequently to enhance interactive operation with active security tools [3]. By active security we mean dynamic measures that track grid interactions and respond to unwanted security events in an adaptive way. It requires both extended security monitoring and tracking, and a good deal of automatic coordination of this across a number of security tools and all the sites involved in the events. The intent is to track and resolve security issues as they occur. Although it must facilitate the various different types of grid access patterns, it will be particularly cognizant of interactive jobs.

4. INTERACTIVE APPLICATION EXAMPLE USAGE – FUSION APPLICATION USE CASE

The fusion application (Computation and Visualization of Plasma Particles in Fusion Devices) has been chosen as a good example of data-intensive, interactive application that requires powerful computing resources, advanced visualization and high network throughput. It is the most representative use case that makes usage of most interactive services that will be developed in the int.eu.grid project. This concerns a user-friendly desktop environment – the Migrating Desktop and Roaming Access Server, CrossBroker, advanced visualization services – GVid, CrossBroker, usage of PACX-MPI on top of Open MPI and glogin.

4.1 Selected Application – Computation and Visualization of Plasma Particles in Fusion Devices

The purpose of porting this application to the grid was exploiting the combined power of a high number of CPUs interactively in order to visualize the behavior of plasmas inside fusion reactors [2]. Numerically the method consists in integrating the equations of movement of a charged particle in the background of the magnetic field inside the reactor.

For every individual particle, and at every step of the algorithm, the position, velocity and other magnitudes of the particle are stored. This is the simulation core of the program. The 3D visualization of those particle trajectories is a very interesting tool for researchers. In particular many structures developing inside the plasma are three dimensional, for example magnetic islands. Also, the possibility of directly observing what happens inside the reactor in a "real-time" simulation provides an insight impossible to obtain by just measuring average values of the magnitudes defining the plasma. For visualization purposes, all CPUs need to transmit the information related with the position and velocity of the particles in a continuous and synchronized way to the central machine, which after rendering will end the output as a video stream to the clients located at the researcher desktops. The user should then be able to interact with the application roughly once per minute. The interaction consists of changing the simulation parameters, such as electromagnetic fields, or the number of particles being simulated, and also zooming in certain parts of the reactor.

In order to show requirements of this application let us go through example usage for the calculations that have been done for the Stellerator TJ-II of the CIEMAT in Madrid. In the TJ-II Stellerator every particle has a physical life of 0.03 seconds. In a workstation with a Xeon at 3 GHz the calculation of a single particle trajectory till it dies is approximately 3.8 seconds of CPU. The memory consumption of the computation itself is rather modest in this respect (order of 500 MB).

In a realistic working scenario we would be simulating about 4000 particles per CPU, which means about 4 hours of simulation. During this time the parameters describing every particle position and velocity have to be transmitted over the network for visualization purposes. The graphical display should be such that 20 frames of data are transmitted every second to give the impression of continuity. Every second each CPU sends 20 frames, each containing the information of 4000 points. Therefore every CPU will be producing 1 Mbyte/second of information. If we assume we spread the calculation over 3 farms, each with about 30 CPUs, the central cluster node will handle a data rate of 30 Mbytes/second, which has to be transmitted for visualization preparation.

4.2 Proposed Use Case

This detailed use case shows technical details and flows of data in the proposed solution. It concerns all interactive services and components that exist in the int.eu.grid architecture (Figure 2):

1. The user runs the Migrating Desktop using Java WebStart

2. User prepares inputs for fusion application. The design of the coals and cavities, together with the magnetic fields they induce, are an input to the simulation.

3. The user specifies initial parameters for his/her application using the Migrating Desktop submission tool called Job Submission Wizard. It concerns choosing the appropriate application-specific parameter input and output files and needed resources. The user selects glogin as the transport layer. The user submits the job.

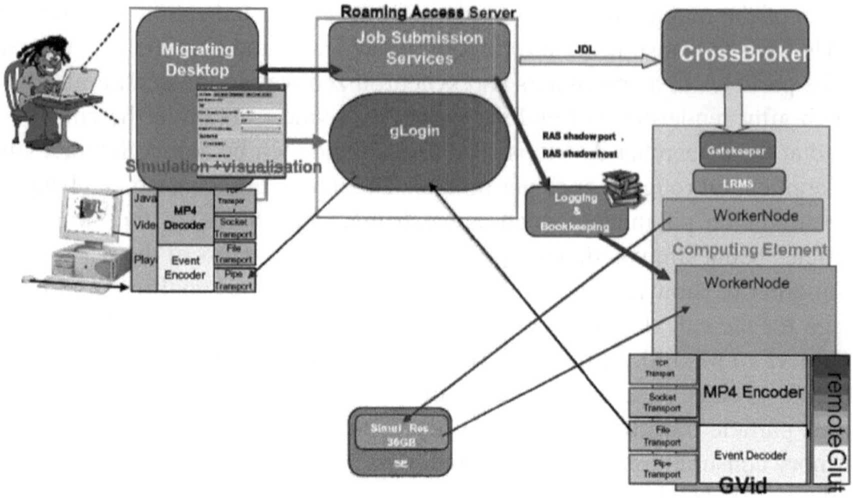

Figure 2. Example use case

4. The job is submitted to the Roaming Access Server (RAS) (using the above data). The appropriate Job Description Language (JDL) file is being created basing on input parameters and restrictions. The JDL includes glogin and application itself.

5. Roaming Access Server receives the request and starts the glogin client locally.

6. The user submission request is sent to the CrossBroker component.

7. The CrossBroker finds an available Computing Element (CE) that matches the requirements (including the capability to run interactive jobs) and submits the job there to be run on CE's Worker Nodes (WN).

8. glogin starts on the Worker Node. glogin starts the application.

9. For PACX-MPI the configuration files are created on the actual startup server.

10. On each cluster the job is run by means of the underlying Open MPI through a standard MPI-startup mechanism – the main startup server containing the contact information of all other clusters participating in the run. One of the sub-jobs is a remote visualization task, the rest are simulation tasks. MPI mechanisms are used for communication.

11. Simulation sub-jobs write an output that is later used by the visualization task. The individual cluster central nodes send the output to the central

computer, which takes care of the processing, rendering and video transmission.

12. Visualization tasks use GVid input adapters, which act as an interface to the visualizing applications and perform the video encoding.
13. Migrating Desktop starts the Java Video Player client with an embedded GVid mechanism (the output adapters), which is in touch with the glogin client on the Roaming Access Server machine that will be in charge of reading/writing the content of the relevant pipes. The output adapters are the final interface to the user displaying the decoded video data and capturing the user's interaction, feeding it back to the rendering application.
14. The trajectory of every particle is drawn as an orbit (blue lines) circling around the stellerator. The user interacts with the application roughly once per minute. The interaction consists in changing the simulation parameters, such as electromagnetic fields, or the number of particles being simulated, and also zooming in certain parts of the reactor. The graphical display should be such that 20 frames of data are transmitted every second to give the impression of continuity.
15. In case of any problem Migrating Desktop takes care of killing the launched process.

This use case is the first major target of our works. We are in the phase of the integration of all above-mentioned components and the first test of presented interactive services and we hope to publish the results and obtained speed-up of the research as soon as possible.

If we combined this detailed description of use case with the application requirements, we can notice that this mechanism enables handling a huge amount of computation, demanding advanced visualization and high network throughput. This proves great value of the proposed solution for supporting interactive applications in grid environments.

5. RELATED WORK

There is some interest in running interactive applications in the grid. These works are very often limited to the specific application or specific aspect of interactive behavior and are not general solutions for interactivity. We should mention two interesting projects that attempt to employ a general scheme for interactivity.

The Interactive Grid Architecture for Application Service Providers, I-GASP work [1] is more focused on graphics and multimedia-rich interactive applications. They have adopted a comprehensive access control and account management system for interactive environments, including the enforcement of QoS guarantees. They proposed components like controlled shell and desktop, dynamic accounts, admission control and monitoring and management agents. They also proposed an affinity scheduling algorithm for an interactive session. The remote display solution used in their implementation is VNC.

Globus Toolkit 4 (GT4) gives interactivity support, with usage of the secure shell -SSH, that has been expanded to support grid-related authentication (GSI-SSH). However, executing jobs with GSI-SSH results in access to the host without usage of grid job policies. Therefore, GSI-SSH cannot be used as a replacement for the usual job submission mechanism.

There are many grid portals and graphical user interfaces that can be compared to the Migrating Desktop. They offer services like job submission, monitoring services and some other functions that can be found also in the MD. However, the interactivity offered usually only refers to the portal middleware itself (in, e.g., the "Grid ENabled Interactive Environment" (GENIE) [6]). In contrast to this, the int.eu.grid provides interactivity for the grid jobs and not only for a grid environment.

6. CONCLUSIONS

In this article we have presented the general approach of supporting interactivity in the int.eu.grid project. We have defined what we mean by the interactivity and explained why such support is needed. Next we have briefly presented the general architecture and the main interactive services, namely the Migrating Desktop Platform, Complex Visualization Services, Advanced Scheduling Services and MPI Grid proxy. We have also put emphasis on the added values of these components. As an example of usage of this complex architecture we took into account a very computing demanding application with a need of advanced visualization and large requirements concerning network throughput. We described the application itself as well as its requirements and also the technical proposition of porting and usage in the int.eu.grid architecture. This application (Computation and Visualization of Plasma Particles in Fusion Devices) use case showed that even such complex and complicated applications can be handled by our approach. We believe that the proposed services are flexible enough for most interactive scenarios of applications usages.

It is worth mentioning that the int.eu.grid infrastructure is going to support several interactive applications from different disciplines of science. Also any new applications and users not yet identified by the project are welcomed in order to test the infrastructure, bring their interactive application in and make profit of their usage.

As it was mentioned, the int.eu.grid project is complementary to EGEE, and it enhances pan-European grids to the interactive and MPI field. This is a clear added value that can have big impact on the usage of e-Infrastructures by researchers.

REFERENCES

[1] Agarwalla B, Basu S, Talwar V, Kumar R (2003), "Interactive grid architecture for application service providers", *Proceedings of International Conference on Web Services (ICWS '03)*, Las Vegas, Nevada, pp. 365–374.

[2] Castejón F, Reynolds JM, Fontdecaba JM, López-Bruna D, Balbín R, Guasp J, Fernández-Fraile D, Fernández LA, Martín-Mayor V, Tarancón A (2006), "Ion orbits and ion confinement studies on ECRH plasmas in TJ-II stellarator", *Fusion Science and Technology*, Vol. 50, pp. 412–418.

[3] Coghlan B, Kenny S (2005), "Towards a grid-wide intrusion detection system", *Proceedings of European Grid Conference*, *LNCS*, Vol. 3470 (Amsterdam, The Netherlands), pp. 275–285.

[4] Fernández E, Heymann E, Senar MA (2006), "Resource management for interactive jobs in a grid environment", *Proceedings of IEEE International Conference On Cluster Computing (Cluster 2006)*, Barcelona (Spain), IEEE CS Press.

[5] Fernández E, Heymann E, Senar MA (2006), "Supporting efficient execution of MPI applications across multiple sites", *Proceedings of 12th International Euro-Par Conference*, Dresden (Germany), *LNCS*, Vol. 4128, pp. 383–392.

[6] GENIE: Grid ENabled Interactive Environment, http://www.npaci.edu/online/v5.5/genie.html.

[7] Gabriel E, Fagg GE, Bosilca G, Angskun T, Dongarra JJ, Squyres JM, Sahay V, Kambadur P, Barrett B, Lumsdaine A, Castain RH, Daniel DJ, Graham RL, Woodall TS (2004), "Open MPI: goals, concept, and design of a next generation MPI implementation", *Proceedings of 11th European PVM/MPI Users' Group Meeting*, Budapest, Hungary.

[8] Keller R, Krammer B, Müller MS, Resch MM, Gabriel E (2003), "MPI development tools and applications for the grid", accepted for publication at the *Workshop on Grid Applications and Programming Tools*, held in conjunction with the GGF8 meetings, Seattle, WA, USA.

[9] Köckerbauer T, Polak M, Stütz T, Uhl A (2006), "GVid – video coding and encryption for advanced Grid visualization", in J. Volkert, T. Fahringer, D. Kranzlmüller, W. Schreiner, editors, *Proceedings of the 1st Austrian Grid Symposium*, volume 210 of books@ocg.at, pp. 204–218, Schloss Hagenberg, Austria, Austrian Computer Society.

[10] Kupczyk M, Lichwała R, Meyer N, Palak B, Płóciennik M, Stroiński M, Wolniewicz P (2004), "The Migrating Desktop as a GUI framework for the application on demand concept", *Proceedings of the ICCS*, ser. *LNCS*, Vol. 3036.

[11] Rosmanith H, Volkert J (2005), "Traffic forwarding with GSH/GLOGIN", *PDP*, 13th Euromicro Conference on Parallel, Distributed and Network-Based Processing (PDP'05), pp. 213–219.

[12] Senar MA, et al. (2006), "Definition of requirements and design document for first release, including QoS and SLA for interactivity, public deliverable", int.eu.grid project.

APPLICATION OF PERFSONAR ARCHITECTURE IN SUPPORT OF GRID MONITORING

A. Binczewski[1], M. Lawenda[1], R. Łapacz[1], and S. Trocha[1]

[1] *Poznan Supercomputing and Networking Center, Noskowskiego 12/14, 61-616 Poznań, Poland*

Abstract In order to satisfy all the requirements of systems based on distributed resources, computer networks to connect them have to be monitored. These tools have to provide network performance indicators for setting optimal data transfers between project nodes. In this chapter we describe a monitoring framework able to perform measurements, which is applied for supporting the GRID-based infrastructure of the e-VLBI network.

1. INTRODUCTION

During the recent years computer networks have dramatically increased the core bandwidth. Advanced networks motivate interest in deploying new services that serve applications from areas of high-energy physics or radio astronomy. This also led to the deployment of status and performance monitoring applications [7, 9].

This software ensures availability and efficiency within the infrastructure, by monitoring various performance metrics. To unify the often distributed task of monitoring, the concept of a Service Oriented Architecture (SOA) offers the ability for specialized, autonomous services to join under a common access scheme. Thus, it is possible to separate the roles of monitoring, storage, processing, and visualization of data into specialized service instances.

The subproject Joint Research Activity 1 of the GN2 project [3] aims at providing a framework for performing multi-domain measurements in the European Research Network (GÉANT2) and the connected National Re-search and Education Networks (NRENs). It is carried out in close cooperation with

Internet2 End-to-End performance initiative (E2Epi) [5] and will result in a common system called PerfSONAR (Performance-focused Service-Oriented Network monitoring ARchitecture) [1, 10]. The name reflects the choice of a SOA for the system implementation.

With the widespread use of multi-domain applications the quality of service – latency, jitter, packet loss – becomes important. Computer networks have to be monitored and managed at every layer of the ISO/OSI model to satisfy all the requirements and expectations and to provide service consistent with an appropriate service level agreement (SLA). On the other hand, research communities using specialized applications based on distributed computations (Grid-based) require proper resource brokering.

In Europe, the Very Long Baseline Interferometer (VLBI) radio telescope array conducts observations known as e-VLBI, in which the telescopes are connected to the central data processor [4]. In order to automate and make the process faster the data is transmitted over a high-speed reliable network and correlated using Grid resources. The limited number of computational and network resources, as well as the need to monitor the service, resulted in the idea of designing and developing a model of distributed radio telescopes' data correlation with a Grid architecture. This objective is implemented within the scope of the EU-funded EXPReS project [6], where the demand for a Grid-based processing system and network monitoring meet together.

In this paper we illustrate the technology for a monitoring framework for collecting network performance data and we present the application of such framework in the Grid environment. The paper is organized as follows. In Section 2 our PerfSONAR is presented, while in Section 3 we detail its service-oriented architecture. Section 4 presents the solution to the problem of Grid monitoring in the EXPReS project. The current status of its proto-typical implementation is the subject of Section 5. Conclusion and future work can be found in the last section.

2. PERFSONAR OVERVIEW

PerfSONAR is a distributed performance monitoring architecture designed to aid in the discovery, troubleshooting and solution of network performance problems, potentially among numerous networks. This framework is comprised of collaborating entities meant to collect, store, analyze and deliver data using well-defined protocols.

The general monitoring infrastructure implemented by perfSONAR is illustrated in Figure 1. The Measurement Points (MPs) are the base layer of this framework and have the responsibility of measuring, as well as storing, different network metrics. The Service Layer is the middle layer of the system that is divided across administrative boundaries. This layer may aggregate, filter or

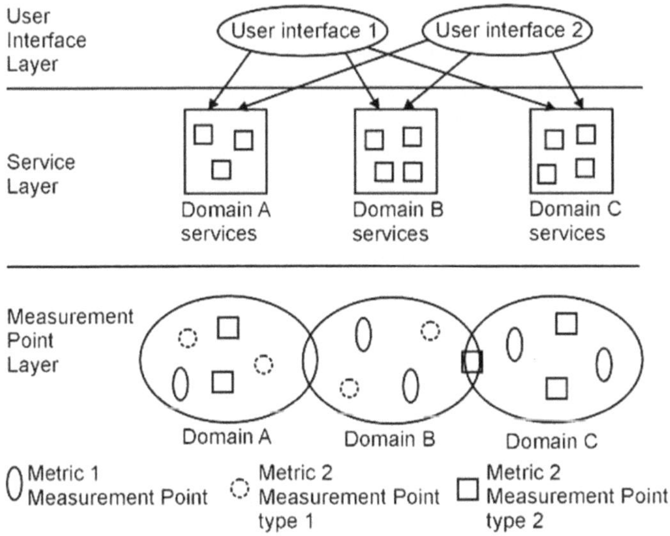

Figure 1. perfSONAR monitoring infrastructure

buffer data and can be seen as a set of data archives and transformers. On top of it the User Interface Layer consists of visualization tools, which present the data in customizable ways. In addition, this layer allows users to perform tests using services available at the lower layers of the framework.

3. MONITORING FRAMEWORK

The Measurement Points are the lowest layer in the perfSONAR and are responsible for measuring and storing network characteristics, as well as for providing basic network information. The measurements can be carried out by active or passive monitoring techniques [2]. The Measurement Point Layer of a network domain consists of different monitoring components or agents deployed within the domain. The domain could be an NREN, metropolitan network or another network under single administration. A set of monitoring tools acting as Measurement Points provides information on specific network performance metrics (e.g., one-way delay, jitter, loss, available bandwidth) through a common interface. Other data consumers, like Service Layer components, can retrieve these performance data by accessing the corresponding Measurement Points. The variety of Measurement Points and the possibility to incorporate existing measurement tools, already used within a domain, allow each network domain implementing a wide range of measurement metrics and reusing existing data.

The Service Layer is the middle layer of the system that allows the exchange of measurement data and management information between do-mains. As discussed earlier, the Measurement Point Layer can be seen as data producers while the Service Layer components are seen as data trans-formers and data archives. In order to discover domain capabilities, the perfSONAR architecture includes a dedicated directory service to help in locating data and other Service Layer components. This layer also considers existing specific functionalities to help resource management or framework elements' protection. A services-based perfSONAR framework implements each of these roles as independent services: Lookup (LS) for service discovery, Authentication (AS) for system elements and data protection, Measurement Archive (MA) for data storage, Topology (ToS) for network topology discovery and Resource Protector (RP) for resource management. The interaction of the entities between the Service Layer and Measurement Points Layer, within the Service Layer, as well as between the Service Layer and User Interface Layer, happens through a web service interface, making use of XML messages defined by the Open Grid Forum Network Monitoring Working Group.

Due to the non-monolithic character of the perfSONAR architecture, independent entities (services) may be added or removed from the system, in-creasing its flexibility and protection against components' failures.

Some of the perfSONAR entities contain an interface, which can be accessed by the User Interface Layer. These are data consumers that are implemented within different analysis tools to help the user in retrieving the data, submitting tests and presenting results in an easily understandable way.

4. GRID PERSPECTIVE

The European VLBI network is spread over European countries and beyond; thus, data transmission will cross several administrative domains, including National Research Networks as well as the pan-European GÉANT2 network. The development of the GRID system and e-VLBI data transmission between different geographically distributed partners under the EXPReS project will require proper resource management and network monitoring, in order to allow for automatic resource distribution and Quality of Service provisioning. This concept led us to the design of the integration of the perfSONAR system with Grid-based e-VLBI.

Networks of radio telescopes are used to produce detailed radio images of stars and galaxies. The resolution of the images depends on the overall size of the network (the maximum separation between the telescopes) and the sensitivity depends on the total collecting area of all the telescopes involved and, crucially, on the bandwidth of the connection between the telescopes.

In this technique, called Very Long Baseline Interferometry (VLBI) the signals between all telescope pairs are combined in a data processor. Typically, this processor must be able to cope with data rates up to 1 Gbps per telescope. At the moment this functionality has been implemented by constructing a massively parallel, purpose-built "supercomputer" – usually referred to as a data processor or correlator. As the processor capabilities continue to develop, the networking infrastructure is also getting more capable of dynamically handling very large data transfers in long periods of time. Taking all these factors into consideration, we can think about designing and implementing e-VLBI solution where the entire system is immersed into the Grid environment. This means that all operations, beginning from the preparation of the measurement scenario, through gathering measurement data, and finishing on visualization, are automated and performed using the functionality of the eInfrastructure.

The e-VLBI requirements for network monitoring fit best into the deployment of perfSONAR, which may be adopted for GRID resource management and QoS provisioning. Figure 2 depicts the concept of integration of perfSONAR into the EXPReS project infrastructure.

The current perfSONAR is the basis for further development. The open architecture of perfSONAR allows external applications making use of data collected by the perfSONAR system through the Network Measurements Working Group XML [12] schema. The XML schema is used to represent measurement data and will be integrated with the GRID Resource Broker environment, willing to interact with the framework. The measurement framework will collect performance data (e.g., available bandwidth, delay, throughput) using perfSONAR services installed in each involved domain. These could be a chain of regional networks, GÉANT2, NRENs and metropolitan area networks. The level of coverage of the e-VLBI network by perfSONAR measurements will significantly impact the ability to take optimal decisions. Ideally, all used domains should deploy the system. The services of the framework will receive monitoring data requests from the Network performance analysis module, co-operating with the Grid Resource broker through a standardized interface. During queries, perfSONAR instances – the Measurement Archive or Measurement Point – will examine their content or perform on-demand tests, in order to provide the requested performance data about links between adjacent domains. By using data returned, we will find the optimal connection between the given radio telescope and file server that should assure the fastest possible and reliable data transfer within e-VLBI Gird environment. Methods for the selection of the best path must be defined.

In addition to path selection, the network performance analysis module will be able to deliver network performance metrics that can visualize e-VLBI network usage.

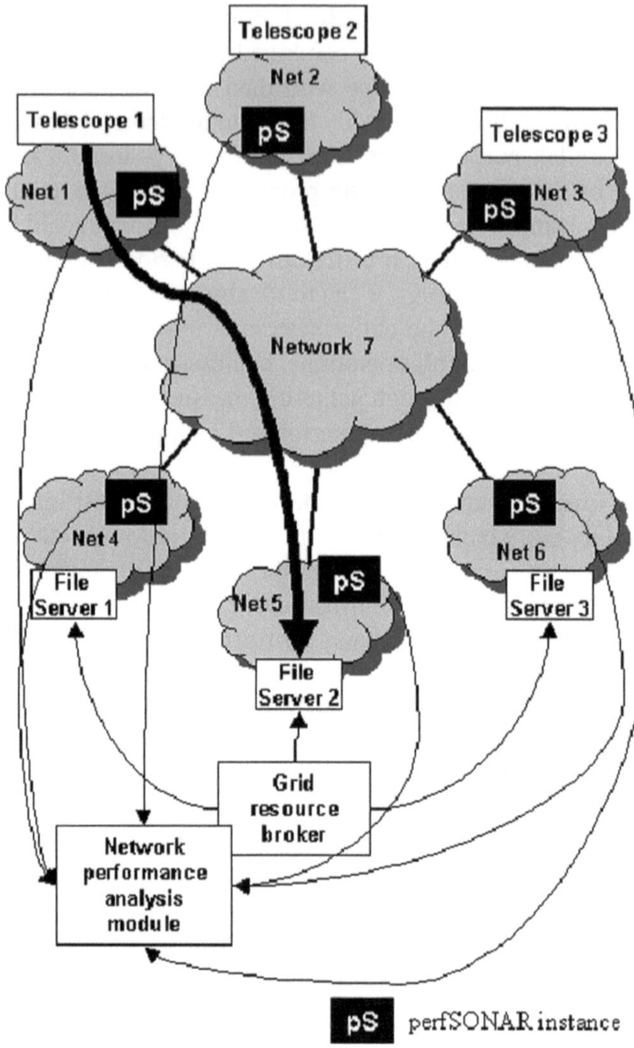

Figure 2. Integration of perfSONAR into EXPReS project infrastructure

5. PROTOTYPICAL IMPLEMENTATION

A prototypical implementation of the perfSONAR system is currently carried out under GN2 to realize the model as Web Services, aiming at the retrieval of link utilization and end-to-end link status data from several networks. The focus of this implementation is to validate the framework design and the service interactions.

More than a dozen perfSONAR installations are currently available and they enabled a first view of the European networks, making a first step toward effective troubleshooting of the performance problems and access to the available data placed in various administrative domains. The largest number of services deployed is in the category of the MA service, which is a wrapper around commonly used Round Robin Databases (RRD) [11] and provides link utilization statistics through a standardized interface. Additional elements of the monitoring infrastructure enable also capacity, packet loss and delay measurements among several European and North American partners. The availability of measurement data within a particular domain depends on the scope of implementation of perfSONAR as well as data access policy of this domain.

The prototypical implementation is already used by the EGEE project to retrieve link utilization data for enabling its own Grid network monitoring [8].

When integrated under the umbrella of the EXPReS project, the number of perfSONAR services will increase over time while adding additional modules, features and measurement types, so that it will cover all domains involved in the e-VLBI network.

6. CONCLUSIONS

This work has discussed the problems and solutions related to integrating a multi-domain perfSONAR framework into the Grid-based e-VLBI network. Although the perfSONAR system reached its second release, we are still in the early stages of development within the EXPReS project. Deployment of perfSONAR services within the involved domains is necessary, and additional work to provide network analysis functionality will be explored in future iterations.

ACKNOWLEDGMENT

The authors wish to thank all the collaborators in the perfSONAR activity for their efforts in defining and implementing the framework.

REFERENCES

[1] Boote J, Boyd E, Durand J, Hanemann A, Kudarimoti L, Łapacz R, Swany M, Trocha S, Zurawski J (2005), "PerfSONAR: A Service Oriented Architecture for Multi-domain Network Monitoring", *Proceedings of the Third International Conference on Service Oriented Computing*, Springer Verlag, Amsterdam, pp. 241–254.

[2] Chen T, Hu L (2002), "Internet Performance Monitoring", *Proceedings of the IEEE, 90*, pp. 1592–1603.

[3] DANTE homepage including information about GN2 project: http://www.dante.net/.

[4] European VLBI network homepage: http://www.evlbi.org/.

[5] Internet2 End-to-End performance initiative homepage: http://e2epi.internet2.edu/.

[6] EXPREeS – A Production Astronomy e-VLBI Infrastructure project homepage: http://www.expres-eu.org/.

[7] IST-MOME (Monitoring and Measurement Cluster) project homepage: http://www.ist-mome.org.

[8] Joint Research Activity 4, Enabling Grids for E-SciencE (EGEE) project homepage: http://egeejra4.web.cern.ch/EGEE-JRA4/.

[9] Leinen S, Przybylski M, Reijs V, Trocha S (2002), "Testing of Traffic Measurement Tools. Deliverable D9.7", *GN2 project.*

[10] Performance focused Service Oriented Network monitoring Architecture project homepage: http://www.perfsonar.net.

[11] Round robin database tool homepage: http://people.ee.ethz.ch/oetiker/webtools/rrdtool/.

[12] The Network Measurements Working Group (NMWG) of the Global Grid Forum homepage: http://nmwg.internet2.edu/.

NETWORK EMULATION ON GLOBUS-BASED GRIDS: MECHANISMS AND CHALLENGES

R. Canonico[1,2], P. Di Gennaro[1], V. Manetti[1], and G. Ventre[1,2]

[1] *Dipartimento di Informatica e Sistemistica, Università di Napoli Federico II, via Claudio 21, 80125 Napoli, Italy,*

[2] *CINI - Consorzio Interuniversitario Nazionale per l'Informatica Laboratorio Nazionale per l'Informatica e la Telematica Multimediali ITEM "Carlo Savy", via Cinthia 4, 80126 Napoli, Italy*

Abstract In the last few years, many emulation systems have been developed to help researchers evaluate the effectiveness of new protocols and applications in realistic network scenarios. NEPTUNE (*Network Emulation for Protocol TUNing and Evaluation*) is a flexible and scalable system developed at University of Napoli for the emulation of different network scenarios by means of a cluster of workstations. Setting up an emulation experiment in a cluster-based system requires, first, the ability to map virtual resources requested by an experimenter onto available physical resources and, second, the ability to exert a precise control over the allocated physical resources. These two requirements have much in common with resource management issues already addressed by the Grid computing community. Hence, we decided to exploit the *virtual workspace* concept at the foundation of the design of the NEPTUNE architecture. In this chapter, we illustrate the peculiar virtualization requirements of a cluster-based emulation system and discuss how a Globus Virtual Workspace based on Xen virtual machines can be used as the basis for implementing a distributed network emulation system.

1. INTRODUCTION

Design validation and performance evaluation of new distributed applications and protocols is usually performed either by means of real testbeds or through simulation. Simulation presents unquestionable merits [2], such as the possibility to simulate large-scale networks in a controlled and repeatable environment; yet, it has some drawbacks, such as the use of ad-hoc implementations of network protocols and applications and the difficulty of modelling low-level implementation details that may affect a system performance [13]. An alternative approach to simulation consists in reproducing the system under test in a real testbed: sets of computers, communication devices, and other resources that either reproduce real-world networks or are overlaid upon the real Internet. Large-scale geographic testbeds, like PlanetLab [12], have been built in the last few years, thanks to cooperative effort of several research institutions. Nowadays, they represent a valuable tool for testing new planet-scale services, like content distribution networks. Nonetheless, the testbed approach still presents some limitations. The main limits of experimental testbeds are (1) limited scalability, (2) limited software reconfigurability, and (3) difficulty to reproduce the heterogeneity of the real Internet, both in terms of terminal capabilities and networking technologies (e.g. PlanetLab hosts are quite homogenous in terms of hardware configuration, and they have more powerful CPUs and faster Internet connections than a typical end-user system). Thus, results obtained on an experimental testbed, even a large-scale one like PlanetLab, may not necessarily be representative of the behaviour of a real-world system. To overcome these limits, in the last few years several emulation systems have been developed to evaluate the effectiveness of new protocols and applications in more heterogeneous, controllable, and realistic network scenarios.

The basic idea behind network emulation is that simulated network elements should be capable of interacting with real network components and applications. A fundamental difference between simulation and emulation is that while the former runs in a virtual simulated time, the latter must run in real time to allow coexistence of both simulated and real network elements [7, 17]. To support network emulation, several approaches have been pursued.

For simple networking experiments (e.g. testing new multimedia applications under variable network conditions) the *dummynet* tool is a practical and simple solution [16]. Dummynet works by simulating the effects of finite queues, bandwidth limitations, and communication delays in a single workstation.

Another approach consists in extending a packet-based network simulator with some emulation facilities, so that it can be fed in real time with real traffic traces directly extracted from a real network. The well-known ns2 network simulator, for instance, has been extended to support emulation [4].

The new frontier of network emulation is cluster-based network emulation. These systems bring together a large number of network components (links, switches, PCs that are interchangeable as hosts, routers, or WAN emulators), in a common facility that can be remotely accessed by users through a web interface. They support flexible configuration, so that users may perform experiments on a variety of distinct network topologies with programmable delays that emulate the latencies of wide-area networks. Through virtualization and space sharing that fully isolates simultaneous experiments, they efficiently use cluster resources.

Maybe the most complex cluster-based emulation system developed so far is Emulab [17]. Emulab is a free-for-use, web-accessible, time- and space-shared, reconfigurable network testbed, providing integrated access to a wide range of experimental environments.

NEPTUNE is a cluster-based emulation system developed at University of Napoli. Even though many design assumptions made for NEPTUNE were borrowed by Emulab, since from the early stages of design, NEPTUNE has assumed virtualization as a key technology for realizing complex networking scenario. The goal of this chapter is to illustrate the importance of virtualization for network emulation and the role of GRID execution management mechanisms in NEPTUNE.

The rest of the chapter is organized as follows. In Section 2 we briefly describe different approaches for resource virtualization, and, with more details, the Xen para-virtualization system. In Section 3 we describe the general architecture of NEPTUNE. In Section 4 we describe the importance of virtualization for cluster-based network emulation systems. In particular, we describe two different kinds of resource multiplexing techniques: one for computational resources (that we call *node multiplexing*) and another for communication resources (that we call *link multiplexing*). In this section we also describe the virtualization techniques we adopted in NEPTUNE to solve the node multiplexing and the link multiplexing problems. In Section 5, we describe the use of grid virtual workspaces as a new instrument for dynamically deploying Xen virtual machines in the NEPTUNE network emulation system. Finally, we describe the current state of the implementation and present some concluding remarks.

2. RESOURCE VIRTUALIZATION TECHNIQUES

The main purpose of virtualization techniques is to hide the physical characteristics of computing resources from the upper layers of a computing system. One of the reasons for virtualization is making a single physical resource appear to function as multiple logical (or "virtual") resources.

There are many approaches to virtualization. Virtualization can be applied to single physical resources of a computing system (e.g. a single device) or to a complete computing system. When applied in this latter sense (a.k.a. *platform virtualization*), virtualization allows the coexistence of multiple "virtual machines" in the same computing "host".

Platform virtualization, is implemented by means of an additional software layer, called virtual machine monitor (VMM) (or *Hypervisor*), that acts as an intermediary between the system hardware resources and the operating system. The so-called *full virtualization* approach implements in software a full virtual replica of the emulated system's hardware, so that the operating system and user applications may run on the virtual hardware exactly as they would in the original system. An alternative, more recent, approach is called *para-virtualization* and consists in implementing a software interface that is similar but not identical to the underlying hardware. Such an approach requires the operating systems to be explicitly ported to run on top of the VMM.

Xen [1] is a para-virtualization system developed at the University of Cambridge. Xen provides a Virtual Machine Monitor for x86 processors that supports execution of multiple guest operating systems at the same time.

3. THE NEPTUNE EMULATION SYSTEM

NEPTUNE is a cluster-based network emulation system developed at University of Napoli [3] that can be used to assess new networking technologies and protocols (e.g. to test new QoS routing protocols and traffic engineering schemes in MPLS-based networks), as well as new distributed applications (e.g. multimedia peer-to-peer applications).

At the time of this writing, the NEPTUNE emulation system runs on a cluster of workstations consisting of 28 biprocessor nodes ProLiant DL380, each equipped with two Intel Pentium IV Xeon 2.8 GHz CPUs, 5 GB of PC-2100 RAM, one 100 Mbps Ethernet NIC, and one Gigabit Ethernet NIC. Each node is equipped with a 34.6 GB SCSI disk. A 700 GB centralized disk array is also available to the whole cluster. The cluster nodes are connected to each other through a set of 100/1000 Ethernet switches.

One of the cluster nodes, the NeptuneManager, provides the fundamental services (like dhcp, dns, tftp, nfs, and so on) needed to properly configure at boot time the physical cluster nodes and the virtual machines participating in the emulation experiments. A web-based system is used to manage and configure the whole system.

Setting up an emulation experiment in NEPTUNE consists primarily in defining a "virtual topology" made of emulated intermediate network nodes (routers) and end-system nodes (user terminals). A testbed mapping module (much like the one used in Emulab [14]) is responsible for mapping the "virtual" topology onto the cluster physical resources. Virtual network nodes are implemented in NEPTUNE as Xen virtual machines. The main advantage of the use of virtualization techniques to instantiate virtual network nodes is the significant reduction in equipment and management costs. Virtual machines allow the creation of customized execution environments, where customization consists in selecting the operating system, installed software packages, and user access policies. Furthermore, virtual machines can be paused or shut down at any time, and later resumed, even at a different physical location (migration). Finally, virtual machines support fine-grained mechanism for resource usage control, allowing to define (and even change at run-time) precise limits to the the amount of usable RAM and disk space.

In NEPTUNE, a complex networked system is reproduced by allocating multiple "virtual" network nodes (both routers and end systems) on each of the cluster physical nodes.

4. USE OF VIRTUALIZATION IN EMULATION SYSTEMS

In a typical cluster-based network emulation system, users submit to the system an experiment request. An experiment request contains a "virtual" network description to be reproduced with the available cluster resources. An experiment description usually contains at least the following information:

- The "virtual" network topology, including a list of relevant parameters for each of the nodes and links

- Operating systems and software packages to be loaded in each of the emulated nodes

When the experiment is activated (swap-in), a system module maps the logical topology of the experiment onto actual testbed hardware, loads the requested operating system and software onto the allocated devices, and establishes the network links.

Traditional network emulation systems use conservative resource allocation and map the emulated "virtual" nodes onto dedicated PCs and emulated links onto switched Ethernet links, thus severely limiting scalability. Nowadays, with increasing computational power made available at low cost, it is possible to exploit virtualization techniques to map multiple "virtual" nodes onto a single CPU. There are many good reasons for doing that. For example many applications need to be evaluated on large topologies, yet they are not resource hungry. Moreover, multiplexing provides a more efficient use of communication resources as the bandwidth of the emulated geographic links is usually much less than that available in the local interconnect used in a modern cluster [8]. These reasons motivate the use of small-scale clusters to emulate medium/large size topologies in an inexpensive manner [6].

4.1 Node Multiplexing

Node multiplexing is the problem of emulating more than a network node on the same physical cluster node. This problem is inherently a problem of machine virtualization, as has been described in Section 2. Hence, it can be solved with one of the many available virtualization techniques. Aspects to be taken into account to select the proper one for a cluster-based emulation system are efficiency, scalability, flexibility, isolation, and operating system customization.

In Emulab, node multiplexing is implemented by means of a modified version of FreeBSD Jail. In NEPTUNE we decided to implement node multiplexing by means of Xen. We decided that Xen suits our needs because

- It is highly scalable
- It potentially supports different kinds of operating systems
- It is transparent to the applications running in the virtual machines
- It provides good isolation among different virtual machines running concurrently
- It supports virtual machine migration, allowing dynamic re-allocation of experiments on the cluster nodes
- It implements different optimization techniques in the communication mechanisms, allowing good communication performance among virtual machines implemented within the same physical node

4.2 Link Multiplexing

The nodes of a cluster are connected by means of one or more switched Ethernet LANs. Each cluster node may be equipped with one or more (Giga or Fast) Ethernet NIC. These NICs, in turn, may be connected to the same switch or to different switches. In theory, it would be possible to connect the cluster nodes in pairs, by means of crossed Ethernet cables, so to physically reproduce the desired topology. However, such a solution is not viable, for at least two reasons. First, because changing the network topology would be extremely impractical, time consuming, and error prone. Second, because this would make it impossible to emulate network topologies with a number of links greater than half of the number of Ethernet NICs. Hence, practical solutions are required to emulate multiple point-to-point connections on top of one or more shared Ethernet LANs. This is usually performed by means of virtual LANs (VLANs) [8, 11]. Such a solution is implemented by properly configuring the Ethernet switches and does not require any configuration and processing in the cluster nodes. This makes, however, the system configuration software extremely dependent on the characteristics of the network switches.

For the above reasons, we decided not to use VLANs in NEPTUNE and we adopted two network device-independent solutions for link multiplexing:

- IP-aliasing and destination MAC address filtering
- Virtual NICs

The first technique consists in activating multiple "aliased" logical network interfaces for each of the node's NICs. Figure 1 shows a virtual network composed of five nodes connected by means of six "virtual links". Each of these links is associated to a point-to-point IP subnet. The IP-level network topology is implemented by setting static IP routes. To emulate links of assigned capacity, outgoing packets are handled by the CBQ scheduler implemented in the Linux Traffic Control kernel module. Outgoing packets are assigned to the corresponding queue by the *ebtables* kernel module that marks packets according to the MAC address of the destination node.

The second technique is implemented by exploiting the virtualization mechanisms implemented in Xen. Every time a new virtual machine (*domU* in Xen terminology) is instantiated, Xen creates new "connected virtual Ethernet interfaces", with one end of each pair within the domU and the other end within the Hypervisor (*dom0* in Xen terminology). Virtualized network interfaces are given different Ethernet MAC addresses. These addresses may be either assigned at random in the range of unicast "locally assigned" MAC addresses (i.e. addresses of the form XY:XX:XX:XX:XX:XX, where X is any

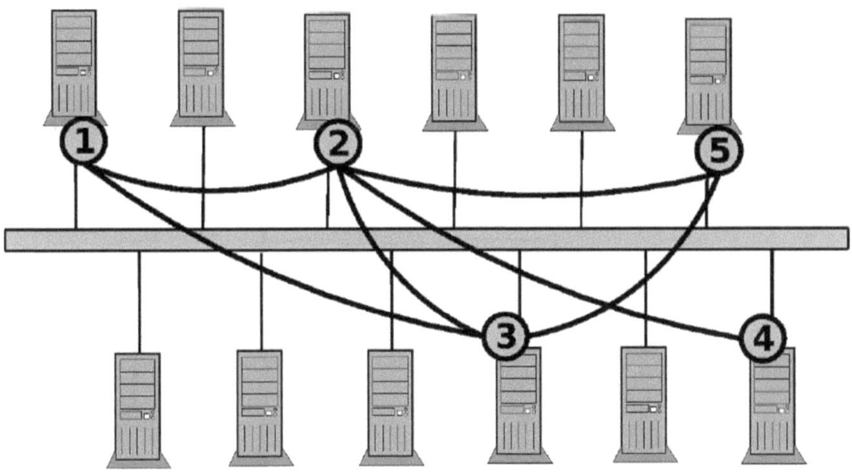

Figure 1. A virtual network with IP aliasing and destination MAC filtering

hexadecimal digit, and Y is one of 2, 6, A, or E) or they can be assigned a given value at domU creation time. When using such an approach, it is not necessary to share a single interface in order to emulate several links, since each virtual network interface is assigned to one of the two ends of an emulated link in the virtual topology.

Figure 2 shows a virtual network in which link multiplexing is implemented by means of this latter technique. The figure shows how a network of six nodes has been emulated by activating three virtual machines in Physical Host A and three virtual machines in Physical Host B. Six virtual links have been created, by properly instantiating one, two, or three virtual Ethernet interfaces in each of the virtual machines and configuring their IP addresses so to create six different point-to-point IP subnets.

In order to assign a given capacity value to each of the emulated links, a queuing discipline and a traffic shaper are associated with both ends of an emulated link. We refer to this technique as the *one link per interface* technique. Figure 3 shows with more details the network configuration of a host, in which two virtual machines (domU_1 and domU_2) have been setup and five different

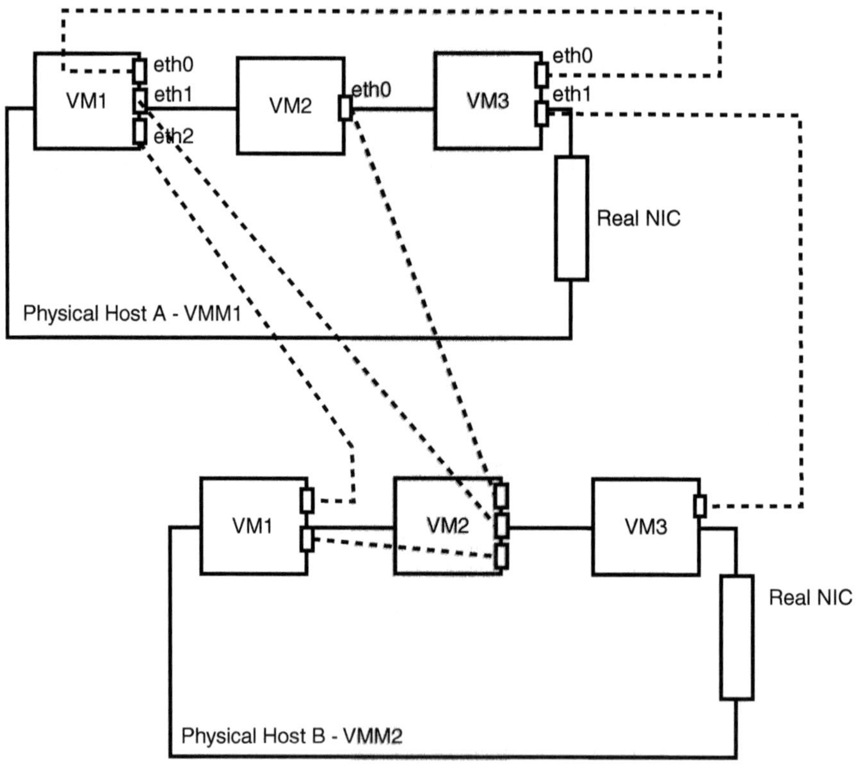

Figure 2. A virtual network implemented by means of virtual machines

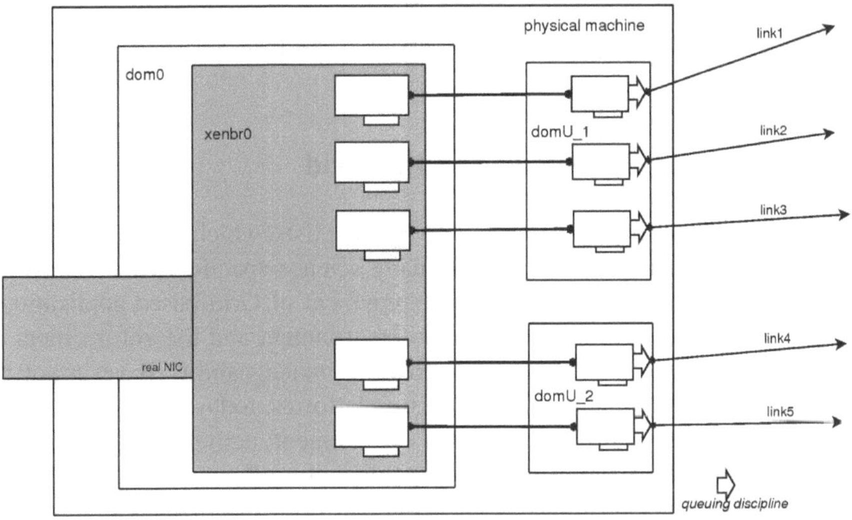

Figure 3. *One link per interface* technique

links have been emulated, three of them connected to node domU_1 and the remaining two of them to node domU_2.

5. NETWORK EMULATION AS AN ON-DEMAND GRID APPLICATION

Resource allocation is a key aspect of shared testbed infrastructures such as PlanetLab and Emulab [15]. While designing our system, we recognized that many resource management issues have already been addressed by the GRID community.

Traditional cluster-based network emulation systems have based their design on very specific and customized hardware, in order to reproduce the behaviour of real-life communications equipments.

Grid computing, on the other hand, is evolving on a different path, i.e. it aims at building a reliable and efficient infrastructure to satisfy different user requirements by means of shared powerful general-purpose computing facilities. In a Grid system, a middleware like Globus has the role of managing users requests and of customizing the computational resources available in a cluster of general purpose computers in order to match the users demands. Such an approach gives the system a higher degree of flexibility and manageability.

To emulate large-scale heterogeneous networking scenarios, the above-mentioned features of a Grid infrastructure are also desirable for a cluster-based network emulation system. In the following, we describe what strategies we are currently pursuing in order to bring some of the technologies developed for Globus-based Grids into the next version of NEPTUNE.

5.1 Virtual Workspaces in the Grid

Nowadays several production-quality Grids exists, which offer powerful execution infrastructures to help solve many science-specific problems. Much progress has been achieved with the deployment of Grid-based applications, but the preparation of remote execution environment and the enforcement of quality of service (QoS) while still providing manageability is yet a hot research topic. One problem with most Grid platforms, today, is the lack of performance isolation and resource usage enforcement: activities relative to one user or virtual organization [5] can unpredictably influence the performance of other processes executing on the same platform. Another issue is the impossibility to dynamically customize the execution environments according to the users requirements, thus compromising their *quality of life* (QoL) [9] in interactions with Grid software.

To satisfy such a requirement, the *virtual workspace* concept has been recently introduced in the Grid computing model. A virtual workspace (VW) is an abstraction of an execution environment that can be made dynamically available to authorized clients by using well-defined protocols [10]. The abstraction captures resource quota assigned to such execution environment on deployment (such as CPU, memory share, disk size, etc.) as well as software configuration aspects of the environment (such as operating system installation or provided services). This environment abstraction can be implemented by using various technologies. The Globus Toolkit supports at least two different techniques: one based on dynamically created user accounts (a.k.a. *dynamic accounts*) and one based on dynamic deployment of virtual machines. Virtual workspaces implemented as virtual machines offer a high degree of isolation, fine-grained enforcement and configuration: VM image configuration, in fact, can reflect a workspace's software requirements, while a hypervisor, such as Xen, can ensure the enforcement of hardware properties.

In Globus, the so-called workspace service allows a Grid client to dynamically create and manage workspaces. The workspace service implementation based on virtual machines takes as input a VM image wrapped in metadata providing critical deployment information and deploys the VM on one of the physical hosts administered by it.

An atomic workspace, implementing a single execution environment, contains both the data (VM image) and some metadata describing deployment information and prerequisites. Atomic workspaces can be combined to form aggregate workspaces such as virtual clusters.

5.2 NEPTUNE as a Virtual Workspace

We are currently working at a next release of NEPTUNE in which nodes participating in a network emulation experiment are part of a virtual cluster implemented according to the virtual workspace concept. This allows the cluster physical resources to be used concurrently with other Grid applications. In our very first implementation we are interested in demonstrating the feasibility of our proposal: a portable cluster-based network emulator easily deployable as a Grid application. Our goal is to not make use of any hardware-dependent solution, so that our system may be replicated on any cluster.

The NEPTUNE cluster emulator can be seen as composed of two workspace sets: the first set contains the head node (*NeptuneManager*), while the second set contains worker nodes. NeptuneManager, the head node of the virtual cluster, is instantiated as a virtual workspace by itself. The virtual network deployment is made by NeptuneManager.

The NEPTUNE worker nodes may be deployed on-demand by authorized users. Each worker node could have a specialized configuration, but for simplicity's sake we can assume in our experiments that all worker nodes share same virtual workspace image, which can be easily cloned.

The service node of the physical cluster runs a Xen 3.0 virtual machine monitor, Globus Toolkit 4 (GT4), and the workspace service. We are currently using the TP1.2.1 version (*technology preview*) of the virtual workspace service. The workspace infrastructure implemented in Globus Toolkit 4 uses the *web services resource framework* (WSRF) model and is composed of

- The *workspace factory service* that allows a Grid client to deploy a Xen-based workspace according to a deployment request specifying resource allocation and length of deployment
- The *workspace service* that allows a Grid client to manage a workspace by starting, stopping, pausing, or destroying it.

Figure 4 shows an overlay network created by means of virtual machines. After the head node (NeptuneManager) has been instantiated by the workspace service, we have performed the following steps:

- Definition of the network topology to emulate using the NeptuneManager services
- Configuration of the virtual node's features like RAM and virtual interfaces
- Definition of the software image to be loaded into the virtual machines
- Selection of the physical cluster's worker nodes (equipped by Xen 3)
- Deployment of the virtual machines

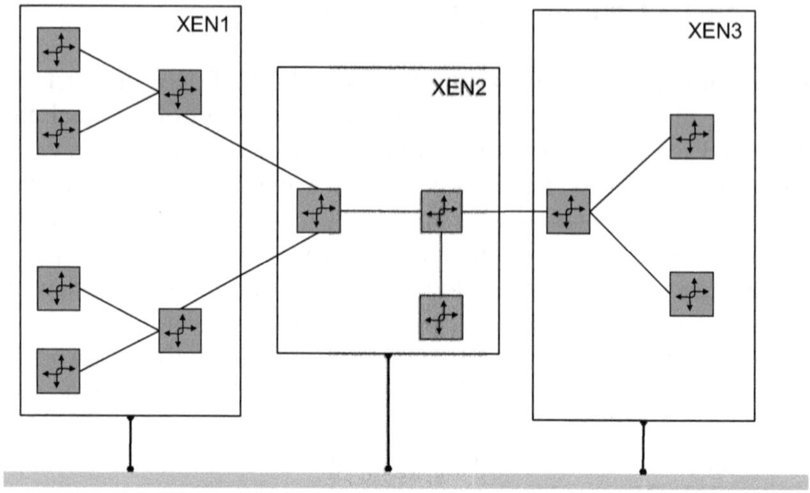

Figure 4. An emulated network implemented by means of virtual machines

As soon as the virtual nodes are available, NeptuneManager configures the overlay network and activates any services. The link multiplexing shown in Figure 4 was realized by the *one link per interface* technique described above. In order to test the implemented configuration, we enable OSPF routing by GNU/quagga on the emulated network; finally, we generate and trace several traffic flows between the virtual nodes.

6. CONCLUSIONS

Synthetic creation, configuration, deployment, and management of medium to large-scale virtual networks by means of a cluster-based system requires fine-grained control of resource utilization and manageability of the instantiated execution environments. These requirements have much in common with the challenges addressed in the context of Grid computing: a single virtual node participating in an emulation experiment, its computational resources, and its software configuration, can be seen like a Grid's virtual workspace. Recently, in the Grid community, the use of virtualization techniques to support on-demand instantiation of virtual workspaces has been proposed. In particular, para-virtualization techniques appear as the most promising, due to the high-degree isolation between the execution environments obtained with low performance overhead. In the next future, para-virtualization will also benefit from hardware support implemented by both Intel and AMD in their latest CPUs, allowing OSs to run natively (i.e. unmodified) and with no overhead penalties on these CPUs.

In this chapter, we have illustrated how Xen can be used to realize the virtualization of both computational and network resources, justifying the use of this instrument in the network emulation context. We have also discussed about how network emulation can be implemented as an on-demand application in a Grid computing environment. Finally, we have described the strategies we have pursued in order to bring some of the technologies developed for Globus-based Grids into the next version of NEPTUNE.

In the future, our main goal is to realize a further integration between our emulation environment (NEPTUNE) and Globus virtual workspaces.

ACKNOWLEDGMENTS

This work has been partially funded by the FP6 IST-2005-034819 project "OneLab: an open networking laboratory supporting communication network research across heterogeneous environments". Pasquale Di Gennaro has been funded from the Italian Ministry of University and Research (MIUR), in the form of a grant in the framework of the S.Co.P.E. Project.

REFERENCES

[1] P. Barham, B. Dragovic. K. Fraser, S. Hand, T. Harris, A. Ho, R. Neugebauer, I. Pratt and A. Warfield, "Xen and the art of virtualization", In *Procs of the 19th ACM Symposium on Operating Systems Principles (SOSP'03)*, pp. 164–177, 2003, ACM Press.

[2] L. Breslau, D. Estrin, K. Fall, S. Floyd, J. Heidemann, A. Helmy, P. Huang, S. McCanne, K. Varadhan, Y. Xu and H. Yu, "Advances in network simulation", *IEEE Computer*, vol. 33, no. 5, pp. 59–67, May 2000.

[3] COMICS research group at University of Napoli Federico II, "NEPTUNE network emulation project", http://neptune.comics.unina.it/

[4] K. Fall, "Network emulation in the VINT/NS simulator", In *Procs of the 4th IEEE Symposium on Computers and Communications (ISCC 1999)*, pp. 244–250, Jul. 1999.

[5] I. Foster, C. Kesselman and S. Tuecke, "The anatomy of the grid: enabling scalable virtual organizations", *International Journal of Supercomputer Applications*, vol. 15, Pages: 200–222, no. 3, 2001.

[6] S. Guruprasad, R. Ricci and J. Lepreau, "Integrated network experimentation using simulation and emulation", In *Procs of the 1st IEEE International Conference on Testbeds and Research Infrastructures for the DEvelopment of NeTworks and COMmunities (TridentCom 2005)*, Trento (Italy), Feb. 2005.

[7] D. Herrscher and K. Rothermel, "A dynamic network scenario emulation tool", In *Procs of the 11th International Conference on Computer Communications and Networks (ICCN 2002)*, Miami, FL (USA), pp. 262–267, Oct. 2002.

[8] M. Hibler, R. Ricci, L. Stoller, J. Duerig, S. Guruprasad, T. Stack, K. Webb and J. Lepreau, "Feedback-directed virtualization techniques for scalable network experimentation", University of Utah, *Flux Group Technical Note 2004-02*, May 2004.

[9] K. Keahey, I. Foster, T. Freeman and X. Zhang, "Virtual workspaces: achieving quality of service and quality of life in the grid", In *Scientific Programming Journal*, vol. 13, no. 4, pp. 265–276, Oct. 2005.

[10] K. Keahey, I. Foster, T. Freeman, X. Zhang and D. Galron, "Virtual workspaces in the grid", In *Procs of Euro-Par 2005, Springer Lecture Notes in Computer Science*, vol. 3648, pp. 421–431, Lisbon (Portugal), Aug./Sept. 2005.

[11] S. Maier, D. Herrscher and K. Rothermel, "On node virtualization for scalable network emulation", In *Procs of the 2005 International Symposium on Performance Evaluation of Computer and Telecommunication Systems (SPECTS 05)*, pp. 917–928, Philadelphia, PA (USA), July 2005.

[12] L. Peterson, D. Culler, T. Anderson and T. Roscoe, "A blueprint for introducing disruptive technology into the Internet", In *Procs of the 1st Workshop on Hot Topics in Networks (HotNets-I)*, Princeton, NJ (USA), Oct. 2002.

[13] R. Ramaswamy, N. Weng and T. Wolf, "Considering processing cost in network simulations", In *Procs of MoMeTools '03, ACM SIGCOMM Workshop on Models, Methods and Tools for Reproducible Network Research*, Karlsruhe (Germany), Aug. 2003, ACM Press.

[14] R. Ricci, C. Alfeld and J. Lepreau, "A solver for the network testbed mapping problem", *SIGCOMM Computer Communication Review*, vol. 33, no. 2, pp. 65–81, 2003, ACM Press.

[15] R. Ricci, D. Oppenheimer, J. Lepreau and A. Vahdat, "Lessons from resource allocators for large-scale multiuser testbeds", *SIGOPS Operating System Review*, vol. 40, no. 1, pp. 25–32, 2006, ACM Press.

[16] L. Rizzo, "Dummynet: a simple approach to the evaluation of network protocols", *SIGCOMM Computer Communication Review*, vol. 27, no. 1, pp. 31–41, 1997, ACM Press.

[17] B. White, J. Lepreau, L. Stoller, R. Ricci, S. Guruprasad, M. Newbold, M. Hibler, C. Barb and A. Joglekar, "An integrated experimental environment for distributed systems and networks", In *Procs of the 5th Symposium on Operating Systems Design and Implementation (OSDI'02)*, pp. 255–270, Boston, MA (USA), Dec. 2002, USENIX Association.

FOOD ENGINEERING INSTRUMENTATION ON LEMDIST WORKSPACE

J. de Jesús Cruz Guzmán[1,2], J.L. Garza Rivera[1], M. Hernández Duarte[1], and J.L. Arjona Roman[1]

[1] *Facultad de Estudios Superiores Cuautitlán, Universidad Nacional Autónoma de México, Cuautitlán Izcalli*

[2] *Dirección General de Servicios de Cómputo Académico Universidad Nacional Autónoma de México, México City*

Abstract Implementing applications with remote access to laboratory equipment on a grid environment requires a design model, focused to solve the interoperability challenge among an enormous diversity of components, endowed with the required functionality. Taking advantage of Grid infrastructure and experimental laboratory equipment, we drive our effort to develop a workspace, called LEMDist, based on a categorical approach for the design application. The principal aim for LEMDist is to build a technological platform to support different pedagogical approaches in natural science teaching and e-science applications. In this work we present a Grid implementation for remote access to laboratory equipment in food engineering that is operated via standard serial and USB computer interfaces, and we present some applications to a freezing process in a blast freezer. In particular, we focus the approach on the remote instrumentations requirements and the interaction with modeling resources and digital image processing capabilities for ice crystal growth.

1. INTRODUCTION

During the recent years many efforts have been made for developing remote access to the experimental infrastructure for science and teaching purposes. The need for quality of service (QoS) capabilities presents a big challenge, because many applications impose a diversity of real-time constraints, with different time scale requirements, together with another kind of quality requirements for instrument control [1]. Many different instrument architectures, experimental data types, security requirements, and user access policies impose strong complexity for a standard-based access design [4, 10, 16, 17].

The deployment of a workspace for e-learning and e-science with remote access to laboratory equipment capabilities for different application fields requires multi-scale analysis for complex systems. Taking into account instrumentation in many specialized areas, together with the highly specialized equipment characteristics, owner conditions of use, user communities requirements, and services, needs guidelines for the application design, and tools for the components' interoperability.

The categorical approach for formal description in open distributed processing systems is promising. In the design and refinement of hierarchical and multilevel systems whose components [20] are attribute dependent, graph analysis [6, 12, 14, 15] can be used.

Using the generalized graph theory's concepts of multi-graph and multi-graph of multi-graphs [3, 13], here we use the term peculiarity, associated with the external structure of graphs, for the definition of a service, and the term preciseness, associated with the internal structure of graphs of graphs, for the definition of an abstraction level in the architecture.

Based on graph theory, the structured systems can be described with a categorical approach, where the architecture definition for the workspace, endowed with a diversity of services, is based on generic interfaces that provide a generalized way for resources' incorporation.

1.1 LEMDist Architecture

The LEMDist [7] architecture is designed like a component integration system in four abstraction levels: Access layer, service layer, grid layer and resource layer, considered like preciseness concepts, and, at the moment, five kinds of service: research repository for the experimental work, distributed enhanced instrumentation, distributed data analysis, computational modeling and simulation, and interactive visualization, considered like the peculiarities [8], all of those integrated in an e-science workspace (Figure 1). The planned characteristics for each service are

- Repository for research project and experimental work. It provides
  - Access to global collaboration communities
  - Scientific data repositories
  - Scientific experiment repositories
  - Workspace
- Distributed enhanced instrumentation needs
  - Discovering laboratory resource service
  - Schedule service

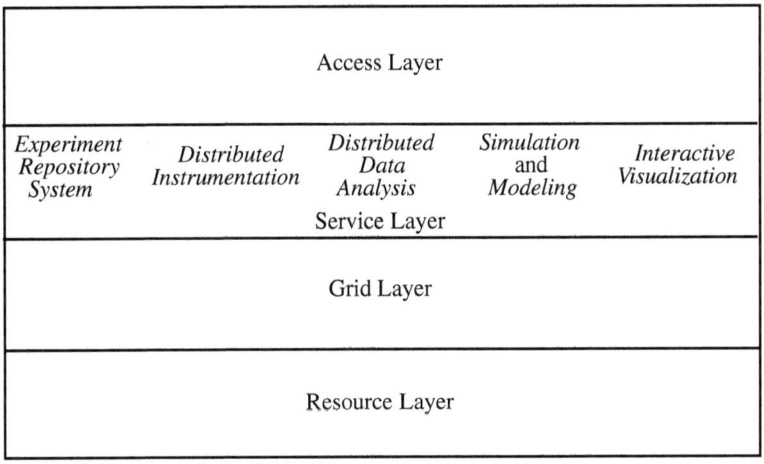

Figure 1. LEMDist architecture

  - Interaction environment with instrument operator
  - Quality of service for control and monitoring in most of the laboratory equipment

- Simulation and modeling. The main requirements for this service depend on access to software resources for simulation and modeling that are
  - CPU intensive
  - Communication intensive (MPI or other)

- Interactive visualization. It is
  - CPU intensive
  - Communication intensive (MPI or other)
  - Needs broadband communication for virtual reality capabilities

2. REMOTE ACCESS TO LABORATORY EQUIPMENT USING GLITE MIDDLEWARE

In the architecture design remote access to laboratory equipment plays a fundamental role. Based on the GLite middleware architecture, we build an instrument element like a Grid node. The generic Grid services for planning and management, discovering resources, job submission, and data transfer are used (Figure 2).

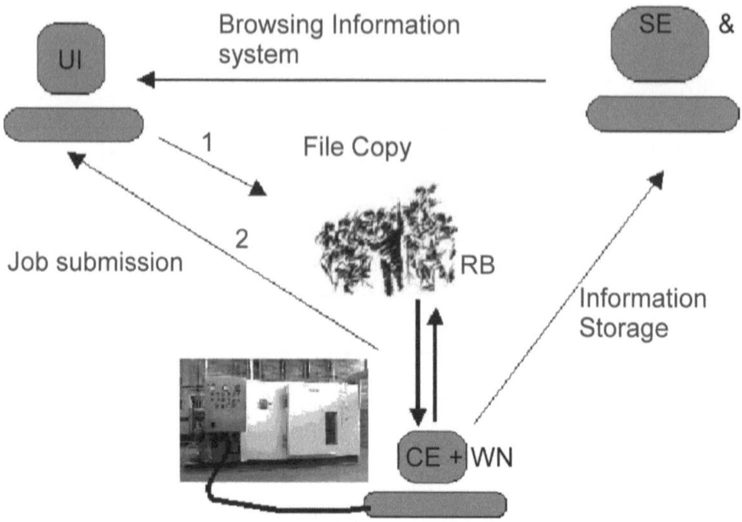

Figure 2. Remote instrument access over the Grid

Using the grid-mapfile mechanism, the user requests an instrument service and that request is mapped, depending on the user-specific privileges to the host which has the specific resource. The job description language (JDL) is used to describe the job and its dependence.

For laboratory instrument access, we registered the application interface (AI) in the computing element (CE), where a dedicated worker node (WN), like a Grid instrument element, was attached. The AI is installed in the WN. From the user interface (UI) an interactive job is submitted using the "Requirements" parameter looking for the AI. The interactive job enables a kind of shell where the user (in the UI) can interact with the device, just using the commands defined in the application interface (Figure 3).

2.1 Application to Freezing Process in Food Engineering

In food engineering teaching, research, and production, it is important to get access to a diversity of equipments. For example, in production environments it is not possible to run experiments: the process equipment is not appropriate for testing, and installing a test laboratory can be too expensive. Access to remote laboratory infrastructure with scaled industrial conditions results in an interesting option.

To foretell freezing times in food engineering is important for many practical applications; for example, designing freezing equipment and production plans, impact estimation when quality of products or process parameters have modifications, etc.

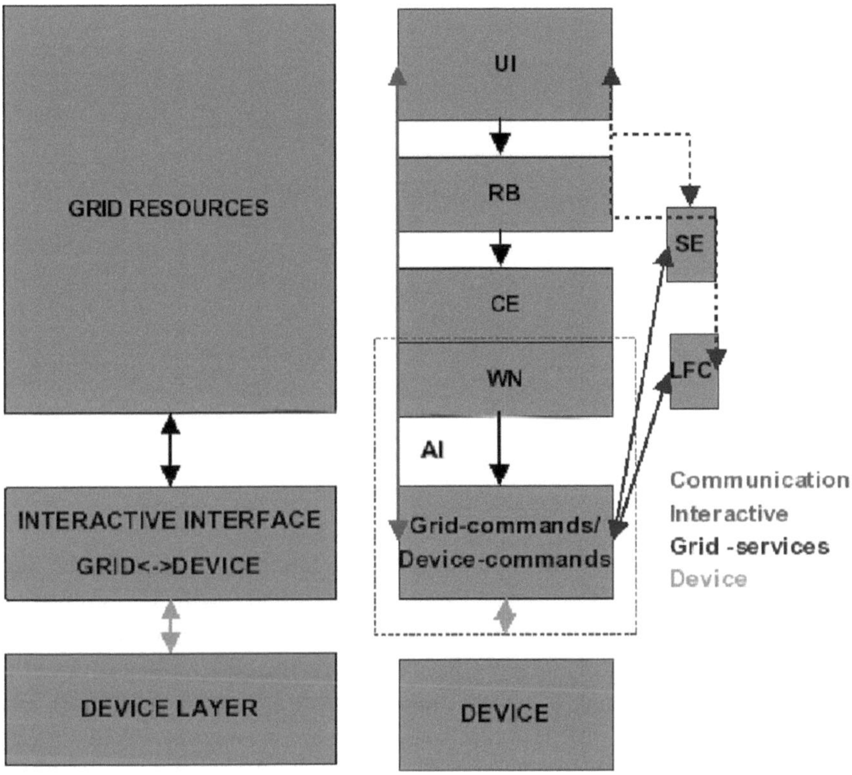

Figure 3. Remote instrument access architecture

Unsteady heat conduction, transition phase, parameter variation with temperature change, geometry impinge on the freezing process. The modeling technique requires experimental verification, and a good understanding of the behavior of temperature time profiles is necessary. For long-time meat preservation, for example, it is necessary to get freezing at −18°C; this temperature is required to inhibit microbial population increase and enzymatic activity. The kinetics for temperature diminution and water crystallization varies from point to point in a product. Establishment of a global freezing rate is complex and there are many criteria in the study of the impact over the product quality. The experimental results here are fundamental.

In Figure 4, we present the temperature behavior related to time for freezing a meat sample in a blast freezer system at −25°C temperature and 4.5 m/s air speed. The purpose is to study the effects of freezing time on the mechanical properties like resistance to compression, where a range of 500–1000 s for ice formation in this sample is observed, beginning in the region −1.5 to −5°C.

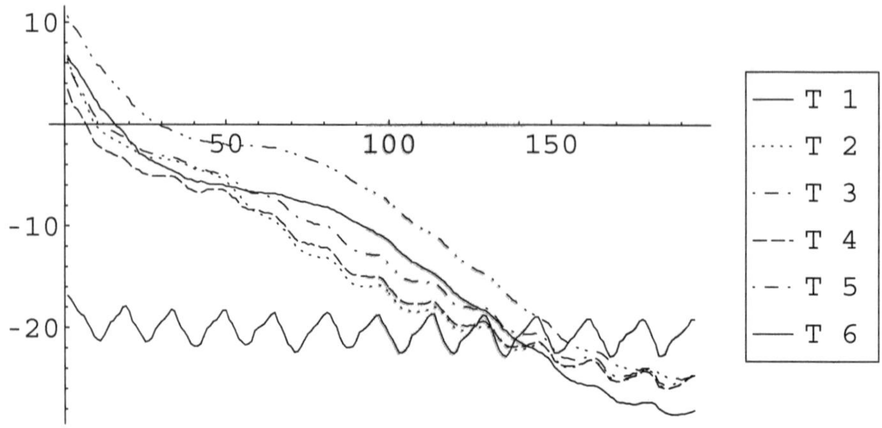

Figure 4. Time temperature profiles in a blast freezer

The implementation of different observation technologies requires the identification and incorporation of all the necessary pieces with the appropriate communication resources between components and the adaptation all of those for grid environment work. The prediction of freezing and cooling times in order to ensure food quality plays an important role in food engineering. Heat and mass transfer phenomena in the freezing process take place as a complex process with transition phase, sudden changes in thermal properties, and other physical–chemical characteristics, where crystallization of water plays a fundamental role. Models available for describing the freezing process are based on non-linear unsteady heat conduction, with variable thermal properties, surface convective cooling, and mass transfer. They are useful for development of simple prediction methods, but in any case verification of thermal data sets is necessary, because of the complicated structures in the food sample and the diversity of physical mechanisms taking place in the freezing process.

2.2 Interoperability Requirements

The components design for freezing process in food engineering is based on the LEMDist architecture, where the required interoperability is found in a composition process between multi-graphs of multi-graphs.

The LEMDist architecture is defined with four abstraction levels and five kinds of service: A collaborative environment connects users to instruments; users can access scientific databases, high-performance computing resources, and high-end graphic capabilities (Figure 5). Interoperability had to be built for each application field. The grid infrastructure provides the integration of services and resources, like remote access to laboratory equipment, low-level

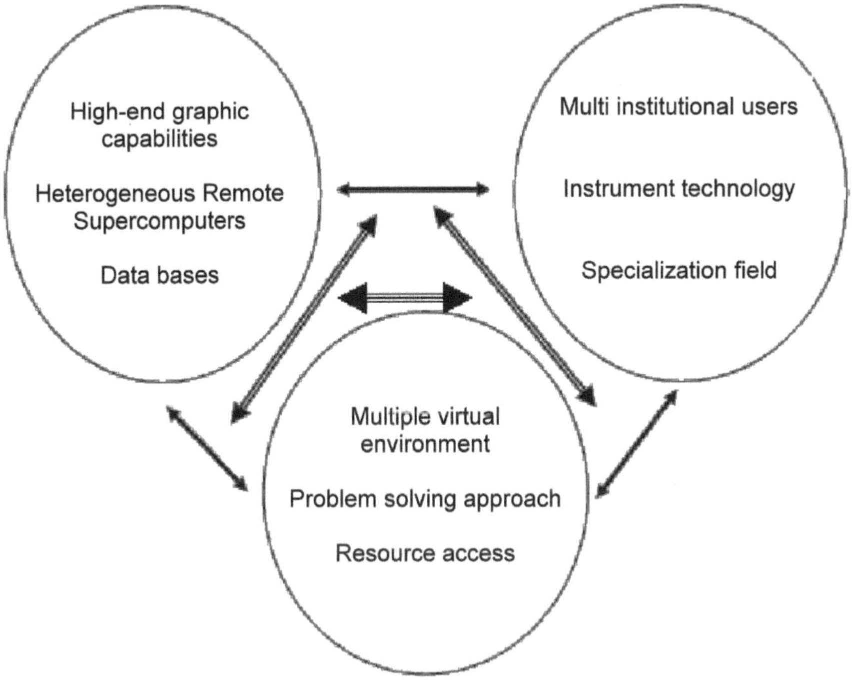

Figure 5. Food engineering architecture

mechanisms for communication authentication, network information, and data access; high-level mechanisms for meta-computing services like parallel programming tools and schedulers are also provided. We use the GLite middleware and the LEMDist is implemented on the EELA virtual organization. For food engineering we have implemented in the grid environment a blast freezer tunnel equipped with a programmable logic controller (PLC) for regulating the belt speed of the tunnel freezer and scan of the temperature in the sample. Remote access has been implemented for the data acquisition process, based on an architecture of reliable scan programming component interfaces between the grid layer and the resource layer. The thermocouples are placed in the thermal center and other locations of the sample, depending on the experimental design and the model's geometrical characteristics.

For the prediction of freezing times we are implementing a heat transfer model, starting with a simple case based on the Stefan problem [9, 18, 19]:

$$\varrho c \frac{\partial T}{\partial t} = \nabla(\kappa \nabla T) + q \tag{1}$$

Discretizing the space domain we obtain a set of ordinary differential equations, and then we use a finite difference method, considering shapes with a nearly regular geometry. The temperature reduction process is divided into three phases: pre-cooling, phase change, and tempering phase. The phase change becomes the difficult point for the heat transfer process. For this we consider first the application of an empirical expression for the region of phase change of the most important physical properties, like thermal diffusivity, heat capacity, density, heat conductivity, enthalpy, and, finally, freezable water fraction (Figure 6). Another interesting alternative is to include in the model the data for heat capacity and enthalpy issued from modulated differential scanning calorimetry. These present an interesting challenge, because it is necessary to focus on the problems for on-line monitoring process generated by Grid process latencies and on the limitations in the definition of differentiate net services for the interconnection among user, instrument and equipment operator in a remote supervision scenario.

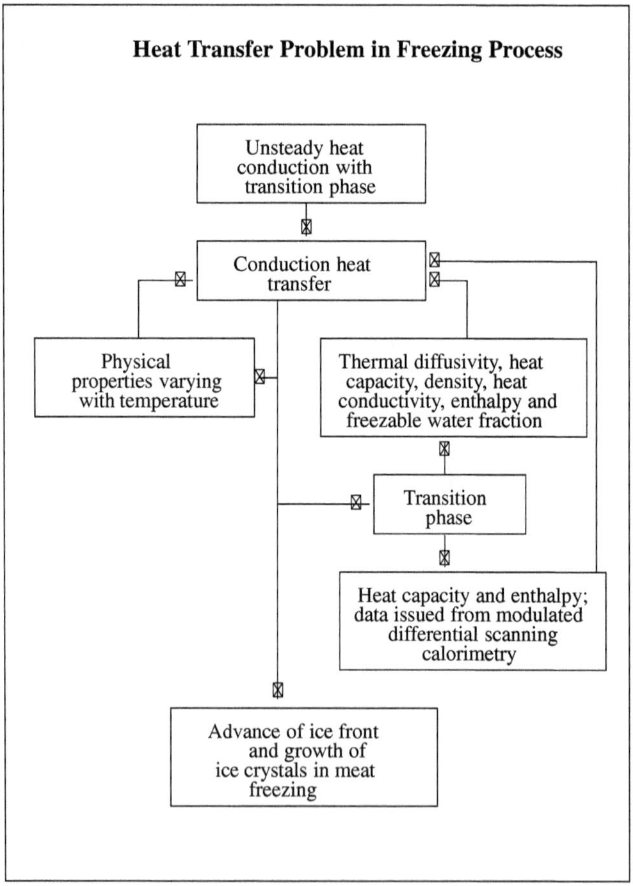

Figure 6. Freezing time modeling

Another component of the systems is based on digital image processing of electronic microscopy of samples, directed to the study of the size of ice crystals' formation that induce morphological changes, responsible for moisture migration [5, 11, 21]. Ice crystals in frozen materials are difficult to observe and the study of the freeze zone is highly imprecise. The measurement techniques used to model the very thin layer near the surface of the sample are adopted to describe the moisture movement on the evaporating surface, which is the cause of change in boundary conditions. We consider texture analysis and ice crystal formation.

3. CONCLUSIONS

LEMDist is a dynamic project that has the aim of building a working environment for e-science and e-learning with access to remote laboratory equipment. So far, we have access to remote laboratory equipment through the Grid layer middleware GLite, when laboratory equipment has a standard serial interface. The implemented commands at the interface level hide the Grid and control device commands, avoid damages, and provide different execution command levels. The main advantage for this design is that we can ever work on the GLite sandbox and it is not necessary to delegate part of the process to another kind of environment.

For the Freezing Tunnel, we need interaction with human operators for initial instrument configuration, sample mounting, and, successively, for starting, stopping, and monitoring processes that are characterized by interaction at different process times, ranging over:

- Real time scales: Interaction with the operator and monitoring process.
- Intermediate delay: Interaction between different instruments or between instruments and simulation process.
- Long delay: Parameter determinations in the instrument initial configuration process.

Not all the process types have been implemented. The main problems presented for the online monitoring process are Grid process latencies and the limitations in the definition of differentiate net service for the interconnection among user, instrument, and equipment operator in a remote supervision scenario.

The remote access to laboratory equipment requires more flexibility on network service and its corresponding middleware implementation for a good performance on real operational conditions.

REFERENCES

[1] Baxevanidis K., Davies H., Foster I., Gagliardi F., "Grids and research networks as drivers and enablers of future Internet architectures", *Computer Networks*, vol. 40, 2002, pp. 5–17.

[2] Burke S., Campana S., Delgado P. A., Donno F., Méndez L. P., Santinelli R., Sciabà A., "GLite 3 user guide", *Manual Series, Worldwide LHC Computing Grid*, CERN-LCG-GDEIS-722398, 2007.

[3] Burroni A., "Algébres graphiques", *Cahiers de topologie et géométrie différentielle*, XXII, no. 3, 1981, pp. 249–265.

[4] Casaca A., Sarma A., "Towards a new Internet architecture", *Guest editorial, Computer Networks*, vol. 40, no. 1, 2002, pp. 1–4.

[5] Chandraratne M.R., Kulasiri D., Frampton C., Samarasinghe S., Bickerstaffe R., "Prediction of lamb carcass grades using features extracted from lamb chop images", *Journal of Food Engineering*, vol. 40, no.1, pp.1–4, 2002.

[6] Cruz G. J. J., Oziewicz Z., "Networks of Networks", *International Conference on Computational Science ICCS'2004*, LNCS 3036, Springer-Verlag.

[7] Cruz G. J., Díaz G. L., Enríquez L., Garza R. J. L., Hernández D. M., Pineda M. C., Álvarez P. M., Ángeles A. E., Arjona R. J. L., Flores M. J., García M., Navarrete B., "LEMDist: An e-learning and e-science workspace", *1st EELA Grid School*, Itacuruca, Brazil, 2–4 Dec. 2006.

[8] Cruz G. J. J., Lewandowicz E., Oziewicz Z., "Multiscale geographic information with multigraph of multigraphs", *The International Archives of the Photogrammetry, Remote Sensing and Spatial Information Sciences, The 12th International Symposium on Spatial Data Handling*, July 10–12, 2006, University of Vienna, Austria.

[9] Delgado A. E., Sun D.-W., "Heat and mass transfer models for predicting freezing processes – a review", *Journal of Food Engineering*, vol. 47, no.3, 2001, pp. 157–174.

[10] Devadithy T., Chie K, Huffman K, McMullen D. F., "The common instrument middleware architecture: overview of goals and implementation", *Proceedings of the First International Conference on e-Science and Grid Computing (e-Science'05)*, Computer Society, IEEE.

[11] Do G.-S., Sagara Y., Tabata M., Kudoh K.-I., Higuchi T., "Three-dimensional measurement of ice crystals in frozen beef with a micro-slicer image processing system", *International Journal of Refrigeration*, vol. 27, no.3, 2004, pp. 302–308.

[12] Korczynski W., Cruz G. J. J, Oziewicz Z., "Multilevel systems as multi-graph", *International Conference on Computational Science, ICCS'2003*, LNCS 2658, Springer-Verlag.

[13] Mac Lane S., "Categories for the working mathematician", volume 5 of *Graduate Texts in Mathematics*, Springer-Verlag, 2nd ed., 1998.

[14] Obtulowicz A., "On n-dimensional graphs and n-graphs", *Miscellanea Algebraicae (Kielce)*, Rok 5, no. 1, 2001, pp. 87–96, ISBN 83-87798-22-3.

[15] Obtulowicz A., "A note on the connections of automata and weak double categories", *Miscellanea Algebraicae (Kielce)*, Rok 5, no. 1, 2001, pp. 75–86, ISBN 83-87798-22-3.

[16] Peterson L., Anderson T., Culler D., Roscoe T., "A blueprint for introducing disruptive technology into the Internet", *First ACM Workshop on Hot Topics in Networks (HotNets-I)*, Oct. 2002.

[17] Robinson J. M., Frey J. G., Stanford-Clark A. J., Reynolds A. D., Bedi B. V., "Sensor networks and grid middleware for laboratory monitoring", *Proceedings of the First International Conference on e-Science and Grid Computing (e-Science'05)*, Computer Society, IEEE.

[18] Rubiolo de Reinick A. C., "Average and center temperature vs time evaluation for freezing and thawing rectangular foods", *Journal of Food Engineering*, vol. 30, no.3, 1996, p. 299–311.

[19] Tuan Pham Q., "Modelling heat and mass transfer in frozen foods: a review", *International Journal of Refrigeration*, vol. 29, no.6, 2006, pp.878-888.

[20] Turner K. J., "Relating architecture and specification", *Computer Networks and ISDN Systems*, vol. 29, no. 4, 1997, pp. 437–456.

[21] Zheng C., Sun D.-W., Zheng L., "Recent applications of image texture for evaluation of food qualities – a review", *Trends in Food Science & Technology*, vol. 17, no.3, 2006, pp. 113–128.

GRID SERVICES FOR 3D DATA ANALYSIS IN VIRTUAL LABORATORIES

A. Clematis[1], A. Corana[2], D. D'Agostino[1], V. Gianuzzi[3], and A. Merlo[2]

[1]*IMATI-CNR, Via De Marini 6, 16149 Genova, Italy,* [2]*IEIIT-CNR, Via De Marini 6, 16149 Genova, Italy,* [3]*DISI-Università di Genova, Via Dodecaneso 35, 16146 Genova, Italy*

Abstract We present a Grid-aware application for the extraction of isosurfaces from volumetric data sets. This is an important issue in the design of a virtual laboratory since it permits to analyze data produced by different types of scientific instruments. After a short description of the basic functionalities we focus on the issues for the efficient use of Grid resources. Our implementation is designed following a Grid service approach and adopts Globus Toolkit 4 as middleware. The main advantages of the use of the Grid are the possibility of processing huge data sets and of satisfying more requests simultaneously.

1. INTRODUCTION

The technological evolution permits the analysis of physical phenomena with an unprecedented accuracy, by means of distributed and remote instruments connected to Internet. To be effectively useful, the huge amount of data produced has to be thoroughly analyzed and understood.

Volumetric data sets are 3D matrices of values produced by different types of devices, e.g., medical instruments as X-ray equipments for computed tomography, or by simulations, e.g., by fluid dynamics or bioinformatics applications. The *isosurface extraction* [3, 17] is a basic operation that permits to implement many kinds of queries on these data. In particular, it permits to determine the shape of tissues or molecules, on the basis of a given threshold value (called *isovalue*).

Isosurfaces are usually represented as triangulated irregular networks and constitute useful visual information that has to be transmitted to remote devices.

3D data analysis, based on isosurface extraction and transmission, represents a key functionality in a virtual laboratory. It can be, however, an exacting task, since many problems, arising from the characteristics of the data and the execution environments, have to be taken into account.

The most important are the following:

- The heterogeneity of the client devices used to analyze the results, since the size of computed isosurfaces may overcome their graphic capabilities. This problem can be tackled by providing a *simplification operation* [9, 10], that produces a good approximation of the original surface using fewer triangles.
- The performance of the interconnection network, since the transfer of the isosurface may be the most costly part of the processing. At this aim a *compression operation* [10, 20] that reduces the size of the data to transfer can be adopted.
- The possible huge amount of data to process, since users would like to receive the result as fast as possible. The design of *parallel algorithms* for the considered operations would improve performance. Clearly, the achieved performance will depend on the characteristics of the available computing resources, but the parallel algorithms may be designed with portable performance.

In this context we implemented a set of programs for distributed 3D data analysis and visualization. The tool includes algorithmic components for the extraction of isosurfaces, their simplification and compression. For each component we implemented at least one sequential and one parallel algorithm. The tool also comprises an application that permits (i) the efficient composition of sequential and parallel operations in a pipeline assembly, taking into account the user request and the execution environment; (ii) the dynamic selection of computational resources, such as single machines or clusters, for the pipeline execution; (iii) the transfer of the resulting mesh to a remote user device.

The algorithmic part of the system is widely presented in [7], while a synthesis is provided in [6], therefore interested readers may refer to them.

So far we focused on the algorithmic issues of our tool, considering dedicated and local computational resources (single machines and local clusters). In this chapter we describe **3DAGOS (3D Data Analysis through Grid-Oriented Services)**, a new Grid-aware version for Globus Toolkit 4 (GT4) [8] that is able to exploit remote and shared resources and can represent a useful tool for distributed cooperative laboratories over the Grid. 3DAGOS includes **GrISO (Grid Isosurface extraction and management Suite of**

cOmponents), a collection of Grid Services corresponding to the above-cited components for the isosurface management and **PLIC_GS (PipeLine Intelligent Composer Grid Service)**. The client, after the authentication, interacts, through a graphical interface, with PLIC_GS that selects a suitable instance of pipeline and a set of Grid resources to minimize the cost function defined by the client.

In order to access and select the resources of the Grid, PLIC_GS needs a suitable support to simplify this essential but complex job. To this aim we make use of **GEDA (Grid Explorer for Distributed Applications)** [5], a Grid Service designed to provide a simplified but non-transparent access to Grid resources. GEDA is a general-purpose resource broker service that can be used to allow a fine-grain resource selection for any kind of grid service.

A visualization tool such as 3DAGOS finds a natural collocation inside projects such as GRIDCC [25] and RINGrid [19], the main goal of which is to provide tools to access to and control remote instrumentations using the distributed computational and storage resources of a Grid, even interactively. Through 3DAGOS, in fact, it is possible to process and visualize the results of experiments selecting the most performant configuration.

The chapter is organized as follows: the second and the third sections present some hints related to the algorithmic components of GrISO, with performance figures for different data sets, and the main features of the PLIC_GS component. Considerations about the design of the related Grid Services are presented in Section 4. In Section 5 the Grid Service GEDA is introduced, while in Section 6 some related works are discussed.

2. ALGORITHMIC COMPONENTS OF GRISO

2.1 The Pipeline Building Blocks and Configurations

GrISO is structured as a set of Grid Services, representing the basic building blocks through which the isoextraction service is provided. They can be linked to one another in a pipeline, composed of different steps, as shown in Figure 1, that presents all the meaningful configurations.

GrISO includes four *computational components* corresponding to the three operations of isosurface extraction, simplification and compression (that comprises both the compression and the decompression components) and three *flow components*. The pipeline composition is performed by PLIC_GS, on the basis of application- and architecture-related parameters.

We introduced the flow components in order to make efficient the composition of the parallel algorithms. In fact, while the composition of sequential algorithms is straightforward, the composition of parallel algorithms involves

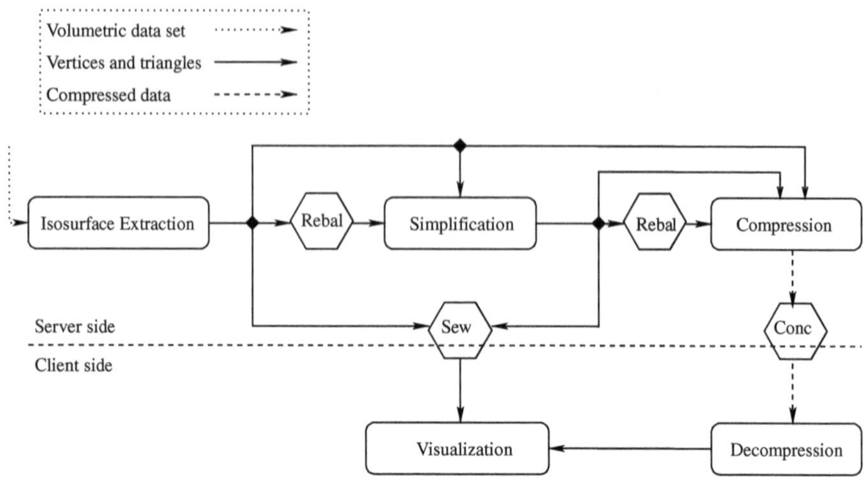

Figure 1. The logical schema of the pipelines that may be built using GrISO. *Rectangles* represent computational components, while *hexagons* represent flow components

to consider the possible different parallelization paradigms, the parallelism degree and load balancing strategies.

The *Rebal* component modifies the data partitioning produced by a component on the basis of the parallelism degree and load balancing requirements of the following one. The *Sew* component merges the parts in which an isosurface is subdivided to produce a single file that may be visualized or further processed. The *Conc* component is similar to the previous one, but it merges the encoding of the parts in which an isosurface is subdivided after compression to produce a single file for the decompression component.

We can consider these components as filters that transform raw volumetric data into triangular meshes representing the required isosurfaces. The data flow among the components is presented in Figure 1. The Sew and the Conc flow components may be executed both on the server and on the client side, while the decompression module must be executed on the client device.

The system is able to acquire volumetric data sets stored in files organized following one out of different possible formats, included the *Digital Imaging and Communications in Medicine* (DICOM) format [23], that represents the standard for medical image storing, handling and transmission. We provide a set of specialized functions, one for each format, that can be easily extended to process other standard formats.

Although we decided to leave rendering aspects outside of our scope, we provided a set of specialized functions also for the output production, for example we consider the production of data in the OFF format [26]. Visualization of isosurfaces is not provided directly by the system, but we exploit the various suitable rendering tools available for most architectures. In particular, we considered the use of the VRML [27] format, as VRML viewers are freely available and widely adopted. So the "Visualization" component in Figure 1 represents a generic VRML viewer.

2.2 The Pipeline Performance

Let us briefly present the performance of the computational components, in order to show how our system may represent an effective analysis tool for the processing of large amount of data.

We considered for the experiments the computerized tomography (CT) scans of a christmas tree (XT) [1] and of a female cadaver (FC).[2] Furthermore we considered a data set resulting from a simulation that represents a fragment of immunoglobulin attacking a rhinovirus. The resulting macromolecule is called the FAB complex. The characteristics of the three data sets are shown in Table 1. For each data set an isosurface (respectively related to the isovalues 180, 0 and 1230) is shown in Figure 2.

Performance evaluation is carried out using a Linux-based Beowulf Cluster of 16 PCs equipped with 2.66 GHz Pentium IV processor, 512 MB of RAM and two EIDE 80 GB disks interfaced in RAID 0. The nodes are linked using a dedicated switched Gigabit network, and they share a PVFS file system [16]. We implemented the sequential components in C and the parallel ones in C and MPICH2 [11]. The performance figures of both versions are presented in Table 2.

For the simplification operation we considered the reduction of 90% of the triangles. The parallel algorithms are executed using 8 and 16 nodes. With up to 8 nodes it is not possible to simplify nor compress the FC-1230 isosurface because of the insufficient amount of aggregate memory. The speed-up reported for the simplification and the compression of the isosurface FC-1230 has been

Table 1. Characteristics of the data sets used in the experiments

Data set	*Volume*	*Size*	*Isovalue*	*#Triangles*
XT	$512 \times 512 \times 999$	499.5 MB	180	4,514,539
FAB	$773 \times 721 \times 746$	793 MB	0	8,799,673
FC	$512 \times 512 \times 1734$	867 MB	1230	12,778,122

[1]Generated from a real-world christmas tree by the Department of Radiology, University of Vienna and the Institute of Computer Graphics and Algorithms, Vienna University of Technology.
[2]Courtesy of the Visible Human Project [28] of the National Library of Medicine (NLM), Maryland.

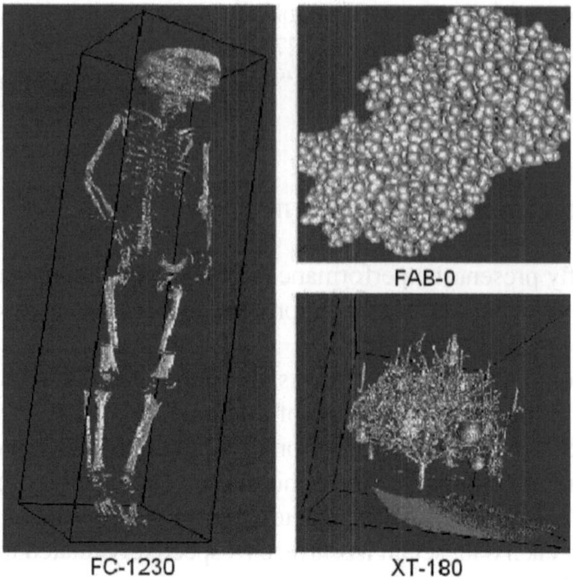

Figure 2. The three isosurfaces considered for the experiments

Table 2. Execution times and speed-up for the considered isosurfaces (* denotes estimated speed up)

	XT-180	*FAB-0*	*FC-1230*
Time sequential iso (sec.)	23.1	34.7	42.6
Speed-up parallel iso (8 proc.)	6.8	7.1	7.5
Speed-up parallel iso (16 proc.)	11.7	12.1	12.2
Time sequential simpl (sec.)	166.3	336.7	N.A.
Speed-up parallel simpl (8 proc.)	6.4	6.7	N.A.
Speed-up parallel simpl (16 proc.)	11.2	12.3	13.1*
Time sequential compr (sec.)	10.8	19.3	N.A.
Speed-up parallel compr (8 proc.)	7.9	7.7	N.A.
Speed-up parallel compr (16 proc.)	15.4	15.3	14.8*

theoretically evaluated, using a validated cost model [7]. We can see that the parallel components are able to provide high performance figures.

For the isosurface extraction and the simplification components we reported the computational times, that is, the execution times without I/O operations. These components in fact may be executed in a pipeline, so they may receive and leave I/O data in main memory. If we consider I/O times also performance is about 20% lower. For the compression component, on the contrary, the output time is taken into account, since the compressed representation of the isosurface is normally transferred as a file.

3. THE PLIC_GS BEHAVIOUR AND ITS USER INTERFACE

The original contribution represented by our tool lies also in the strategies we designed to provide satisfactory response time to the user.

The goals considered when designing a server are usually two: (i) to satisfy at the best each user request and (ii) to maximize the throughput of the system. Using a limited set of resources, as our local cluster, it is possible to achieve satisfactory results only for the first goal. Actually, if many users simultaneously ask for processing large data sets, the computational power may quickly run out, and the waiting time will become too long.

So far we concentrated our effort on designing suitable strategies to satisfy the first goal, while the integration of our tool in the Grid environment allows us to pursue the second goal.

Adaptivity is a key point for providing an efficient and satisfactory service. In fact, the selection of the components to activate, their efficient composition and the parallelism degree have to take into account the parameters that characterize both the execution environment and the user request.

As regards the execution environment, let us consider a user requiring to extract the isosurface denoted as FC-1230. The visualization of such triangular mesh is not possible on a domestic PC, owing to the high number of triangles. This problem can be solved by activating the simplification module, in order to produce an isosurface made up of less triangles.

A similar consideration has to be made for the transmission time of the isosurface. The optimized non-compressed representation of this isosurface has a size of about 220 MB, so the download time will be huge in the presence of a low bandwidth link. In this case the system should activate the compression module that is able to reduce the size of the isosurface to about 20 MB.

As regards the request parameters, when a user requires a service, he/she normally wishes that

- the request is satisfied in the minimum time;
- the quality of results is the highest one.

It is often very difficult to conciliate both these aspects. The parallel version of an algorithm may lower the quality of results with respect to the sequential version. This is true for our parallel simplification operation owing to the management of edge points in the mesh. A lower parallelism degree better preserves the quality of the isosurface, at the cost of lower performance, and vice versa.

Furthermore the execution of the compression operation must be carefully considered, since it requires the execution of the decompression on the client device. If the device, e.g., a personal digital assistant (PDA), cannot provide a sufficient computational power, the compression cannot be performed.

On the basis of the previous considerations the user has to choose between the minimum response time and a higher quality of the mesh.

PLIC_GS satisfies the user request composing a pipeline and selecting the resources that better fit the requirements. Remote accessibility is provided by means of a GUI interface (see Figure 3).

Through the interface the user can or cannot require the execution of the simplification and compression operations or he can delegate this decision to PLIC_GS. In the second case the user has to provide a threshold on the resulting number of triangles, which represents the maximum allowable size of an

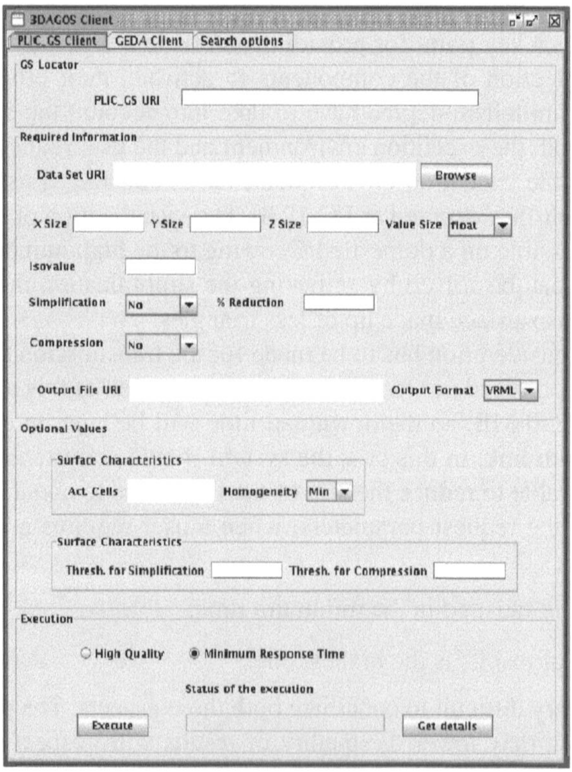

Figure 3. The first tab of the 3DAGOS Java client related to PLIC_GS. It permits users to specify the parameters of the isosurface extraction operation and the optional execution of the simplification and/or compression operation

isosurface to be visualized and/or transmitted without long waiting time. For the visualization this size depends on the characteristics of the used device.

After the posting of the request, PLIC_GS determines the pipeline composition and the parallelism degree. A complete description of these aspects is presented in [7]. Here we briefly describe the decision process through the following example.

Let us suppose the request of the XT-180 isosurface using a PDA, preferring the quality of the result. The user determined that triangular meshes made up of more than 800,000 triangles are difficult to visualize, therefore he requires the simplification step if that threshold is overcome. Moreover, because of the low computational capabilities of his device, he is not able to perform the decompression or any other operation.

On the basis of these parameters PLIC_GS at first executes the sequential isosurface extraction, then it considers the size of the isosurface. Since the number of triangles is higher than the given threshold, the simplification module with a reduction of 80% will be activated. The parallelism degree has to be the minimum possible, because high quality of results is chosen. With our cluster we need the aggregate memory of eight PCs to manage this mesh. In conclusion, PLIC_GS will execute the Rebal component to partition the isosurface into eight parts, then the parallel simplification module using eight processes. At the end the result will be processed by the Sew component that produces a VRML file that the user can download and directly visualize.

4. GRISO AND PLIC_GS GRID SERVICE INTERFACES

GrISO and PLIC_GS allow the user to obtain high-throughput figures exploiting a wider set of remote computational resources to satisfy requests. A key point of their implementation is the careful decision of moving the computation where it is more convenient, considering the interconnection network and the available computational resources.

PLIC_GS makes use of the **GEDA** Grid Service, a kind of Resource Broker Grid Service, to search and select the computational resources needed to execute the pipeline. GEDA explores the GRID through WS MDS, proposes to the client a selection of resources satisfying some filter criteria, and finally reserves resources and activates a distributed program through WS GRAM.

In the following sections, some hints are given about the implementation of GrISO and PLIC_GS in GT4, while GEDA is presented in Section 5.

We developed all the components of 3DAGOS using the Grid Development Tools for Eclipse [24] that represents a new and easy-to-use tool that manages the creation of all the interfaces and the configuration files required to create and deploy a Grid Service.

4.1 The Implementation of GrISO

We implemented the computational and flow components by wrapping the C and MPI code within specific Grid Services written in Java. Only the decompression component of GrISO remains a legacy application, because it may be executed on a device outside the Grid. Each Grid Service provides a single method representing all the implementations of the operation. It collects the parameters of the execution, then the following operations are carried out:

- the staging of the input data and of the selected code on one or more provided locations, if needed;
- the execution of the application on the provided resources;
- the staging of the results on a provided location, if needed;
- the indication of the correct termination or the encountered errors.

To perform these operations each Grid Service produces an RSL2 file that is submitted to the WS GRAM service of one of the resources to use. As regards the code to execute, a Grid Service is associated with all the executables of the operation. In the present version of the system we considered only the use of 32 bit Linux architectures.

These Grid Services require three kinds of parameters:

- The operation-specific parameters. For example, the isovalue is only needed by the isosurface extraction operation, while the reduction percentage is needed by the simplification. In our system these parameters are provided by the user through the interface shown in Figure 3.
- The selection of the implementation and the parallelism degree. In our system these parameters are determined by PLIC_GS on the basis of the logic sketched in Section 3.
- The URIs of the resources to use and the possible files to transfer. In our system these parameters are determined by PLIC_GS interacting with GEDA and the user.

As an example, let us consider the method provided by the Isosurface Extraction Grid Service:

```
String Isoextraction(int xdim, int ydim, int zdim, char valtype,
   String datafile, float isovalue, float isosize
   boolean homogeneity, String impl, int pardegree
   String compresources, String fileIn, String fileOut);
```

The selection of the implementation and the parallelism degree are represented by the *impl* and *pardegree* parameters, the resources to use and the possible files to transfer by the *compresources*, *fileIn* and *fileOut* parameters, while the remaining ones are parameters specific of the operation.

Performance of the computational kernels of the Grid-enabled operations is presented in Section 2.2. The Grid environment adds some unavoidable overheads, mainly due to the transfer of executable and I/O files and to the interaction with the Grid middleware. Such overheads are difficult to quantify, since they depend on various factors at the different layers of the Grid architecture, but in any case they are largely compensated by the availability of remote computational resources.

4.2 The Implementation of PLIC_GS

The aims of this Grid Service that is completely implemented in Java are the following:

- the selection of the pipeline composition and of the parallelism degree;
- the selection of the computational resources;
- the management of the interactions among the Grid Services and the management of the data flow.

The first point, that is the logic of PLIC_GS, has been described in Section 3. The remaining interesting aspects are then the selcction of the computational resources and the management of the Grid Services.

The selection of the computational resources is made considering the parameters provided by the user and the response obtained by GEDA that is described in the next section.

In general the most important research projects [2, 13, 15, 22] aim to hide the resource selection from the users. In our opinion this approach is suitable for a large class of embarrassing parallel applications, but not for all parallel applications.

In many cases the suitability of the scheduling strategy depends on the knowledge of the application structure, so it has to be made on the basis of a specific set of characteristics. In our case we consider for the execution of the various Grid Services of GrISO only the clusters available in the Grid. Furthermore, if we found several clusters that provide the same computational power, we should prefer the one which provides a large amount of aggregate memory, which is made up by few nodes that have to be connected through sufficiently fast links. In this manner it is more probable that we will be able to simplify and/or compress the resulting isosurface using it. Such kind of criteria may be difficult to formalize, therefore in our opinion the final decision should be delegated to the user.

Once the computational resources to be used have been determined, PLIC_GS activates each selected Grid Service by providing it the proper parameters. In particular, it is responsible for the management of the data flow, supplying the URIs of the input data and the destination of the resulting files. This is an important aspect as each Grid Service of the pipeline might be executed on a different cluster.

5. GEDA GRID SERVICE AND THE USER INTERFACE

Grid middlewares like GT4 offer mechanisms that permit the discovery and the selection of resources and the execution of jobs on such resources. The main problem in performing such operations is that there are different middleware components to interact with (e.g., in GT4 there are WS MDS for discovering, brokers for selection and WS GRAM for execution); obtaining and using resources through a direct interaction with such a set of different services can be quite difficult or inefficient. The aim of GEDA is precisely to provide an aid for the execution of parallel jobs in Grid environments [5].

GEDA provides three methods:

- `String discoverResources(String XMLConstraints)`:
  this method permits to perform a research among the computational resources of a Grid to find those that satisfy the provided requirements. It returns a list of all the resources fulfilling the requirements.

- `String setResources(String XMLChosenResources)`:
  it allows to select a subset of the computational resources of a Grid provided as list. The returned value indicates the success or the error code.

- `String executeJob(String XMLExecutionRules)`:
  it grants the execution of a job on the selected resources. The parameters contain all the execution information (e.g., the location of executables, input and output files and staging operations), while the returned value indicates also in this case the success or the error code.

These three methods allow the Grid service to be queried and used by other Grid Services or applications. When the request is received, an instance of GEDA is created by GT4 and a set of resources is allocated to allow it to work. The instance takes into account the requirements provided by the client and starts to query the WS MDS services to search for the set of resources satisfying the request. After the research has been correctly made, an XML string containing the ID of the instance and the set of fulfilling resources is returned; a copy of this information is also kept in the state of the instance.

At this point, the client chooses the subset of resources for the execution and calls the method `setResources: XMLChosenResources`, is an XML string containing the list of the selected resources[3] and a set of conditions that define constraints over them (e.g., the estimated time of usage of a resource or the need to have a certain number of the selected resources available at the same time). GEDA checks these parameters and starts an interaction with the WS GRAM services that manage the requested resources to verify if they are still available; if so, GEDA tries to make a reservation[4] of the resources. Finally, it returns an ack to the client if all is ok, otherwise it sends back the encountered errors, for example, because some resources are not available or some constraints are not satisfiable.

After receiving the ack from GEDA, the client can use the third method to execute the application, otherwise it can directly contact WS GRAM to launch it by itself. In this case the client has to produce an RSL2 description of the job that otherwise is automatically created by the executeJob method. If a negative answer is got, the client tries a new selection.

In our system GEDA is used by PLIC_GS to find the resources that best satisfy the computational requirements of the pipeline that has to be executed. The selection is made by a cooperation among the two Grid Services and the user. The interaction of the user with the two Grid Services is performed through a client we implemented in Java. From a functional point of view, the client is subdivided into three parts.

The first one is the Parameter Specification, and it is represented by the tab presented in Figure 3. In this tab the user specifies the parameters of the request, that is the data set, the isovalue and the optional execution of the simplification and compression steps.

The second one is the Resource Selection, and it is represented by the two tabs presented in Figure 4. In these tabs the user interacts with PLIC_GS and GEDA in order to determine the computational resources to use for the execution of the pipeline. Using the third tab it is possible to restrict the list of considered resources by specifying further parameters.

The third one is the Execution. By clicking the Execute button (Figure 3) all the information collected in the two previous parts are provided to PLIC_GS that determines the pipeline configuration and manages the execution on the selected resources.

The interactions among the entities of 3DAGOS are presented in Figure 5.

[3]That is a subset of the previously sent list.

[4]If there are mechanisms on the resources that allow reservation, as in Platform LSF [18].

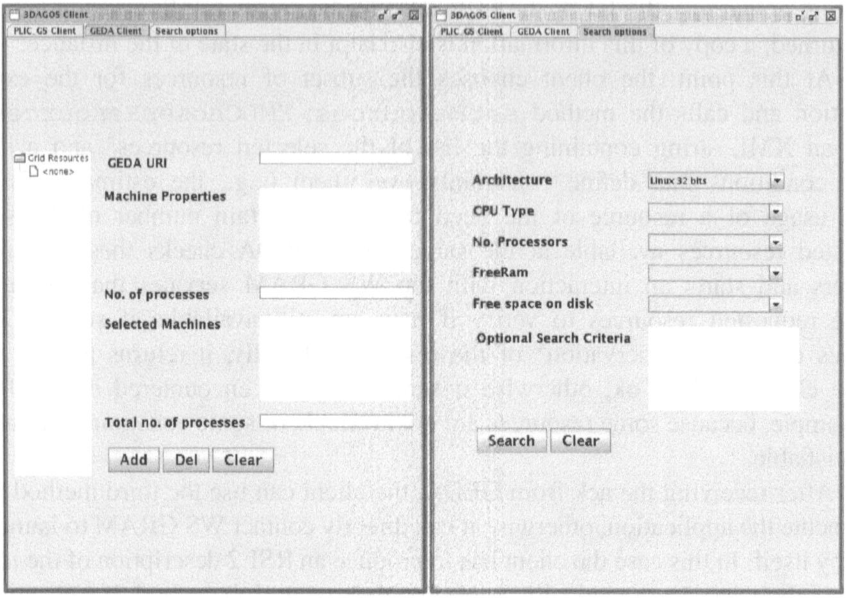

Figure 4. The tabs of the 3DAGOS Java client that permit the users to interact with PLIC_GS and GEDA for the selection of the resources

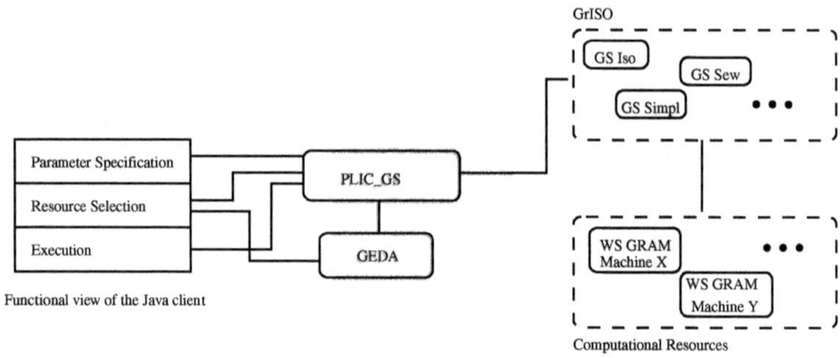

Figure 5. The interactions among the various entities involved in 3DAGOS

6. RELATED WORKS IN GRID ENVIRONMENT

Many current research projects aim to provide suitable visualization tools in the Grid environment. The most remarkable projects are the Grid Visualization Kernel (GVK) [12], funded within the EU CrossGrid Project, and the Distributed Visualization Architecture (DiVA) [4].

Both of them aim to define a way to efficiently map a visualization pipeline on the resources provided by a Grid. The ideas are very similar: they consider the possibility of moving parts of the visualization pipeline, composed of the filtering, visualization and rendering operations, from the client to the Grid server. The solutions proposed to this problem are slightly different.

GVK aims to support near real-time interaction between the visualization client and the server, incorporating in the pipeline operations like occlusion culling or reference-rendering in order to reduce the size of the data to be transmitted [14]. In DiVA the efforts are devoted to the design of an accurate performance prediction system and of new algorithms for remote visualization, to obtain the best time performance in a dynamically changing computing environment [21].

Although these two works have been considered, we designed our pipeline with quite different goals, as presented before. At first, our system is not intended to simply provide a rendering to a remote client; instead, it produces and sends data that can be visualized or further processed, satisfying an even complex client request, not necessarily to have a minimal response time.

Great attention has also been devoted to the definition of the guidelines for the pipeline composition, the selection of the most performant algorithms and the management of the data flow among the pipeline components, in order to obtain the best match with the client request.

In our purpose the integration of the tool in the Grid environment allows us to realize a high-throughput system. In fact, we may exploit a wider set of non-local computational resources to satisfy requests. A key point to carefully consider is the possibility to move the computation where it is more convenient, taking into account the interconnection network. For example, we may find that the repository of the data set we have to process is linked with one or more computational nodes through a fast and dedicated network. Depending on the size of the data, the selection of one of these resources rather than a more powerful cluster linked through a shared network will often represent the best solution.

GEDA is a general-purpose Grid resource manager and can be compared with similar middleware components. For example, considering the GridLab project, GEDA can be intended as a component of the service layer in the capability space for a GT4-based Grid. It can be compared to the Grid Resource Management and Brokering Service (GRSM), working on GT2-based Grid, but with a greater attention to the management of parallel hardware and software resources, since it uses a structured description of these resource (as described, for example, in [15]). An essential part of GridLab is the Grid Application Toolkit (GAT) that provides a single interface for requesting and using remote resources, independently of the underlying infrastructure. GEDA is the service layer component able to interface with the GAT resource and job description functions and the GAT resource broker function.

7. CONCLUSIONS

In this chapter we presented 3DAGOS, a set of Grid Services that permits to perform efficiently the isosurface extraction operation in the Grid environment.

The proposed approach represents a general way to realize effective tools for Virtual Laboratories. In particular, our solution is able to effectively exploit the computational resources shared in a Grid, yielding three main advantages: availability of the application to remote users; better performance, since more powerful computer resources than the local ones can be used; higher throughput, since multiple requests can be accomplished using different computational resources.

The main goal of 3DAGOS is to support the distributed computation over the Grid, even interactively. Until now, Grid technology focused on providing batch access to resources, while interactive behaviour requires some kind of Quality of Service, to be attained by means of modifications to the grid middleware managing the interactions between different grid entities.

PLIC_GS and GEDA are the first steps in this direction. PLIC_GS allows to define pipelines satisfying some cost function optimization criteria and to instantiate them considering the allowable resources. GEDA considers the Grid resources as described by the WS MDS, but it requires also other kinds of description such as network interconnections, topology and performance characteristics and a hierarchical structured description of the distributed resources, to efficiently support the deployment of distributed applications.

Our future work will be devoted to define a suitable architecture over an OGSA-compliant Grid to support QoS, providing new tools, e.g., distributed schedulers with Soft Real Time constraints, and mechanisms, such as more restrictive Service Level Agreement (SLA).

ACKNOWLEDGMENTS

This work has been supported by FIRB strategic project on Enabling Technologies for Information Society, Grid.it, and by the regional program of innovative actions PRAI-FESR Liguria.

REFERENCES

[1] A Grid Application Toolkit and Testbed, http://www.gridlab.org/.

[2] Andronico G, Ardizzone V, Barbera R, Catania R, Carrieri A, Falzone A, Giorgio E, La Rocca G, Monforte S, Pappalardo M, Passaro G, Platania G (2005). "GILDA: The

Grid INFN Virtual Laboratory for Dissemination Activities". *Proceedings of the First International Conference on Testbeds and Research Infrastructures for the DEvelopment of NeTworks and COMmunities (TRIDENTCOM'05)*, 304–305.

[3] Andújar C, Brunet P, Chica A, Navazo I, Rossignac J, Vinacua A (2005). "Optimizing the Topological and Combinatorial Complexity of Isosurfaces". *Computer-Aided Design* 37:8, 847–857.

[4] Baker P, Bethel W, Clyne J, Fulcomer S, Hathaway B, Kohl J, Moran P, Parker S (2003). "DiVA: Distributed Visualization Framework Component Interface". *Proceedings of Workshop Findings Document.*

[5] Clematis A, Corana A, D'Agostino D, Gianuzzi V, Merlo A (2006). "Resorurce Selection and Application Execution in a Grid: A Migration Experience from GT2 to GT4". *Proceedings of the International Symposium on Grid computing, high-performAnce and Distributed Applications (GADA)*, LNCS 4276, Springer, 1132–1142.

[6] Clematis A, D'Agostino D (2006). "Parallel Remote Visualization for the Grid". Dongarra J, Zima H, Hoisie A, Yang LT, Di Martino B (eds) *Engineering the Grid: Status and Perspective*, American Scientific Publishers, 521–536.

[7] D'Agostino D (2006). "Designing Parallel Programs for 3D Data Processing on Distributed Architectures". *PhD Thesis, DISI-TH-2006-03*, Università di Genova, Genova.

[8] Foster I (2006). "Globus Toolkit Version 4: Software for Service Oriented Systems". *Journal of Computer Science and Technology* 21:4, 513–520.

[9] Garland M, Heckbert PS (1997). "Surface Simplification Using Quadric Error Metrics". *Computer Graphics* 31, 209–216.

[10] Gotsman C, Gumhold S, Kobbelt L (2002). "Simplification and Compression of 3D Meshes". Iske A, Quak E, Floater MS (eds) *Tutorials on Multiresolution in Geometric Modelling*, Springer-Verlag, 319–361.

[11] Gropp W, Lusk E, Doss N, Skjellum A (1996). "High-Performance, Portable Implementation of the MPI Message Passing Interface Standard". *Parallel Computing* 22:6, 789–828.

[12] Heinzlreiter P, Kranzlmuller D (2003). "Visualization Services on the Grid: The Grid Visualization Kernel". *Parallel Processing Letters* 13:2, 135–148.

[13] Heymann E, Senar MA, Fernández E, Fernández A, Salt J (2004). "Managing MPI Applications in Grid Environments". *Proceedings of the Second European Across Grids Conference (AxGrids 2004)*, LNCS 3165, Springer, 42–50.

[14] Kranzlmuller D, Kurka G, Heinzlreiter P, Volkert J (2002). "Optimizations in the Grid Visualization Kernel". *Proceedings of Workshop on Parallel and Distributed Computing in Image Processing, Video Processing, and Multimedia*, 129–135.

[15] Lacour S, Pérez C, Priol T (2005). "Generic Application Description Model: Toward Automatic Deployment of Applications on Computational Grids". *The 6th IEEE/ACM International Workshop on Grid Computing (Grid2005)*, Springer.

[16] Ligon III WB, Ross RB (2001). "PVFS: Parallel Virtual File System". *Beowulf Cluster Computing with Linux*, MIT Press, 391–430.

[17] Lorensen WE, Cline HE (1987). "Marching Cubes: A High Resolution 3D Surface Construction Algorithm". *Computer Graphics (Proceedings of SIGGRAPH 87)* 21:4, 163–169.

[18] Lumb I, Smith C (2004). "Scheduling Attributes and Platform LSF". *Grid Resource Management: State of the Art and Future Trends*, Kluwer Academic, 171–182.

[19] Remote Instrumentation in Next-Generation Grids (RINGrid), http://www.gup.uni-linz.ac.at/ringrid/.

[20] Rossignac J (1999). "Edgebreaker: Connectivity Compression for Triangle Meshes". *IEEE Transactions on Visualization and Computer Graphics* 5:1, 47–61.

[21] Shalf J, Bethel EW (2003). "The Grid and Future Visualization System Architectures". *IEEE Computer Graphics and Applications* 23:2, 6–9.

[22] Sipos G, Kacsuk P (2006). "Multi-grid, Multi-user Workflows in the P-GRADE Portal". *Journal of Grid Computing* 3:3–4, 221–238.

[23] The Digital Imaging and Communications in Medicine standard (DICOM), ftp://medical.nema.org/medical/dicom/.

[24] The Grid Development Tools for Eclipse, http://ds.informatik.uni-marburg.de/MAGE/gdt/.

[25] The Grid enabled Remote Instrumentation with Distributed Control and Computation (GRIDCC), http://www.gridcc.org/.

[26] The Object File Format (OFF), http://www.dcs.ed.ac.uk/home/mxr/gfx/3d/OFF.spec.

[27] The Virtual Reality Model Language (VRML), http://www.web3d.org/.

[28] The Visible Human Project, http://www.nlm.nih.gov/research/visible/visible_human.html.

NEW PROTOTYPE ARCHITECTURE FOR AUTOMATED IRRIGATION BASED ON POWER LINE COMMUNICATIONS

C. Donciu[1], O. Costea[1], M. Temneanu[1], R. Damian[1], and M. Branzila[1]
[1] *Department of Electrical Measurements, Faculty of Electrical Engineering, Iasi, Romania*

Abstract The realization of measurement and control modules for distributed data transmission is proposed, to be implemented on automatic irrigation systems, so that it may lead to an intelligent system of irrigation combined with automatic nutrients' injection, relying on the sensory investigation of the environment and soil parameters' conditions. The decisional algorithm of the control center prescribes combined irrigation schedules according to the exploited crop and to its development specificity, and the investigation is carried out at the irrigation cell level, modularly. The communication data flow exploits the existent infrastructure of power supply of the automatic irrigation systems as a physical support, and it is developed through the Power Line Communications (PLC) technique. The system architecture is so conceived that it should allow its implementation on the automatic irrigation systems of both circular movement (central pivot) and linear movement. When a monitoring, control and data storage application involves the presence of the sensors and actuators on a large area, the FieldPoint-distributed input–output systems offer an integrated, modular and economical solution for industrial use.

1. INTRODUCTION

The chapter describes a work in progress, which should be completed by August 2008. The existent automatic irrigation systems use aspersion irrigation, a method that has some disadvantages:

- the unfavorable effect of the drops (in the case of average and high-pressure aspersers) upon plants, especially when plants are young and blooming;

- the presence of the wind during the irrigation – if its intensity overcomes 2.5 m/s, up to 10–25% of the crop areas may receive no water;
- high water losses caused by the global irrigation and absence of cell level individualized information;
- increased nutrients' pollution of the water table;
- high-level power consumption.

The intelligent irrigation system architecture proposed by the present project designs a structure of ample complexity that solves all these problems. This way, through the various telecommunication techniques used (radio transmission, data delivery through the PLC power supply network, TCP–IP type data transmission over the Internet, serial and parallel-type local communications), a wide range of devices are interconnected (electrical engines, nutrients' injection batteries, microelectronics specific elements, sensors and transducers, actuators, computers, etc.), whose action is supervised by software programs developed through methods of virtual instrumentation. The final objective is imposing the norms and thresholds specific of modern agriculture and environmental protection.

The system contains a rule-type database, as multi-parametric graphs of the vegetation stage, with which the control center decisional algorithm interacts, having as manifest goal the agricultural productivity growth, with the constraints of keeping a low degree of groundwater pollution by nutrients [12].

Data transmission inside the system is performed via the PLC technique, which represents an important research objective.

In the situation when, between two communication points on the powered lines, there is a transformation post, the problem raises of crossing it with the useful data signal. The PLC key component is the coupling element between the transformer primary and secondary circuits, a complex and expensive device. We note that, since in the online command of the intelligent irrigation system the response time can take seconds, the present project proposes the implementation of a *by-pass* method of the transformers, without using supplementary equipment.

Data transmission will be performed by phase modulation and the signal reading by digital signal processing, making use of a method developed by the members of the research team. In this way, the signal processing will be done by using a variable resolution computational algorithm for the frequency spectrum, which supposes a dependency between the resolution value and the amplitude value of the spectrum component previously calculated; this allows increasing the resolution in the interval of interest, while keeping the imposed minimum resolution outside the interval of interest.

The implementation of this method, due to the reduction of the maximum relative phase error up to 98.9%, compared to the classical discrete Fourier method, allows the quantization of the information on the spectral components at the phase level.

2. THE PROPOSED METHOD

2.1 Intelligent Irrigation System

The intelligent irrigation system architecture is structured on six main levels, as follows:

- the level of irrigation modules, which join the automatic irrigation systems with angular or linear movement and which have as main function the controlled command of the electro valves for the admission of the water-nutrients mixture, in accordance with the rules prescribed by the control;
- the level of sensor modules, which are fixed and placed on the ground, having as main purpose the sampling through sensors of the surrounding cell characterization data and their transmission toward the control center;
- the level of nutrients' injection batteries, with the role of injecting nutrients into the irrigation water, in the concentration prescribed by the control center;
- the level of data flow transmission through PLC, which has the network of the power supply of the engines operating on the automatic irrigation systems and water pumps as physical support, and realizes the data transfer between the control center on the one hand and the irrigation modules and the nutrients' injection batteries on the other;
- the decisional level, based on the data resulting from the sensor modules' level and those extracted from its own database of combined irrigation strategies, which delivers the execution commands to the injection batteries, regarding the irrigation output and the characterization cell geographical localization;
- the level of the FieldPoint module, which allows the communication between the sensor modules and the decisional level.

The general architecture of the intelligent irrigation system is shown in Figure 1. This system consists of two independent circuits, one of water supply and one of power supply. The water supply circuit (CAA) makes the junction between the water supply source (AA) and the nutrients' injection batteries (BIN).

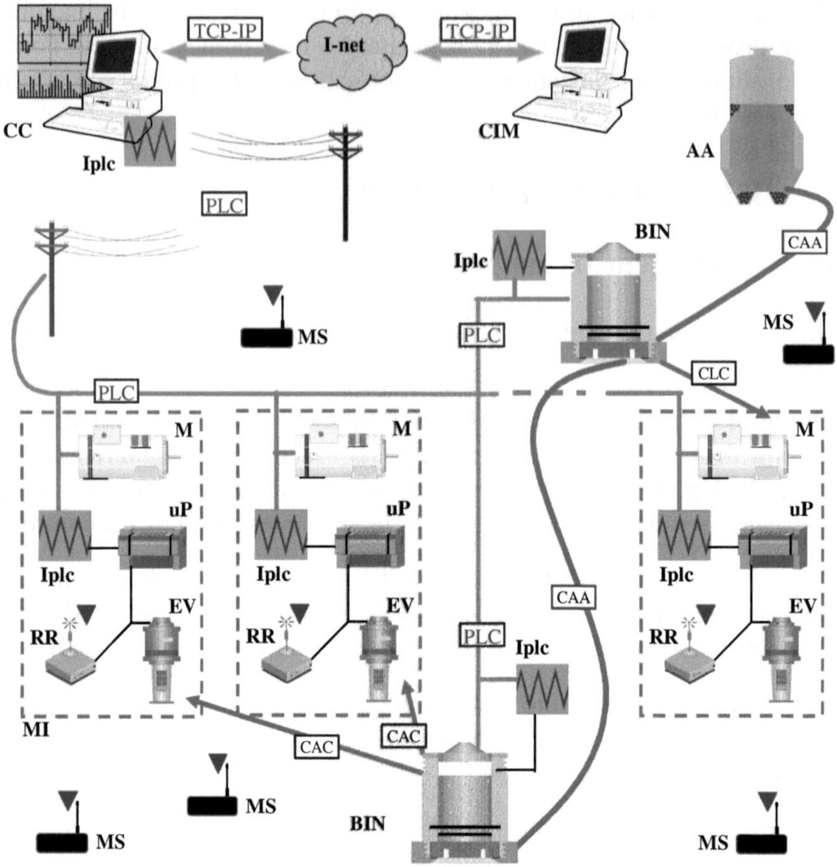

Figure 1. Intelligent irrigation system architecture

This circuit converts into the combined supply circuit (CLC), after the nutrients' injection, and it connects with the pulverization nozzles of the irrigation module (MI). The electric circuit powers up the engines and the pumps afferent to the automatic irrigation system, and it constitutes the data transmission physical environment. The nutrients' injection batteries are installed near the pumps. These batteries have a common power supply.

The irrigation module is set up at the level of the mobile arms of the automatic irrigation system and it is made up of the communication interface (IPCL), the microchip control device (μP), the radio receiver (RR) and the electro valve (EV). At the MI module movement, the identification of the membership to a characterization cell is performed upon entering its range of action. The emission power of the sensor modules (MS) is limited within the cell and its detection is based on the maximum emission criterion.

Following the communication setup between the sensor module transmitter and the irrigation module receiver, data are transmitted by the command device μP to the PLC interface, through the supply circuit of the operative electrical engine.

Power line communication is a method of achieving bi-directional communication by using a power line as a transmission channel and superposing high-frequency signals over the 50/60 Hz electrical signal. Since incoming lines and electric outlets can be used unaltered for PLC, making new wiring unnecessary, it is a reasonable alternative.

Power line communications, or broadband over power lines (BPL), allow transmission of data over power lines. The radio frequency (RF) signal sent over medium- and low-voltage AC power lines allows end users to connect to the Internet. The RF signal is modulated with digital information that is converted by an interface in the home or small business into Ethernet-compatible data.

PLC is based on the idea that any copper medium will transport any electrical signal for a certain distance. Basically, a RF signal is modulated with the data we wish to send. This radio signal is then sent down the copper medium (our power lines) in a band of frequencies not used for the purposes of supplying and managing electricity.

The frequencies and encoding schemes used greatly influence both the efficiency and the speed of the PLC service. Most PLC radio traffic generally occurs in the same bandwidth, roughly 1.6–80 MHz. These frequencies are in the MF (medium frequency; 300 kHz–3 MHz), HF (high frequency; 3–30 MHz) and some of the VHF (very high frequency; 30–300 MHz) spectrum [3]. Various encoding schemes [5, 8] have been used for sending the data along power lines; these include [5]

`GMSK`: Used with the single carrier version of PLC, providing low bandwidths (< 1 Mbit/s);

`CDMA`: Used with the single carrier version of PLC, providing low bandwidths (< 1 Mbit/s);

`OFDM`: Used with the multi-carrier version of PLC, providing a bandwidth of up to 45 Mbit/s.

The advantages of PLC for the end users are the following:

- The equipment needed to set up PLC in the home in the USA is cheaper, on average, than that of other broadband solutions, such as DSL and cable modem, usually costing (depending on equipment) from 45 up to 75 USD.
- The equipment uses existing power outlets in the home, making the setup a lot easier (plug-and-play). There is no need for complicated wiring and additional installations. It is possible to move one's computers and appliances where one wants them.

- Power line communication outperforms its competitors with speeds of up to 14 Mbit/s.
- For users in rural areas, who cannot receive DSL or cable modem services, PLC can be used to provide an all-in-one service, carrying telephone, cable television and high-speed data.

Via the PLC transmission, the useful information is received by the control center CC, on a computer server. As a consequence of the interpretation of the response type data provided by the characterization cells' sensors, control decisions are taken upon the limits of the nutrients' concentrations and relative humidity, which are imposed by the rule-based database.

The quality and quantity commands are transmitted by the PLC system to the nutrients' injection batteries, providing the combined irrigation agent to the pulverization nozzles and to the volume controlling electro valves belonging to the irrigation module.

Useful data are delivered by the PLC communication system, modulated in beams. The signal is of the *broadband* type and the physical environment enables multiple operations running on the same existing infrastructure.

2.2 Sensor Module

The hardware architecture of the sensor module is shown in Figure 2 and is composed of a set of sensors, specialized in the climatic parameters' detection (temperature, humidity, dew point, speed, direction and movement sense of air masses, precipitations) and a set specialized on the soil parameters' detection (conductivity, humidity) [2].

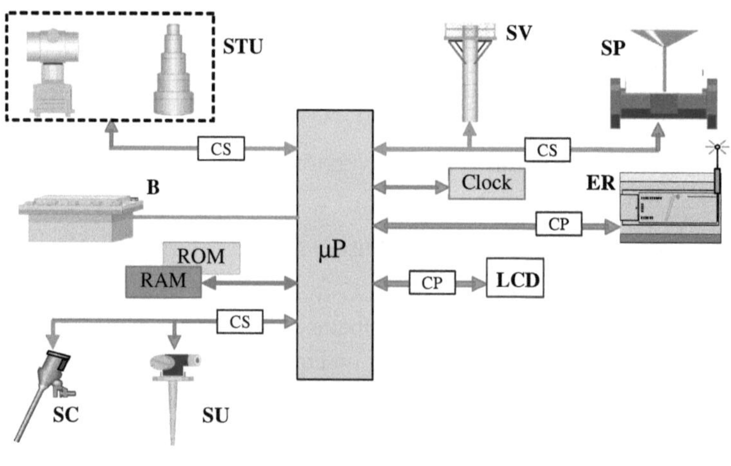

Figure 2. Hardware architecture of the sensors module MS

For the air-relative temperature and humidity measurement, an incorporated sensor with numerical output (STU) is used, made in CMOS technology, which makes it thermally safe and stable. This generates a useful signal of superior quality, it has a very quick response time and insensitivity to external noises (EMC) [7].

The data delivered by this sensor are used to forecast the degree of the water evaporation off the ground. The need to utilize the rain gauge sensor type (SP) comes from the fact that, during precipitations, following the data provided by the soil humidity sensor, the interruption of the irrigation procedure happens late, because of the time needed by water to permeate through the ground, to the depth where the humidity sensor is placed. The data coming from the anemometer (SV) are used to eliminate irrigation unevenness caused by wind blasts.

At the ground level the conductivity sensor (SC) and the humidity sensor (SU) are set, on whose account the input data for the estimation procedure of the relative nutrients' and humidity concentration will be supplied [6]. The bidirectional communication with the sensors' block is performed through the microchip μP, and it is of serial type.

As the sensor module is an independent system, it is equipped with a control keyboard and a LCD screen. The radio data delivery to the irrigation module is done by the transmitter (ER), having the transmission range adjusted in order that it should not surpass the characterization cell by more than half the radius. The assembly power supply is performed by a dry battery (B).

The control center is the physical support of the decisional algorithms, which operate the actions of the executor elements (the actuators), relying on the data resulting from the level of the sensor modules.

The database found at this level contains information referring to the leguminous species' classification, in terms of water consumption and absorption capacity, and it provides irrigation graphs, regarding the cultivated species, regional climatic factors, culture denseness and rows' orientation, plant habits, root system and vegetation stage. Besides, beginning with this level, the time diagram starting point (the culture initiation) and the spells designated for the maintenance works are programmed [1].

The Intervention and Monitoring Centre (CIM) represents the interface with the human operator in the process of supervision and monitoring the control actions elaborated by the Control Center and allows making modifications in the decisional algorithms' development.

Also, starting with this level, the database can be completed or modified (software upgrade), with respect to the irrigation networks. Since the control center is configured by the server, it can be accessed from any geographical area where an Internet connection is available. The TCP–IP communication between CC and CIM is provided by two dedicated virtual instruments, server and client. For security reasons the login to the server is permitted only on a password basis.

Traditional control systems use a point-to-point architecture to communicate, which implies the existence of a permanent communication between the main controller and each sensor or executing element. This approach does not satisfy the requirements of modularity, control decentralization, integrated diagnosis, maintenance and low cost [14]. A new communication architecture, of the "shared bus" type, satisfies all the aforementioned requirements and results in efficiency, flexibility and safety improvement, with ensuing reduction of mounting, reconfiguration and maintenance times and costs.

The practical implementation of distributed control systems relies on the utilization of distributed hardware and software equipment. An advantageous solution is the adoption of the FieldPoint input–output SCADA (Supervisory Control And Data Acquisition) software modules, as provided by National Instruments' LabView [4, 9–11, 17].

The FieldPoint modules are easy to use and configure, reliable and robust, and allow the conditioning and conversion components (A/D and D/A) to be placed near the process. The FieldPoint module used in data acquisition is connected straight away to the sensors and transducers, and filters, calibrates and scales the signals, which can then be transmitted through serial interfaces (RS-232 and RS-485), Ethernet, CAN, FOUNDATION Fieldbus, wireless or radio.

3. TESTS AND COLLECTION OF EXPERIMENTAL RESULTS

The FieldPoint module can be connected to a large variety of sensors with current outputs in the range 4–20 mA and receive a variety of digital signals from 5–30 VDC to 0–240 VAC. By using the FieldPoint modules for Ethernet communications, LabView real-time applications can be directly interfaced. In addition to this, the system has the possibility to work independently, without computer connection.

Figure 3 presents the block diagram of the virtual instrument, which assures the communication between the FieldPoint module and the workstation's computer. The functions implemented by the main instrument are merged in a While structure.

The front panel of the user interface is shown in Figure 4 and contains four numeric indicators, which indicate the values registered by the sensors. The virtual instrument has a stop button, which allows stopping the recording process.

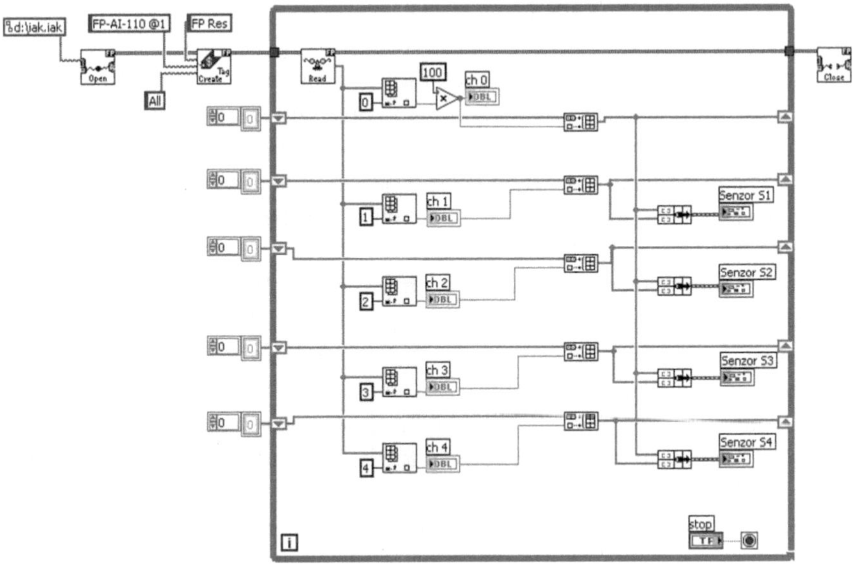

Figure 3. The block diagram of the communication between the computer and the FieldPoint module

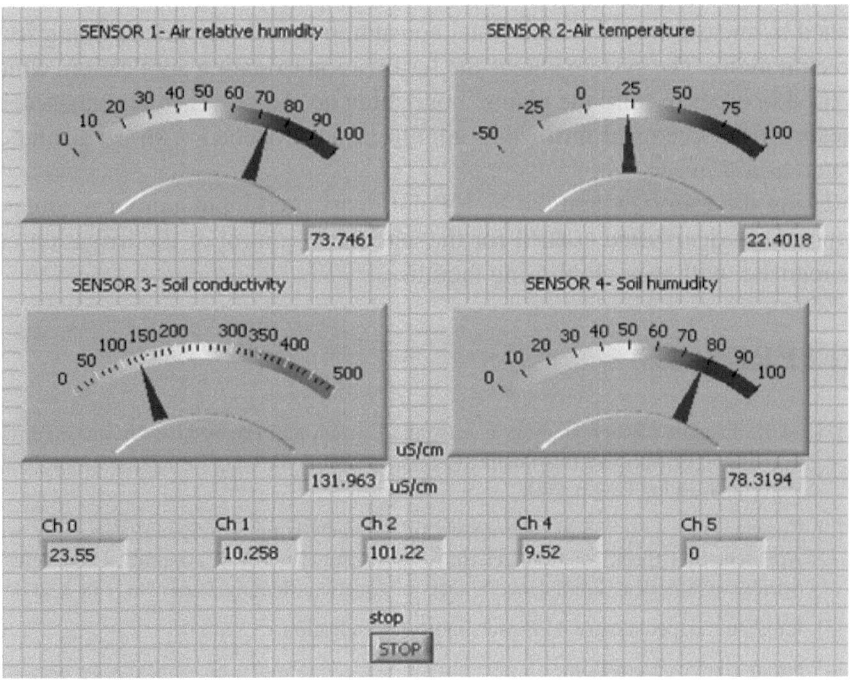

Figure 4. The front panel of the user interface

4. CONCLUSIONS

Irrigation is essential for the vegetables demanding in terms of ground humidity, especially when cultivated on light soils, with great water permeability. Considering the high prices of the leguminous production, very large economical losses can be incurred, when such production is obtained in droughty circumstances. This is one more reason pleading for the "precision irrigation" of the leguminous crops [13].

As the leguminous plants' needs, with respect to the water factor, vary substantially during the vegetation period, the proposed system allows the assurance of humidity necessary in the ground and air to an optimum degree, according to the vegetation stage. This way, knowing at every moment the plants' water consumption, their sweating quotient, as well as the water losses through evaporation and infiltration into the ground, the system permits to overcome the watering stress.

The proposed system allows the spatial and temporal monitoring of water consumption, in terms of the water used by plants, plus the water lost by evapo-transpiration. After the spatial and temporal monitoring, water consuming graphs can be drawn up, which allow the maximum rigorous determination of the intervals of maintenance and ingathering work [16]. Moreover, the water balance in the soil can be determined with high precision; the balance is established by measuring the whole water consumption and the water coming from precipitations. The system allows the fertilization irrigation application, which is used in order to fulfil the plants' nourishing needs, on vegetation phases, by means of the chemical fertilizers' administration together with the irrigation water, as a solution.

Using the LabView and the FieldPoint-distributed input–output modules a structure of distributed control for the automatic control of the irrigation can be realized, with web monitoring facilities.

REFERENCES

[1] ACIL Tasman, "Water Trading in Australia Current and Prospective Products: Current Trends and Prospective Instruments to Improve Water Markets", Prepared for the Water Reform, Working Group. 2003.

[2] Corwin, D.L., Lesch, S.M., "Protocols for Characterizing Soil Spatial Variability with Apparent Soil Electrical Conductivity", Agronomy Society of America, Crop Science Society of America, Soil Science Society of America Meeting, Paper No. 168-3, 2005.

[3] http://en.wikipedia.org/wiki/Power_line_communication.

[4] http://wiki.lavag.org/Home.

[5] Kutz, T.C., "Broadband over PowerLine: A New Technology for the Future", *The Telecommunications Review* 2005, 41–46.

[6] Lesch, S.M., Corwin, D.L., "Field Surveying Techniques/Assessment and Diagnosis of Salinity Problems", Usda-Nrcs Conservation Program Book #2, National Employment Development Center, Washington, DC, pp. 1–41, 2005.

[7] Lesch, S.M., "Sensor-Directed Spatial Response Surface Sampling Designs for Characterizing Spatial Variation in Soil Properties", *Computers and Electronics in Agriculture* 2005, 46:153–179.

[8] LINK Personal Communications, Programme by ERA Technology Ltd, Cleeve Road, Leatherhead, Surrey, KT22 7SA, England, 1999.

[9] National Instruments, LabVIEW User Manual and G Programming, 2001.

[10] National Instruments, Measurement and Automation Catalogue, 2004.

[11] V. Nayak, "Lookout and FieldPoint SCADA System Provides Increased Efficiency for Pilot Plant", Third Quarter, 2000.

[12] Nijbroek, N., Hoogenboom, G., Jones, J.W., "Predicting Soybean Yield Variability Under an Irrigated Environment", Presented at the July ASAE International Meeting, Milwaukee, WI, Paper No. 003038. ASAE, 2950 Niles Rd., St. Joseph, MI, 2000.

[13] Sadler, E.J., Evans, R.G., Stone, K.C., Camp Jr, C.R., "Opportunities for Conservation with Precision Irrigation", *Journal of Soil and Water Conservation* 2005, 60(6):371–379.

[14] Torre-Neto, A., Schueller, J.K., Haman, D.Z., "Networked Sensing and Valve actuation for Spatially-Variable Microsprinkler Irrigation", ASAE International Meeting, Milwaukee, WI, Paper No. 001158. ASAE, 2950 Niles Rd., St. Joseph, MI, 2000.

[15] Viterbi, A.J., *CDMA: Principles of Spread Spectrum Communication* (1st edition), Prentice Hall PTR. ISBN 0201633744, 1995.

[16] Wraith, J.M., Robinson, D.A., Jones, S.B., Long, D.S., "Spatially Characterizing Bulk Electrical Conductivity and Water Content of Surface Soils with Time Domain Reflectometry", *Computers and Electronics in Agriculture* 2005, 46:239–261.

[17] www.ni.com – Solutions.

THE USE OF PIMA(GE)^2 LIBRARY FOR EFFICIENT IMAGE PROCESSING IN A GRID ENVIRONMENT

A. Clematis[1], D. D'Agostino[1], and A. Galizia[1]

[1] *IMATI-CNR, Via De Marini 6, 16149 Genova, Italy*

Abstract Grids enable the creation of virtual laboratories for the collaborative use of sophisticated instruments producing large amount of data that have to be processed in order to extract knowledge. In this context, a very important task is related with image processing, since most of the data are images that have to be adequately analyzed. We present a Grid-aware version of the Parallel IMAGE processing GEnoa library (PIMA(GE)^2 Lib). The major features of our approach are the preservation of the ease in the development of parallel image processing applications and the possibility to efficiently exploit the Grid resources for their executions. The resulting tool, called PIMA(GE)^2 Grid, is based on the Grid Service technology; from the user point of view, it acts as an intermediate layer between Grid resources and parallel image processing applications. PIMA(GE)^2 Grid represents a feasible solution to exploit multiple and computationally intensive image processing applications in a virtual laboratory.

1. INTRODUCTION

Nowadays Grids have been established as a new opportunity in the scientific community. Their development is largely due to the need of cooperation among scientists who are eager to share resources, such as scientific laboratory equipment, computing facilities, software tools and data [13].

For example, the hundreds of gigabytes of genome sequence and the terabytes of data of digital observatories require large data repositories and a large computational power to be processed. The Grid permits to exploit distributed resources to accomplish this task [28, 30].

The main consequence of producing huge amounts of data is the need of availability on the Grid of software libraries and tools, which scientists employ to extract knowledge from data. With this approach, the heaviest computations may be done using remote resources, while the lower amount of resulting data may be moved for further processing on local facilities.

Image processing is a key computational aspect in a virtual laboratory. Indeed, many works are aimed to enable the use of existing image processing libraries and tools on the Grid [14, 15, 17]. However, it is not an easy task, since many different aspects have to be considered, including the following:

- the preservation of the ease of use of the original tools;
- the possibility of efficiently processing multiple and computationally intensive requests exploiting suitable facilities available in the Grid;
- the management of remote application executions, including the I/O and data transfer.

In this chapter we present our experience in the development of a Grid-aware version of PIMA$(GE)^2$ Lib, our parallel image processing library. The resulting tool, called PIMA$(GE)^2$ Grid, allows to develop and execute parallel image processing applications on the Grid. It represents a useful computational resource that can be employed in a virtual laboratory. From the user point of view, it acts as an intermediate layer between Grid resources and parallel image processing applications. We developed it considering the abstraction model proposed in [2]. Actually, Grid services aimed at image processing and to the exploitation of the architecture have been deployed in advance, and through PIMA$(GE)^2$ Grid, they are coordinated and provided to the users.

However, to execute the applications on Grid resources, the computational nodes to be used need to be specified by the users. To accomplish this task, we make use of *GEDA*, *Grid Explorer for Distributed Applications* [5], a Grid Service designed to provide a simplified but non-transparent access to Grid Resources. GEDA is a general-purpose resource broker service that can be used to allow a fine-grain resource selection for any kind of Grid Service.

The parallel image processing services are obtained through the porting of the native library code. The Application Programming Interface (API) of the library is a standard and sequential one, and the aspects inherent to parallel programming are hidden to the user. These points have been inherited by PIMA$(GE)^2$ Grid; they represent a further step to simplify the development and the execution of efficient image processing applications on the Grid.

The chapter is organized as follows: in the next section an overview of the library integration in a Grid environment is given. In Section 3 PIMA$(GE)^2$ Lib is described, while in Section 4 we provide the overall organization of PIMA$(GE)^2$ Grid, the Grid-aware version of the library. In Section 5 related works are discussed, and in the last section conclusions and future directions are outlined.

2. MOVING TOWARD HETEROGENEOUS DYNAMIC IMAGE PROCESSING

The use of sequential and parallel libraries is well assessed for numerical applications. The software evolution history provides examples in this direction through the diffusion of different packages, in particular for numerical applications [4], but also for the field of image processing [18, 25].

A sequential library, for numerical computation or image processing, represents a user-friendly tool that provides optimized and ready-to-use operations in its specific domain. A parallel library is a more complex tool. In fact, with respect to the sequential library, it has to manage in a proper way the burdens related to the setup of the parallel environment, the management of communication, data exchange, input/output handling and parallel processing.

The exploitation of parallel libraries is a common practice for the development of computationally demanding applications, which improves the development process and still provides optimized performance on a wide range of target architectures [3, 11].

In this context, we developed *PIMA*$(GE)^2$ *Lib*, the Parallel IMAGE processing GEnoa Library [9], with the purpose of providing robust and efficient implementations of the most common low-level image processing operations, according to the classification provided in Image Algebra [23]. We briefly describe the library in Section 3.

We also considered the problem of making the library interoperable and available in distributed and heterogeneous environments. To achieve this goal we designed and implemented *PIMA*$(GE)^2$ *Server* [6], a version of the library obtained by wrapping the original code using CORBA [31]. The result is a client–server architecture that is able to satisfy more clients at a time.

The main drawback of this approach is that the server is coupled with a fixed set of computational resources, such as a specific cluster. If many users require to process large images, which is a common situation in a virtual laboratory, the available computational power will be quickly run out, with the consequence of a long waiting time.

Designing a scheduler that uses a set of distributed servers is possible, but this approach would result in an ad hoc and complex solution. Moreover, these servers have to be deployed in advance. Considering that the most effective way to obtain the required computational power is using the Grid, we decided to adopt different technologies. In particular we considered the use of a Globus-compliant middleware and of the Grid Service technology for producing a Grid-aware version of PIMA$(GE)^2$ Lib; we called it *PIMA*$(GE)^2$ *Grid*.

In this way we maintained the interoperability and availability level of PIMA$(GE)^2$ Server, and we added a high level of flexibility and a dynamic behavior, as well. In fact, the Grid Service interface allows the user interaction with PIMA(GE)^2 Grid through heterogeneous clients in order to run image processing applications on Grid resources. The user has to provide to PIMA(GE)^2 Grid the source code of the application, where all the functions made available by PIMA(GE)^2 Lib can be employed. A Grid Service compiles the application linking it to the library, and in this way we are able to preserve the whole management, load balancing and optimization policies we defined in the library. The result is then executed on the computational resources that the user selected among those available in the Grid.

We do not need to install or deploy anything on these resources to exploit them, and this is the most important difference with PIMA(GE)^2 Server. PIMA(GE)^2 Grid represents a general solution in the porting and execution of parallel legacy code on a Grid. However, it is to note that the performance of the executions depends on the selections made by the user.

3. A BRIEF OVERVIEW OF THE PIMA(GE)^2 LIB

The PIMA(GE)^2 Lib, the Parallel IMAGE processing GEnoa Library, provides a robust and efficient implementation of the most common image processing low-level operations. It has been implemented using C and MPICH2 and allows the development of compute-intensive applications, achieving good speed-up values and scalability.

The library elaborates 2D and 3D data sets and allows to perform operations both in sequential and in parallel, depending on the user requirements. In the second case the parallelism is hidden from the users and completely managed by the library. This behavior is mainly obtained through the definition of an effective and flexible interface that appears completely sequential [7].

The main advantage of this approach is that PIMA(GE)^2 Lib shields users from the intrinsic complexity of a parallel application. On the other hand this approach requires a great effort in designing effective solutions for the efficient management of underlying architectures. In the library we tackled this issue at different levels: an optimization policy is applied to perform a suitable management of communications and memory operations; the data distribution is oriented toward obtaining load balance; other efforts are dedicated to the exploitation of memory hierarchies. Another important point to improve application performance is the adoption of a parallel I/O [8] in PIMA(GE)^2 Lib that permits to also speed up the I/O time.

3.1 The PIMA(GE)^2 Lib Hidden Parallelism

The major issue in the development of PIMA(GE)^2 Lib is the definition of a robust and effective parallelization strategy. It has to permit the automatic management of all the aspects related to the data distribution and communications.

In this way we are able to hide the parallelism of the computation and still ensure efficient execution.

The image processing operations are implicitly parallelizable; they usually require a spatially localized portion of the input image data to compute the output image. Moreover it is possible to use a similar methodology to implement the computational engines of many operations. Thus we were able to group together similar image processing operations, defining an *algorithmic pattern* that corresponds to each group. The algorithmic patterns may correspond to the execution, in parallel, of sequential operations or it may also require one or more communication phases. They provide a restricted number of parallel behaviors and are useful to avoid redundant data distributions and communications when a sequence of operations is applied to the same data. This approach has already been adopted in the image processing community [24].

The parallel behavior of the library is the following: data are distributed among the parallel processes when a user instantiates an image. In particular data are subdivided in contiguous row regions with the same granularity during the I/O phases or during the memory management operations. Each process performs the appropriate functions on its own chunk of the image, following the adequate parallelizable patterns. Similar to what happens in the distribution phase, an implicit collection phase is performed when necessary, e.g., at the end of computation. This behavior is depicted in Figure 1. We implemented the parallelization strategy using a dedicated software layer, transparent to the users. The functions of this layer are activated through the sequential interface of the operations; they start up the algorithmic patterns and manage the corresponding parallel schemes. The parallelization strategy leads to high speed-up values in the execution of each operation, because of the proper parallelization of the algorithmic patterns. The same result is achieved considering a pipeline of operations, since we avoid redundant communications, as demonstrated by speed-up values presented in the next subsection.

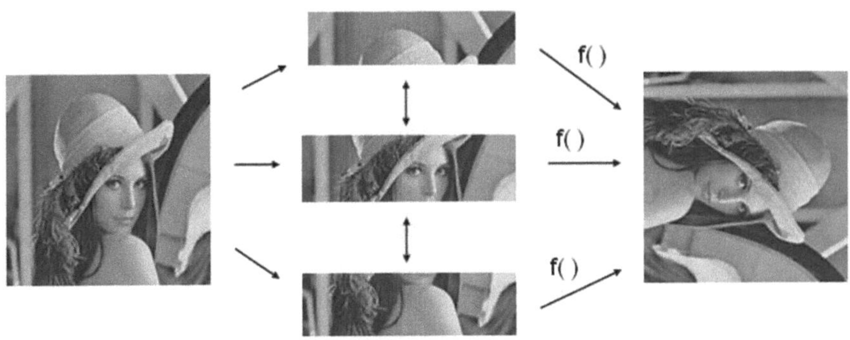

Figure 1. An intuitive example to show how images are distributed, elaborated and collected among the MPI processes; in this case we considered three processes

Table 1. Experimental results of the edge detection algorithm implemented using PIMA(GE)^2 Lib. We indicated the execution times in seconds, and we make explicit the obtained advantages through the speed-up values achieved

Processes	*1*	*2*	*4*	*6*	*8*
Execution time	1037.6 s	564.7 s	326.1 s	243.5 s	208.8 s
Speed-up	–	1.8	3.2	4.3	5

3.2 Experimental Results

In order to evaluate the speed-up achievable using PIMA(GE)^2 Lib we briefly present the experimental results obtained with the implementation of an edge detection algorithm using our library. We consider an image representing human tissue, whose size is 1 MB. This image and the resulting one are presented in Figure 2. Table 1 reports execution times in seconds and the corresponding speed-up values for the edge detection algorithm using up to eight nodes of a Linux cluster. The interconnection is provided by a Gigabit switched Ethernet and each node is equipped with a 2.66 GHz Pentium processor, 1 GB of RAM and two EIDE disks interface in RAID 0. We can see that the algorithm requires about 17 min to process the image sequentially, while using 8 nodes we can speed up the execution about 5 times.

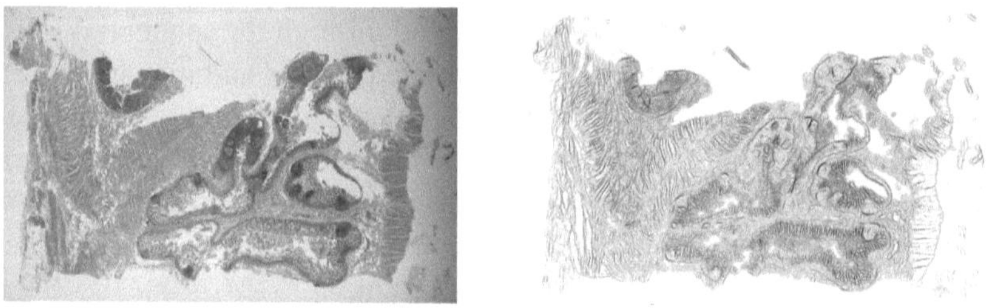

(a) (b)

Figure 2. An example of edge detection application developed using the PIMA(GE)^2 Lib. In **(a)** a tissue image is represented before the elaboration; in **(b)** there is the resulting image that represents the edge structure of the tissue

4. USING GRID SERVICES TO INTEGRATE PIMA(GE)^2 LIB IN A GRID

The present version of the library can be employed to develop applications that may be executed on Grid resources. To do this a user has to

1. select the architecture he/she wants to use;

2. write the source code of the application;

3. compile it with the compiler and the library version corresponding to the selected architecture;

4. select computational resources;

5. transfer the input data and the executable on the selected resource(s);

6. launch and monitor the execution;

7. transfer the output data on the machines he/she specifies;

8. remove all the files from the selected resource(s).

The aim of our work is to leave to the user only the second step of the previous list of operations, while the others are managed by the library and the Grid Service. In particular the management of the other steps may be completely delegated to an automatic tool, but we decided to let expert users cooperate with the Grid Service, at least in order to select the computational resources to use. This choice differs from the usual behavior of most of the job submission tools [1, 26], that is to exclude the users from the resource selection step. However, in our opinion this is, in many cases, a good way to achieve high performance executing parallel applications.

Actually the efficient execution of a parallel application is a complex task in the Grid environment. A first screening of the computational resources is easy to obtain on the basis of common application requirements, such as the requested amount of computational power, aggregate memory or disk space. Instead, the effective selection of the resources that are able to provide the highest performance is complex to achieve.

The major issue is to limit the unavoidable overhead due to the Grid environment, in particular considering the need to move data among the repositories, the computational resources and the users using low bandwidth links. Furthermore, new problems may arise, for example if computational resources belonging to different organizations are coupled to perform a parallel job [21].

In the Grid environment, the library represents a service to develop image processing applications. For this reason it is natural to develop it using the Grid Service technology. PIMA(GE)^2 Grid is composed of three Grid Services:

- *PIMA(GE)^2 GS*, aimed to couple and execute the parallel image processing applications developed using PIMA(GE)^2 Lib;
- GEDA, a resource selection service already mentioned;
- the GridFTP file transfer service, provided by the middleware.

We briefly present the first prototype of the system.

4.1 The User Interface and the Functional Description

The user interacts with PIMA(GE)^2 Grid using the graphic interface of a Java client application, which is presented in Figure 3. We describe it considering the list of operations presented before.

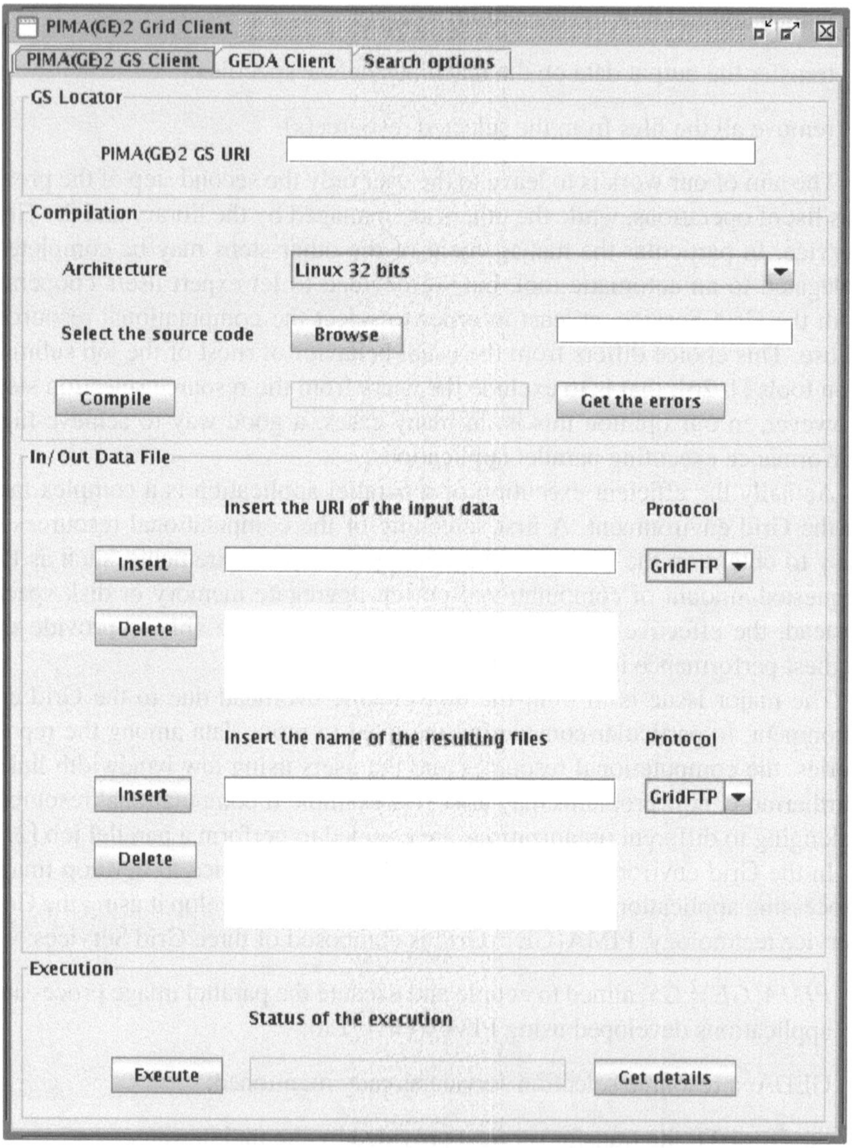

Figure 3. The interface of the java client that permits to interact with PIMA(GE)^2 Grid

The first parameter to provide is the Uniform Resource Identifier (URI) of PIMA(GE)^2 GS. Then, the user has to specify the C source code of the application and the platform to use. The corresponding file is uploaded using GridFTP, then it is compiled and linked with the library. Presently, we consider only the use of 32-bit Linux architectures. Thus the upload of the file and its compilation are made on the machine where PIMA(GE)^2 GS is running. In a future version, we will extend the system considering other possibilities.

After the compilation, the user receives a feedback which may be "Success" or "There are errors." In the second case the output is provided to the user that has to properly modify the code and re-upload it.

The next step is the selection of the computational resources to use. This is a common operation in many Grid applications, therefore we preferred to exploit a separate Grid Service rather than to implement an ad hoc solution. In particular we used *GEDA*, which represents the Grid Service version of a selection tool we already developed to aid the execution of parallel jobs. To simplify its use, we implemented in the PIMA(GE)^2 Grid Java interface a client for GEDA, whose graphic interface is presented in Figure 4.

The graphic interface permits users to browse the list of computational resources available in the Grid. It may present all the available resources or they can be limited by the user providing one or more requirements using the last tab. In particular in the current version of the system the research is limited by defaults to only 32-bit Linux architectures.

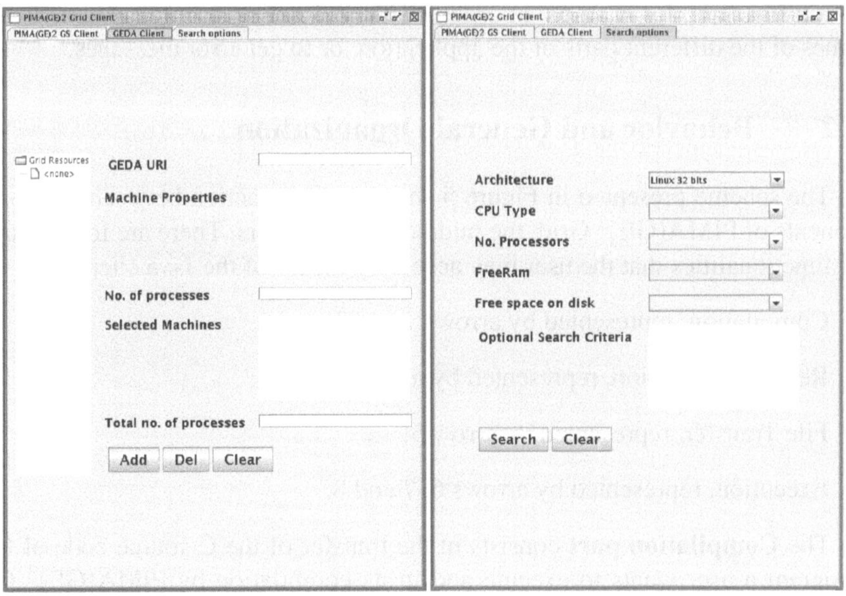

Figure 4. The interface of the java client that permits to interact with GEDA

Selecting each computational resource, its characteristics are presented in the corresponding text area of the first tab.

When a resource is chosen, by default the number of processes that will be launched on it is equal to the number of processors, but it is possible to modify it. The resulting parallelism degree of the application and the selected machines are also presented. Considering that a computational resource in Globus middleware is represented by the URI of its WS GRAM, the result of the selection step is a list of pairs *(URI, parallelism degree)*.

The last step to perform before the execution of the application is the indication of the URIs of input and output files. This step is performed using the first tab of the graphic interface corresponding to the PIMA(GE)^2 GS client. Notice that we provide the possibility to select the protocol to use. The selection of a specific protocol is an issue arising by the availability of several data replication systems, each of them using its own protocol, even if there are many proposals aimed to develop a uniform interface [16]. In the present version, however, we provide only the possibility to transfer the I/O data using the GridFTP protocol.

When all the parameters are specified the user has to press the *Execute* button, which corresponds to the execution of the application on the selected resources. During the execution a feedback is provided by PIMA(GE)^2 GS. The status of the job may be one of the following: "Submission," "Staging of Input files," "Computing processing," "Staging of output files," "Cleaning," "Done" or "There are errors." It is also possible to require the output of the program as a further file, for example in order to analyze statistical information, as the times of the different parts of the application, or to get error messages.

4.2 Behavior and General Organization

The scheme presented in Figure 5 shows the interaction between the components of PIMA(GE)^2 Grid, the middleware and users. There are four groups of functionalities that the user may access by means of the Java client:

- Compilation, represented by arrows 1 and 2;
- Resource Selection, represented by arrows 3 and 4;
- File Transfer, represented by arrow 5;
- Execution, represented by arrows 6, 7 and 8.

The **Compilation part** consists in the transfer of the C source code of the program a user wants to execute and in its compilation by PIMA(GE)^2 GS using PIMA(GE)^2 Lib and MPICH2. The resulting executable is stored on the disk and will be transferred on the resources selected for the execution.

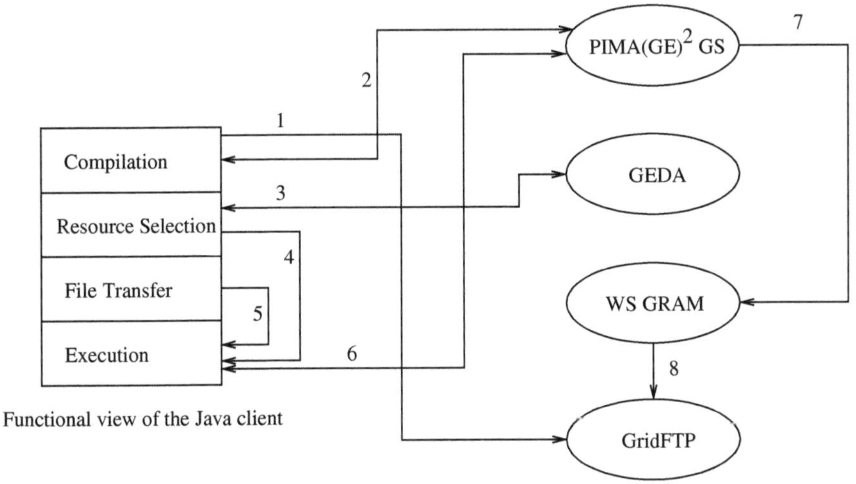

Figure 5. Interaction among the various entities involved in the execution of a user application using PIMA(GE)^2 Grid. *Numbers* represent the order in which they take place

The **Resource Selection** is provided by GEDA. In the present version of the system PIMA(GE)^2 GS and GEDA do not cooperate directly. The user selects the machines, then the resulting resource list will be provided to PIMA(GE)^2 GS by the client, and not through a direct interaction between the two Grid Services. In a future version we plan to provide a default selection of resources. It will be based on an estimation of the computational requirements of the application, and on a possible execution time threshold provided by the users. In both cases we will maintain the possibility to modify the proposed selection of machines, in order to exploit possible further information, for example the presence of a particular optimized version of some software.

It is to notice that the best performance will be obtained if all the selected machines belong to the same LAN and share a parallel file system, as PVSF. Indeed the parallel operations provided by the library are mostly based on the data parallel approach, therefore parallel processes may need to frequently communicate and exchange data. If the machines belong to the same LAN, the communication time is lower than we select machines belonging to different organizations. Another advantage is represented by the presence of a parallel file system. We implemented a version of the I/O operations that is optimized for the use of MPI-IO, and we collect satisfactory results using PVFS.

The **File Transfer part** is represented by the specification of the URIs of the I/O files of the program. These files will be transferred to/from each computational resource or only to/from the parallel file system using the GridFTP protocol. These transfers are made in conjunction with the execution of the application, managed by the **Execution part**.

All the information collected in the two previous parts in fact are provided to PIMA(GE)^2 GS by clicking the Execute button. The execution of the application is performed by PIMA(GE)^2 GS that produces an RSL2 file containing all the necessary parameters. The resulting file is then submitted to the first WS GRAM of the list of computational resources to start the execution.

4.3 The Performance

We considered the use of the Grid we built with other research institutes in Genoa for the execution of the same edge detection algorithm presented in Section 3.2. We selected, through GEDA, the same cluster than in the previous case, while both user and the PIMA(GE)^2 GS exploit different workstations.

We noticed that the performance achieved by the application by itself was the same than using PIMA(GE)^2 Lib, while the overall execution time increases by many seconds because of the need to transfer the I/O files and to set up the execution environment.

Those overheads do not depend on the sequential or the parallel nature of the application, but on the size of the images and the network characteristics.

5. RELATED WORKS

In the porting of existing software tools in different environments, there are two common solutions: the re-engineering and re-development of the code or the encapsulation of the code with a proper technology [27].

Also in the case of the Grid environment, it is possible to adopt both strategies. Most of the projects aimed at the integration of parallel libraries or tools in the Grid environment follow the second one. In particular they use the Grid Service technology, because it permits to achieve the required interoperability level with few modifications to the code [10, 29].

These projects can be further subdivided into two classes. In the first class, the legacy code functions are called in a Grid Service, e.g., using the Java Native Interface [20]. In the second class, the solution is the execution of the whole legacy parallel application by a Grid Service using external commands. In particular the MPI-based applications are executed using the *mpirun* command.

In the first class one of the most important projects is the GrADS project [29]. The aim of GrADS is to develop a framework that incorporates scheduling policy based on the application requirements and on the performance dynamic monitoring of resources. It is obtained using various specific Grid components, such as Globus MDS, NWS and Autopilot. Each application running in the GrADS system is responsible to select the resources and monitor the execution on the basis of its performance contract.

GrADS performs a set of experiments toward the porting of specific functions of parallel libraries such as ScaLAPACK [11] and PETSc [3]. These experiments are described in [22] and belong to the first class. Their approach considers each library operation separately, integrating it in an atomic way and adding proper scheduling policies in order to improve its performance. Each legacy function is wrapped in the GrADS framework with the definition of a new GrADS function. It has the purpose to execute the native code, to dynamically select and monitor the resources and also to manage the aspects related to the fulfillment of the performance contracts.

In our case this solution is infeasible since we want to inherit the library optimization policy that requires to know the sequence of function calls.

An example of the second class is proposed in [12]. This solution is fast and efficient, because the Grid Service is independent of the implementation and execution details of the MPI application. Considering that this means the possibility to maintain the policies of our library we decided to follow this approach. This solution represents a way to encapsulate parallel code that could be assumed more in general in service-oriented and object-oriented approaches, like Web Services and CORBA. For example, this way of integrating parallel code in the CORBA framework is proposed in [19].

At the best of our knowledge PIMA(GE)^2 Grid is one of few examples where an entire native library is ported on the Grid. Furthermore, its ease of use in the development of parallel applications and in the exploitation of Grid resources for their execution represents a relevant feature of the work. PIMA(GE)^2 Grid represents a general solution in the porting and execution of parallel legacy code on a Grid.

6. CONCLUSIONS

In this chapter the integration of the Parallel IMAGE processing GEnoa library (PIMA(GE)^2 Lib) in a Grid environment is presented. The proposed approach represents a general way to produce effective tools for virtual laboratories and presents many advantages.

First of all, users have the same degree of freedom than using the original version of the library to develop applications. We did not port only a particular set of functions, instead we made available the whole API of the library. In this way we allowed the integration of the generic applications that is possible to develop using the library, achieving also the goal of inheriting the optimization policy and the flexibility of PIMA(GE)^2 Lib.

Furthermore, we allow the exploitation of the whole amount of computational power provided by the Grid, and at the same time, providing Grid Services deployed in advance, we hide the burdens related with the execution of parallel applications and with the use of the Grid as well. PIMA(GE)^2 Grid acts as an abstraction layer that coordinates Grid Services, data and image processing applications and as a consequence enables a simplified exploitation of the architecture.

However, it is to note that the performance of the applications depends on the selected resources. A careful resource selection also permits the reduction of the data transfer time. This aspect is of particular importance when PIMA(GE)^2 Grid is used in virtual laboratories. Normally the data repository, corresponding to a particular instrument, is connected with one or more computing centers with a dedicated large bandwidth network. If the user application is executed using the computational capabilities of one of these centers, the transmission time and consequently the whole execution time will be shorter.

We plan to implement many improvements to the present version of PIMA(GE)^2 Grid. We want to allow users to extend the operations provided by the library with their applications. In this manner a user that has often to execute the same algorithm on newly produced data will recall the stored application without the need to re-upload it. The perspective is to produce an extensible library of software components, representing higher-level functions.

At the same time we plan to add an intelligent scheduler module to PIMA(GE)^2 Grid in order to provide a default resource selection based on a given temporal threshold for the production of results. This solution requires the development of a cost model for each library function. On the basis of the information provided by GEDA, we plan to implement a scheduler to propose a matching between the user requests and a set of computational resources.

ACKNOWLEDGMENTS

This work has been supported by FIRB strategic project on Enabling Technologies for Information Society, Grid.it, and by the regional program of innovative actions PRAI-FESR Liguria.

REFERENCES

[1] G. Andronico, V. Ardizzone, R. Barbera, R. Catania, A. Carrieri, A. Falzone, E. Giorgio, G. La Rocca, S. Monforte, M. Pappalardo, G. Passaro, G. Platania (2005). "GILDA: The Grid INFN Virtual Laboratory for Dissemination Activities". *Procs of TRIDENTCOM'05*, 304–305.

[2] K. Amin, G. von Laszewski, M. Hategan, R. Al-Ali, O. Rana, D. Walker (2006). "An Abstraction Model for a Grid Execution Framework". *Journal of Systems Architecture*, vol. 52, no. 2, 73–87.

[3] S. Balay, W. Gropp, L. Curfman McInnes, B. Smit (1996). "PETSc 2.0 Users' Manual". *Technical Report ANL-95/11 revision 2.0.17*, Argonne National Laboratory.

[4] R.F. Boisvert (2002). "Mathematical Software: Past, Present and Future". *Computational Science, Mathematics, and Software*, Purdue University Press.

[5] A. Clematis, A. Corana, D. D'Agostino, V. Gianuzzi, A. Merlo (2006). "Resource Selection and Application Execution in a Grid: A Migration Experience from GT2 to GT4". *Procs. of the International Symposium on Grid computing, high-performAnce and Distributed Applications (GADA)*, LNCS 4276, Springer, 1132–1142.

[6] A. Clematis, D. D'Agostino, A. Galizia (2005). "A Parallel IMAGE Processing Server for Distributed Applications". *Parallel Computing: Concurrent & Future Issues of High-End Computing*, John von Neumann Institute for Computing (NIC) series, vol. 33, 607–614.

[7] A. Clematis, D. D'Agostino, A. Galizia (2005). "An Object Interface for Interoperability of Image Processing Parallel Library in a Distributed Environment". *Procs. of ICIAP 2005*, LNCS 3617, Springer, 584–591.

[8] A. Clematis, D. D'Agostino, A. Galizia (2007). "Parallel I/O Aspects in $PIMA(GE)^2$ Lib". Parallel Computing: Architectures, Algorithms and Applications, Proceedings of the ParCo 2007, Advances in Parallel Computing, vol.15, IOS Press, 441–448.

[9] A. Clematis, D. D'Agostino, A. Galizia (2006). "The Parallel IMAGE processing GEnoa Library: PIMA(GE)^2 Lib". *Technical Report IMATI-CNR-Ge*, no. 21/2006

[10] T. Delaitre, T. Kiss, A. Goyeneche, G. Terstyanszky, S. Winter, P. Kacsuk (2005). "GEMLCA: Running Legacy Code Applications as Grid Services". *Journal of Grid Computing*, vol. 3, no. 1–2, 75–90.

[11] J. Dongarra, L.S. Blackford, J. Choi, A. Cleary, E. D'Azeuedo, J. Demmel, I. Dhillon, S. Hammarling, G. Henry, A. Petitet, K. Stanley, D. Walker, R.C. Whaley (1997). *ScaLAPACK user's guide.* Society for Industrial and Applied Mathematics, Philadelphia.

[12] E. Floros, Y. Cotronis (2004). "Exposing MPI Applications as Grid Services". *Procs. of EuroPar 2004*, LNCS 3149, Springer, 436–443.

[13] I. Forster, C. Kesselman (2004). *The Grid: Blueprint for a New Computing Infrastructure*, 2nd edition. Morgan Kaufmann.

[14] S. Hastings, T. Kurc, S. Langella, U. Catalyurek, T. Pan, J. Saltz (2003). "Image Processing or the Grid: A Toolkit or Building Grid-Enabled Image Processing Applications". *Procs. of the 3rd International Symposium on Cluster Computing and the Grid*, 36–43.

[15] P. Heinzlreiter, D. Kranzlmuller (2003). "Visualization Services on the Grid: The Grid Visualization Kernel". *Parallel Processing Letters*, vol. 13, no. 2, 135–148.

[16] T. Ho, D. Abramson (2006). "A Unified Data Grid Replication Framework". *Procs. of the 2nd IEEE International Conference on e-Science and Grid Computing*, IEEE Computer Society, 52.

[17] X. Huang, L. Huang, M. Li (2006). "Grid-Enabled Medical Image Processing Application System Based on OGSA-DAI Techniques". *APWeb Workshops 2006*, LNCS 3842, Springer, 460–464.

[18] J. Lebak, J. Kepner, H. Hoffmann, E. Rudtledge (2005). "Parallel VSIPL++: An Open Standard Library for High-Performance Parallel Signal Processing". *Proceedings of the IEEE*, vol. 93, no. 2, 313–330.

[19] M. Li, O.F. Rana, D.W. Walker (2001). "Wrapping MPI-Based Legacy Codes as Java/CORBA Components". *Future Generation Computer System*, vol. 18, no. 2, 213–231.

[20] S. Liang (1999). *Java Native Interface: Programmer's Guide and Specification.* Pearson Education.

[21] M.S. Müller, M. Hess, E. Gabriel (2003). "Grid Enabled MPI Solutions for Clusters". *Procs of the 3rd IEEE International Symposium on Cluster Computing and the Grid (CCGrid 2003)*, 18–25.

[22] A. Petitet, S. Blackford, J. Dongarra, B. Ellis, G. Fagg, K. Roche, S. Vadhiyar (2001). "Numerical Libraries and the Grid". *The International Journal of High Performance Computing Applications*, vol. 15, no. 4, 359–374.

[23] G. Ritter, J. Wilson (2001). *Handbook of Computer Vision Algorithms in Image Algebra*, 2nd edition. CRC Press.

[24] F.J. Seinstra, D. Koelma (2001). "The Lazy Programmer's Approach to Building a Parallel Image Processing Library". *Procs. of the 15th International Parallel & Distributed Processing Symposium (IPDPS 2001)*, IEEE Computer Society, 115.

[25] F. Seinstra, D. Koelma, J.M. Geusebroek (2002). "A Software Architecture for User Transparent Parallel Image Processing". *Parallel Computing*, vol. 28, no. 7–8, 967–993.

[26] G. Sipos, P. Kacsuk (2006). "Multi-grid, Multi-user Workflows in the P-GRADE Portal". *Journal of Grid Computing*, vol. 3, no. 3–4, 221–238.

[27] H.M. Sneed (2000). "Encapsulation of Legacy Software: A Technique for Reusing Legacy Software Components". *Annals of Software Engineering*, vol. 9, no. 1–4, 293–313.

[28] A. Szalay, J. Gray (2001). "The World-Wide Telescope". *Science*, vol. 293, no. 5537, 2037–2040.

[29] The Grid Application Development Software Project (GrADS), http://hipersoft.rice.edu/grads/index.htm.

[30] The Human Proteome Folding Project, http://www.grid.org/projects/hpf/.

[31] N. Wang, D.C. Schmitdt, D. Levine (2000). *An Overview of the CORBA Component Model*. Addison-Wesley.

REMOTE OPERATIONS OF AN ACCELERATOR USING THE GRID

M. Prica[1], R. Pugliese[1], C. Scafuri[1], L. Del Cano[1], F. Asnicar[1], and A. Curri[1]

[1]*Sincrotrone Trieste S.C.p.A., Strada Statale per Basovizza 14, km 163.5, 34012 Trieste, Italy*

Abstract GRIDCC [7] is a 3-year project funded by the European Commission. Its goal is integrating instruments and sensors with Grid computing and storage resources. Its middleware is being designed bearing in mind use cases from a very diverse set of applications and, as a result, the GRIDCC architecture provides access to the instruments in as generic a way as possible. GRIDCC is also developing an adaptable user interface and a mechanism for executing complex workflows in order to increase both the usability and the usefulness of the system. The new middleware is incorporated into a few significant applications that will allow software validation in terms of functionality and quality of service. The pilot application this chapter focuses on is applying GRIDCC to support remote operations of the ELETTRA synchrotron radiation facility. We describe the results of implementing complex GRIDCC workflows involved in both routine operations and troubleshooting scenarios. In particular, the implementation of an orbit correction feedback shows the level of integration of instruments and traditional Grid resources that can be reached using the GRIDCC middleware.

1. INTRODUCTION

Grid research has concentrated mainly on the marshaling of computation and data and their interactions at hubs of analysis and synthesis. This focus has spawned the notion of computational and data Grids, respectively. As the technology for these has matured, increasing attention is being directed toward the actual sources of data: instruments and sensors. Scientific instruments and sensors provide the raw observations used to develop, falsify and verify scientific theories and thus drive the scientific progress. The scientific process begins not with the data, but rather with the preceding data collection. This collection interacts profoundly with interpretation and analysis, whether by human or by machine. Another important trend is the increasing use of in silico experiments and the integration of model-based simulation with physical sensors. These projects further couple the data collection with analysis and simulation. In these scenarios, a unified paradigm for Grid-enabling both physical instruments and virtual instruments will result in flexible, extensible systems for exploiting the tremendous potential of information technology. The above considerations suggest that we seek to push the Grid to the edges of the practice of science: to the point where data are collected and to the point where the scientists use data; and thus create an end-to-end Grid environment for scientific activity. Many instruments and sensors are already digitally accessible, but they are poorly integrated into the Grid. In this chapter we define an instrument to be any piece of equipment controlled through a computer interface such as telescopes, particle accelerators or power stations. Instruments work in real time and their successful operation often requires rapid interaction with conventional computing and storage resources as well as with other instruments. The control of instruments is often an interactive process. The real time and interactive nature of instrument control provides a critical requirement for the definition of acceptable quality of service (QoS) constraints for interactions between the different Grid components. In order to meet these defined QoS constraints it is often necessary to provide mechanisms for reserving the instruments [8], the required computing and storage elements and network bandwidth. While running an experiment, an application scientist will inevitably need to perform more than one task. A common scenario involves running an experiment using a piece of equipment, analyzing the obtained results and then in the light of this analysis eventually conducting further experiments. Hence, the need to provide the best possible support for workflows, defined as the execution and management of sets of dependent and potentially heterogeneous tasks. (In this chapter, the term "task" is used to indicate a general piece of work, while the term "job" refers specifically to a computational task.) Workflow support must be powerful enough to deal with a myriad of possible workflows that the application scientist may wish to describe, but also presented in simple and intuitive manner. It should not be assumed that application scientists are competent software engineers, nor should they be forced to act as such.

Application scientists should be presented with a similar interface to the Grid and to the instruments they are already familiar with. For this purpose, the GRIDCC project is developing a virtual control room (VCR) [10] environment that presents to its users a tailored interface both to the Grid and the instruments. One of the barriers to placing instruments onto the Grid in the past has been the risk of damage to the instruments caused by (unintentional) misuse by inexperienced users. Protection of the instruments through appropriate checks and measures becomes thus an essential requirement. These checks and measures include local problem solvers, which are capable of diagnosing situations that would cause damage to an instrument, e.g., by monitoring the temperature of the positioning motors on a telescope and shutting them down in the case of overheating. Likewise, there is a need for global problem solvers required to identify problems that cannot be detected or remedied by a single instrument. Considerable effort has already been made within the e-science community to develop Grid middleware through a number of projects [4, 9, 13, 14]. Whenever possible, GRIDCC builds on these projects and the community standards and extends them with the added functionality. The GRIDCC architecture implementation is being developed within the gLite [6], a Grid middleware built within the EGEE [3] project. Among the many workflow description languages available for use within the Grid, BPEL [1] has been selected, having a number of commercial and open source engines becoming available.

2. THE GRIDCC ARCHITECTURE

In this section we present the overall architecture [15] for the GRIDCC project. The project is composed of several service groups (Figure 1), each of which may be decomposed into a number of individual services.

- Virtual control room (VCR): The VCR is the application scientists' interface into the GRIDCC system. VCR offers a customizable, user-friendly interface to the interactive real-time control of instruments and a complete collaborative environment for support of the group work.
- Information service (IS): The IS provides a repository for information collected within the architecture. It may be queried to discover information.
- Global problem solver (PS): The PS uses information from the IS in order to diagnose potential situations which may lead to damage of the instrument or other elements. It is also responsible for taking whatever actions are appropriate to remedy these situations.
- Security services (SS): These services are omnipresent throughout the GRIDCC architecture and provide the mechanisms for performing authentication and authorization of users.
- Execution service (ES): The ES is responsible for taking a workflow defining a sequence of tasks to be performed and ensuring that it is executed with all existing QoS requirements.
- Instrument element (IE): The IE [5] is an abstraction of an instrument (or a group of instruments) into a standard interface that can be used within the rest of the GRIDCC architecture.

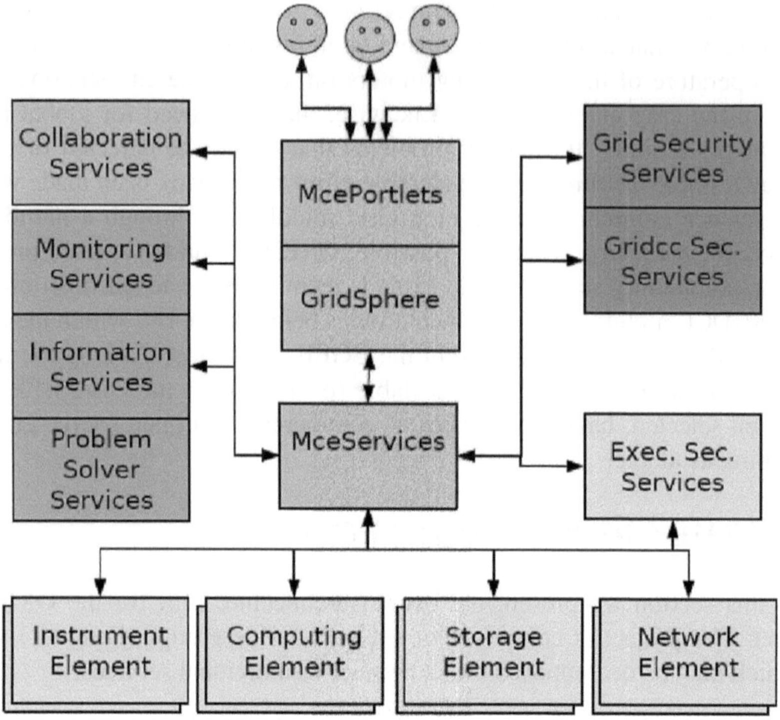

Figure 1. GRIDCC architecture

- Compute element (CE) and storage element (SE): The CE and SE are usual Grid abstractions of computational resource(s) and storage resource(s).

3. APPLYING GRIDCC TO REMOTE OPERATIONS OF AN ACCELERATOR

Remote operations of an accelerator (e.g., ELETTRA) involve planning of the accelerator operations, maintenance of the accelerator and its troubleshooting, repair of delicate equipment; understanding and pushing performance limitations; performing commissioning and set ups and various routine operations. All these activities are based on large amounts of information, which are at present accessible only at the accelerator site. All the high-level software (HLS) of ELETTRA, a third-generation synchrotron radiation facility located in Trieste (Italy), is suitable to be implemented using the GRIDCC testbed (Figure 2). ELETTRA HLS consists of application programs like OneBM (One Button Machine) and OrbitFB (Orbit FeedBack). The implementation of the OrbitFB application via a GRIDCC workflow, a real feedback via Grid, shows

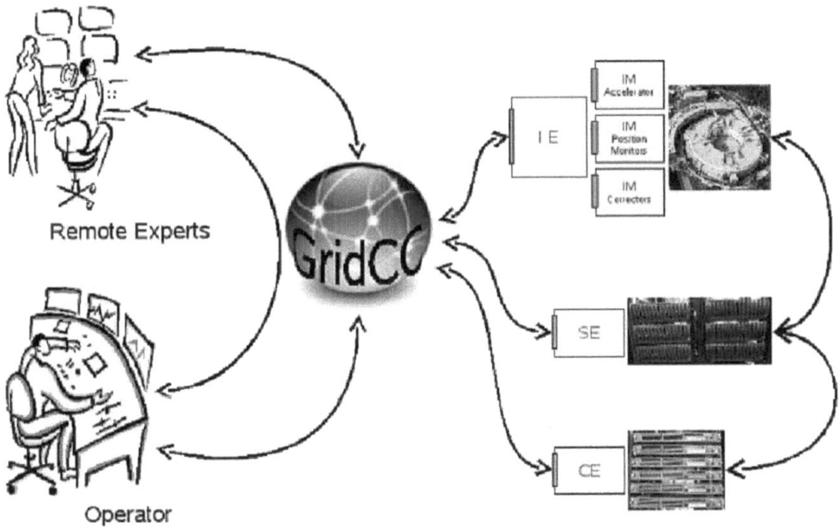

Figure 2. GRIDCC applied to far remote operations of ELETTRA

the level of integration of instruments and traditional Grid resources, which can be reached using the GRIDCC middleware. The machine operators and remote experts in a virtual control room can collaboratively operate the accelerator in a routine and in a troubleshooting scenario. In the OrbitFB application, beam positions are sent via Grid to a remote computing center where orbit correction code is running and the corrections are fed back to the corrector elements, thus improving the orbit of the storage ring. Remote control of an accelerator facility has the potential of revolutionizing the mode of operation and the degree of exploitation of large experimental physics facilities. Grid technologies facilitate the sharing of these resources across a distributed computing environment. In particular, the Grid handles issues of authentication, authorization, resource description and location, data transfer and resource accounting, all of which make Grid technologies extremely useful for implementing highly distributed control systems such as those of an accelerator. Grid technologies can also be used to integrate operations with computing farms where complex machine physics models can run.

The implementation of the OneBM and the OrbitFB programs using the GRIDCC middleware and the involved complex workflows is a demanding task. The OneBM application controls different parts of the accelerator and turns them on gradually executing a set of tasks (Figure 3), monitoring the outcome of each task and taking corrective actions in case of malfunctions in order to have the machine up and running at the desired working parameters. In this case, different components of the GRIDCC architecture are involved

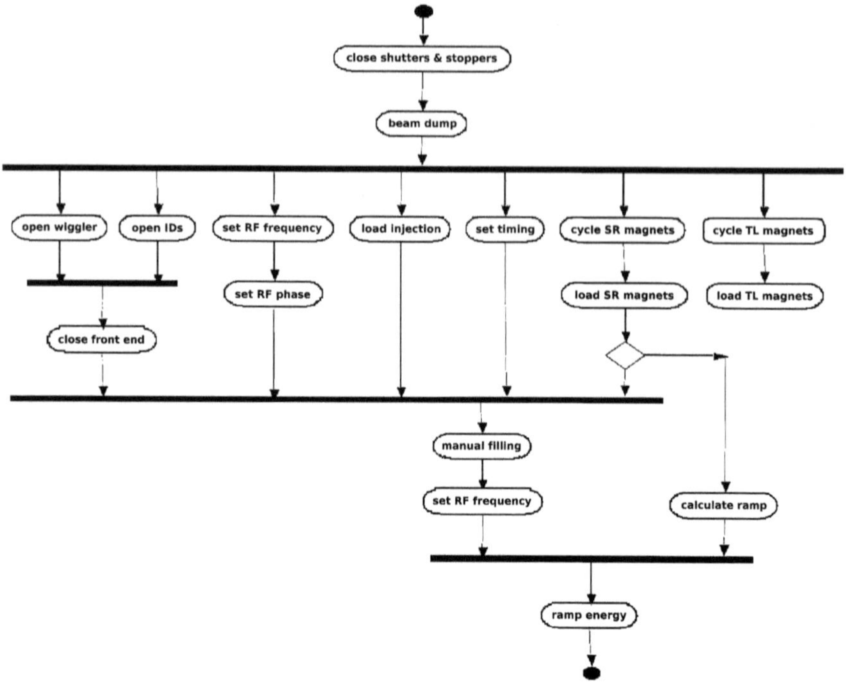

Figure 3. OneBM workflow flowchart

and interact frequently, in particular the instrument elements and the execution services. In the OrbitFB (Figure 4) application the data coming from the machine are fed to a simulator to correct and optimize machine parameters in a loop. Implementing this challenging application required the development of a new machine simulator that can be controlled via Tango [12], the control system used at ELETTRA. As we are developing a generic bridge between the IE middleware and the Tango control system it will be very simple to move from a simulated environment to the real accelerator. The simulator may thus be effectively used to develop the pilot application. Moreover, the OneBM and the OrbitFB have to be coded as GRIDCC workflows and the corrector algorithm has to be coded as a separate program to be executed on a computing resource. A very useful side effect of porting the HLS to the GRIDCC testbed is the possibility for ELETTRA to set up a powerful training tool by which a user can learn to control the accelerator using the simulator without affecting the status of the real machine. We have implemented an IE with three instrument managers (IMs). The first IM is the accelerator, which allows to control the general parameters of the machine and implements commands to initialize, start and stop the machine and inject current into the storage ring. Also, it allows injection of the beam orbit distortions used in simulations. The PositionMonitors

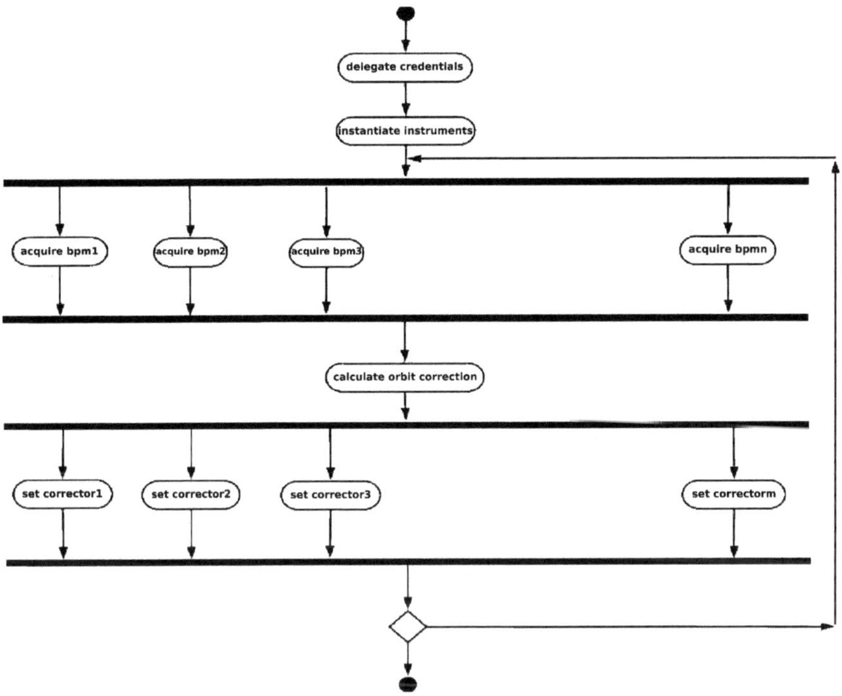

Figure 4. OrbitFB workflow flowchart

IM allows the acquisition of the beam orbit in the storage ring of the accelerator through beam position sensors and saving the obtained data either locally or on a remote Grid storage element. The Correctors IM allows the correction of the beam orbit acting on a set of corrector magnets distributed along the storage ring. The values applied to the corrector magnets are the result of the orbit correction calculation and may be uploaded from the local file system or from a remote Grid storage. The IMs are all interfaced with a set of Tango devices that implement the machine simulator and in the future they could be connected to the Tango devices that control the real machine. An additional IM, the OrbitCorrector, has been developed to interact with a local computing resource that executes the orbit correction code.

4. RESULTS

In the current setup a remote operator can log in the virtual control room and launch the OneBM workflow that starts the machine simulator and gradually injects current into the storage ring of the accelerator, reaching the normal operating conditions. The operator may then interact with the machine, modifying

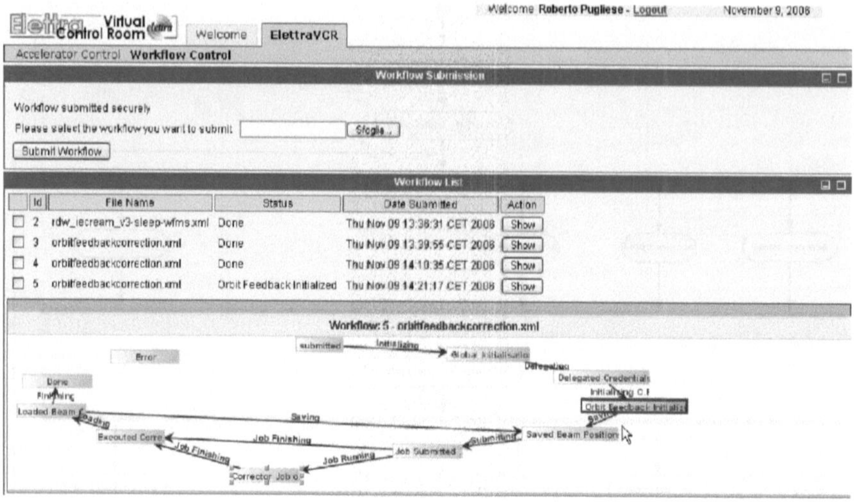

Figure 5. Workflow submission and monitoring

the orbit by injecting a random distortion and thus simulating a problem. VCR offers the operator a set of collaboration tools such as Skype [11] and VRVS [16] that allow contacting remote experts and discussing operating procedures or seek advice in case of possible malfunctions. To deal with an orbit distortion, an expert would likely suggest launching the OrbitFB workflow. The operator can monitor the workflow execution (Figure 5) from the VCR and at the same time interact with the PositionMonitors IM to visualize the orbit (Figure 6) and

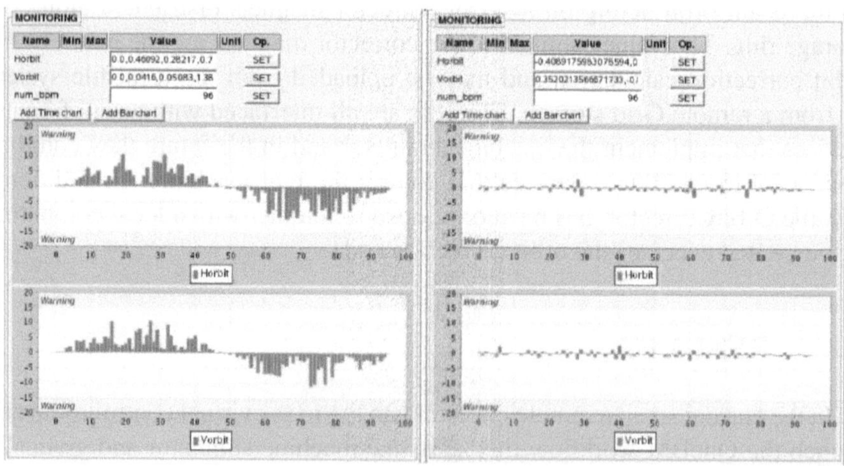

Figure 6. Interacting with an IM: beam before and after running the OrbitFB

evaluate the quality of the beam. The OrbitFB has been tested in three different setups: The first one uses an IE to interact with a local CE, an approach similar to the one used by the NEESgrid [8] project. With this configuration we managed to run the OrbitFB loop at 1 Hz. The second setup uses CREAM [2] as a web service interface to the specific remote CE. We managed to run the OrbitFB loop every 5 s with this setup. In the final setup we used the WMProxy [17] as a web service interface to the gLite broker. As expected, it turned out to be the slowest setup, with OrbitFB loops running every 20 s.

5. CONCLUSIONS

The far remote operations of the ELETTRA accelerator pilot application shows the potential of the GRIDCC middleware in accessing and controlling distributed instrumentation in a large experimental physics facility. It also demonstrates how Grid technologies can be used to integrate operations with computing farms where complex machine physics models can run. VCR, the GRIDCC user interface, provides operators and experts with a set of tools that allow collaborative efforts in coping with everyday tasks as well as handling exceptional events. This pilot application will evolve with the further development of the GRIDCC middleware. GRIDCC will use the agreement service layer between the workflow management system and the grid resources, providing the resource reservation service. An integrated problem solver will monitor the running tasks in order to identify error conditions and provide support for fault-tolerant operation.

ACKNOWLEDGMENTS

We would like to thank the GRIDCC team and the ELETTRA machine physics and machine operators groups. The GRIDCC project would not have been possible without the contribution of the European Commission (FP6, IST-511382).

REFERENCES

[1] BPEL4WS. ftp://www6.software.ibm.com/software/developer/library/ws-bpel.pdf.

[2] CREAM. http://grid.pd.infn.it/cream/.

[3] ELETTRA Sincrotrone Trieste. http://www.elettra.trieste.it/.

[4] Enabling Grids for E-sciencE. http://public.eu-egee.org/.

[5] E. Frizziero, M. Gulmini, F. Lelli, G. Maron, A. Oh, S. Orlando, A. Petrucci, S. Squizzato, S. Traldi. "Instrument Element: A New Grid Component that Enables the Control of Remote Instrumentation". http://www.gridcc.org/getfile.php?id=1496.

[6] GLite: Lightweight Middleware for Grid Computing. http://glite.web.cern.ch/glite/.

[7] GRIDCC Grid Enabled Remote Instrumentation with Distributed Control and Computation. http://www.gridcc.org/.

[8] NEESgrid project. http://it.nees.org/.

[9] Platform LSF. http://www.platform.com/products/LSF.

[10] R. Pugliese, F. Asnicar, L. Del Cano, L. Chittaro, R. Ranon, L. De Marco, A. Senerchia "Collaborative Environment for the Grid – Multipurpose Collaborative Environment". *Distributed Cooperative Laboratories*. Springer 2006.

[11] Skype communication tool. http://www.skype.com.

[12] Tango Control System. http://www.tango-controls.org/.

[13] The Condor Project. http://www.cs.wisc.edu/condor/.

[14] The Globus Alliance. http://www.globus.org/.

[15] The GRIDCC Architecture Document. http://www.gridcc.org/getfile.php?id=1382.

[16] VRVS Virtual Room Videoconferencing System. http://www.vrvs.org/.

[17] WMProxy. http://egee-jra1-wm.mi.infn.it/egee-jra1-wm/wms.shtml.

E-VLBI USING A SOFTWARE CORRELATOR

N. Kruithof*
Joint Institute for VLBI in Europe (JIVE), Postbus 2, 7990 AA Dwingeloo, The Netherlands
Kruithof@jive.nl

Abstract Very long baseline interferometry (VLBI) is a technique in astronomy where several distant radio telescopes observe the same source and the received signals are recorded and later correlated to produce an image with a larger resolution. An *e*-VLBI experiment is a VLBI experiment where the data are transferred directly over the Internet and correlated in real time.

The primary task of JIVE is the correlation of data from the EVN telescopes [2] all around Europe. A special purpose-built hardware correlator is used for data processing. We are currently investigating the capabilities of a next-generation software correlator using Grid processing power. In this chapter we describe the design of this software correlator.

1. INTRODUCTION

Very long baseline interferometry (VLBI) [8] is a type of interferometry used in radio astronomy, in which data received at several telescopes are combined to produce an image with very high resolution. VLBI can be used for both astronomy and geodesy. For astronomy, VLBI provides high-resolution images of radio sources in the sky, whereas in geodesy VLBI measures the location of the telescopes and the Earth orientation parameters (EOP).

Astronomical research aims to study the sky and requires high angular resolution. The resolution of the image increases linearly with the size of the telescope dish. However, it is not possible to build telescope dishes of arbitrarily large size. Instead, measurements of several telescopes can be combined using VLBI to simulate a telescope as large as the Earth. By measuring the

*On behalf of the FABRIC team.

cross-correlation of the signals between all telescope pairs, one is able to measure the angular Fourier components of the image in the sky.

VLBI

In order to approximate a telescope with a larger dish, multiple telescopes can observe the same object, and the data can be combined using interferometry. A pair of telescopes forms a baseline. The baseline projected to the plane orthogonal to the source direction defines the spatial frequency for a given observing frequency. So the maximal frequency is defined by the two telescopes farthest apart. Due to the rotation of the earth, these projected distances change during an experiment, giving the possibility to measure a range of spatial frequencies with a limited number of telescopes. The angular resolution of the VLBI array depends on the maximum projected baseline length, while the sensitivity depends on the number of telescopes and the bandwidth. Because the radio emission has a broad white noise spectrum, it is most efficient to use two-bit (four-level) sampling of the signal. The data rate is a limiting factor for the total bandwidth, as Nyquist sampling is required.

In practice, the data are recorded at the telescopes on disk packs during a VLBI experiment. After the experiment the disks are shipped to a central institute, the Joint Institute for VLBI in Europe (JIVE), for correlation. At JIVE, the data from the different telescopes are played back and correlated by a dedicated hardware correlator [6]. The maximal capacity of this hardware correlator is 16 telescopes at a data rate of 1 Gb/s each. In practice, there can be several weeks between the experiment and the time when the correlated data become available.

e-VLBI

In an electronic VLBI (*e*-VLBI) experiment [7], data from the telescopes are transferred directly over the Internet to JIVE, where they are streamed in real time into the correlator. The data transport from the telescopes to JIVE goes over several networks like local connections, paths provided by NRENs and the GÉANT backbone in Europe.

The sensitivity achievable using interferometry is proportional to the square-root of the data rate and the number of telescopes, whereas the angular resolution is proportional to the maximal distance between two antennas. Hence, heavy requirements are put on both the network connections and the computing power to achieve a good sensitivity, see also Table 1.

Transporting the data over the network has several advantages over a traditional experiment. Obviously, the results of the experiments are almost immediately available. This opens up the possibility to change the course of an

Table 1. Network bandwidths and computing power needed for an *e*-VLBI experiment based on a XF architecture

Description	# telescopes	# sub-bands	Data-rate (Mb/s)	Spect/prod	Tflops
Fabric-demo	4	2	16	32	0.16
1 Gb/s, full array	16	16	1024	16	83.39
Future VLBI	32	32	4096	256	~21457

experiment based on earlier findings. Also, *e*-VLBI allows for real-time analysis of the data and helps to identify and resolve minor technical problems in the data collection during the course of the experiment.

Several experiments in the past have shown that real-time *e*-VLBI is possible. The EC funds the EXPReS project[1] [3] which aims at building a production-level *e*-VLBI instrument of upto 16 intercontinental telescopes connected in real time to JIVE and available to the general astronomy community.

Correlation

Correlation is the process by which data from multiple telescopes are collected and combined to measure the spatial Fourier components of the image of the sky. The high data rates and the optimizations complicate the process.

Assume that we are correlating the signal of two telescopes. First, both signals are delayed to account for the different time at which the signal arrives at the telescopes, see Figure 1. Next, a phase shift is performed to compensate for the Doppler effect produced by the rotation of the Earth. This process requires very accurate timing information in the data and a very detailed model of the geometry of the experiment. The signals are now ready to be correlated.

During correlation, the first signal is delayed with discrete steps and each delayed signal is multiplied with the second signal and then integrated. The output is a summation per delay step.

For more than two stations, each station is correlated with itself (autocorrelation) and every other station (cross-correlation). Note that the complexity is quadratic in the number of telescopes.

[1]EXPReS is made possible through the support of the European Commission (DG-INFSO), Sixth Framework Programme, Contract #026642.

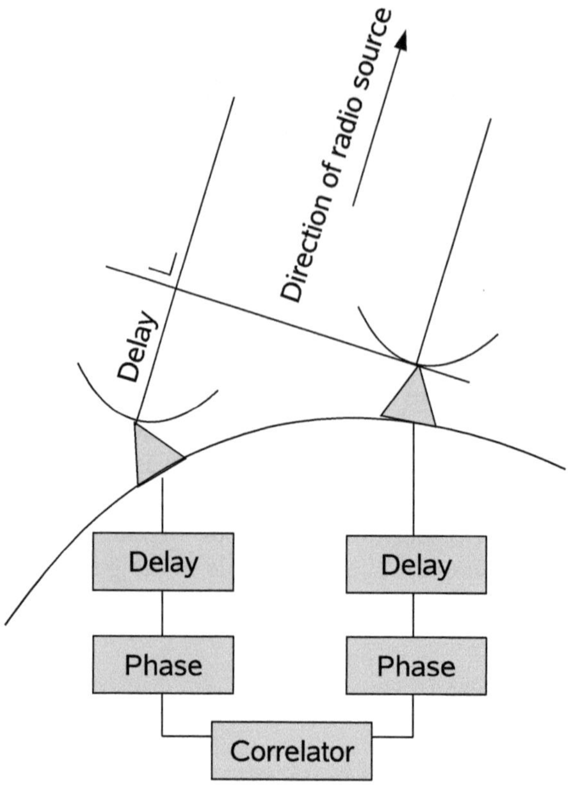

Figure 1. Block diagram of the correlation

2. SOFTWARE CORRELATOR

If the data in an *e*-VLBI experiment can be streamed over the Internet to JIVE, they can also be sent to another correlator. We are currently investigating the possibilities of a next-generation correlator using a computing Grid. The advantages of a software correlator over a new dedicated hardware correlator lie in its flexibility and accuracy. The software correlator can be tuned for special experiments. The main advantage of a dedicated hardware correlator is the greater performance. The advance of general-purpose computing is making software correlation a cost-effective solution for a range of applications. A similar approach is presented in [1].

The flexibility of its design allows the software correlator to change with the needs of researchers. In fact, the first version of the software correlator was developed to track the Huygens spacecraft during its descent through the atmosphere of Saturn's moon Titan. Due to the nature of this experiment,

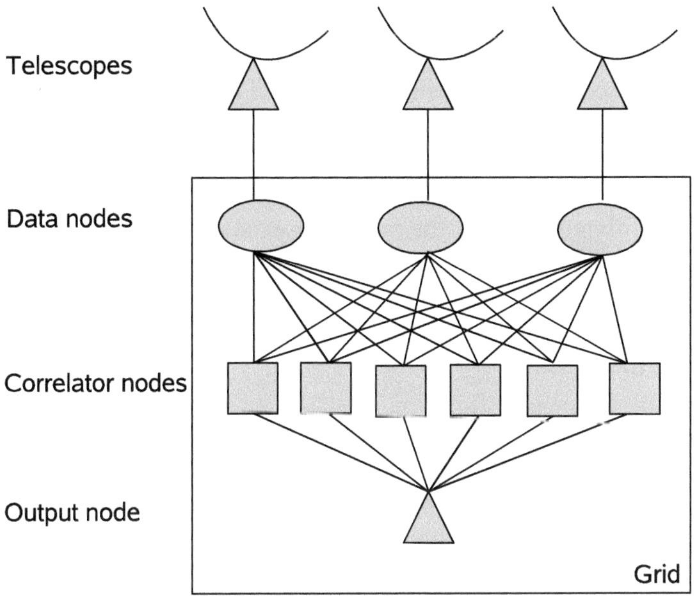

Figure 2. Outline of the network connections between different components in the software correlator

special requirements were put on the correlator, which the current hardware correlator was not able to provide. Moreover, we expect that the costs of developing a software correlator are much lower than the costs for a hardware correlator.

The correlation is done by splitting the signal in time slices that are processed in parallel (see Figure 2). The signal from a telescope is received by a single, so-called data node. The data node sends slices of data to an available correlate node. The correlate node receives data from all telescopes for a certain time slice and performs the correlation. The size of the output of the correlation is much smaller than the input size and can be collected and stored by a single output node. There is one manager node that assigns data to available computing nodes.

The correlation is not computationally expensive, in the sense that it requires only few operations per transferred bytes. However, due to the high data rates, the absolute number of clock cycles required by the application is still extremely high. Moreover, the problem is quadratic in the number of telescopes participating in the experiment since it is linear in the number of channel pairs that have to be correlated. The huge need for networking and computing power makes a computing Grid an ideal platform for this application.

Design

The software correlator is written in `C++` and uses several standard libraries like `fftw` [4], `mpich` [5] and the standard template library.

At initialization, each node is assigned a role: data node, correlator node, output node or manager node. The node then creates a number of controllers that manage different tasks of the node. For example, there are controllers for reading input, processing the data and writing the output. A single type of controller can be used by different kinds of nodes. If a node receives a message, it delegates the message to the different controllers. The proper controller can then process the message.

Data Node. The data node opens a controller for reading data and a controller for forwarding the data to the correlate nodes. It then connects the two using a buffer. The data node will receive a message from the manager node specifying how to obtain the input: from file or over the network using one of various types of transfer protocols. It will also receive messages containing a start and stop time and the correlate node to send the data to.

Correlate Node. A correlate node will initialize the correlation process and connect to the output node. It can receive messages from a data node asking to open an input connection and from the manager node to process a time slice. After the slice is processed the node will send a message to the manager node saying it is available for a next job.

Output Node. The output node will receive a message from the manager node where to store the data, and it allows connections from the correlate node to be opened. The node sorts the received data from the correlate nodes and stores it for further processing. The output node has to make the received data available to the user, and it should be archived in a proper way.

Manager Node. The manager node is the most complicated node in the software correlator. It sends messages to initialize the other nodes and tells them how to connect to each other. After the initialization, it maintains a list of available correlate nodes and delegates time slices to available correlate nodes. General error messages can also be sent from any node to the manager node.

The interface to the user will communicate with the manager node to send commands to the correlator and obtain status information from the correlator. The user interface will be based on virtual lab.

Network Connections. Since correlation is mainly a networking problem, testing and optimizing the data flows is of vital importance for the performance of the software correlator. We distinguish two types of data flows: containing control messages and the signals from the telescopes.

Control Messages. The control messages are sent between different nodes to regulate the correlation. They form a low bandwidth stream and MPI is used to send these messages. The network is mainly star shaped around the control node, but there are some connections during the initialization between input nodes and the correlate nodes and between correlate nodes and the output node. Since the messages control the correlator, the delivery of the messages has to be guaranteed.

Signal of the Telescopes. The signal from the telescopes requires far more bandwidth than the control messages. The constant throughput for these connections is very important as we are dealing with real-time data. Some packet loss is even acceptable if a data stream at a constant rate can then be maintained. The connections for these data streams are set up to be exchangeable such that it is possible to test different network protocols like TCP (using jumbo frames) or protocols used for streaming media. The network consists of three layers: from the telescopes to the data nodes, then to the correlate nodes and finally to the output node, as shown in Figure 2.

3. CONCLUSIONS

Within Future Arrays of Broadband Radio-telescopes on Internet Computing (FABRIC), which is a joint research activity in the EXPReS project, we are currently implementing the software correlator. The work flow management is designed by the the Polish Supercomputing and Networking Center in Poznan.

In order to be able to guarantee data transfer, it might be necessary to add an extra node near the telescopes with guaranteed bandwidth to the telescope. This node could buffer the input data in case the transfer rates to the Grid drop temporarily. When the software correlator is run on a cluster of super computers, this node can also send large time slices directly to alternating super computers. In this way the network connectivity is optimized by introducing an additional layer of input nodes.

Other research areas that need attention are found in resource leveling, dealing with delays in data transfers and managing insufficient computing power.

We also want to test several networking protocols to optimize the data transfer. For example, we could imagine using UDP for the main data stream, where some data loss is acceptable and a TCP connection for data-headers to guarantee their proper delivery.

REFERENCES

[1] A. T. Deller, S. J. Tingay, M. Bailes, and C. West. "Difx: a software correlator for very long baseline interferometry using multi-processor computing environments", 2007.

[2] EVN, Website: `http://www.evlbi.org`.

[3] EXPReS, Project website: `http://expres-eu.org`.

[4] M. Frigo and S. G. Johnson. "The design and implementation of FFTW3". *Proceedings of the IEEE*, 93(2):216–231, 2005. Special issue on "Program Generation, Optimization, and Platform Adaptation".

[5] W. Gropp, E. Lusk, N. Doss, and A. Skjellum. "A high-performance, portable implementation of the MPI message passing interface standard". *Parallel Computing*, 22(6):789–828, September 1996.

[6] R. T. Schilizzi, W. Aldrich, B. Anderson, A. Bos, R. M. Campbell, J. Canaris, R. Cappallo, J. L. Casse, A. Cattani, J. Goodman, H. J. van Langevelde, A. Maccafferri, R. Millenaar, R. G. Noble, F. Olnon, S. M. Parsley, C. Phillips, S. V. Pogrebenko, D. Smythe, A. Szomoru, H. Verkouter, and A. R. Whitney. "The EVN-MarkIV VLBI data processor". *Experimental Astronomy*, 12:49–67, 2001.

[7] A. Szomoru, A. Biggs, M. A. Garrett, H. J. van Langevelde, F. Olnon, Z. Paragi, S. Parsley, S. Pogrebenko, and C. Reynolds. "From truck to optical fibre: the coming-of-age of evlbi", 2004.

[8] J. A. Zensus, P. J. Diamond, and P. J. Napier, editors. "Very long baseline interferometry and the VLBA", 1995.

DIGITAL SCIENCE LIBRARY IN REMOTE INSTRUMENTATION SYSTEMS

M. Lawenda[1], N. Meyer[1], M. Stroiński[1], D. Kaliszan[1], T. Rajtar[1], M. Okoń[1], and D. Stokłosa[1]

[1] *Poznań Supercomputing and Networking Center, ul. Z.Noskowskiego 10, 61-704 Poznań, Poland*

Marcin.Lawenda@man.poznan.pl,

home page: http://www.ringrid.eu/

Abstract This chapter is devoted to the analysis of data management in remote instrumentation systems. Storage and management of significant amounts of data are crucial issues in every system where data need to be collected from different scientific equipment and next used to analysis or further processing. The authors propose their own solution, called digital science library (DSL), and analyze a specific implementation of DSL designed for nuclear magnetic resonance (NMR) spectroscopy. DSL, based on the data management system, is a distributed data-access environment. Its main function is to store and present data in the grid environment. Moreover, results of the conceptual work are presented in this chapter, starting with the requirements specification and finishing with the architectural design.

1. INTRODUCTION

There are numerous areas of science, industry and commerce that require broad international co-operation for their success. A number of problems may be addressed by using sophisticated equipment and top-level expertise, which is often locally unavailable. Therefore, the development and dissemination of techniques and technologies that allow virtualized, remote and shared access to industrial or scientific instruments is essential for the progress of society. The possibility of using scientific or industrial equipment independent of their physical location helps in the egalitarian attitude in using expensive facilities and for unification of communities and subsequently opens new opportunities for industry, science and business. That is why a new branch of grid science – remote instrumentation systems (RIS) – becomes more and more popular.

The specific representative of the RIS systems is the virtual laboratory systems (VL). One of them is PSNC Virtual Laboratory (VLab) [14], the research project developed in Poznań Supercomputing and Networking Center [8] since the beginning of the year 2002.

A crucial component in most typical RIS (and VL) systems is the module responsible for data storage and management (DSM). Performing remote experiments on laboratory devices in a multi-user environment is related with the necessity of storing results data, which could be next used in the post-processing and visualization stage. This requirement is especially true for experiments where some number of devices need to work in parallel, e.g., in e-VLBI experiments. The execution of e-VLBI experiments with 16 radio telescopes takes a few dozen terabytes (dependently on measurement resolution). It illustrates the scale of the problem.

The functional requirements which DSM systems should meet are not limited to storage and management only. These systems should provide their users the possibility to publish electronic papers and presenting digital content. Moreover, nowadays users want to have the possibility of using more and more sophisticated tools designed to create, browse and search for electronic documents. These tools perform an important role in the global information infrastructure and can benefit the education and scientific environments. These kinds of tools are called digital science libraries (DSLs).

The DSL which is implemented in the VLab system is created on the basis of the data management system (DMS) [6], which was developed for the PROGRESS project [9]. Its main functionality, which is storing and presenting data in grid environments, was extended with the functions specific to the requirements of the virtual laboratory [14].

This work presents the general assumptions, project design and implementation of the digital science library for the exemplary purpose of nuclear magnetic resonance (NMR) spectroscopy data. The digital science library should allow store and present data used by the virtual laboratories. These data can be input information used for scientific experiments, as well as the results of performed experiments. Another important aspect is the capability of storing various types of publications and documents, which are often created in the scientific process. This type of functionality, which is well known to all digital library users, is also provided by the DSL.

2. PSNC VIRTUAL LABORATORY

In general, VL is a distributed environment, providing remote access to the various kinds of scientific equipment and computational resources. In VLab users can submit their tasks in the form of dynamic measurement scenario –

the sets of experiments and computational tasks of different kind. These tasks form a dependencies' graph – describing the connections between them and all possible flow paths; the actual flow path is determined upon run-time – based on results obtained at each stage. The tasks are scheduled on one-by-one (or on group) basis, after the appropriate results from the preceding tasks are obtained. The virtual laboratory is not a stand-alone system. It was designed to cooperate with many other grid systems, providing only the purpose-specific functionality and relying on well-known and reliable grid solutions. The most important system the VLab cooperates with is the Globus Toolkit [2] – in the scope of scheduling computational tasks, software services and libraries for resource monitoring, discovery and management. All computational tasks submitted in the VLab system are transferred to the Globus via the GRMS [5] module – an important part of the GridLab project [4]. Among other external systems used by the virtual laboratory are the VNC [10] system, SZD [6] (data management system), authentication and authorization modules.

Workflow in EXPReS

The virtual laboratory system has a modular architecture. The modules can be grouped into three main layers. The general diagram is presented in Figure 1.

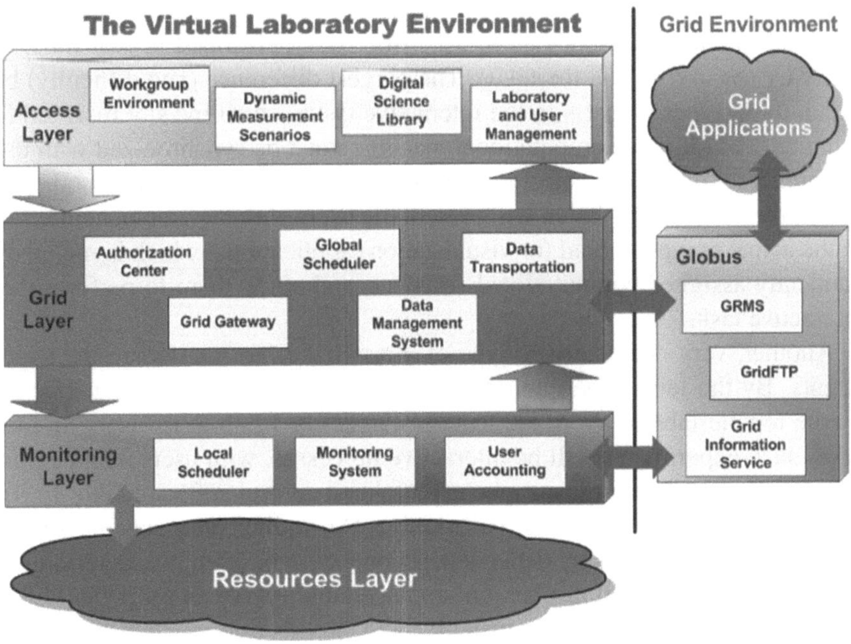

Figure 1. General architecture of the virtual laboratory

An access layer is the top layer of this structure. Basically, it contains all modules and components responsible for a VLab user access and graphical interface to the system (including a web portal) and data input interface. Below there is a grid layer, which communicates with the external grid environment. It is responsible for user authorization and authentication, data management, general task scheduling. Modules from this layer also handle the transfer of the computational tasks to the Globus system and gather feedback data and the computed results. The monitoring layer consists of lower-level modules, such as hardware-dedicated schedulers, system monitoring, gathering accounting data. All the physical scientific devices are located in the resources layer, as well as modules responsible for their direct control. On the grid environment side the most important element is the Globus system, with all its components (GRMS, GridFTP, etc.). Globus also allows execute computational tasks (batch and interactive) on a wide variety of grid applications.

The Peculiar Nature of the Virtual Laboratory Experiments

To understand how the VLab works and why the data management system is so important for its proper working, the peculiar nature of this system is described next in this chapter.

In general, there are two main kinds of tasks in the virtual laboratory: experimental and computational ones. The latter can be divided into regular (batch) jobs and interactive/visualization tasks (the ones performed in real time, directly by the users – via the GUI). The biggest difference (and difficulty) between those types is that – in the interactive tasks – the time slot reserved for running the task on a computational machine must be synchronized with user preferences, considering specific work hours, daily schedule, etc. Another aspect is the mechanism which will present the users with the graphical interface of the actual computational (or visualization) application – which is run on dynamically assigned computational servers – and allow them to perform their interactive task.

Another, very characteristic type of virtual laboratory tasks are the experiments. By the term experiment we mean a task, scheduled to be performed on the remote laboratory equipment, available via VLab to its users. In most cases such experiments will be interactive processes, with users manipulating directly the remote equipment via a specialized control software GUI. Experiments are difficult for formal description. Depending on a specific science domain, there can be many dependencies on external, often non-deterministic factors. The device will not be non-stop available for VLab users but will be shared with local researchers (usually with higher priority than the remote users). There are also maintenance periods, in which the device is unavailable.

Sometimes the presence and assistance of the device operator may be necessary – especially at the beginning of experiments.

The first scientific device incorporated into the VLab system was an NMR spectrometer. The most important problems with scheduling the NMR experiments, apart from those described above, come from the samples management. To perform an NMR experiment, an actual sample containing chemical compound has to be delivered to the NMR spectrometer and inserted into the machine. This causes a number of scheduling problems. At the task (and corresponding sample) submission point the actual NMR device has to be known and chosen, because the sample has to be sent from the remote location to the device site. The exact time of sample arrival is not known as well, making the exact task scheduling impossible until the sample arrives.

3. NUCLEAR MAGNETIC RESONANCE SPECTROSCOPY

Nuclear magnetic resonance (NMR) spectroscopy is a unique experimental technique that is widely used in physics, organic and inorganic chemistry, biochemistry, as well as in medicine.

The analysis of one- or multidimensional homo- and heteronuclear spectra obtained in the course of an NMR experiment can provide information about the chemical shifts of the nuclei, scalar coupling constants, residual dipolar coupling constants and the relaxation times T_1, T_2. All of these data can be stored in the presented database, which also offers tools for performing quick and optimal search through the repository.

Information about each compound that is stored in the database comprises its unique name, a chemical formula, information about its physicochemical properties and a specific description of all its atoms consistent with a HOSE code [1]. Graphical representation of one- and multidimensional NMR spectra, as well as their source files obtained from NMR experiments, and a complete description of parameters used to acquire spectra are also kept in the repository. Additionally, it is possible to place the pulse sequence programs, which can be used by all the database users. In case of bio-molecules – proteins and nucleic acids – the database yields a possibility of depositing atom coordinates, structural constraints used in the computation and the complete structure determination protocols. References to all the stored data in the form of a list of papers are also included. A complete list of record fields and their formats is presented in Table 1.

Compared to the other NMR databases available through the Internet, like BioMagResBank [11], NMR data-sets bank [7], NMRShiftDB [13], SDBS [16] and Spectra Online [12], the presented repository is more suitable for teaching, since it contains an entire range of information about the performance

Table 1. List of attributes present in the NMR database

No	Name	Type	Size	Description
			Information about the compound	
1	Compound name(s)	Text	256 characters	Chemical name(s) of the compound
2	Molecular formula	Text	32 characters	Molecular formula
3	Summary formula	Text	32 characters	Summary formula
4	Chemical structure	Graphics		Molecular formula with atom numbering
5	Atoms name with numbers	HOSE code	Text Array[1] : $N{\cdot}16$ (names, numbers of atoms) Array[1]: $N{\cdot}64$ (HOSE codes)	Atom names and their corresponding HOSE codes
6	Chemical class	Text	256 characters	Chemical class of the compound
7	Molecular weight	Text	16 characters	Molecular weight
8	Boiling point	Text	16 characters	Boiling point
9	Melting point	Text	16 characters	Melting point
10	Form type	Text	2 characters	Type of the form
			Information about NMR experiment	
11	1D NMR spectra	Graphics	~ 0.5 MB	Graphical representation of 1D NMR spectra
12	Correlation type	Text	16 characters	Type of correlation of 2D and 3D NMR spectra: homonuclear, heteronuclear, other
13	2D NMR spectra	Graphics	~ 0.5 MB	Graphical representation of 2D NMR spectra
14	3D NMR spectra	Graphics	~ 0.5 MB	Graphical representation of 3D NMR spectra
15	Pulse sequence	Binary	~ 0.25 MB	Pulse sequence
16	Pulse sequence – description	Text	64 characters	Description of the pulse sequence
17	Spectrometer model	Text	32 characters	Model of NMR spectrometer used in an experiment
18	Spectrometer manufacturer	Text	16 characters	Manufacturer of NMR spectrometer used in an experiment

19	Spectrometer basic frequency	Text	16 characters	Spectrometer basic frequency
20	Temperature	Text	16 characters	Temperature of experiment
21	Solvent	Text	32 characters	Solvent name
22	Concentration	Text	16 characters	Concentration of the compound
23	PH	Text	8 characters	pH of the solution
24	Buffer	Text	128 characters	Buffer contents and its name
25	Reference	Text	32 characters	Reference compound name and its chemical shift
26	Reference Compound	Text	32 characters	Example of reference compounds with their chemical shifts
27	Additional spectrum information	Text	Text file	Additional spectrum information
		Information about spectral data		
28	^{1}H, ^{13}C, … chemical shifts	Text	Array[1]: $N{\cdot}8$	Available chemical shifts
29	Coupling constants – scalar	Text	Array[1]: $N{\cdot}64$	An array of available values of scalar coupling constants
30	Coupling constants – residual dipolar couplings (RDC)	Text	Array[1]: $N{\cdot}64$	An array of available values of residual dipolar coupling constants
31	Relaxation times	Text	Array[1]: $N{\cdot}2{\cdot}16$	Values of relaxation times T_1 and T_2
32	Authors	Text	256 characters	List of authors
33	Institution	Text	128 characters	Institution
34	Contact	Text	128 characters	Contact to the author (e.g., e-mail, phone)
		Bibliography		
35	Authors	Text	256 characters	List of authors
36	Journal	Text	128 characters	Journal name
37	Year	Text	8 characters	Issue year
38	Volume	Text	16 characters	Volume
39	Pages	Text	16 characters	First and last pages of the article
40	Title	Text	256 characters	Title of the article

(continued)

Table 1. (continued)

No	Name	Type	Size	Description
41	PDF, PS	Binary		Paper in PDF or PS file format
42	Link	Text	128 characters	Link to the bibliographic base
			Experimental data	
43	Original NMR data	Binary	~ 10 MB	Compressed NMR data
44	Format of data	Text	64 characters	Format of NMR data
45	Processed data (zipped)	Binary	~ 50 MB	Compressed processed NMR data
46	Format of processed data	Text	16 characters	Format of processed NMR data
			Structural analysis	
47	Protocol (zipped)	Binary	~ 1 MB	Protocols, scripts, ...
48	Protocol description	Text	512 characters	Description of a protocol
49	Restraints (zipped)	Binary	~ 0.2 MB	Experimental restraints
50	Restraints description	Text	256 characters	Description of experimental restraints
51	Structures (zipped)	Binary	~ 5 MB	Atom coordinates
52	Database	Text		Name of the database storing atom coordinates
53	Reference to database	Text	128 characters	Reference to database
54	Code	Text	16 characters	Structure code
55	Additional Data	Text	512 characters	Additional information
			Other information	
56	Other information	Text	512 characters	Additional information

[1] *N*-line array, *N* denotes number of atoms

and analysis of the NMR experiment. Based on the data that are stored in the database such as chemical shifts, coupling constants or relaxation times, the user can quickly identify a molecule.

In case of unknown compounds the knowledge of chemical shifts and coupling constants, together with a set of the HOSE codes, can be used to predict the chemical environment of atoms and to suggest the types of functional groups which are present in the analyzed molecule. The HOSE codes can also be helpful in the simulations of the spectra performed on the basis of the molecular formula of the compound.

4. DIGITAL LIBRARY

The main functional assumptions of the digital science library can be described in reference to the reasons for founding the Digital Library of Wielkopolska (Wielkopolska Biblioteka Cyfrowa – WBC) [15], developed by Poznań Supercomputing and Networking Center. Some of the most important assumptions are

- uniqueness of some digital collections,
- digital library software should make it possible to significantly extend the usefulness of the library, which takes advantage of this software,
- digital library software must cooperate with the library-integrated systems and, in particular, allow using the catalogue database.

Referring to the first assumption, the digital science library will serve students and scientists by storing the results and computation data of the experiment. There is a need to digitize the research in a form further called a digital publication.

The second assumption is related to the usefulness that should be provided by the digital library in comparison with the existing libraries. A digital library is not supposed to constitute a separate unit or to be in competition with the existing libraries. It should complete and broaden the overall library usefulness, expanding its universality and availability on the one hand, and on the other allowing the librarian (or library administrator) to adapt the digital library functionality to its readers' (users) and administrator's needs.

In this context, DSL exemplifies the digital library which, in assumption, will serve the scientific unit providing the selected instrument (e.g., NMR spectrometer) through the virtual laboratory. The scientists should have a possibility of collecting and accessing their results of the experiments and computations. The collected resources related to a defined institution and the provided instrument should have a teaching character, too.

The third issue concerns the semantic description of the stored resources, i.e., metadata. It allows searching the content of the library. DSL realizes these assumptions in a complete manner – the publications can be described by any metadata set, including those renowned as quasi-standards (e.g., Dublin Core). The issue of the scientists' experiment results is similar; in this case the available functionality is broadened by the possibility to store the NMR-data. For this purpose the NMR-dedicated database described in Section 5 has been developed.

To sum up, the digital science library is an example of a digital library dedicated to scientists, which extends the possibility to store the results of experiments and related documents. It should be outlined that, unlike other libraries, e.g., the Digital Library of Wielkopolska [15], DSL realizes a different approach to the representation of digital resources. In WBC an electronic publication represents some separated content (an article, a report, a movie, a book), which can exist in the system in many versions, and the structure of that publication has a form of a single file (e.g., PDF) or multi-files (e.g., HTML). In DSL the electronic publication (an article, results of scientific work and a chemical compound) has a multi-file format and the metadata sets are used to model the possible relations and dependencies.

The Functionality of the Digital Libraries

Over the last several years a few fundamental features which the modern digital library should provide [3] have been established. The list below presents the most important ones:

- Categorization – the ability to group the related publications,
- Catalogue description – detailed information about the digital entries in the library,
- Searching – the possibility of the library content lookup by the user-defined criteria (paper search or the experiment results search),
- Browsing – visualization of the searched out data in the user-friendly interface.

Convenient and Widespread Access to Publications

There are many scientists who find publishing their work results, materials, etc., in the digital library necessary and share them with other users. However, to make it possible, the basic functionality described in the previous section is not sufficient. The digital science library software has to be equipped with some additional features adequately to the requirements of the scientific environment. The list of such additions for the publication of documents and experiment results is presented below:

- Widespread access to the published papers – Internet access to the digital library resources has to be assured,
- Durability – a publication made available should never change its form (i.e., the file format) or the localization (i.e., network localization),
- Version management – the possibility of keeping the document in many versions and managing them,
- Access management – resource access protection. The access cannot be granted to valuable, unpublished resources or experiment results,
- Copying protection – protection from unauthorized resource copying,
- Notifications – the user should be notified about the changes of the library content,
- Credibility – the published document should not be changed by non-trusted or unauthorized users.

Many existing digital libraries (e.g., WBC) can be characterized by the common qualities described above.

The digital science library is built on the basis of the data management system. This implies the list of features, which are not so common in the digital library environments, but on the other hand are valuable and significant to the scientific environment. DMS is a distributed virtual file system, extended by the data storing on the hierarchical storage management (HSM).

5. EXEMPLARY SET OF DATA

The first steps of the DSL preparation assume that all data which will be placed in the library should be gathered, categorized and characterized. The characterization should consist of name, type, size (range), description.

A detailed list of data stored in the exemplary NMR library is presented in Table 1. Information is processed with the use of the following operations: enter, modify and search.

The Use Case Diagrams

After gathering all data which are designed for putting them into the database, the use case diagrams should be prepared for most frequent scenarios. Below one of them is discussed. There we can distinguish the main actor

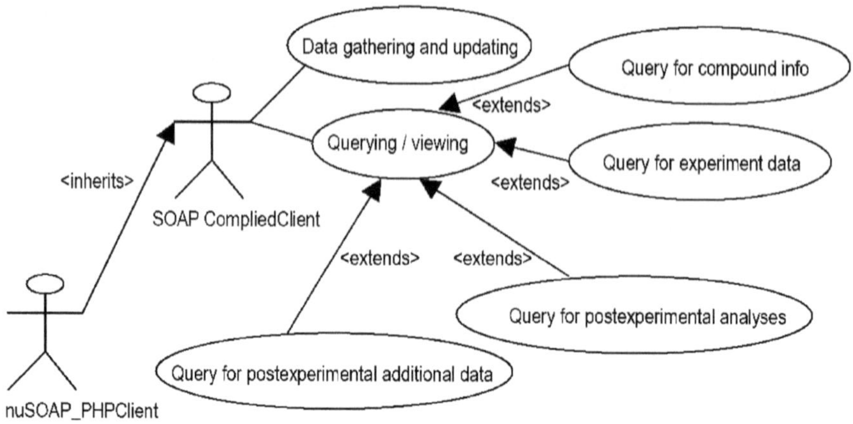

Figure 2. Use case diagram

on the basis of demands defined for the digital science library of nuclear magnetic resonance spectroscopy. It is presented in Figure 2. As an actor we can consider the system placed in the presentation or the indirect (middle) layer, on the condition that it is able to communicate over the SOAP protocol. The figure presents the two actors. The first – SOAPCompliedClient – is identified with a certain pattern of actor with granted rights for executing the specified operations and provided with the ability of intercommunication with the library over the SOAP protocol. The nuSOAP (nuSOAP_PHPClient) is the second actor in the figure. It can be used to create an access service for the NMR library. In this case the access service is based on WWW pages and the PHP programming language.

The figure shows some main examples of using the DSL-NMR library system:

- **Data gathering and updating** – enables introducing and updating information related to the chemical compound and the experiment data.
- **Querying / viewing** – enables getting information on a chemical compound, experiment and its results and the analysis executed on the basis of received data. The course of searching out the data depends on the demands. A standard package of demands consists of
  - Query for compound info – condition: existence of information concerning researched chemical compound; searching out its attribute,
  - Query for experiment data – condition: existence of experiment data; searching out after attributes of an experiment, like type of spectrum, sequence of impulses, spectrometer type, solvent, measurement temperature,

- Query for post-experimental analyses – condition: existence of the analysis description; searching out of chemical compound according to value of chemical shift, coupling constant,
- Query for post-experimental additional data – condition: existence of the research documenting data, e.g., reference for authors of experiments, for articles concerning research or papers documenting the received results.

6. PROJECT DESIGN ASPECTS

The creation process of the database scheme for the DSL-NMR proceeded in several periods and depended on the user demands. An important period in the database project design process was the process of its normalization, allowing the removal of excessive information from tables and increasing the optimization level of the data structure. It is possible to distinguish four logical parts in the database scheme. They are dedicated for

- storage of basic information characterizing the examined chemical compound,
- storage of information and localizations describing the experiment carried on,
- storage of information about the experiment data analysis,
- storage of information and localizations of the research complementary data.

7. CONCLUSIONS

Remote instrumentation systems become more and more popular among users wishing to use very rare and expensive scientific devices remotely. One of the important components of these systems is the module for storage and management of data.

Nowadays, there are many digital libraries available via Internet access. They all have different profiles. They can be devoted to the cultural heritage or regional art, while others could have educational purpose or belong to scientific research projects. That is why it is very important to use open standards in digital libraries information recording. It will allow the collaboration between the digital science library and other projects of this kind (e.g., WBC). Furthermore, it will create mutually complementing environments that will give the users access to the repositories containing digital information of different types and subject matters.

Moreover, it is worth to remember that in the case of remote instrumentation systems the DSL systems should closely cooperate with scientific equipment and application servers to facilitate experiment results, storing and conducting post-processing or visualization stages.

An example interface to the digital science library for NMR data was implemented in PSNC and can be found on the web page of the virtual laboratory project [14].

ACKNOWLEDGMENTS

The authors are very grateful to scientists from the Laboratory of Structural Chemistry of Nucleic Acids, Institute of Bioorganic Chemistry and Computer Centre of the Technical University of Lodz for their suggestions, assistance and support.

REFERENCES

[1] Bremser W.: "HOSE – A Novel Substructure Code", *Anal. Chim. Acta*, 103, 355–365 (1978).

[2] Globus Toolkit, http://www.globus.org/toolkit/.

[3] Górny M., Nikisch J.A., Gruszczyński P., Mazurek C., Stroiński M., Swędrzyński A.: "Zastosowanie oprogramowania dLibra do budowy Wielkopolskiej Biblioteki Cyfrowej", http://ebib.oss.wroc.pl/matkonf/iwb2/dlibra.php (in Polish).

[4] GridLab, http://www.gridlab.org/.

[5] GRMS, http://www.gridlab.org/WorkPackages/wp-9/.

[6] Grzybowski P., Mazurek C., Spychała P., Wolski M.: "Data Management System for Grid and Portal Services. Submitted to Grid Computing", *Infrastructure and Applications special issue of The International Journal of High Performance Computing Applications (IJHPCA)*, Cardiff University, UK, http://progress.psnc.pl/English/DMS.pdf.

[7] NMR Data-sets Bank: A repository for raw NMR data-sets, http://nmrb.cbs.cnrs.fr/ index.html.

[8] Poznań Supercomputing and Networking Center, http://www.man.poznan.pl.

[9] PROGRESS – Polish Research on Grid Environment for SUN Servers, http:// progress.psnc.pl/English/index.html.

[10] RealVNC, http://www.realvnc.com/.

[11] Seavey B.R., Farr E.A., Westler W.M., Markley J.L.: "A Relational Database for Sequence-Specific Protein NMR Data", *J. Biomol. NMR*, 1, 217–236 (1991).

[12] Spectra Online, http://spectra.galactic.com/SpectraOnline/Default_ie.htm.

[13] Steinbeck Ch., Krause S., Kuhn S.: "NMRShiftDB – Constructing a Free Chemical Information System with Open-Source Components", *J. Chem. Inf. Comput. Sci.*, 43, 1733–1739 (2003).

[14] Virtual Laboratory PSNC, http://vlab.psnc.pl/.

[15] Wielkopolska Biblioteka Cyfrowa, http://www.wbc.poznan.pl.

[16] Yamamoto O., Someno K., Wasada N., Hiraishi J., Hayamizu K., Tanabe K., Tamura T., Yanagisawa M.: "An Integrated Spectral Data Base System Including IR", *MS, 1H-NMR, 13C-NMR, ESR and Raman Spectra, Anal. Sci.*, 4, 233–239 (1988).

GRID INTEGRATION OF FUTURE ARRAYS OF BROADBAND RADIO-TELESCOPES – MOVING TOWARD *E*-VLBI

M. Okoń[1], D. Stokłosa[1], R. Oerlemans[2], H.J. van Langevelde[2], D. Kaliszan[1], M. Lawenda[1], T. Rajtar[1], N. Meyer[1], and M. Stroiński[1]

[1]*Poznań Supercomputing and Networking Center, Poznań, Poland*
[2]*Joint Institute for VLBI in Europe, Dwingeloo, The Netherlands*

Abstract The purpose of this chapter is to present the design for the integration of radio-astronomical VLBI experiments with the Grid environment. The current status of VLBI process is described, and the main part of this document presents the proposed architecture for the development of the next-generation e-VLBI system, together with the analysis of potential problems and design limitations. This design was done for the purpose of the EXPReS project [5], partially funded by the European Commission (FP6–IST). It was created with significant help and close cooperation from the Joint Institute of VLBI in Europe (JIVE) [10].

1. INTRODUCTION

Networks of radio telescopes can be used to produce detailed radio images of stars and galaxies. The resolution of the images depends on the overall size of the network (the maximum separation between the telescopes) and the sensitivity depends on the total collecting area of all the telescopes involved and, crucially, the bandwidth of the connection between the telescopes. In this technique, called very long baseline interferometry (VLBI), the signals between all telescope pairs are combined in a data processor. Typically this processor needs to find the correlated response in 2-bit (4-level) sampled signals on every baseline pair with high resolution and must be able to cope with data rates up to 1 Gbps per telescope.

This functionality has been implemented by constructing a massively parallel, purpose-built "supercomputer" – usually referred to as a data processor or correlator. A decade ago, a 20,000-node PC cluster would have been required to match the processing power of the EVN (The European VLBI Network) [4] MkIV Data Processor at JIVE. However, this number is falling rapidly as processor capabilities continue to develop. The networking infrastructure is also getting more capable of dynamically handling very large data transfers in long periods of time. There is a growing trend in research communities to design and develop a number of "virtual laboratories" – or remote instrumentation environments [1, 2, 13], which benefit from the increasing power of available computational systems and high-bandwidth networks. A good example here is the RINGrid project [17] (co-funded by the EC under the 6th Framework Programme), which aims at trend analysis and recommendations for designing next-generation remote instrumentation services and synergy between remote instrumentation and next-generation high-speed communications networks and grid infrastructures [19].

2. THE CURRENT STATUS OF VLBI OPERATIONS

As was mentioned in Section 1, the data correlation is currently being done by a dedicated hardware. The radio-telescope data are recorded locally, put on magnetic tapes or hard disks at the telescope sites and sent to the central processing facility using traditional methods like couriers or post office.

In the past some effort was dedicated to introducing the e-VLBI [4] (electronic VLBI) concept, very similar to the "regular" VLBI, with the main difference being data sent via the network to the central processing facility at JIVE.

Currently, JIVE can guarantee for regular science experiments e-VLBI at 128 Mbps for six radio-telescopes (Torun, Onsala, Westerbork, Medicina, Jodrell Bank, Cambridge). EVN aims at one regular e-VLBI session every 6 weeks. In practice this schedule may vary. Occasionally, 256 Mbps for 6 and 512 Mbps for 3 radio-telescopes were achieved in research experiments. Also the Arecibo telescope, which is the largest telescope in the world (305 m antenna, run by National Astronomy and Ionosphere Center, Arecibo, USA), has participated at 32 Mbps.

For the data transfer e-VLBI relies on dedicated light-paths (Wb, Jb, Cam) and the classical IP switched network (Med, Ons, Tor).

In detail, the VLBI process can be described as follows (Figure 1):

- The radio astronomer (user) is given a research grant for performing the observation. The exact time slot is scheduled and the information is returned to the user.
- The user uses the SCHED program to produce a single VEX file [20], which contains the complete description of the VLBI experiment. The file is of the

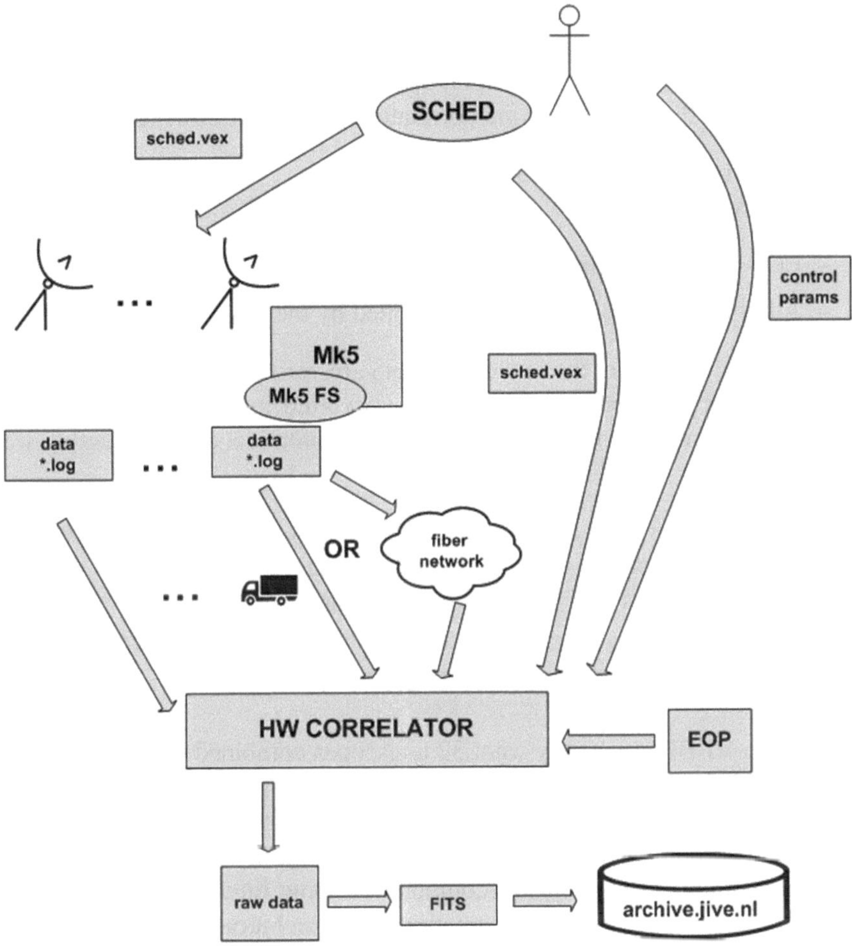

Figure 1. VLBI correlation process

ASCII text format. SCHED uses a text file as an input and has a command-line interface and some graphical output.

- The VEX file is transferred to the radio-telescopes, via an FTP system. The telescope operators are responsible for loading the file and setting up the observation.

- During the observation the Field System controls the telescope using settings obtained from the VEX file. Data are recorded in the Mk4 data format on Mk5A disks. These disks are inside a Mark5 computer and can only be accessed through special software. More information on these subjects can be found at the Haystack website [9].

- After the observation the collected data are sent to the central JIVE processor at Dwingeloo, via dedicated connections as is done for e-VLBI or still in most cases by physical transport of disks or tapes. In order to perform the correlation, the JIVE hardware correlator needs the following:
  - Data from all the telescopes
  - User VEX file
  - EOP (earth orientation parameters)
  - Additional control parameters, supplied by the user
- After the correlation, the raw output data are verified by the JIVE staff, and if no errors are detected, the data are converted into the FITS [7] format. This file format can be read by standard astronomical data processing software.
- Finally, the converted correlation results are put in the JIVE archive where the radio astronomer who asked for the observation can access the data. After 1 year others will gain access to the data.

3. THE IDEA OF GRID-BASED E-VLBI

Future e-VLBI arrays may want 32 telescopes combined with 4 GHz bandwidth (generating at least 16 Gbps of sampled data per telescope), which requires two orders of magnitude more computing power than available in the current data processor. Conceptually, the processing of a single baseline product can be assigned to a cluster computer at some random Grid point. This requires a state-of-the-art implementation of a correlation code, which is completely scalable and easily distributed. Recent research has shown that it is possible to take advantage of floating point processing to implement more accurate and sensitive algorithms that also allow special functionality, such as unlimited spectral zoom. This unique machinery will allow spacecraft tracking with VLBI precision.

It is not just the issue of sheer computing power that drives the design issues of radio astronomy data processors. The connectivity aspects are just as demanding and may in fact be the defining issue in the future. As high-speed data sampling becomes available at the telescope, it will be more and more challenging to bring all data to a single place. Moreover, to form all the baseline pairs a matrix of computing nodes needs to be fed to calculate all data products, requiring massive switching capacity. An attractive idea for future data processors is to map this computational topology onto an international Grid of computing nodes. This will reduce the requirement to have all data converge to one central place; with a large fabric of reasonably fast, interconnected data processor nodes it will be possible to provide the necessary computer power for a wide band data processor, while the international wide band connections themselves provide the complex interconnect. In such a scheme, digital data must be packaged at the telescope and copied for distribution to several data processor nodes, arbitrarily distributed over Europe. The resulting data product must be archived as a single entity, presumably at a central node, as there are proprietary issues associated with its use for scientific research.

With the start of an e-VLBI experiment that employs distributed correlation, a central routing scheme needs to be implemented. Each telescope needs to be instructed where to send data, and computer resources must be allocated accordingly. There are different topologies possible for the implementation; we aim for scalable solutions that can work both with relatively few data processor nodes, as well as implementations in which there are more nodes than baselines between telescopes. To support data management and information flow for the entire e-VLBI experiment in an automated way, a workflow management system will be created. The workflow module is placed on top of the existing applications and steers them to control the measurement process in a Grid environment. The total process must be real time, in the sense that it keeps up with observing, but otherwise processing can be asynchronous.

4. DESIGN LIMITATIONS

The integration of e-VLBI operations with the Grid environment has to cope with certain limitations of currently used Mark5 operational system. It poses some specific problems, which may disappear when other systems can be employed in the future. This part of the VLBI evolution is also covered in the same EXPReS project, in the scope of JRA1: FABRIC (Future Arrays of Broadband Radio-Telescopes on Internet Computing) WP1 [6].

The Mark5 is a disk-based VLBI data recording and playback system capable of handling 1 Gbps data rates. It is widely used at the telescopes participating in the EVN. A software correlator running on some grid node should therefore conform to this interface of the radio-telescopes to the outside world.

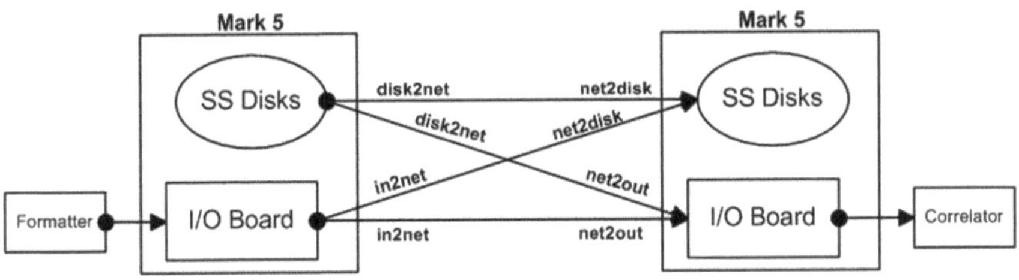

Figure 2. Mark5 to Mark5 transfer through the network

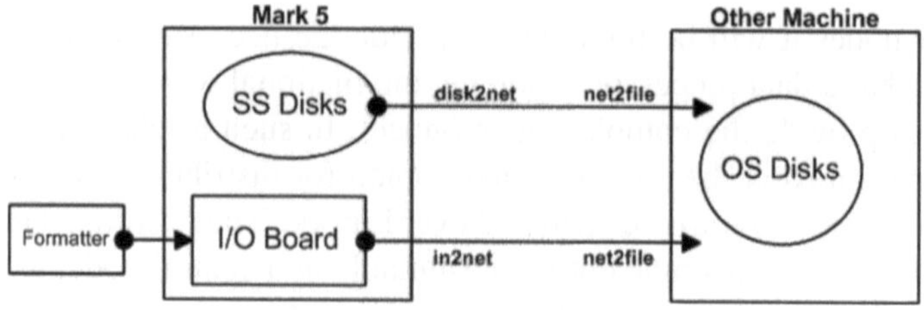

Figure 3. Mark5 to file transfer through the network

Before describing ways to get the observed data onto some grid node for further processing, first the relevant data transfer modes of the Mark5 computer will be described. For the current e-VLBI practice a Mark5 computer is connected to the radio-telescope through a formatter. The data enter the I/O board and are transmitted to another Mark5 computer (Figure 2). At the receiving Mark5 data can be recorded on disks or an I/O board can transmit the data directly to the EVN correlator. The appropriate Mark5A commands have to be activated on both sides.

Usually a grid node does not have a Mark5 computer at the receiving side but UNIX/Linux-like computers. There are two ways of getting the data onto a UNIX/Linux machine. First the data are recorded on disks in the Mark5 computer and later they are played back. In this case the disk2net command puts the data to the network and Net2file gets the data from the network (Figure 3).

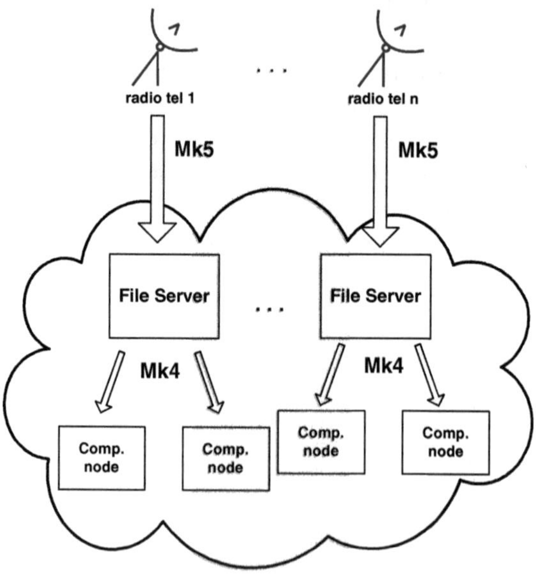

Figure 4. The overview of the solution with file servers

The data stream can also be put on the network by the Mark5 during the astronomical observations by in2net. At the receiving side Net2file gets the data from the network onto a disk.

So, somewhere a UNIX/Linux file server is necessary, either at the telescope site or at a remote grid node. In the first case a powerful file server at the telescope site is required, and it operates as a data buffer. The data are sent over the network to a compute node. The second case requires a point-to-point Internet connection between the Mark5 system and a remote file server. Buffering can optionally be done at the telescope site on the Mark5 system itself. The connection between radio-telescope and appropriate file server must be set up manually by the telescope operator, who receives the relevant information from the experiment description (Figure 4). The connections will be defined by the VLBI operator, using the Workflow Manager application, as is described in more detail here.

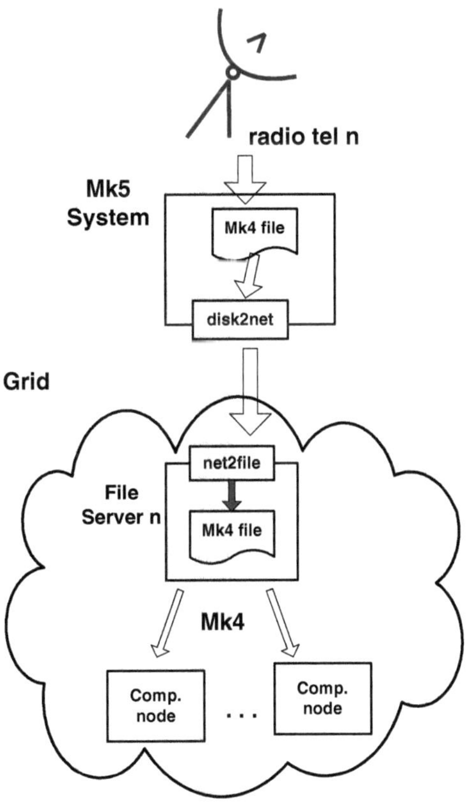

Figure 5. The close-up view on the file server solution

One of the problems lies in the effective switching of the routing path from the telescopes (there is no automated way to do that; it requires manual adjustment of networking equipment). Furthermore, once the data acquisition has started, the data destination point cannot be changed.

This limitation has a significant impact on the design, because the initial design assumptions were based on the idea of direct and dynamic routing of the data streams from the radio-telescopes to the computational nodes. Considering the above, this is technically impossible with the current hardware. Unfortunately, the routing problem is difficult to solve at this point, as it requires hardware modification at the telescopes sites. A workaround is necessary and is presented in this section. The solution is explained in detail earlier in this section.

Figure 5 presents a more detailed view on a single telescope.

5. GRID TECHNOLOGY FOR THE E-VLBI CONCEPT

One of the main tasks of FABRIC was to identify and describe the e-VLBI domain areas that could be addressed by existing or currently developing Grid solutions. These areas include network security, data transport, workflow management and Grid resource management systems.

Security

Security is a very important aspect in every distributed, grid-related system. But this “openness” of grid systems can cause many disturbances. On the one hand, there is the necessity to protect resources like computational servers or laboratory equipment from intruders; on the other hand, these facilities should be easily accessible to authorized users.

The very popular solution for developing distributed and heterogeneous systems, which solves most of the mentioned problems, is Globus Toolkit (GT) [8]. GT is a set of tools, which provides distinct Web Services [21] (WS) and pre-WS authentication and authorization capabilities. Both build on the same base, namely standard X.509 end entity certificates and proxy certificates, which are used to identify persistent entities, such as users and servers, and to support the temporary delegation of privileges to other entities, respectively. Globus Toolkit in the confines of security aspects provides the following functionality:

- message-level security mechanisms – implement the WS-Security standard and the WS-SecureConversation specification to provide message protection for GT’s SOAP messages,

- transport-level security mechanisms – uses transport-level security (TLS) mechanisms,
- authorization framework – allows for a variety of authorization schemes, including a "grid-mapfile" access control list, an access control list defined by a service, a custom authorization handler and access to an authorization service via the SAML (Security Assertion Markup Language) protocol [18].

For non-WS components, GT provides similar authentication, delegation and authorization mechanisms, although with fewer authorization options.

The Grid Security Infrastructure (GSI) is used by the Globus Toolkit for enabling secure authentication and communication in a distributed and heterogeneous environment. It provides useful services for Grids, including mutual authentication and single sign-on, which bases on public key cryptography. The main GSI functionality covers

- secure communication between nodes in the Grid,
- supporting security across organizational boundaries,
- supporting "single sign-on" for users of the Grid.

Every user and service on the Grid is equipped with a certificate and in this way identified. The certificate contains the following information:

- a subject name, which identifies the person or object that the certificate represents,
- the public key belonging to the subject,
- the identity of a Certificate Authority (CA) that has signed the certificate,
- the digital signature of the named CA.

CA is used to certify the link between the public key and the subject in the certificate. GSI certificates are encoded in the X.509 certificate format, a standard data format for certificates established by the Internet Engineering Task Force (IETF).

Mutual authentication is a very important operation in establishing a secure connection. It is used for proving that the two parties are who they say they are. The Secure Sockets Layer (SSL) is used in the Grid to perform mutual authentication.

Data Transport and Management

An enormous amount of data is collected under VLBI experiments. As a consequence, the designed system has to have additional, specialized functionality connected with data transport and management. Data retrieved from radio telescopes are then transferred through many processing stages of experiment scenario to get the final correlation product (e.g., visualization). Due to the huge size of data and assurance of efficient post-processing computation, a set of grid tools must be involved in this process. It is not enough to simply store data on some machine in the grid and download it when the computation will be performed. The retrieving and downloading processes must be optimized for speedy access in different geographic locations, cataloged with descriptive information for easy retrieval and made available to computation jobs running on the Grid.

Data management is one of the core services in the Grid systems. Based on the high-capacity storage systems and broadband networks, the Grid community has produced a set of components for working with and managing data on the Grid.

Several components in the Grid space are aimed specifically at providing uniform Grid interfaces to various types of data:

- GridFTP – A uniform, secure, high-performance interface to file-based storage systems on the Grid,
- OGSA-DAI – An OGSA interface for accessing XML and relational data stores,
- Metadata Catalog Service (MCS) – A stand-alone metadata catalog service with an OGSA service interface.

These tools specialize in moving and transferring data between Grid systems. Each tool meets specialized applications or user needs and some also provides interfaces to specialized storage systems:

- globus-url-copy – a command-line tool for requesting GridFTP transfers,
- Reliable File Transfer (RFT) Service – an OGSA service that allows clients to request data transfers and then "disconnect" while the transfer takes place,
- UberFTP – A text-based interactive client for GridFTP,
- GSI-SCP/SFTP – Popular OpenSSH tools that support Grid authentication,
- Data Management System (DMS) [3,16] – part of Gridge Toolkit developed by PSNC [15] and designed for the creation of a complete and robust data management environment, ready to fulfill even the enterprise requirements. The basic functionality enables the creation of a distributed environment capable of managing large amounts of data. Advanced security solutions ensure the appropriate protection of the collected resources. The universal gateway to the scientific databases may be used to integrate external data

sources. Web Portal facilitates users' work with data and helps to comfortably manage the DMS resources.

Moreover, a set of tools were implemented, which help optimize the use of storage systems for specialized user communities:

- Replica Location Service (RLS) – a distributed mechanism for keeping track of the locations of replicated data on a Grid,
- NeST – a "storage appliance" that provides remote access to local data when computation jobs are running,
- DataCutter – a system that uses data filters and streams to segment data sets in efficient ways on a Grid.

6. NETWORK MONITORING AND MANAGEMENT

The European VLBI network is spread over European countries and beyond; thus, data transmission will cross several administrative domains, including National Research Networks (NRENs), as well as the pan-European GÉANT2 network [GÉANT2]. The development of a GRID system and e-VLBI data transmissions between different geographically distributed partners will require proper resource management and network monitoring, to allow for automatic resource distribution and QoS provisioning.

Network Monitoring

Based on the set-up of the e-VLBI network and initial requirements these are the highlights of the monitoring architecture:

- It must operate inter-domain and end-to-end. There is also a need for intra-network measurements in order to identify the exact amount of resources. Therefore, metrics should be provided on a hop-by-hop basis.
- It should be flexible enough to accommodate different types of metrics (e.g., capacity, available bandwidth, delay).
- The architecture must encompass using different existing monitoring tools to measure various metrics, e.g., MRTG [MRTG], Cacti [Cacti], BWCTL [BWCTL].
- It must standardize interfaces between various architecture components and provide a defined interface to integrate GRID resource brokers or QoS provisioning systems.

- It should carefully cover security issues. The architecture must be able to deal with customized requirements, on what metrics and with whom they can be shared.

Quality of Service

e-VLBI, as an application supposed to transmit huge amounts of real-time data, will impose very strict requirements on the wide-area transmission networks that can be used by e-VLBI. Such requirements are identified in the networking community as QoS (quality of service). There are four major parameters of QoS, which will determine the quality of transmission services for e-VLBI:

a) capacity – the number of bits per second that is available to a flow or an aggregation of flows,

b) one-way packet delay – the time between the transmission of a packet at its source to the moment it is fully received by the destination,

c) one-way packet delay variation – the difference in the one-way packet delay of two successive packets in the same IP flow,

d) packet loss – the ratio of packets sent by the source and not delivered to the destination (or delivered in error).

Capacity seems to be the most crucial QoS parameter for e-VLBI, as it determines the amount of data that may be transmitted over the network. e-VLBI is going to transmit large portions of data, which will require that the network offer high-capacity transmission services with sufficient values of the other QoS parameters.

There are two networking technologies that may be used by e-VLBI, which have adopted different approaches to QoS. Both technologies are used in research networks (GÉANT2, national research and education networks), which can offer transmission services for e-VLBI. In a traditional IP network traffic is transited on a best-effort basis without any guarantee of QoS; QoS can be achieved by reservations of bandwidth and admission control, which requires that some additional mechanisms be introduced into the network. In the other approach, which is adopted in optical networks, every transmission requires prior reservation of bandwidth and the QoS parameters are guaranteed upon the reservation acceptance by the network.

7. SYSTEM ARCHITECTURE

The final system design and architecture is a result of careful problem analysis and requirements specification. It would be impossible to present a successful solution to such a complicated problem without the help and advice

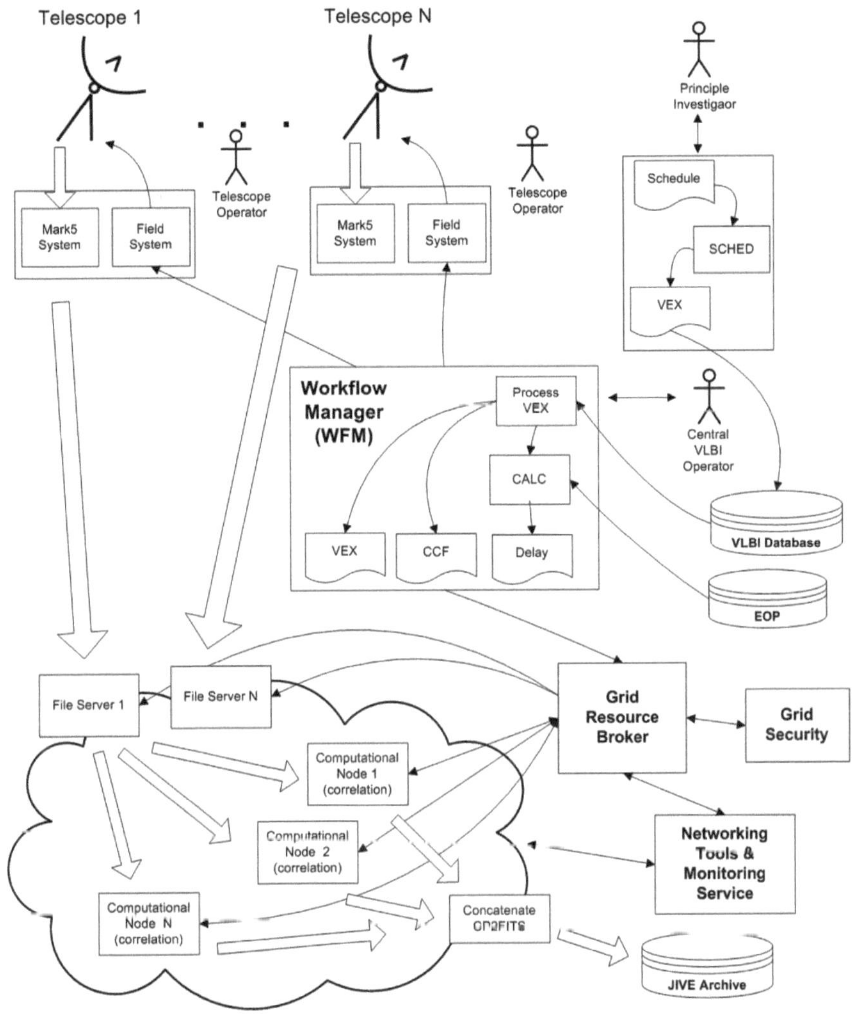

Figure 6. System architecture for a distributed broadband correlation

of radio astronomers (JIVE staff in particular). The proposed architecture for meeting the FABRIC goal is based on the current VLBI and e-VLBI status and is presented in Figure 6.

The whole process can be divided into four major parts.

Experiment Preparation

A principal investigator (PI) will be granted observation time on various VLBI telescopes. Before the actual observation can start, the PI will create a schedule file containing all the details of the observation, like what source to observe, which telescopes to use and when to observe. This schedule file will be used to create a VEX file with the SCHED application. A central operator will be notified of the experiment. Then, the PI will have to wait until the data arrive in the JIVE archive.

Experiment Set-Up

The central operator will use the Workflow Manager (WFM) [11, 12, 14] to process the VEX file. The WFM will be a graphical Java application, downloaded and executed from the VLBI web portal (Figure 7). The information contained in the VEX file will be displayed in the appropriate forms, in which the experiment control parameters can be verified and modified if necessary. The central operator will be able to set up more e-VLBI parameters, according to the operator's domain knowledge and some specific experiment conditions. Another important role of the WFM (and central operator) will be to associate specific file servers with radio-telescope locations and create a workflow for the post-experiment distributed data correlation. The description of connections between file servers and radio-telescopes will be sent to telescope operators together with experiment definition (VEX file). It will be the operators' responsibility to manually set up the routing from their sites to the file servers before the experiment starts. Additionally, the operator will build an experiment workflow graph with virtual correlation tasks, which will divide the correlation process logically, on a number of processing instances. Each

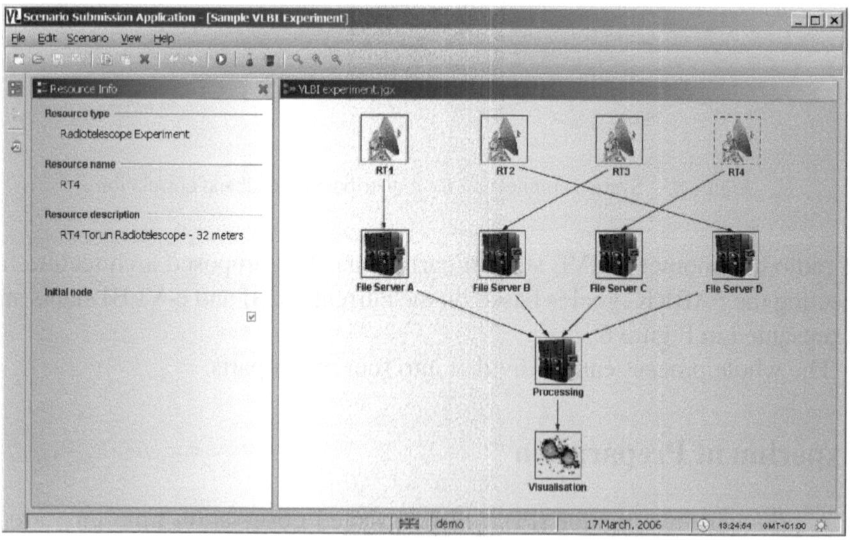

Figure 7. A sample view of the WFM application prototype

instance will be treated as a single and independent sub-task and can be allocated on a different computational node.

The WFM will also calculate the necessary delay tables before the correlation takes place, by calling the external program CALC and Earth orientation parameters (EOP). CALC is a standard application developed by geodesists to accurately determine positions on Earth.

Experiment Start

After the central operator's work is done, a new VEX file will be created and sent to the telescopes participating in the observations. The WFM will also notify the telescope operators that a new experiment is scheduled. As was mentioned before, the WFM will also send the planned routing information between telescopes and file servers and it will also contact the dedicated Grid Resource Broker to allocate the necessary computational nodes for each of the correlation tasks, and for the necessary broadband network connections between the file servers and computational nodes. The resource broker will use the mechanisms described in Section 4 (networking) in order to plan and schedule the optimal route for the data transfer between file servers and nodes.

In the meantime, the telescope operator will load the VEX file in the Field System, which controls the telescope, and in the Mark5 system, which records the data. At the start of the experiment the allocated compute nodes should be ready and waiting for data. The WFM has already sent the Correlator Control File (CCF) and Delay file to the compute nodes. The data recorded by the Mark5 system will be sent to the allocated compute nodes via the file servers (see Section 4 for details). Additionally, the WFM may also provide the operators (as well as principal investigator) with the possibility of verifying the status of the entire e-VLBI experiment (by displaying the current status of all radio-telescopes and the status of the grid broker).

Distributed Correlation

One of the goals of the FABRIC project is to enable distributed correlation using grid facilities. It was concluded that the implementation of the first version of the software correlator will be based on the off-line time-sliced correlation scenario. Other software correlator architectures and implementations will be developed and tested in the FABRIC project.

8. CONCLUSIONS

At this point the project is at the design phase. Therefore, the design presented in this chapter has not yet been verified in the field. However, we trust that this is the optimal solution for this kind of system. It was a result of careful analysis of the problem domain and radio astronomy hardware specifications and state-of-the-art Grid solutions. It was created with great help from JIVE staff. Unfortunately, the nature and complexity of this problem still raise many questions and uncertainties that may have to be solved only by experimenting and creating working models and prototypes. Therefore, in the end the final implemented system architecture might look different in some areas, and some issues may be handled differently. But, in general, the presented approach gives the complete view on the shape of Grid-embedded e-VLBI operations.

We strongly believe that this scientific radio astronomy domain will benefit greatly from the new opportunities, given by this state-of-the-art Grid-wise approach, and that this is a good start to a project that will benefit in the better understanding of the universe and the nature itself.

ACKNOWLEDGMENTS

The work presented in this chapter was carried out with funding from the European Commission under the EXPReS project (6th Framework Programme – IST), contract number 026642.

REFERENCES

[1] A. Atkinson, D. du Boulay, C. Chee, K. Chiu, T. King, D.F. McMullen, R. Quilici, P. Turner, M. Wyatt "CIMA Based Remote Instrument and Data Access: An Extension into the Australian e-Science Environment", *e-Science 2006 Conference*, Amsterdam, The Netherlands, 2006.

[2] A. Bagnasco, A. Poggi, A.M. Scapolla, "A Grid-Based Architecture for the Composition and the Execution of Remote Interactive Measurements", *e-Science 2006 Conference*, Amsterdam, The Netherlands, 2006.

[3] Data Management System (DMS), http://www.gridge.org/content/view/19/100/1/1/.

[4] EVN website, http://www.evlbi.org.

[5] EXPReS project website, http://expres-eu.org.

[6] FABRIC WP1 from description of work, found at http://www.expres-eu.org/docs.html.

[7] FTIS data format, http://heasarc.gsfc.nasa.gov/docs/heasarc/fits.html.

[8] Globus Toolkit website, http://www.globus.org.

[9] Haystack, www.haystack.mit.edu/tech/vlbi/mark5/index.html.

[10] JIVE website, http://www.jive.nl.

[11] M. Lawenda, N. Meyer, T. Rajtar, M. Okon et al., "Workflow with Dynamic Measurement Scenarios in the Virtual Laboratory", *6th CARNet Users Conference*, Zagreb, Croatia, 2004, ISBN 953-6802-04-X.

[12] M. Lawenda, N. Meyer, T. Rajtar, M. Okon, D. Stoklosa, M. Stroinski, "Job Workflow in the Virtual Laboratory", *Global Grid Forum 10*, Germany, March 9, 2004.

[13] M. Okon, M. Lawenda, M. Stroinski, N. Meyer, "Virtual Laboratory as a Remote and Interactive Access to the Scientific Instrumentation Embedded in Grid Environment", *e-Science 2006 Conference*, Amsterdam, The Netherlands, 2006.

[14] M. Okoń, M. Lawenda, N. Meyer, D. Stokłosa, T. Rajtar, Kaliszan, M. Stroiński, "Scheduling Interactive Tasks in the Grid-Based Systems", *TERENA Networking Conference 2005*, Poznan, Poland, 2005.

[15] PSNC website, http://www.psnc.pl.

[16] J. Pukacki, M. Kosiedowski, R. Mikołajczak, M. Adamski, P. Grabowski, M. Jankowski, M. Kupczyk, C. Mazurek, N. Meyer, J. Nabrzyski, T. Piontek, M. Russell, M. Stroiński, M. Wolski, "Programming Grid Applications with Gridge", *Computational Methods in Science and Technology* 12(1), 47–68 2006.

[17] RINGrid website, http://www.ringrid.eu.

[18] SAML, http://www-128.ibm.com/developerworks/xml/library/x-seclay4/.

[19] The presentation: Damian Kaliszan "Mobile aspects in remote instrumentation", *European&Indian e-Infrastructures conference*, New Delhi, India (http://www.beliefproject.org/events/IndiaConference/Session_B4/).

[20] VEX file format, http://lupus.gsfc.nasa.gov/vex/vex.html.

[21] W3C Web Services Activity, http://www.w3.org/2002/ws.

USING THE GRID FOR THE INTERACTIVE WORKFLOW MANAGEMENT IN BIOMEDICINE

I. Porro[1], L. Torterolo[1], M. Fato[1], A. Schenone[1], and M. Melato[2]

[1] *Department of Communication, Computer and System Sciences (DIST), University of Genoa, Genova, Italy,*

[2] *Nice s.r.l. - via Marchesi Roero 1, 14020 Cortanze (AT), Italy*

Abstract According to the virtual physiological human paradigm, integration is often required among different levels and disciplines. Large, distributed and heterogeneous repositories, as well as computationally demanding analysis tools, are more and more involved in biomedical studies. Both for storage of distributed biomedical data and metadata and for access to distributed analysis tools, a Grid-based approach may provide a shared, standardized and reliable solution. To make users able to easily access available services and data, a Grid portal has been implemented in order to hide the complexity of the framework. The portal is intended to act as Grid services provider and as Web interface for managing workflow enactment and management on the Grid. Workflows may be submitted as Grid jobs through a service-based workflow engine and monitored during execution. As a first case study for testing workflow functionalities of the Grid portal, an application is presented for the search and analysis of microarray data.

1. INTRODUCTION

Both in clinical and in research studies a multilevel approach, combining clinical, cellular and molecular data, is more and more required in order to build a patient model according to the virtual physiological human (VPH) paradigm to support personalized diagnoses and therapies [15]. As regards the clinical environment, an increasing number of images are acquired for both morphological and functional medical imaging studies. Moreover, the acquisition of several image data sets along time is usually needed in order to monitor the disease progression and to follow up therapeutic treatments. On the other hand, the scientific approach to the molecular basis of cells and tissues studies has been deeply changed, in both physiological and pathological conditions, by recent data acquisition technologies like high-throughput sequencing equipment and microarray analysis systems, giving a new comprehensive view to the research community.

Extensive computational efforts are required to make sense of the large amount of data generated by these experiments and to produce accurate and biologically consistent predictions. From another point of view, data coming from different instruments and/or repositories have to be managed in a virtual experiment environment. Eventually, the interactive management of acquisition and analysis workflows arises as one of the most challenging issues in the field.

Due to such an integrated approach, clinical and genomic studies tend to converge in more complex multiscale biomedical experiments to open the way to new research goals ranging from personalized medicine to in silico drug design and discovery [7], but advanced tools must be developed to easily manage the amount and the complexity of the information generated every day.

In the clinical field, multiscale studies may combine information coming from different acquisitions (molecular data, cellular data, signals, morphological images, functional images) along temporal sequences of analysis for both diagnosis and therapy. And, more and more, in medical imaging studies data are stored in different and geographically distributed clinical repositories, thus forcing doctors at difficult remote searches (when possible) for collecting the complete information about patients. On the other hand, also in bioinformatics experiments data come from different heterogeneous and distributed data sets, and interoperability among different remote repositories arises as a main concern for researchers. In both cases, the complexity and the interdisciplinarity of experimental setups have to be faced by ontologies and metadata managing tools.

From a computational point of view, the fast timing of the clinical practice may force different clinical institutions to connect their computing resources to provide results for advanced medical applications supporting diagnosis and therapy. In bioinformatics, as well, the use of distributed computing resources running suitable applications on networks of federated research centers is often needed due to difficulties in finding relationships among heterogeneous data sets of large-scale experiments. To widen the range of such analysis tools, the access to geographically distributed services has begun to be a suitable resource in the biomedical field, but has to be correctly managed.

To further improve the results of studies and experiments, tools and data have to be managed through predefined functional sequences of acquisition and analysis. This can be accomplished by transparently managing complex workflows over such sequences and by enabling automatic enactors of such workflows both within the usually heterogeneous environment of research and on clinical networks.

The sharing of such data and metadata repositories as well as of computational resources may be provided by a Grid approach. In a Grid framework, resources may be shared within virtual organizations (VO) in a standardized, secure and reliable way, and computational power and performances may also be scaled in terabytes scenarios. Furthermore, a certificate-based authentication schema coming out for free from the Grid technology [14] may address security and privacy issues. Nevertheless, as regards data security, special requirements are often needed in clinical environments, due to the high sensitivity of medical information. On the other hand, the complexity and heterogeneity of data and services may be hidden from unpracticed users by a Grid portal.

2. MATERIALS AND METHODS

Starting from the aforementioned reasons, a Grid-based platform, named BMPortal, has been built as the core block of different specific Grid environments able to integrate data storage, retrieval and analysis in several fields of biomedicine. This platform provides shared, standardized and reliable tools for managing and analyzing distributed and heterogeneous biomedical data in order to maximize the results of virtual experiments. Users may store their experimental data on distributed and shared storage units and then retrieve their own data or data coming from other experiments. In the same way they may provide and access shared algorithms and tools. Through a simple interface, also third-party resources, such as databank and computing services, may be accessed by users as well as different distributed analysis tools.

The BMPortal follows a traditional model view controller pattern. The interface (View layer) includes a Web GUI browsable through common standard browsers and a number of Web services available through a WSDL description. The portal (Controller layer) transforms interface (user) requests into commands to server-side components, invokes server-side components to interact with resources and returns the result of processing to users. The resources (Model layer) include databases, storage areas, system commands and library functions that may be exposed via API or command line.

The BMPortal is implemented over the EnginFrame Grid Gateway developed by Nice [16]. The EnginFrame Grid Gateway is composed by two main components, a server and one or more agents. The server part of the platform relies on a standard J2EE Servlet container (e.g., Tomcat) to provide both static and dynamic contents. The EnginFrame agents manage both local and Grid resources on behalf of users and publish services to be accessed through EnginFrame Servers, extensively using XML for service definition (Figure 1).

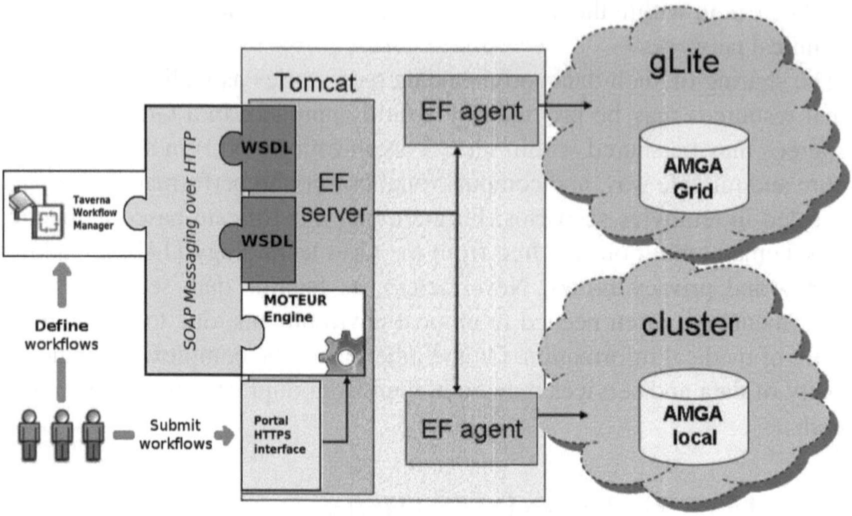

Figure 1. Architecture of the platform

The BMPortal supports gLite version 3.0, which is the official Grid middleware for the European Grid infrastructure [1]. gLite provides some fundamental functions as (i) job scheduling and submission, (ii) data storage management, (iii) metadata access, (iv) monitoring and (v) accounting services. As metadata container, AMGA [11] has been used, both for local and Grid domains. gLite has been chosen as the reference middleware but other middlewares may be linked as well, through specific plugins. The plugin to gLite has been developed within the activities carried out for the development of the Genius Grid Portal [2] (also based on EnginFrame).

With particular attention to the definition and the execution of complex workflows for acquisition and analysis, the BMPortal has also been designed to expose Grid services that can interact with workflow designer applications and to support workflow enactment and life-cycle management [3].

To these goals, the BMPortal is intended to act as

1. Grid services provider. Grid services exposed as Web services by EnginFrame can be used in the workflow as standard nodes.

2. Web interface for managing workflow enactment and management on the Grid.

As regards the first function, defined Grid services should be exposed as WSDL/SOAP standards Web services and made available to in-house or ISV applications like workflow enactors, hiding the complexity of the underlying

Grid infrastructure and actually providing an abstraction layer on the Grid technologies. Thus, the exposed Web/Grid services could be used as nodes of a workflow graph by workflow designer applications and could be triggered for running during the workflow execution.

For the second function, the BMPortal Web interface should allow users to submit workflows as Grid jobs and to monitor/control them as standard jobs. Moteur (hoMe-made OpTimisEd scUfl enactoR) has been selected as workflow enactor. Moteur is a service-based workflow engine developed in the AGIR project [4], optimized for dealing with data-intensive applications. Moteur has been selected as workflow enactor for our prototype due to its ability of implementing parallelization at different levels: intrinsic workflow parallelism, data parallelism (multithread) and services parallelism (pipeline), in order to better exploit the resources available on a grid infrastructure. In order to make the workflow enactment more scalable, the Moteur engine has been configured properly on a LSF-based cluster. LSF is an industry cluster scheduler provided by the Platform Company. Therefore the "Moteur jobs" can be launched like other jobs on the local computing cluster, according to cluster CPU's current load. This approach discharges the Portal server CPU from executing the workflows and thus from incurring, in memory, CPU or bandwidth limitations.

In both of these roles the computational power and data access capabilities provided by the backend Grid infrastructure may be exploited.

The BMPortal Web interface is designed to provide the following functionalities:

1. *Upload of workflows and related inputs:* users can upload their own workflows and insert input data for execution.

2. *Submission of Moteur as Grid job:* workflows are executed by Moteur on the Grid infrastructure as standard Grid jobs.

3. *Monitoring of workflows:* both Moteur submission and workflow processing status may be checked together with data produced by intermediate results.

4. *Results visualization:* when jobs are terminated, workflow results are staged in the spooler area and are made available through the BMPortal interface for visualization, post-processing or download.

Thus, by simply using a standard Web browser, researchers can run and monitor their own workflows, exploiting both the computational power and the data access capabilities provided by the backend Grid infrastructure. Moreover, typical features of server-side services (reliability, fault tolerance, load balancing, etc.) may also be exploited for a more efficient approach.

3. A CASE STUDY

The BMPortal has been designed for supporting a wide range of biomedical experiments. Nevertheless, in order to test its functionalities, a simple case study involving use of workflows in the field of bioinformatics has been set up.

Microarray techniques have been successfully used for gene expression profiling in a variety of genomic analyses such as gene identification, drug discovery and clinical diagnosis, providing a genome-wide system-level understanding of cellular processes and transcriptional networks. Due to this large extension of microarray studies we focused on the analysis of microarrays data in order to carry out tests about the reliability and efficiency of our platform.

Several systems have been presented in the last years in order to manage the complexity of big microarrays experiments. The amount of data, the necessity of standardization, the lack of leading software packages have driven laboratories to the pursuing of the perfect system.

The MARS system [6] provides a comprehensive suite for storing, retrieving and analyzing multicolor microarrays data. This system can be used to keep track of information that accumulates during microarray production, with emphasis on quality management and process control. Several applications can be connected through a Web service infrastructure in a tightly integrated environment. Although it uses SOAP as interconnection protocol, there is no mention of remote access to applications or of the use of a Grid middleware.

Another interesting work has been done with the eXpression Profiling System (XPS) [12], which is a distributed system for microarray storage and analysis. XPS is intended to run in a Grid context and is entirely based on the ROOT data analysis framework. XPS has extended R toolkit to access Root TFiles data storage. The most important feature of the system is also its main drawback: data are stored as *Root trees*. Therefore, the system is fully dependent on the availability of Root all around the Grid infrastructure.

The ODD-Genes system [13] has been developed in order to demonstrate several Grid technologies. They are mostly provided by Sun Microsystem components. The system provides some basic microarray analysis tools based on R and a knowledge discovery module. The users may also select one or more analysis workflows for microarray data. Notwithstanding its robustness, ODD-Genes is heavily dependent on the Sun Technology framework and only one supercomputing facility has been involved.

From these considerations it emerges that, at least in this domain, the BM-Portal can provide better interoperability and standardization for advanced functionalities in a full Grid environment.

For testing the platform, a data set with 448 patients has been used concerning the gene expression of breast cancer tissue in a large population-based cohort of patients from both Sweden and Singapore [8]. Acquisitions had been performed on HG-U133 microarray chips with the GeneChip^{TM} Affymetrix platform and had been uploaded into the GEO database.

The simple case study defined for testing our prototype is defined by an operative sequence including

- the search of suitable microarrays data for the study of gene expression profiling of breast cancer tissues in two different databases (GEO, KEGG)
- the downloading of data sets from the databases
- the normalization of image sets
- a model-based gene expression profiling based on a PM/MM difference model
- the definition of differentially expressed genes

All these operations are being exposed as Web services on the BMPortal.

As regards the analysis of data, all steps are performed by a parallel version of the dChip software [5]. dChip is a commonly used Windows application for microarray analysis. It has been ported to a parallelized version running on Linux and on the Grid, using MPI for parallel job definition and GFAL APIs for remote data access through the Grid. A very good performance speedup has been achieved [9] by parallelizing this application that was intrinsically sequential in its original version. The module providing the definition of differentially expressed genes is planned but not available yet.

Bioinformatics scientists need to orchestrate the integration in complex workflows of services provided by different computational tools and data coming from information repositories [10]. With the increasing number of bioinformatics databases and processing tools exposed as Web services, a workflow managing system able to operate according to the SOA standards is an essential tool for e-Scientists for taking full advantage of such resources. In this experiment, the Taverna application has been adopted as workflow designer while most of the building block tasks are provided by the BMPortal. Taverna is a workflow tool, which allows users to construct complex analysis workflows from Web services standards components located on both remote and local machines.

The sequence of the implementation includes

- the definition through Taverna of a workflow describing the above-listed steps of analysis
- the exposition of such functionalities as Web services on the BMPortal
- the enactment of the workflow through Moteur by using the services of the BMPortal

- the use of Grid technologies for authorization procedures as well as for accessing distributed resources both for data management and for computation.

Moreover, complementary services for microarray quality control, based on the R/Bioconductor execution environment, will be implemented.

The workflow shown in Figure 2 is intended to build a data set suited for analysis by searching for data on different repositories, starting from some search parameters (patient Ids, studies, genes, etc.) that can be identified even through other previously performed workflows. Data are downloaded from repositories, selected and refined through some selection parameters and eventually merged in a final data set to be submitted to an analysis workflow. The execution status of the workflow is continuously checked and displayed on the portal interface.

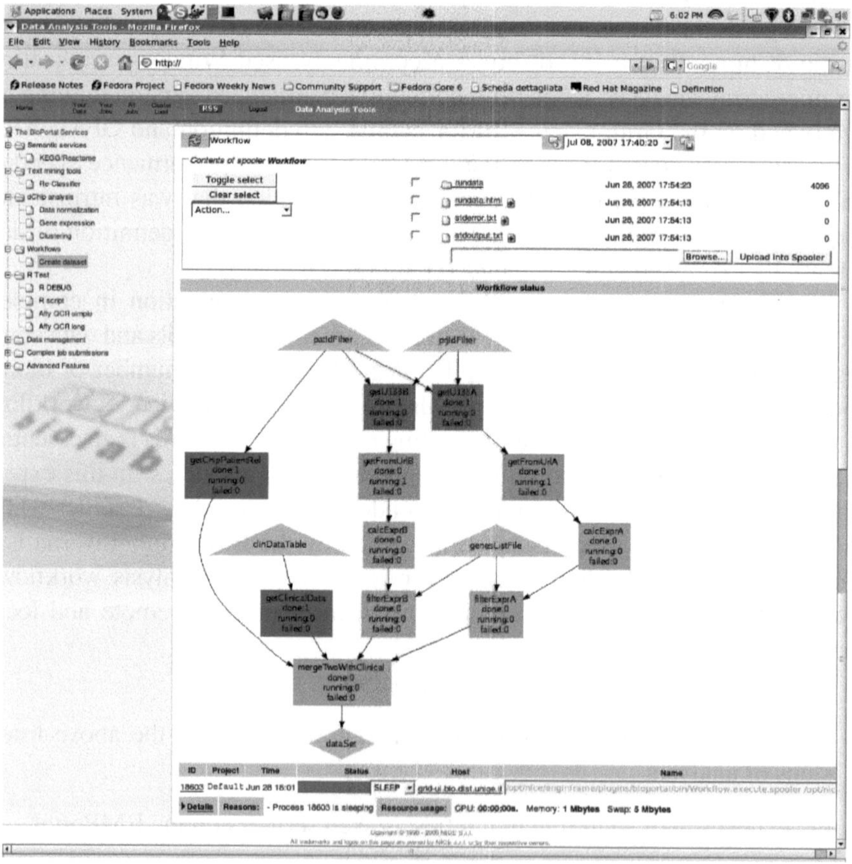

Figure 2. Example of workflow: construction of a data set

4. CONCLUSIONS

The BMPortal has been built to provide shared, standardized and reliable tools for managing and analyzing distributed and heterogeneous biomedical data in order to maximize the results of virtual experiments. The BMPortal is intended to act both as provider of Grid services to be used in workflows as standard nodes and as Web interface for performing workflow enactment and management on the Grid. The complexity and heterogeneity of data and services are hidden from users and results may be produced using distributed resources both for data and for computation. The execution status is continuously checked while performing workflows and is displayed to the user through the portal interface. The portal has been tested on a case study concerning the analysis of microarray data for breast cancer studies and satisfying results have been achieved.

ACKNOWLEDGMENTS

This platform is currently part of the Italian FIRB project LIT-BIO (Laboratory for Interdisciplinary Technologies in BIOinformatics) (http://www.litbio.org).

We thank the technical staff at Nice (http://www.nice-italy.com) and at National Institute of Nuclear Physics in Catania (GILDA Support Team) for their invaluable support to our work.

REFERENCES

[1] Alfieri R, Barbera R, Belluomo P, Cavalli A, Cecchini R, Chierici A, Fiaschini V, Dell'Agnello L, Donno F, Ferro E. "The INFN-Grid testbed", *Future Gen. Comput. Syst.* 2005, 21(2):249–258.

[2] Andronico A, Barbera R, Falzone A, Kunszt P, Re GL, Pulvirenti A, Rodolico A. "Genius: a simple and easy way to access computational and data grids", *Future Gen. Comput. Syst.* 2003, 19(6): 805–813.

[3] Gil Y, Deelman E, Blythe J, Kessleman C, Tangmunarunkit H, "Artificial intelligence and grids: workflow planning and beyond", *IEEE Intell. Syst. special issue on e-science* 2004, 19(1):26–33.

[4] Glatard T, Montagnat J, Pennac X. "MOTEUR grid-enabled data-intensive workflow-manager", *Proc. Healthgrid 2006*, Valencia, Spain, June 7–9, 2006.

[5] Li C, Wong WH. "Model-based analysis of oligonucleotide arrays: expression index computation and outlier detection", *Proc. Natl.Acad. Sci. USA* 2001, 98:31–36.

[6] Maurer M, Molidor R, Sturn A, Hartler J, Hackl H, Stocker G, Prokesch A, Scheideler M, Trajanoski Z. "MARS: microarray analysis, retrieval, and storage system", *BMC Bioinform.* 2005, 6(1):101.

[7] Muggleton SH. "2020 computing: exceeding human limits" Commentary, *Nature*, 22 March 2006, 440: 409–410.

[8] Pawitan Y, Bjöhle J, Amler L, Borg AL, Egyhazi S, Hall P, Han X, Holmberg L, Huang F, Klaar S, Liu ET, Miller L, Nordgren H, Ploner A, Sandelin K, Shaw PM, Smeds J, Skoog L, Wedrén S, Bergh J. "Gene expression profiling spares early breast cancer patients from adjuvant therapy: derived and validated in two population-based cohorts", *Breast Cancer Res.* 2005, 7(6): R953–64.

[9] Porro I, Torterolo L, Corradi L, Fato M, Papadimitropoulos A, Scaglione S, Schenone A, Viti F. "A Grid based solution for management and analysis of microarrays in distributed experiments" *BMC Bioinform.* (accepted).

[10] Romano P, Bertolini G, De Paoli F, Fattore M, Marra D, Mauri G, Merelli E, Porro I, Scaglione S, Milanesi L. "Oncology over Internet: integrating data and analysis of oncology interest on the net by means of workflows", *International Workshop on Integrative Bioinformatics Complex Metabolic Networks*, Bielefeld, Germany, July 4–5, 2005.

[11] N. Santos and B. Koblitz, "Metadata services on the grid", in *Proc. of Advanced Computing and Analysis Techniques (ACAT'05)*, Zeuthen, Berlin, May 2005.

[12] Stratowa C. "XPS, a novel framework for distributed storage and analysis of microarray data in the terabyte range: an alternative to bioConductor", *Proceedings of the 3rd International Workshop on Distributed Statistical Computing*, 2003.

[13] Thorsten F. "ODD genes – from microarray data to discovery, using the Grid and high performance computing", *Scottish Center for Genomics Technology and Informatics (SCGTI) 3rd E-BioSci/ORIEL Annual Workshop*, Hinxton Hall Conference Centre, Hinxton, England, October 12–15, 2004, www.gti.ed.ac.uk.

[14] Tuecke S, Welch V, Engert D, Pearlman L, Thompson M. "Internet X.509 public key infrastructure (PKI) proxy certificate profile", *IETF RFC 3820*, June 2004.

[15] "Towards Virtual Physiological Human: multilevel modelling and simulation of the human anatomy and physiology – White Paper", edited by *European Commission – DG INFSO & DG JRC*, Bruxelles, December 2005.

[16] http://www.enginframe.com.

AUTHOR INDEX

SUBJECT INDEX

(continued from page ii)

EM Modeling of Antennas and RF Components for Wireless Communication Systems
F. Gustrau, D. Manteuffel
ISBN 3-540-28614-4

Interactive Video Methods and Applications
R. I Hammoud (Ed.)
ISBN 3-540-33214-6

ContinuousTime Signals
Y. Shmaliy
ISBN 1-4020-4817-3

Voice and Speech Quality Perception
Assessment and Evaluation
U. Jekosch
ISBN 3-540-24095-0

Advanced ManMachine Interaction
Fundamentals and Implementation
K.-F. Kraiss
ISBN 3-540-30618-8

Orthogonal Frequency Division Multiplexing for Wireless Communications
Y. (Geoffrey) Li and G.L. Stüber (Eds.)
ISBN 0-387-29095-8

Circuits and Systems Based on Delta Modulation
Linear, Nonlinear and Mixed Mode Processing
D.G. Zrilic
ISBN 3-540-23751-8

Functional Structures in Networks
AMLn—A Language for Model Driven Development of Telecom Systems
T. Muth
ISBN 3-540-22545-5

RadioWave Propagation for Telecommunication Applications
H. Sizun
ISBN 3-540-40758-8

Electronic Noise and Interfering Signals
Principles and Applications
G. Vasilescu
ISBN 3-540-40741-3

DVB
The Family of International Standards for Digital Video Broadcasting, 2nd ed.
U. Reimers
ISBN 3-540-43545-X

Digital Interactive TV and Metadata
Future Broadcast Multimedia
A. Lugmayr, S. Niiranen, and S. Kalli
ISBN 3-387-20843-7

Adaptive Antenna Arrays
Trends and Applications
S. Chandran (Ed.)
ISBN 3-540-20199-8

Digital Signal Processing with Field Programmable Gate Arrays
U. Meyer-Baese
ISBN 3-540-21119-5

Neuro-Fuzzy and Fuzzy Neural Applications in Telecommunications
P. Stavroulakis (Ed.)
ISBN 3-540-40759-6

SDMA for Multipath Wireless Channels
Limiting Characteristics and Stochastic Models
I.P. Kovalyov
ISBN 3-540-40225-X

Digital Television
A Practical Guide for Engineers
W. Fischer
ISBN 3-540-01155-2

Speech Enhancement
J. Benesty (Ed.)
ISBN 3-540-24039-X

Multimedia Communication Technology
Representation, Transmission and Identification of Multimedia Signals
J.R. Ohm
ISBN 3-540-01249-4

Information Measures
Information and its Description in Science and Engineering
C. Arndt
ISBN 3-540-40855-X

Processing of SAR Data
Fundamentals, Signal Processing, Interferometry
A. Hein
ISBN 3-540-05043-4

Chaos-Based Digital Communication Systems
Operating Principles, Analysis Methods, and Performance Evalutation
F.C.M. Lau and C.K. Tse
ISBN 3-540-00602-8

Adaptive Signal Processing
Application to Real-World Problems
J. Benesty and Y. Huang (Eds.)
ISBN 3-540-00051-8

Multimedia Information Retrieval and Management Technological
Fundamentals and Applications
D. Feng, W.C. Siu, and H.J. Zhang (Eds.)
ISBN 3-540-00244-8

Structured Cable Systems
A.B. Semenov, S.K. Strizhakov, and I.R. Suncheley
ISBN 3-540-43000-8

UMTS
The Physical Layer of the Universal Mobile Telecommunications System
A. Springer and R. Weigel
ISBN 3-540-42162-9

Advanced Theory of Signal Detection
Weak Signal Detection in Generalized Obeservations
I. Song, J. Bae, and S.Y. Kim
ISBN 3-540-43064-4

Wireless Internet Access over GSM and UMTS
M. Taferner and E. Bonek
ISBN 3-540-42551-9

MIX
Papier aus verantwortungsvollen Quellen
Paper from responsible sources
FSC® C105338

If you have any concerns about our products,
you can contact us on
ProductSafety@springernature.com

In case Publisher is established outside the EU,
the EU authorized representative is:
Springer Nature Customer Service Center GmbH
Europaplatz 3, 69115 Heidelberg, Germany

Printed by Libri Plureos GmbH
in Hamburg, Germany